Time-Dependent Effects in Disordered Materials

NATO ASI Series

Advanced Science Institutes Series

A series presenting the results of activities sponsored by the NATO Science Committee, which aims at the dissemination of advanced scientific and technological knowledge, with a view to strengthening links between scientific communities.

The series is published by an international board of publishers in conjunction with the NATO Scientific Affairs Division

A	**Life Sciences**	Plenum Publishing Corporation
B	**Physics**	New York and London
C	**Mathematical and Physical Sciences**	D. Reidel Publishing Company Dordrecht, Boston, and Lancaster
D	**Behavioral and Social Sciences**	Martinus Nijhoff Publishers
E	**Engineering and Materials Sciences**	The Hague, Boston, Dordrecht, and Lancaster
F	**Computer and Systems Sciences**	Springer-Verlag
G	**Ecological Sciences**	Berlin, Heidelberg, New York, London,
H	**Cell Biology**	Paris, and Tokyo

Series B: Physics

Time-Dependent Effects in Disordered Materials

Edited by

Roger Pynn

Los Alamos National Laboratory
Los Alamos, New Mexico

and

Tormod Riste

Institute for Energy Technology
Kjeller, Norway

Plenum Press
New York and London
Published in cooperation with NATO Scientific Affairs Division

Proceedings of a NATO Advanced Study Institute on
Time-Dependent Effects in Disordered Materials,
held March 29–April 9, 1987,
in Geilo, Norway

Library of Congress Cataloging in Publication Data

NATO Advanced Study Institute on Time-dependent Effects in Disordered Materials
  (1987: Geilo, Norway)
   Time-dependent effects in disordered materials /   edited by Roger Pynn and
Tormod Riste.
      p.   cm.—(NATO ASI series. Series B, Physics; v. 167)
   "Proceedings of a NATO Advanced Study Institute on Time-dependent Effects in
Disordered Materials, held March 29–April 9, 1987, in Geilo, Norway"—T.p. verso.
   "Published in cooperation with NATO Scientific Affairs Division."
   Bibliography: p.
   Includes index.
   ISBN 978-1-4684-7478-7   ISBN 978-1-4684-7476-3 (eBook)
   DOI 10.1007/978-1-4684-7476-3
   1. Order–disorder models—Congresses. 2. Fractals—Congresses. 3. Fluid
dynamics—Congresses. 4. Crystals—Growth—Congresses. 5. Spin glasses—
Congresses. 6. Solid state physics—Congresses. I. Pynn, R. II. Riste, Tormod,
1925–   . III. North Atlantic Treaty Organization. Scientific Affairs Division. IV. Title.
V. Series.
QC173.4.073N3   1987
530.4′1—dc19   87-28562
                                                            CIP

© 1987 Plenum Press, New York
Softcover reprint of the hardcover 1st edition 1987

A Division of Plenum Publishing Corporation
233 Spring Street, New York, N.Y. 10013

PREFACE

This volume comprised the proceedings of a NATO Advanced Study Institute held in Geilo, Norway between 29 March and 9 April 1987. Although the principal support for the meeting was provided by the NATO Committee for Scientific Affairs, a number of additional sponsors also contributed. Additional funds were received from:

Institutt for Energiteknikk (Norway)
The Norwegian Research Council for Science and Humanities
NORDITA (Denmark)
VISTA (Norway)

The organizing committee would like to take this opportunity to thank all sponsors for their help in promoting an exciting and rewarding meeting.

This Study Institute was the ninth of a series of meetings held in Geilo on subjects related to phase transitions and was a natural successor to the 1985 meeting on Scaling Phenomena in Disordered Systems. Many of the subjects discussed at the latter meeting were revisited in 1987, with time dependence as an added feature. Often the common theme was the concept of fractals first introduced into statistical physics some six years ago. However, by no means all disordered systems can be forced into a fractal framework, and many of the lectures reinforced this lesson.

One of the main lines of research on fractals has been concerned with the study of the physical mechanisms responsible for the manifestation of fractal geometry in nature. The color plates at the front of this volume show two pertinent examples: fluid flow in a porous medium and electrodeposition. In the former case, the study is motivated by a practical concern to describe the displacement of a viscous fluid (oil) by a less viscous one (water) in a porous medium (rock). Under suitable conditions the patterns of displacement generated bear a remarkable resemblance to computer simulations of diffusion limited aggregation (DLA). The same remark applies to the color pictures of electrodeposition and to many other phenomena, suggesting the existence of universality among growth patterns. At Geilo, the emphasis was not on those patterns *per se* but rather on their time-dependent properties. The original DLA model involves a time-ordered sequence of (growth) events but is not explicitly time-dependent. However, the kinetics of this and other growth processes have been explored extensively by computer simulation and many of the results are discussed in this volume. Experimental studies of growth kinetics are also reported for DLA-like systems in which the distributions of growth probabilities on the cluster surface can be described in terms of a multifractal distribution. Such distributions occur whenever a measure is placed on a fractal structure and have been used, for example, to analyze the onset of turbulence.

Both of the structures shown in the color plate involve growth in a field which obeys the Laplace equation. As for models which describe thermal or geometric critical phenomena, the Laplace equation insures that local fluctuations propagate over a much larger length scale. Although the analogy with phase transitions is still somewhat ephemeral, the properties of several variants of Laplace-equation growth have been studied in detail. This compilation is probably a prerequisite for the recognition of universal behavior. One model discussed at the meeting is that of dentritic growth, apt in view of the snowflake density at Geilo in March. It is quite clear both from the computer studies of this problem and from experiments on colloidal aggregation that the patterns formed depend sensitively on the local growth probabilities. The work on colloidal aggregation also demonstrates that the rate of accretion plays a crucial role in determining the ramification of resulting aggregates. In keeping with the "time-dependent" theme of the Study Institute, results for colloidal aggregation and electrodeposition were presented as films, allowing local details of growth processes to be observed first-hand. A real-time experiment involving a vertical Hele-Shaw cell in a slide projector nicely demonstrated viscous fingering of a "Newtonian fluid" of micron-sized polystyrene spheres in water.

The study of the dynamics of pattern formation, in which ordered structures develop from an initially disordered state, is by no means limited to models of aggregation. First order phase transitions involve kinetic processes such as nucleation and growth in addition to spinodal decomposition and coarsening. Universality classes for this dynamical behavior are imperfectly understood but the scaling behavior of several models has been determined. Computer situations presented at the Study Institute demonstrated that growth kinetics, microstructure and topology can be understood in terms of the Potts model.

The phenomenon of 1/f noise is ubiquitous: it can be associated with the flow of sand in an hour glass, with the spectral distribution of starlight and with the flow of water in the river Nile. It is generally argued that this diversity precludes a universal origin for the noise. Nevertheless, progress in localization theory seems to provide some insight and recent results were presented at the School. It was argued that a comparative study of magnetic and electrical noise might help in the understanding of the physics of spin glasses.

Various aspects of spin glass models were presented at Geilo and connection was made to the notion of hypermetricity and the study of neural networks. Although hypermetricity has been a familiar concept to mathematicians since the turn of the century it only appeared in physics three years ago. In an ultrametric set there are no points intermediate between any pair of points. A classical example of this feature occurs in the taxonomy of living species. Such classifications are often represented as inverted hierarchical trees in which the branching points, representing the discrimination between two species, occur at successive times. Any natural definition of the distance between two extant species is proportional to the age of their most recent common ancestor, that is, to the time elapsed since branching occurred. This distance is found to obey the ultrametric inequality stated above for pairs of points. The fascinating lectures on hypermetricity presented at Geilo by Gerard Toulouse were based on a recent review of this subject by Rammal, Toulouse, and Virasoro (Rev. Mod. Phys., $\underline{58}$, 765-788 (1986)). We have chosen not to contribute to the proliferation of duplicated papers by including this article in these proceedings, but advise interested readers to consult the very well-written original paper.

The theory of neural networks was invaded by physicists about five years ago and the subsequent emphasis has been on the application of techniques of statistical mechanics to memory nets which perform pattern recognition tasks. Various models for memory have been proposed with different learning rules and degrees of error tolerance. The ultimate memory capacity of a neural system of a given size depends on the model, as does the occurrence of such tragedies as the confusion catastrophe. In spite of its deficiencies, statistical mechanics has contributed to a wider understanding of neural networks and has provided surprising insight related, for example, to the existence of spurious states and phase transitions. Correspondingly, neural-network theory has broadened the scope of traditional statistical mechanics by spurring interest in non-Hamiltonian systems, multi-valley analysis, sizes of attraction basins, etc. Von Neumann's forty-year-old prediction that the theories of computing would move closer to thermodynamics seems to be verified.

Although the relationship between spin glasses and real (window) glass is not at all clear, the dynamics of both systems are relevant to a school on time-dependent effects in disordered systems. The glass transition itself is not limited to special systems since any class of materials can be prepared in amorphous form if the cooling rate is adjusted to the dynamics of the material. In this proceedings, a discussion of the glass transition in polymers is presented with particular emphasis on an understanding of the Vogel-Fulcher law, which describes the relaxation time in such systems. The subject of polymer dynamics is also treated in relation to the sol-gel transition and the dynamics of sol clusters. In this field, light-scattering measurements have provided detailed data which can be compared with scaling predictions. The combination of inelastic light scattering and dielectric relaxation measurements also yields information on the non-linear response of dipolar glasses, once again yielding the ubiquitous Volker-Fulcher behavior.

At this school, there was an unprecedented number of contributed papers, most of which were presented as posters in order to respect the allotted 10 day schedule. In many cases, these shorter contributions served to illustrate or amplify points made by invited lecturers, and they have been included in this volume.

The organizing committee would like to take this opportunity to thank the invited speakers and participants for contributing so enthusiastically to yet another successful and thought-provoking Geilo school.

R. Pynn
T. Riste

CONTENTS

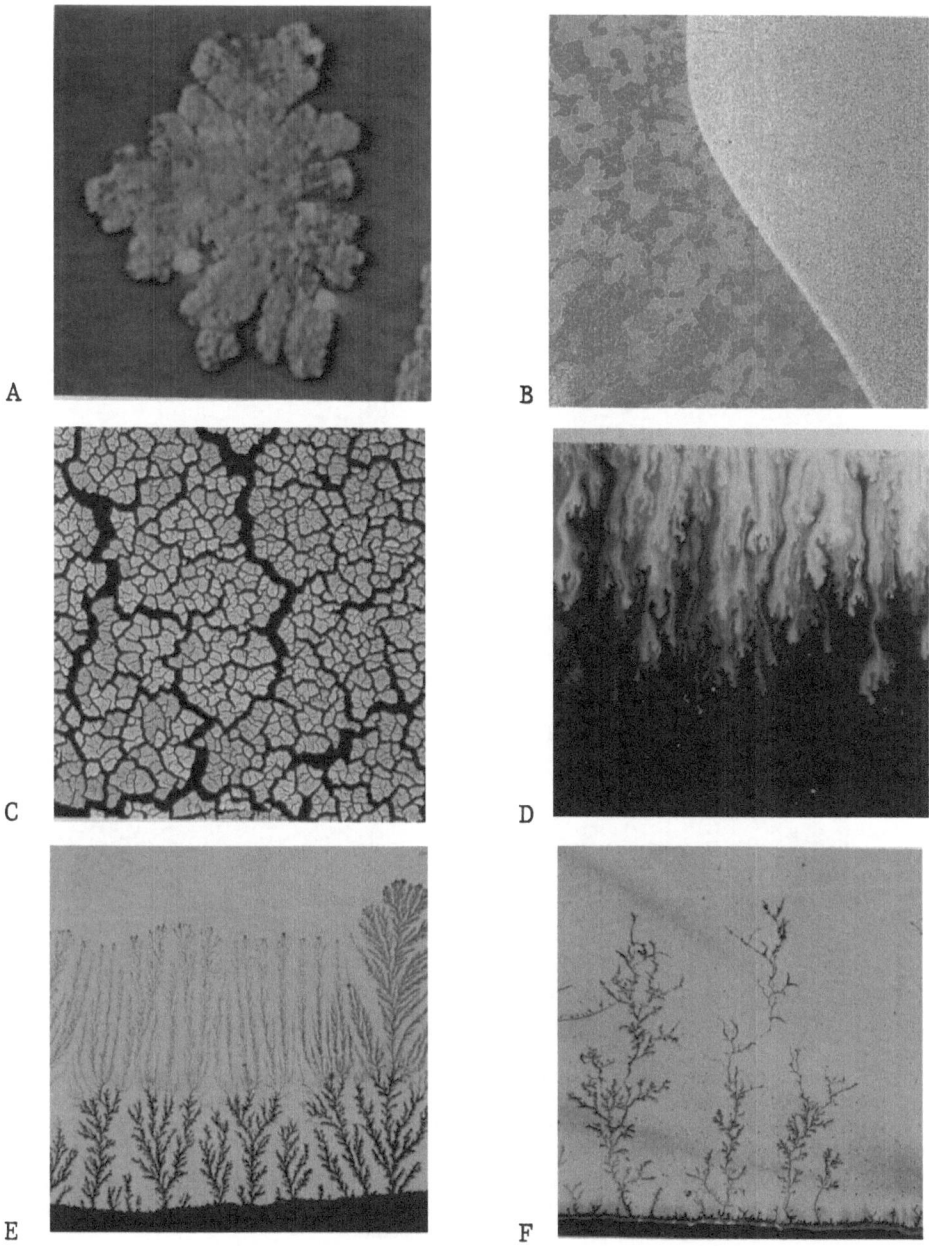

Examples of uniformly sized microspheres used for model studies of disordered materials. (A) Slow diffusion-limited aggregation of 1.1-μm spheres in a plane producing a "snowflake"-like crystal with dendrites after several months. (B) 1.1-μm spheres dispersed in water (gray) packing into a planar polycrystalline lattice. The different shadings in blue indicate different orientations of the grains. This systems may be used for model studies of annealing and interfacial melting. (C) Ramified fracture pattern resulting from a uniform contraction of 3.5-μm polystyrene spheres in a plane. This process takes place as a regular lattice of water-dispersed spheres dries out. (D) Colloidal flow of 1.9-μm water-dispersed spheres induced by gravitation in a vertical Hele-Shaw cell (glass plates separated by 25 μm). This exemplifies a chaotic process for fluid flow in the case of zero interfacial tension between two fluids. (A.T. Skjeltorp)

(E) Ramified growth in electrodeposition of silver at high ion concentration. The change in morphology in the middle of the growth pattern is resulting from an increase in voltage between the electrodes. (F) DLA-like growth patterns in electrodeposition of silver at low ion concentration. (G. Helgesen)

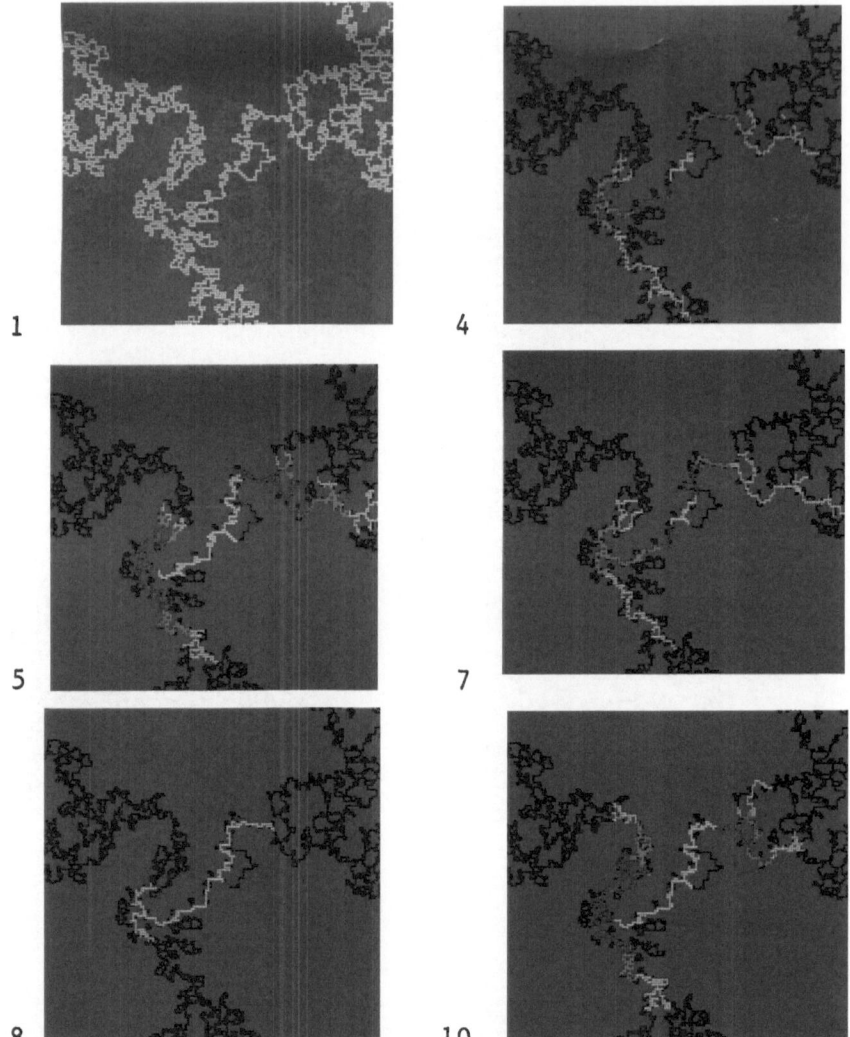

Fig. 1. A spanning percolation cluster (light blue) on a lattice of 147 × 147 bonds. The cluster shown consists of 6261 sites. The concentration of occupied sites is $p_c = 0.593$, but only sites on the spanning cluster are shown. Occupied neighboring sites are connected by conducting bonds. The backbone of the spanning percolation cluster is shown in white, and consists of 3341 sites.

Fig. 4. A digitized version of the "fast" experiment shown in Fig. 3. The backbone is black. The sites added during the time interval between two consecutive pictures are colored differently. The masses are 30, 86, 213. The center of injection is marked by a light blue point.

Fig. 5. A digitized version of the "slow" experiment at 1 ml/h ∿ 0.33 pore/sec. The backbone is black. The sites added during the time interval between two consecutive pictures are colored differently. The masses are 145, 379, 543. The center of injection is marked by a light blue point. The mass at breakthrough is 753.

Fig. 7. A sequence of growth stages in a typical computer simulation at the same aggregate masses as in Fig. 4.

Fig. 8. The "fast" experiment of Fig. 4 and the simulations from Fig. 5 superimposed. The backbone is black, the experimental result is yellow, the result of a simulation is red, The overlap, i.e., sites found both in the simulation and in the experiment, is white. The cluster mass is 213.

Fig. 10. Simulation of "slow" flow with $\eta \to 0$. The backbone is black. The sites added during later growth stages are colored differently. The masses are 145, 379, 543. The center of injection is marked by a light blue point. (U. Oxaal et al.)

AGGREGATION AND CRYSTAL GROWTH*

Arne T. Skjeltorp

Institute for Energy Technology
P.O.B. 40, N-2007 Kjeller, Norway

ABSTRACT

Dispersions of monosized microparticles are used to study

- Diffusion-controlled aggregation
- Field-induced aggregation
- Quenching, annealing and interfacial melting

A film has been produced from direct microscopic observations to
illustrate the various processes.

I. INTRODUCTION

The interplay of small subunits like atoms and colloidal particles to
form macroscopic objects like crystals, dendrites, glasses and aggregates
represents some compromise between crystalline order and the lack of it.
Order predominates if the many-body system is close to equilibrium. For
example, single crystals are grown if added particles are given sufficient
time to find a favourable place on the perimeter. The resulting structures
may be classified quite straightforwardly by using the conventions of
crystallography.

However, disorder predominates for structures formed far from equilibrium like glasses and aggregates. Recent advances over a wider field -
chaos, percolation theory, fractals - may provide a new angle to unify
many of the problems related to the physics of disordered systems[1-3]. The
fundamental equations governing nonlinear processes are often easily
stated but are usually hard to solve.

On the experimental side, traditional techniques like scattering of
beams of electrons, X-rays, light or neutrons produce only an averaged
picture of spatial and temporal effects from results in reciprocal space.
This does not give any direct information about local structure and movement of individual particles or atoms. Even for moderate disorder like
dislocations and grain boundaries, scattering techniques are not capable
of mapping the dynamics on the atomic level. One of the reasons for increased use of computer simulations in these areas is to bridge the gap
between theory and experiment.

Another alternative to exploring these problems is to let nature play

---

*See figures A-C on color page.

the rules with macroscopic analogues. Classical examples are the first visualization of dislocations in a two-dimensional soap bubble raft[4] and the use of steel ball bearings to obtain conceptual ideas of the instantaneous structures of liquid and glassy materials[5]. Even closer analogues of atomic systems are offered by monosized colloidal particles. The size is typically in the range 0.1 - 5 μm, bridging the gap between atoms and macroscopic objects. By changing the dispersions and particle size it is possible to tune the interactions relative to the disruptive forces created by the Brownian motion.

The purpose of this presentation is to use monosized microparticles to visualize a wide range of growth patterns[6], dynamic aggregation, polycrystalline behaviour and melting. The results are compared with various models mostly investigated using computer simulations.

II. EXPERIMENTAL

The particles used in the present experiments consisted of very uniformly sized polystyrene spheres of diameter typically d = 1- 5 μm produced using the Ugelstad swelling technique[7]. By special methods the spheres could also be made magnetic[8] by introduction of magnetic iron oxides. The spheres were dispersed in water and confined to a monolayer between glass boundaries, Fig. 1. The separation between the plates could be adjusted evenly by using a small fraction of larger spheres as spacers. The structures were observed directly in light microscopes. A video camera attachment and recorder allowed long term observations and digital analysis of the structures using a frame grabber with 512 x 512 pixels resolution.

Fig. 1. Schematic experimental set-up.

III. DIFFUSION CONTROLLED AGGREGATION AND CRYSTAL GROWTH

The models used to describe aggregation, dendritic and crystal growth have aimed partly at reproducing the characteristic patterns of the resulting structures while following essentially two approaches. One is the simple diffusion-limited aggregation (DLA) computer simulation model initiated by Witten-Sander (WS)[9]. Here, randomly diffusing particles (Brownian motion) stick irreversibly to the growing cluster on contact. The computer experiments produce randomly ramified, self-similar structures which may be characterized by a Hausdorff fractal dimension D below the Euclidian dimension d. There is now a wide range of DLA related models (see presentations by Meakin and Stanley in these proceedings). The other approach relates to modelling of dendritic crystal growth taking place in supercooled or supersaturated melts. The socalled boundary layer model[10] and the geometrical model[11] are able to produce qualitative agreement with observations. However, these models may also produce nonphysical results[12].

At present, there are considerable efforts to find connections between fractal aggregation and dendritic crystal growth, both analytically and experimentally. In particular, Viscek[12] introduced surface tension and lattice anisotropy into the DLA model and simulated a crossover between DLA and dendritic growth. Nittmann and Stanley[13] have shown the importance of noise and anisotropy in dendritic growth from simulations based on a generalized version of the socalled dielectric breakdown model. This model is again closely related to the DLA model (see H.E. Stanley these proceedings).

There have been few experiments showing the transition from ramified aggregation to dendritic growth in the same system. Perhaps the only examples are the experiments by Sawada et al.[14] and Grier et al.[15] of zinc electrodeposits in a plane. The results showed crossover from fractal to dendritic growth by increasing the voltage between the electrodes. However, subsequent simulations[16] have shown that a subtle competition between diffusion and electric current field may generate patterns similar to those obtained in the experiments. It is therefore not clear whether this may classify as DLA growth (see also G. Helgesen, these proceedings)

A closer approximation to physical realizations of typical diffusion controlled aggregation of colloidal particles is based on the diffusion-limited cluster aggregation (DLCA) model[17,18]. Here, there is a clustering of single particles as well as clustering of clusters. The aggregation rate is limited by the time taken for clusters to collide via Brownian motion. Recent simulations have also allowed for rotation of clusters on contact[19]. Another limiting kinetic regime has been introduced through the chemically-limited cluster aggregation (CLCA) model[2]. In this case the aggregation rate is limited by a small probability of forming a bond upon collision of two clusters.

The models for colloidal aggregation discussed above predict a limited set of fractal dimensions, but the experimental results show a relatively large spread of values. It is in fact doubtful whether diffusion-controlled aggregation of the type discussed above has been realized in two dimensions, until now. In one reported experiment[21], 0.3 μm diameter spheres trapped on a water-air interface showed D = 1.20 ± 0.15 which is well below the simulated value D = 1.4 - 1.5 for DLCA. This was attributed to anisotropic electrostatic forces around the growing tips and prompted a socalled tip-to-tip model[22] which produced a simulated value D = 1.26.

The experimental results in three-dimensional colloidal systems also show a spread of values from D = 1.65 for proteins[23] to 2.5 and 2.6 for silica[24] and immunoglobulin[25], respectively. However, the results by Weitz et al.[26] on gold colloids appear to be consistent with the simulated values D = 1.78 and 2.0 for DLCA and CLCA, respectively.

Using the present type of monodisperse colloidal particles it has been possible to provide a new level of detailed observations in two dimensions. In the following, a wide range of morphologies will be visualized and confronted with various aspects of the models outlined above.

A. DLA Growth from Ramified to Faceted Crystals

In these experiments 1.1 μm spheres were stabilized with a surfactant (0.1 % sodium dodecyl sulphate) and ionic strength of 0.02 M. The DLVO model[27] shows that for this case the interaction potential for two spheres will have a "secondary minimum" (attractive forces) of approximately 0.3 kT. This indicates that the spheres will need roughly three or more neighbours to obtain a total bonding energy $\geq$ kT to overcome the disrupting

Brownian motion. In one case (Fig. 2a) salt was added (0.05 M) to the dispersion to reduce the Coulomb repulsion and produce strong van der Waals bonding.

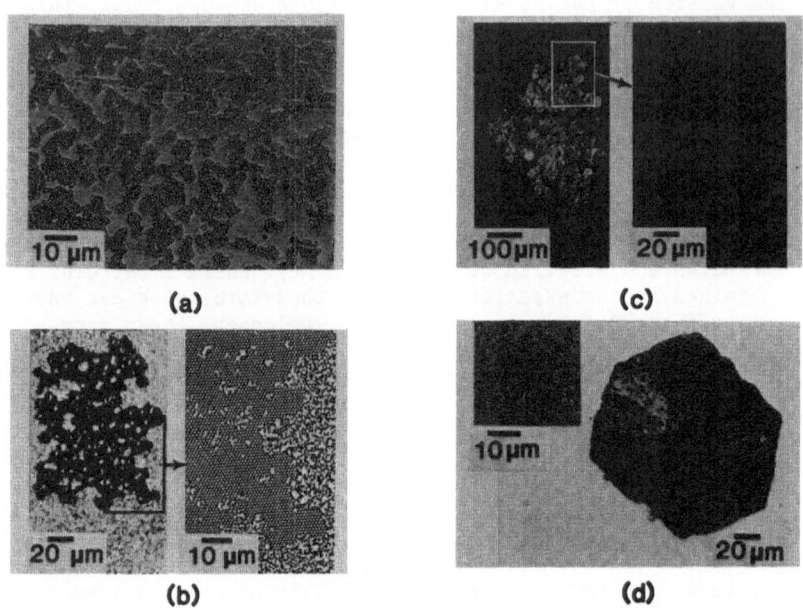

Fig. 2. Microscope pictures of aggregated 1.1 μm spheres for various growth periods (t), initial concentration (ϱ) and average growth velocity (v_g) as discussed in the text: (a) Ramified clusters (t = 20 min., ϱ ≃ 0.3, v_g ≃ $2 \cdot 10^{-2}$ μm/sec); (b) "porous" cluster with rough surface (t ≃ 100 hrs., ϱ ≃ 0.7, v_g ≃ $3 \cdot 10^{-4}$ μm/sec; (c) dendritic crystal (t ≃ 4000 hrs., ϱ ≃ 0.15, v_g ≃ 10^{-5} μm/sec) at two different magnifications; (d) faceted hexagonal crystal (t ≃ 4500 hrs., ϱ ≃ 0.1, v_g ≃ $5 \cdot 10^{-6}$ μm/sec) with small seed crystal shown to the left.

Fig. 2 displays a series of growth patterns for samples containing typically 10^7 - 10^8 dispersed spheres. Fig. 2a shows the usual DLCA situation with relatively fast irreversible clustering for a sphere concentration ϱ ≃ 0.30 defined relative to a compact lattice (for which ϱ = 1). A quantitative analysis of this case will be presented later in this paper.

Figs. 2b - 2d show examples of successively slower DLA growth. Here, the growing clusters were thus so far apart that only single spheres stuck to the aggregates and there was no clustering of clusters.

Fig. 2b shows a snapshot of a slowly growing aggregate 100 hrs after initiation of the growth process with an average growth velocity of v_g ≃ $3 \cdot 10^{-4}$ μm/sec (defined as the average growth of radius of gyration from the initiation) with a fairly high particle density ϱ ≃ 0.7.

Fig. 2c shows a situation of dendritic growth with an average growth velocity of approximately v_g = 10^{-5} μm/sec during 5 months. The starting concentration in this case was relatively low (ϱ = 0.15). The overall pattern is "snowflake-like" with branching dendrites as shown in the magnified inset.

Finally, Fig. 2d shows a near to perfect hexagonal single crystal (some dislocations) which has grown for a period of about 6 months from an initial concentration of $\varrho \approx 0.1$. The growth velocity here was so low ($v_g \approx 5 \cdot 10^{-6}$ µm/sec) that a complete equilibrium situation was achieved. Observations during growth showed that there was an even, slow layer by layer faceted growth from the seed crystal shown in the inset.

It is possible to obtain a quantitative characterization of isolated clusters examplified in Figs. 2b - 2d based on the determination of a bulk and perimeter fractal dimension, D_b and D_p, respectively, vs. growth velocity. For this, D_b was determined in the usual way[2,3] from the relationship between the total number of particles N in the cluster and the radius of gyration R_g:

$$N \propto R_g^{D_b}. \tag{1}$$

Here, R_g was calculated using

$$R_g^2 = \frac{1}{N} \sum_{i=1}^{N} (\vec{r}_i - \vec{r}_0)^2 \tag{2}$$

with $\vec{r}_0$ the center of mass for the cluster and the sum is taken over all N particles positioned at sites $\vec{r}_i$. A log-log plot of N vs R_g at successive stages of growth thus has slope D_b. The perimeter fractal dimension D_p was obtained using the box counting technique[2,3]. The number of L x L square boxes $N_b(L)$ needed to cover the perimeter was counted and averaged over different center points. For a fractal structure it is expected that

$$N_b(L) \propto L^{-D_p} \tag{3}$$

and D_p may be determined from a log-log plot of $N_b(L)$ vs L.

On the basis of approximately one hundred growth patterns examplified in Fig. 2, Fig. 3 thus shows the phase diagram of DLA for the 1.1 µm spheres expressed as D_b and D_p vs growth velocity v_g. The bands in the figure reflect the uncertainties involved in determining the fractal dimension. It may be seen that both D_b and D_p appear to approach the simulated DLA value 1.7 for the fastest growth in the present experiments. As v_g decreases, there is a crossover to the compact values $D_b = 2$ and $D_p = 1$ as expected for single crystal growth.

From the experiments visualized in Figs. 2b - 2d, the overall physical picture is as follows:

The random walk by the spheres ensures a DLA process ("nonlocal diffusion field"). In the limit of extremely slow growth (Fig. 2d), the global features (hexagonal facets) reflect the local anisotropy (triangular close packing of spheres). A condition for this is that the spheres coming into contact with the growing cluster are able to relax to neighbouring sites with the largest number of bonds (lowest potential energy).

5

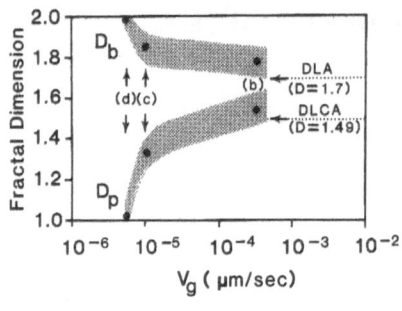

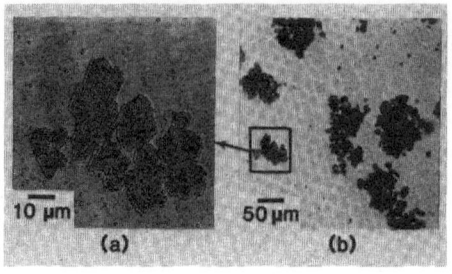

Fig. 3   Fig. 4

Fig. 3. Bulk (D_b) and perimeter (D_p) fractal dimension vs
growth velocity, v_g, for DLA of 1.1 μm spheres. The bands
reflect the uncertainties in the determination of D_b and
D_p for a wide range of clusters. Points labeled b-d
correspond to the photographs in Fig. 2. Also marked in the
figure is the bulk simulated value for DLA and the DLCA value
obtained in the present work (Fig. 6).

Fig. 4. Slow two-stage aggregation of 1.1 μm spheres into partly faceted
single crystals (a) which form clusters throughout the sample
(b). (t = 100 hrs, $\varrho \simeq 0.2$, $v_g \simeq 3 \cdot 10^{-5}$ μm/sec for individual crystals)

For successively faster growth the surface becomes more irregular
with patterns changing from dendritic (Fig. 2c) to rough (Fig. 2b). It
should be noted that the "surface tension effect" is still active here
(close packing of the bulk cluster). The "roughening" therefore appears to
come from an increased number of dendrites or "tip splitting" of dendrites. These results, demonstrating the competition between surface tension and screening effects, seem also to come out from Vicsek's computer
simulations[17]. He used a modification of the DLA process with site dependent sticking probabilities.

For relatively high concentration and slow growth, there are also
other growth schemes as shown in Fig. 4. This demonstrates a slow two-step
growth process for an intermediate concentration $\varrho = 0.2$ with an initial
formation of partly faceted single crystals (Fig. 4a) forming clusters
throughout the sample (Fig. 4b). This would classify as a crossover between slow DLA and DLCA growth.

B. DLCA Growth

In the following we will analyze in detail a prototypical DLCA process believed to be very close to a physical realization of recent computer simulations by Meakin and Jullien (MJ)[19]. Their model allows for
cluster-cluster aggregation as well as cluster rotations around points of
contact. For these experiments 4.7 μm spheres were used. The clustering
was slow due to small Brownian motion but finite due to the secondary potential minimum discussed earlier. Fig. 5 shows the aggregates formed
after 24 hrs from an initial concentration of $\varrho = 0.14$ dispersed spheres.
Direct observations during the growth process showed that the spheres
stuck fairly close to where they came in close contact with the aggregate.
There was a tendency for some rearrangement of the spheres by migration
from the first nearest-neighbour site it reached to an energetically more
favourable neighbour site. During the clustering-of-clusters process,

there was also a rotation of clusters around the point of contact. The rotations predominantly went in the direction of the smallest angle to make closed loops with a second point of contact for the clusters. The combination of these processes thus produced crystalline behaviour on small length scales as shown in Fig. 5a, but ramified on larger length scales as shown in Fig. 5b. This is still only a tiny fraction of the total sample containing about 10^6 spheres.

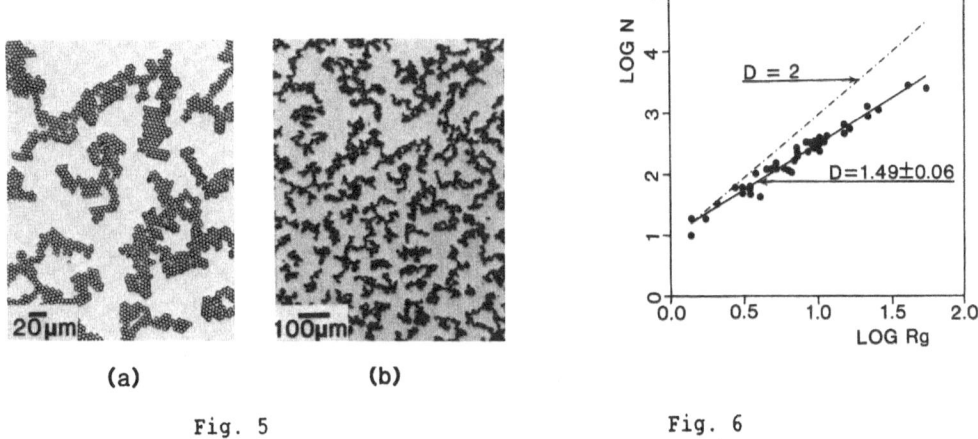

(a)   (b)

Fig. 5   Fig. 6

Fig. 5. DLCA of 4.7 μm spheres ($\rho = 0.14$) showing crystalline behaviour on short length scales (a) and ramified clusters on large length scales (b).

Fig. 6. Determination of fractal dimension D for DLCA clusters in Fig. 5 using radius of gyration method as discussed in the text. R_g is expressed in units of sphere diameter (4.7 μm).

    The quantitative analysis of the DLA structures was carried out by using the two methods discussed earlier to determine the fractal dimension D. In Fig. 6, the radius of gyration method produces a log-log plot of N vs R_g with slope D = 1.49 ± 0.06. Here R_g is expressed in units of sphere diameter a = 4.7 μm. An independent check of D was obtained using the box counting method for which the number of boxes $N_b(L)$ containing any part of the cluster was counted and averaged over different center points. A log-log plot of $N_b(L)$ vs L thus produced a slope D = 1.48 ± 0.05 for box sizes L ~ 2-100 sphere diameters. From these estimates we therefore conclude that D ~ 1.49 ±0.05. This result may be compared with the MJ simulation[19]. In particular, the aggregates in fig. 5a show a striking resemblance to Fig. 4 in MJ's paper. Their model in this case corresponds to the observations in the present system with cluster-cluster aggregation and rigid rotation of clusters around points of contact as discussed above. The MJ fractal dimension for this case was found to be D = 1.485 ± 0.015 using the radius of gyration method and thus very close to the value found for the present system. As noted in MJ's paper[19], it is possible that in the limit of large clusters (N → ∞), the value for D allowing for rotation could decrease to their cluster-cluster-aggregation value D = 1.438 ± 0.005 without rotations. Thus, rotations have a compacting effect on relatively small lengths scales but could be insignificant on large length scales.

C.   Dynamical Scaling of DLCA

It is also of interest to analyze the temporal evolution of the cluster size distribution for the DLCA discussed above. There have been essentially two approaches reported in the literature (see presentation by Meakin): One uses the classical Smoluchowski equation[28] and the other uses scaling forms postulated from phase transition work[29]. We will choose to analyze the results using the scaling form method proposed by Vicsek and Family[29] for the cluster size distribution with time t:

$$n_s(t) \sim t^{-\omega} s^{\tau} f(s/t^z) \tag{4}$$

Here, $n_s(t)$ is the normalized cluster size according to

$$n_s(t) = N_s(t)/N_0 \tag{5}$$

with $N_s(t)$ the number of clusters with s particles and N_0 the total number of particles. The reduced density used earlier is given by

$$\varrho = N_0/N \tag{6}$$

where N is the number of particles for a densely packed sample. The critical exponents ω, τ and z in Eq. (4) thus characterize the static as well as the dynamic behaviour for DLCA. The scaling function in Eq. (4) is expected to be valid in the limit $\varrho \to 0$ at large s and t.

Vicsek and Family also went on to use normalization to obtain the following scaling relationship

$$\omega = (2-\tau)z. \tag{7}$$

It is also possible to use the mean cluster size S(t) defined by

$$S(t) = [n_s(t)s^2/[n_s(t)]_s \tag{8}$$

to show that

$$S(t) \sim t^z \tag{9}$$

for $t \to \infty$. An alternative form of the scaling in Eq. (4) is[29]

$$n_s(t) \sim s^{-2} g(s/S(t)). \tag{10}$$

These predictions will be analyzed in the following from aggregation of 4.7 µm spheres. Fig. 7 shows a series of snapshots during the growth process for increasing times t from the initation of the experiments. As may be seen, there is an initital formation of small clusters due to particle-particle aggregation. The clusters and single particles continue to diffuse and form a succession of larger clusters due to cluster-cluster

8

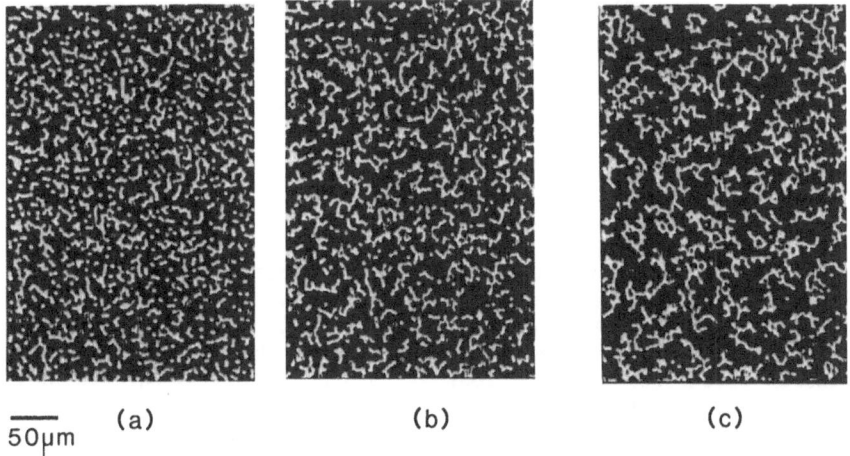

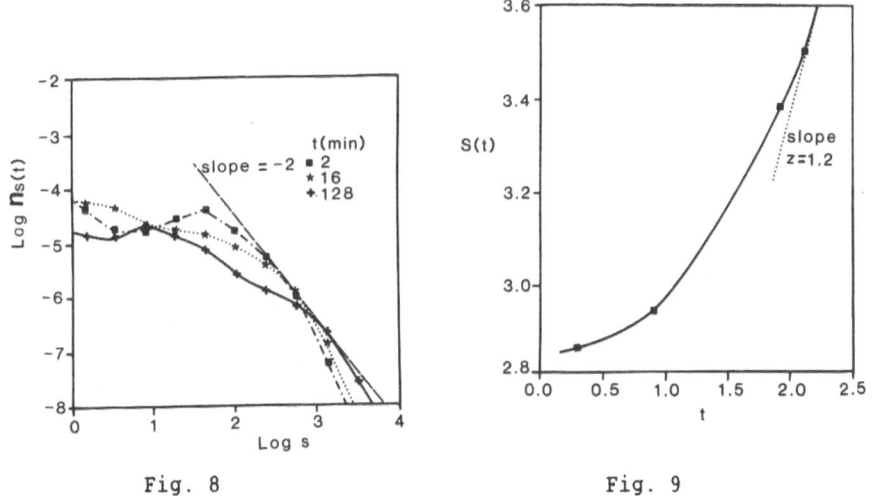

Fig. 7. Aggregation of 4.7 µm spheres versus time: (a) t = 2 min;
(b) t = 16 min; (c) t = 128 min.

Fig. 8   Fig. 9

Fig. 8. The cluster size distribution $n_s(t)$ vs the number
of particles s at different times t with limiting slope
- 2 as discussed in the text.

Fig. 9. The mean cluster size vs time with asymptotic slope z as
discussed in the text.

aggregation. Fig. 8 shows typical log-log plots of the number of clusters $n_s(t)$ versus the number of particles s. It may be seen that the limiting slope for large values of s is -2, consistent with the scaling in Eq. (10). The mean cluster size $S(t)$ vs time in Fig. 9 approaches the scaling form in Eq. (9) only at long times. The asymptotic slope is $z \leq 1.2$.

The results presented here are only preliminary. More extensive analysis is now in progress to probe dynamical scaling for a wider range of particle sizes and densities.

IV. MAGNETIC COLLOIDAL AGGREGATION

This section will visualize some unusual cases of aggregation of non-magnetic and magnetic microspheres dispersed in water and ferrofluid producing quite different morphologies from those discussed earlier. Theoretical calculations by Jacobs and Bean[30] showed that single-domain magnetic spheres could form rings of four or more particles as the lowest energy state. De Gennes and Pincus[31] pointed out that such spheres might self-organize to form chains, open loops and rings of various sizes and shapes like a polydisperse polymer melt. Recent Monte Carlo simulations[32] have shown open loop structures which break up into chains for increasing applied fields.

The magnetic particles used in the experiments consisted of d = 3.5 µm porous spheres containing 20% (volume) magnetic crystals distributed evenly in the pores. The spheres possessed a net magnetic moment

$$m = \pi d^3 M/6 \qquad\qquad (11)$$

with magnetization M = 0.8 emu/g found in independent measurements[32].

Fig. 10 shows the aggregation vs time for the same sample before and after the spheres were magnetized. As may be seen, the magnetized spheres have a pronounced tendency to form chains and loops as predicted, whereas the unmagnetized spheres form the usual aggregates discussed earlier. It is possible to obtain estimates of the fractal dimensions for the fully developed "magnetic" clusters in Fig. 10, but a better approach to analyze the chaining and looping may be to make comparisons with polymer melts[31,33]. This work is now in progress.

Another interesting aggregating situation is the use of magnetic fluids or ferrofluids as the dispersing fluid. Magnetic fluids are colloidal dispersions of single domain magnetic particles like magnetite in carrier fluids like water or kerosene[34]. The size of the magnetite particles is typically 100 Å and thus much smaller than the microspheres used in the present experiments.

Fig. 11a and 11b show some unusual aggregation phenomena in ferrofluid[35] of 1.9 µm nonmagnetic and 3 µm magnetic microspheres, respectively. In Fig. 11a, the nonmagnetic 1.9 µm spheres tend to form chains and loops as well as compact clusters. The pattern is strikingly similar to the backbone of a percolating cluster. For the magnetic spheres in Fig. 11b, there are practically no compact clusters and the patterns have an even more "chainy" appearance than for the magnetic spheres in Fig. 10. The tendency to chaining of the nonmagnetic spheres in ferrrofluid may be due to a dynamic steric effect induced by the magnetic particles in the ferrofluid. The pronounced chaining of the magnetic spheres is probably

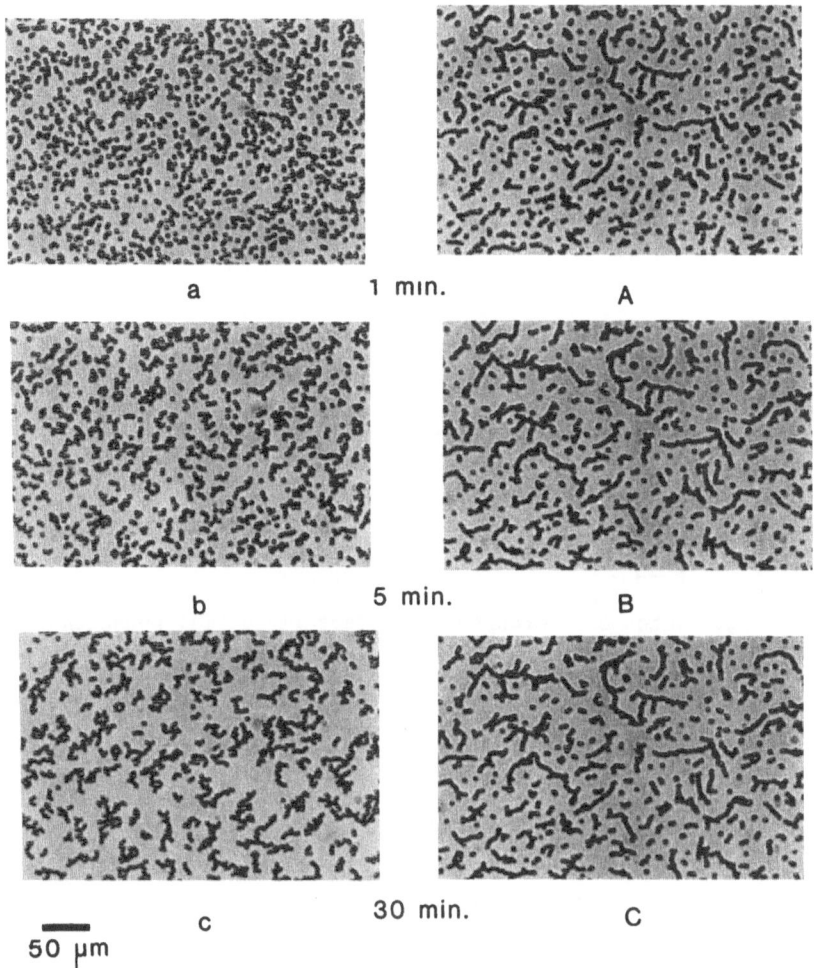

a   1 min.   A

b   5 min.   B

30 min.

c   C

50 μm

Fig. 10. Aggregation of 3.5 μm spheres versus time before (a) - (c)
and after (A) - (C) the spheres have been magnetized.

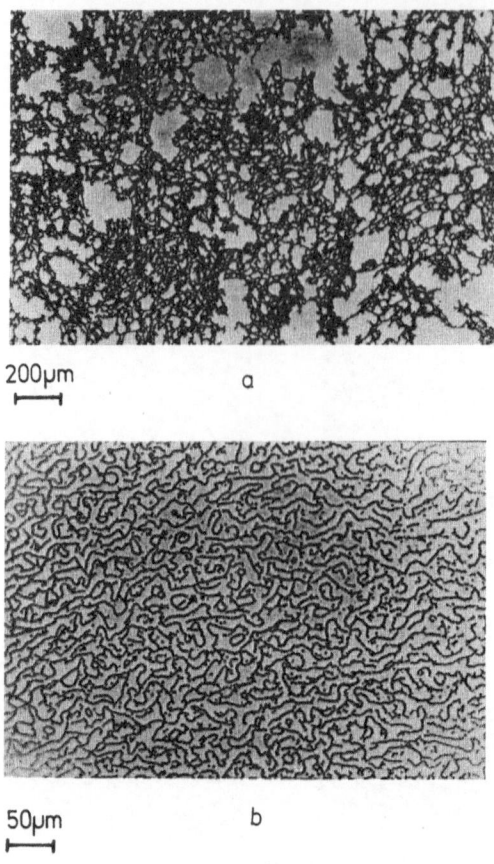

200μm   a

50μm   b

Fig. 11. Aggregates of microspheres in ferrofluid: (a) non-magnetic
1.9 μm spheres; (b) magnetic 3 μm spheres.

greatly enhanced by a reaction field: The net moment of the spheres polar-
izes the surrounding ferrofluid particles which in turn influence the
moment of the spheres.

Yet another field induced aggregation phenomenon involves the use of
the magnetic hole effect[36]. The basis for this is that non-magnetic
spheres dispersed in a magnetized fluid (external field H different from
0) will create holes which appear to possess magnetic moments. The
effective interaction between spheres confined to a plane may thus be
varied from attractive to repulsive by turning the field from a parallel
$(H_{||})$ to a vertical $(H_{\perp})$ direction relative to the layer.

An interesting situation occurs by reducing $H_{\perp}$ to zero while
maintaining a rotating field h_{xy} in the plane. In this case the spheres
quickly make 2D aggregates as shown in Fig. 12. It is possible to adjust
the concentration to a gelation situation where the whole lattice will
suddenly "freeze" to one aggregate. By increasing $H_{\perp}$ there will be a
sudden dis-aggregation into a rotating lattice phase and eventually to a
repulsive triangular lattice[37].

It is possible to determine numerically the fractal dimension of the
aggregate in Fig. 12b (box counting), and it is found that D = 1.7 ± 0.1.

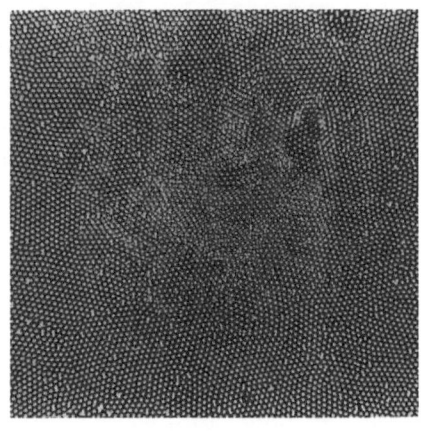

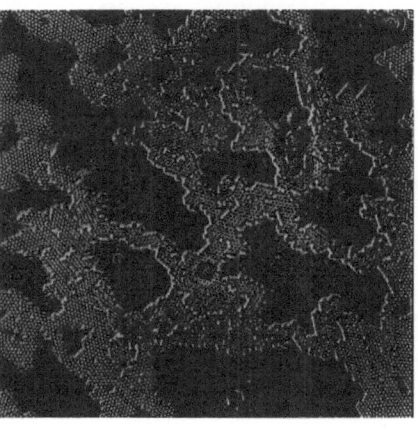

10μm   b

Fig. 12. Monolayer of 1.9 μm spheres in ferrrofluid. (a) static field
$H_\perp$ = 100 Oe; (b) aggregation of the lattice in (a) by a
rotating field h_{xy} = 30 Oe in the plane.

V. ANNEALING AND MELTING OF MICROSPHERE POLYCRYSTALS

Domain formation, growth and melting are of considerable interest in
metallurgy and surface science[38] . For example, various rates of quenching
of molten metals below the melting temperature, T_m, produce various
levels of disorder in the form of grains with random boundaries and orientations. This constitutes typically a metastable situation even for T <<
T_m. With time, a pure system (same type of atoms or particles) will
thus evolve towards larger grains or a single crystal. As the temperature
is increased towards T_m from below, there are two competing mechanisms taking place: (i) melting of the material near GB, and (ii) annealing towards larger grains. Both simulations and experiments indicate that
process (i) is a precursor effect in the form of a continuous transition
before the bulk first-order melting transition at $T = T_m$.

It is possible to demonstrate certain aspects of these phenomena using microspheres. Fig. 13 thus shows a "quenched" monolayer of 1.9 μm
spheres dispersed in water and confined between two glass plates as discussed in Sec. II. The reduced temperature is T/T_m = 0.8 with $T_m \simeq$ 330 K
estimated below. As may be seen, there are randomly oriented grains separated by random grain boundaries. Fig. 13a is a snap-shot whereas Fig. 13b
is the averaged and thresholded picture taken from 30 snap-shots about 10
secs. apart. The difference between Fig. 13a and 13b shown in Fig. 13c
thus shows the "liquidlike" spheres performing Brownian motion between the
grains.

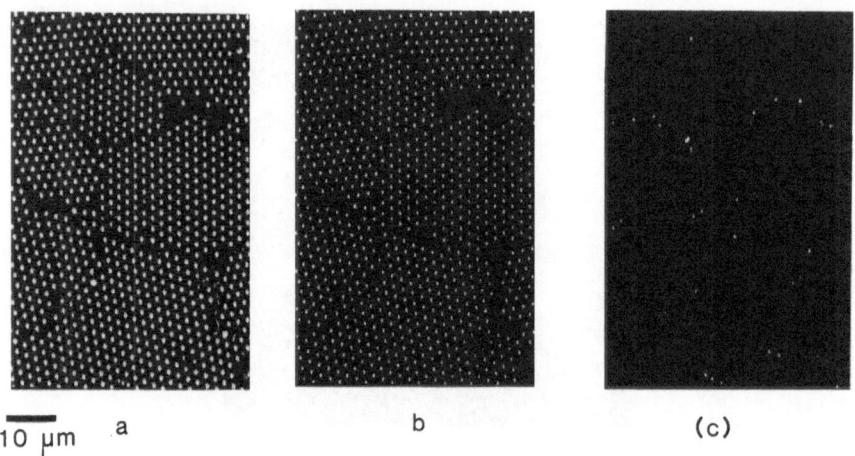

10 μm   a   b   (c)

Fig. 13. "Quenched" monolayer of 1.9 μm spheres forming a polycrystalline lattice: (a) snap-shot; (b) averaged over 5 min;
(c) difference between (a) and (b) showing the "liquid-like"
spheres.

A slight heating of the system from T/T_m = 0.8 to T/T_m = 0.9
produced some annealing after a period of about 12 hrs as shown in Figs.
14a and 14b. The dynamics of this annealing is demonstrated in Fig. 14c.
This picture is the difference between the pictures in Figs. 14a and 14b.
As may be seen the largest grains have moved laterally whereas there have
been rotations around the grain boundaries, and the smaller grains have
annealed into the larger grains.

Figs. 15a - 15c demonstrate the melting of the polycrystal as the
system is heated through T_m. During this process, the melting clearly starts
at the grain boundaries without further annealing. The solid grains are
thus separated and continue to melt around the perimeter. The qualitative
picture is quite similar to that which was observed in simulations[38,39].

14

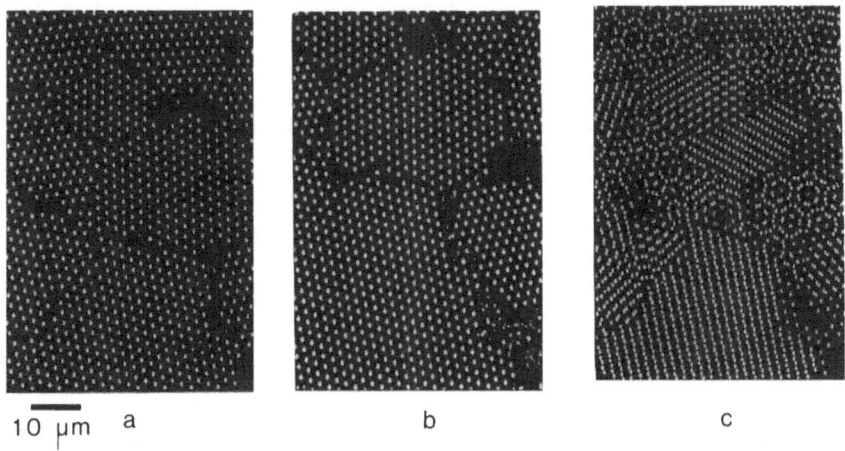

10 µm   a   b   c

Fig. 14. Annealing of a polycrystalline 1.9 µm lattice: (a) before
annealing; (b) after 12 hrs; (c) pictures (a) and (b)
superimposed showing the dynamics in the annealing process.

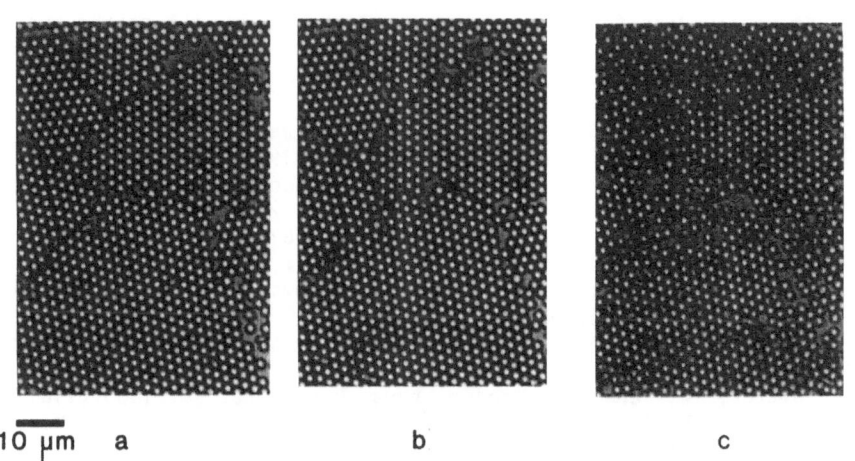

10 µm   a   b   c

Fig. 15. "Melting" of the 1.9 µm polycrystal as the system is heated
to $T/T_m \simeq 1.1$: (a) 10 min; (b) 20 min; (c) 30 min.

VI. CONCLUSIONS

The main objective of this presentation has been to show that colloidal microspheres may be quite useful in order to visualize non-equilibrium many-body processes like aggregation and crystal growth. It has
thus been possible to demonstrate colloidal growth from disordered to dendritic to faceted stuctures. The observations support the importance of
small sticking probability or low growth velocity combined with surface

15

tension effects to reach the limit of ordered structures. Considering the large number of computer simulations in this field, experimental realizations on physical systems are clearly much in demand as a corrective feedback. It has thus been possible to show that recent modifications of the DLA model including surface tension (Vicsek[12]) and cluster-cluster aggregation with rotations (Meakin and Jullien[19]) compare well with experiments. The postulated dynamic scaling by Vicsek and Family[23] is also consistent with the observed temporal evolution of the cluster size distribution. The introduction of magnetic interactions between the particles and use of magnetic carrier fluids (ferrofluids) shows quite different morphologies than the ordinary aggregation processes. Finally, the use of microspheres to study quenching, annealing and interfacial melting for polycrystals shows dynamical behaviour also seen in recent computer simulations.

VII. ACKNOWLEDGEMENTS

The research was supported in part by Dyno Particles A/S, the Norwegian Research Council for Science and Humanities (NAVF) and the Nansen Foundation. I also wish to thank John Ugelstad and collaborators at SINTEF for useful advice and for providing monodisperse spheres as well as Geir Helgesen for helpful discussions and programming assistance.

REFERENCES

1. Recent reviews are found in e.g. refs. 2, and 3 as well as the present proceedings.
2. "Scaling Phenomen in Disordered Systems", eds. R. Pynn and A. Skjeltorp, (Plenum, 1985).
3. "On Growth and Form: Fractal and Non-Fractal Patterns in Physics", eds. H.E. Stanley and N. Ostrowsky (Nijhoff, 1986).
4. W.L. Bragg and J.E. Nye, Proc. Roy. Soc. (London) $A190$, 474 (1947).
5. J.D. Bernal, Proc. Roy. Soc. (London) Ser. A, 284, 299 (1964).
6. A.T. Skjeltorp, Phys. Rev. Lett. 58, 1444 (1987).
7. J. Ugelstad et al., Adv. Colloid Interface Sci. 13, 101 (1980). The spheres are commercially available from Dyno Particles A/S, POB 160, N-2001 Lillestrøm, Norway.
8. J. Ugelstad, L. Søderberg, A. Berge and J. Bergstrøm, Nature 303, 95 (1983).
9. T.A. Witten, Jr., and L.M. Sander, Phys. Rev. Lett. 47, 1400 (1981).
10. E. Ben-Jacob, N. Goldenfield, J.S. Langer, and G. Schon, Phys. Rev. Lett. 51, 1030 (1983).
11. R.C. Brower, D.A. Kessler, J. Koplik and H. Levine, Phys. Rev. Lett. 51, 1111 (1983).
12. T. Vicsek, Phys. Rev. $A32$, 3084 (1985).
13. J. Nittmann and H.E. Stanley, Nature 321, 663 (1986).
14. Y. Sawada, A. Dougherty, and J.P. Gollub, Phys. Rev. Lett. 56, 1260 (1986).
15. D. Grier, E. Ben-Jacob, R. Clarke, and L.M. Sander, Phys. Rev. Lett. 56, 1264 (1986).
16. Ou-Yang Zhong-can, Yue Gang, and Hao Bai-lin, Phys. Rev. Lett. 57, 3203 (1986).
17. P. Meakin, Phys. Rev. Lett. 51, 1119 (1983).
18. M. Kolb, R. Botet, and R. Jullien, Phys. Rev. Lett. 51, 1123 (1983).
19. P. Meakin and R. Jullien, J. Physique 46, 1543 (1985).
20. R. Jullien and R. Botet, J. Phys. $A17$, L639 (1984).
21. A.J. Hurd and D.W. Schaefer, Phys. Rev. Lett. 54, 1043 (1985).
22. R. Jullien, J. Phys. $A19$, 2129 (1986).
23. H.J. Stapleton et al., Phys. Rev. Lett. 45, 1456 (1980).

24. S.K. Sinha, T. Freltof, and J. Kjems in "Kinetics of Aggregation and Gelation", eds. F. Family and D.P. Landau (Elsevier, Amsterdam 1984)
25. J. Feder, T. Jøssang, and E. Rosenqvist, Phys. Rev. Lett. 53, 1403 (1984).
26. D.A. Weitz, J.S. Huang, M.Y. Lin, and J. Sung, Phys. Rev. Lett. 54, 1416 (1984).
27. J.M. Victor and J.P. Hansen, J. Physique Lett. 45, L-307 (1984).
28. D.A. Weitz and M.Y. Lin, Phys. Rev. Lett. 57, 2037 (1986).
29. T. Vicsek and F. Family, Phys. Rev. Lett. 52, 1669 (1984).
30. I.S. Jacobs and C.P. Bean, Phys. Rev. 100, 1060 (1955).
31. P.G. de Gennes and P.A. Pincus, Phys. Konden. Mater. 11, 189 (1970).
32. Magnetization measurements using a vibrating sample magnetometer
33. For sol-gel transitions, see e.g. H.J. Herrmann et al., J. Phys. A.: Math. Gen. 16, 1221 (1983).
34. R.E. Rosensweig, Sci. Am. 247, 124 (1982).
35. Water based ferrofluid type SP2 supplied by S.W. Charles at UCNNW, Bangor with saturation magnetization M_s = 320 Gauss.
36. A.T. Skjeltorp, Phys. Rev. Lett. 51, 2306 (1983).
37. A.T. Skjeltorp, J. Magn. Magn. Mat. 65, 195 (1987).
38. O.G. Mouritsen and M.J. Zuckermann, Phys. Rev. Lett. 58, 389 (1987) and references therein.
39. O.G. Mouritsen et al., this proceedings.

ROLE OF FLUCTUATIONS IN FLUID MECHANICS AND DENDRITIC SOLIDIFICATION

H. Eugene Stanley

Center for Polymer Studies and Department of Physics
Boston University, Boston, MA 02215 USA

Abstract:

Our purpose is to review certain recent advances in understanding the role of fluctuations in fluid mechanics and dendritic solidification; many of these represent joint work of the author and J. Nittmann. If one understands completely the simple Ising model, then one understands virtually all systems near their critical points— although the detailed descriptions of many such systems requires a suitably-chosen variant of the Ising model (such as the XY or Heisenberg model). By analogy, we shall argue here that if one understands completely the simple diffusion-limited aggregation (DLA) model or the closely-related dielectric breakdown model (DBM), then one understands the role of fluctuations in a range of fluid mechanical systems, as well as in dendritic solidification. The detailed descriptions of some such systems requires suitably-chosen variants, such as DBM with anisotropy and noise reduction.

The overall theme I'll develop is that recent work on relatively simple *non-deterministic* models has some utility for describing experimentally-observed phenomena in fluid mechanics and dendritic growth. I'll first make the case that we can approach these experimental subjects of classic difficulty with the same spirit that has been used in recent years to approach problems associated with phase transitions and critical phenomena. This approach is to carefully choose a microscopic model system that captures the essential physics underlying the phenomena at hand, and then study this model until we understand "how the model works." Then we reconsider the phenomena at hand, to see if an understanding of the model leads to an understanding of the phenomena. Sometimes the original model is not enough, and a variant is needed, and we shall see that this is the case here also. Fortunately, however, we shall see that the same underlying physics is common to the model and its variants.

We begin, then, with the classic Ising model.

Table 1

A "Rosetta stone" connecting the physics underlying (a) an electrical problem (dielectric breakdown), (b) a fluid mechanics problem (viscous fingering), and (c) a diffusion problem (dendritic solidification).

(a) electrical	(b) fluid mechanics	(c) dendritic solidification
electrostatic potential:	**pressure:**	**concentration:**
$\phi(r,t)$	$P(r,t)$	$c(r,t)$
electric field:	**velocity:**	**growth rate:**
$E \propto -\nabla\phi(r,t)$	$v \propto -\nabla P(r,t)$	$v \propto -\nabla c(r,t)$
conservation:		
$\nabla \cdot E = 0$	$\nabla \cdot v = 0$	$\nabla \cdot v = 0$
Laplace Equation:		
$\nabla^2 \phi = 0$	$\nabla^2 v = 0$	$\nabla^2 v = 0$

(a) The Ising Model and Its Variants

The first time I heard a lecture on the Ising model, the speaker apologized for having what was termed "the Ising disease" (an appellation attributed to Montroll). The Ising model was proposed 67 years ago (Lenz 1920) and its solution for a one-dimensional lattice occurred 62 years ago (Ising 1925). However, at that time no one knew that the Ising model describes a wide range of materials near their critical points. Over 1000 papers have been published on this model, but only since 1977 have we known that if one understands the Ising model thoroughly, one understands the essential physics of virtually all 3-dimensional materials systems near thermal critical points. This is because other systems are simply variants of the Ising model. For example, most systems are related to special cases of the n-vector model, which in turn is a simple Ising model in which the spin variable s has not one component but rather n separate components s_j: $s \equiv (s_1, s_2, \ldots, s_n)$.

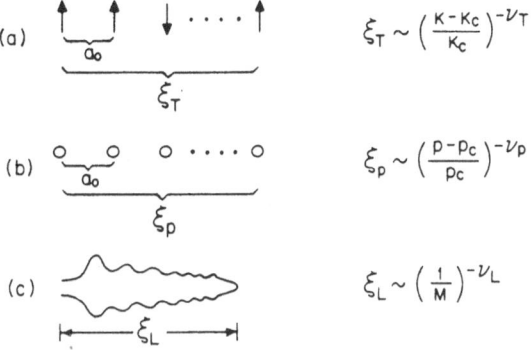

$$\xi_T \sim \left(\frac{K - K_c}{K_c} \right)^{-\nu_T}$$

$$\xi_p \sim \left(\frac{p - p_c}{p_c} \right)^{-\nu_p}$$

$$\xi_L \sim \left(\frac{1}{M} \right)^{-\nu_L}$$

Fig. 1: Schematic illustration of the analogy between (a) the Ising model, which has fluctuations in spin orientation *on all length scales* from the microscopic scale of the lattice constant a_o up to the macroscopic scale of the thermal correlation length ξ_T, (b) percolation, which has fluctuations in characteristic size of clusters *on all length scales* from a_o up to the diameter of the largest cluster—the pair connectedness length ξ_p, and (c) the DLA/DBM problem, whose clusters have fluctuations *on all length scales* from the microscopic length $d_o = \gamma/L$ (γ is the surface tension and L the latent heat) up to the diameter of the cluster ξ_L. Also shown, on the right side, is the analogy between the scaling behavior of the three length scales ξ_T, ξ_p, and ξ_L.

The Ising model solves the puzzle of how it is that nearest-neighbor interactions of **microscopic** length scale 1Å "propagate" their effect cooperatively to give rise to a correlation length ξ_T of **macroscopic** length scale near the critical point (Fig. 1a). In fact, ξ_T increases without limit as the coupling $K \equiv J/kT$ increases to a critical value $K_c \equiv J/kT_c$,

$$\xi_T \sim A \left(\frac{K - K_c}{K_c} \right)^{-\nu_T}, \tag{1.1a}$$

The "amplitude" A has a numerical value on the order of the lattice constant a_o. A snapshot of an Ising system shows that there are fluctuations on all length scales from a_o ($\cong 1$Å) to ξ_T (which can be from $10^2 - 10^4$Å in a typical experiment).

Attempts to simplify the essential problem of propagation of order from one spin to its neighbors by making mean-field type of truncations (such as the Weiss approximation, the Bethe approximation, and the Kasteleyn-van Kranendonk constant coupling approximation) fail to describe 3-dimensional systems near their critical points.

To describe the specific heat near the λ point of ^{4}He, one finds that the Ising model is not appropriate. This is because the order parameter in ^{4}He is not a one-dimensional variable with only two values (up or down), but rather a two-dimensional object with an amplitude and a phase. Accordingly, the Ising model has to be replaced by a "variant" for which the one-dimensional Ising spins are replaced by two-dimensional XY spins.

(b) Random Site Percolation on a Lattice, and Its Variants

In its simplest form, one randomly occupies a fraction p of the sites of a d-dimensional lattice (the case $d = 1$ is shown schematically in Fig. 1b). Again, phenomena occurring on the local 1Å scale of a lattice constant are "amplified" near the percolation threshold $p = p_c$ to a macroscopic length ξ_p.

Here p plays the role of the coupling constant K of the Ising model. When p is small, the characteristic length scale is comparable to 1Å. However when p approaches p_c, there occur phenomena on all scales ranging from a_o to ξ_p, where ξ_p increases without limit as $p \to p_c$

$$\xi_p \sim A \left(\frac{p - p_c}{p_c} \right)^{-\nu_p}. \tag{1b}$$

Again, the amplitude A is roughly 1Å.

It is by now a well-known piece of "magic" that each phenomenon of thermal critical phenomena has a corresponding analog in percolation, so that the percolation problem is sometimes called a geometric or "connectivity" critical phenomenon. Any connectivity problem can be understood by starting with pure random percolation and then adding interactions, or whatever. Thus, e.g., we understand why the critical exponents describing the divergence to infinity of various geometrical quantities (such as ξ_p) are the same regardless of whether the elements interact or are non-interacting (Kertész et al 1985; Geiger and Stanley 1982b). This has been predicted on theoretical grounds and confirmed by detailed numerical simulations. Similarly, the same connectivity exponents are found regardless of whether the elements are constrained to the sites of a lattice or are free to be anywhere in a continuum (see, e.g., Gawlinski and Stanley 1981 for $d = 2$, and Geiger and Stanley 1982a,b for $d = 3$).

(c) The Laplace Equation and Its Variants

Is there some lesson to be learned for fluid mechanics from our experience with thermal and geometric critical phenomena? We don't know the answer to this question, but J. Nittmann and I have been exploring this possibility in recent months. Just as variations in the Ising and percolation problems were found to be sufficient to describe a rich range of thermal and geometric critical phenomena, so we have found that variants of the original Laplace equation are useful in describing puzzling patterns in fluids mechanics and dendritic growth.

In the Ising model, we place a spin on each pixel (site) of a lattice. In percolation we allow each pixel to be occupied or empty. In fluid mechanics, we assign a number—call it ϕ—to each pixel. Generally we shall understand ϕ to be the pressure at this region of space.

The spins in an Ising model interact with their neighbors. Hence the state of one Ising pixel depends on the state of all the other pixels in the system—up to a length scale given by the thermal correlation length ξ_T. The "global" correlation between distant pixels in an Ising simulation arises from the fact that neighboring pixels at i and j have a "local" exchange interaction J_{ij}. Similarly, the correlation in connectivity between distant pixels in the percolation problem arises from the "propagation" of local connectivity between neighboring pixels. In fluid mechanics, the pressure on each pixel is correlated with the pressure at every other pixel because the pressure obeys the Laplace equation.

One can calculate an equilibrium Ising configuration by "passing through the system with a computer" and flipping each spin with a probability related to the Boltzmann factor. Similarly, one can calculate the pressure at each pixel by "passing through the system" and re-adjusting the pressure on each pixel in accord with the Laplace equation.* If we were to arbitrarily flip the configuration of a single pixel in the Ising problem (from $+1$ to -1), we would significantly influence the equilibrium configuration of the system out to a length scale on the order of ξ_T. Similarly, if we were to arbitrarily impose a given pressure on a single point of a system obeying the Laplace equation, we would drastically change the resulting pattern out to a length scale that we shall call ξ_L.

Does ξ_L obey a "scaling form" analogous to Eqs. (1a) and (1b) obeyed by the functions ξ_T and ξ_p for the Ising model and percolation? We believe that the answer to this question is "yes," although our ideas on this subject remain somewhat tentative and subject to revision.

---

* There is an intimate connection between the diffusion equation and the random walk problem (see, e.g., Chandrasekhar 1943).

The best way to see the fluctuations inherent in structures grown according to the Laplace equation is to first introduce some specific models. There are two models that were at once thought to be fully equivalent, although it is now recognized that the actual patterns produced by each have a different "susceptibility to lattice anisotropy" (Ball 1986). The first of these models is diffusion limited aggregation (DLA). Here one releases a random walker from a large circle surrounding a seed particle placed at the origin. When the random walker touches a perimeter site of the seed, it "sticks" (i.e., the perimeter site becomes a cluster site), and we have a cluster of mass = 2. A second random walker is then released. This process continues until a large cluster is formed. Initially the "mass" M of clusters was typically 10^3 to 10^4. However it has become possible to make very fast algorithms, and the largest cluster to date has a mass of 4×10^6 (Meakin 1986a).

The dielectric breakdown model (DBM) differs from DLA in that nothing happens until the random walker touches a cluster site, at which time the perimeter site it was just on at the previous step is transformed into a cluster site. Not surprisingly, this tiny local change in boundary conditions does not affect the "critical exponents" of this problem—DLA and DBM have the same value of the fractal dimension d_f describing how the cluster mass depends on cluster diameter L: $M \sim L^{d_f}$.† In both thermal critical phenomena (or percolation) the length L introduced when we have a finite system size scales the same as the correlation lengths ξ_T (or ξ_p). Hence for DLA we expect that there will be fluctuations on length scales up to ξ_L, where ξ_L itself increases with the cluster mass according to

$$\xi_L \sim A\left(\frac{1}{M}\right)^{-\nu_L} \qquad [\nu_L = 1/d_f]. \qquad (1c)$$

Here the amplitude A is again on the order of 1Å. Note that (1c) is analogous to (1a) and (1b) if we think of $M \to \infty$ as being analogous to $K \to K_c$. This reasoning is common in polymer physics, where we relate the radius of gyration R_g of a polymer to the mass through an equation of the form of (1c), $R_g \sim (1/M)^{-1/d_f}$. Note that $\nu_L = 1/d_f$ plays the role of the critical exponents ν_T and ν_p of (1a) and (1b). Suppose we test this idea, qualitatively, by examining the largest DLA clusters in detail. We find that indeed there are fluctuations in mass on length scales less than, say, the width W of the side branches. If one makes a log-log plot of W against mass M, one finds the same slope $1/d_f$ that one finds when one plots the diameter against M.

---

† The difference in boundary conditions *does* affect the rate at which the asymptotic behavior shows up (Ball 1986). For example, for DLA the screening will be more severe: as soon as a random walker steps on a perimeter site, the walker is stopped and the perimeter site becomes a cluster site. However for the DBM a random walker is free to walk on perimeter sites with impunity: only when the walker steps on a **cluster** site does the walker stop walking. Hence in the DBM the walkers can better penetrate the fjords of the system, so in overall appearance DBM clusters appear to have thicker branches and to be more "compact." The critical exponent $\nu_L = 1/d_f$ is not changed since it depends not on the density but on the *rate* at which the density decreases as the mass increases.

Evidence for Similarity of Viscous Fingering Patterns and Laplace Equation (DLA/DBM) Patterns

In the remainder of this talk, we'll describe in some detail the sorts of results we obtain from variants of the Laplace equation. First, it is necessary to describe the simplest system that produces patterns resembling interesting objects found in nature. Consider, e.g., the classic Saffman-Taylor viscous fingering problem. Here one injects a low-viscosity fluid into a medium filled with high viscosity fluid. In the limit that the viscosity ratio between the high and low viscosity fluids can be taken to be zero, we can assume that the pressure everywhere inside the low viscosity fluid is a constant: $P(i) = 1$ for $i \in$ [cluster of pixels occupied by low-viscosity fluid]. The pressure everywhere else in the system will have a value given by the solution of the Laplace equation, (2). This problem is modelled by the dielectric breakdown model or DBM (Niemeyer et al. 1984) or diffusion-limited aggregation model or DLA (Witten and Sander 1981). These two models have in common that both are solutions to Laplace equation for the case in which the pressure is zero at infinity and $P = 1$ on an object called the cluster.

Daccord has made accurate measurements on the fractal dimension of viscous fingers in both lateral (Nittmann et al 1985) and radial (Daccord et al 1986) geometries (Fig. 2). He reduced the length scale normally imposed by surface tension by using liquids with zero interfacial tension—the two fluids were water and a viscous aqueous solution of polysaccharide (Fig. 3). He found that the resulting patterns are indeed fractal, with a fractal dimension identical to that of DLA/DBM (Fig. 4). Måløy et al (1985) found analogous behavior where the cell itself introduced the randomness: he accomplished this by placing glass beads inside the cell at random. Chen and Wilkinson (1985) imposed the randomness by studying viscous fingering inside a network of glass tubes whose diameter L was randomly chosen from a probability distribution $\pi(L)$.

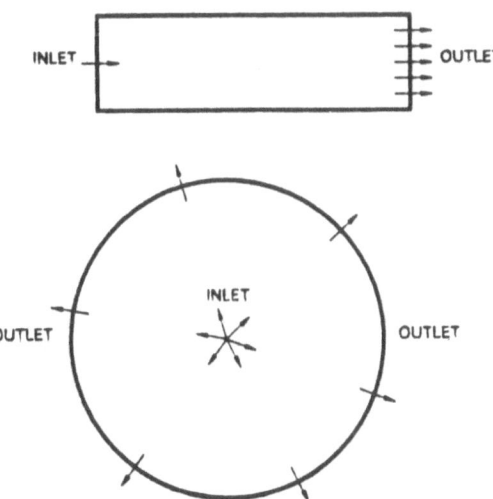

Fig. 2: Schematic illustration of the lateral and radial Hele-Shaw cells. Shown are top views. The spacing between the plates is typically 1 mm or less. From Daccord et al. (1986).

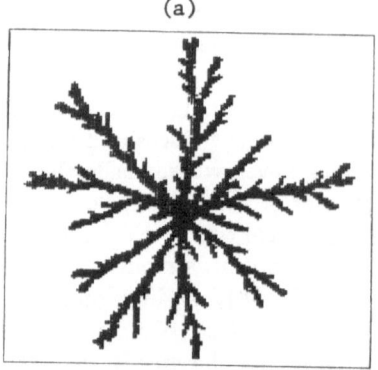

(a)

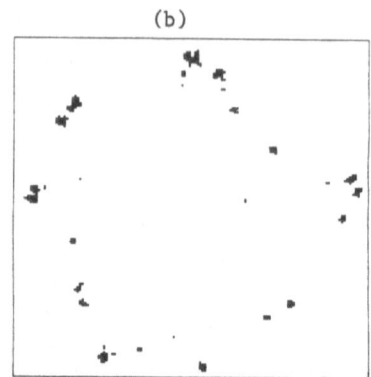

(b)

Fig. 3: The growth region of a radial viscous finger, a typical experimental pattern for which DLA is the appropriate model. The finger at time $t = t_o$ is shown in (a), while (b) displays the difference between the pattern at $t = t + \Delta t$ and $t = t$, obtained experimentally by simply subtracting the images of the same finger photographed at slightly different times. After Daccord et al. (1986).

(a)

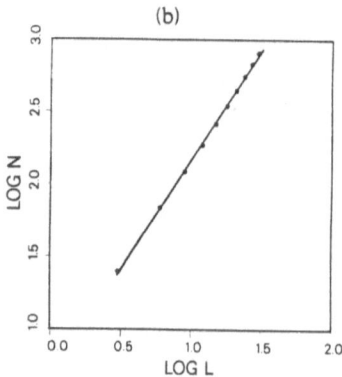

(b)

Fig. 4: Analysis of the fractal dimension typical of a radial viscous finger by the sandbox method (N is the number of occupied pixels in a $L \times L$ sandbox whose center is on an occupied pixel). The slope of the straight line shown is $d_f = 1.70 \pm 0.05$, while for DLA d_f is believed to be about 1.71 (from Daccord et al. 1986).

Not only is the fractal dimension the same for the fluid mechanics problem and for the Laplace patterns, but **so also are the multifractal properties the same**. Multifractals arise when one defines some quantity on all the pixel sites. Perhaps the simplest example is that of a charged needle: if we assign to every pixel a number equal to the electric field, then the set $\{E_i\}$ of field values for the perimeter sites of the needle form a multifractal set. The distribution $n(E)$ giving the number of perimeter pixels with electric field E is characterized, like all distribution functions, by its moments

$$Z(q) = \sum_E n(E) E^q. \tag{2}$$

As might be anticipated for a self-similar system, these moments scale with the mass M (or with the diameter L)

$$Z(q) \sim M^{\sigma(q)} \sim L^{-\tau(q)}. \tag{3a}$$

Since $M \sim L^{d_f}$, the exponents σ and τ are related by the fractal dimension d_f,

$$\sigma = \frac{\tau}{d_f}. \tag{3b}$$

For thermal and geometric critical phenomena, exponents analogous to the $\sigma(q)$ and $\tau(q)$ can be defined by considering a large $L \times L$ system at the critical point $[K = K_c]$ (or $p = p_c$). One finds that the ratio of two successive exponents is a constant "gap," so that there is no new information obtained by studying higher moments of the distribution. Connected with this simplicity is the fact that there is only one independent exponent in finite size scaling at the critical point (a second exponent arises if we wish to relate quantities that describe the approach to the critical point).

In percolation, these exponents have geometric interpretations:

(i) $y_h = d_f$, the fractal dimension of the incipient infinite cluster (the largest cluster found in a box of edge L at $p = p_c$), and

(ii) $y_T = d_{\text{red}}$, the fractal dimension of the red bonds that occur inside the largest spanning cluster (red bonds are singly connected bonds: when cut, the cluster falls into two pieces).

Relation (i) was noted by Stanley (1977) while (ii) was proved by Coniglio (1982).

In the case of the moments Z_q, there is an infinite hierarchy of exponents in the sense that the ratio $\tau(q+1)/\tau(q)$ depends on q:

$$\frac{\tau(q+1)}{\tau(q)} = D(q). \tag{4}$$

For the case of a long thin needle, the exponent $D(q)$ sticks at the value $3/2$ for small q, but for q above a critical value $q = q_c$, $D(q)$ becomes "unstuck" and varies continuously with q.

The same considerations apply to the fluid mechanics problem. Here the analog of the electric field $E \propto \nabla V$ is the growth probability $p_i \propto \nabla P$, where the index i runs over all perimeter sites i. Thus p_i is the probability that site i is the next to be added to the cluster. If we think of random walkers (Fig. 5), then p_i is the hit probability (the probability that site i is the next to be hit by a random walker). Clearly the set p_i play a vital role in determining the dynamics of growth, since if we know all the p_i for every perimeter site i at a given time t, then we can predict (in a statistical sense) the state of the system at time $t + 1$.

Recently, considerable attention has focussed on the question of how a DLA aggregate grows. Such growth phenomena are **completely** characterized by assigning to each perimeter site i the number p_i, the probability that site i is the

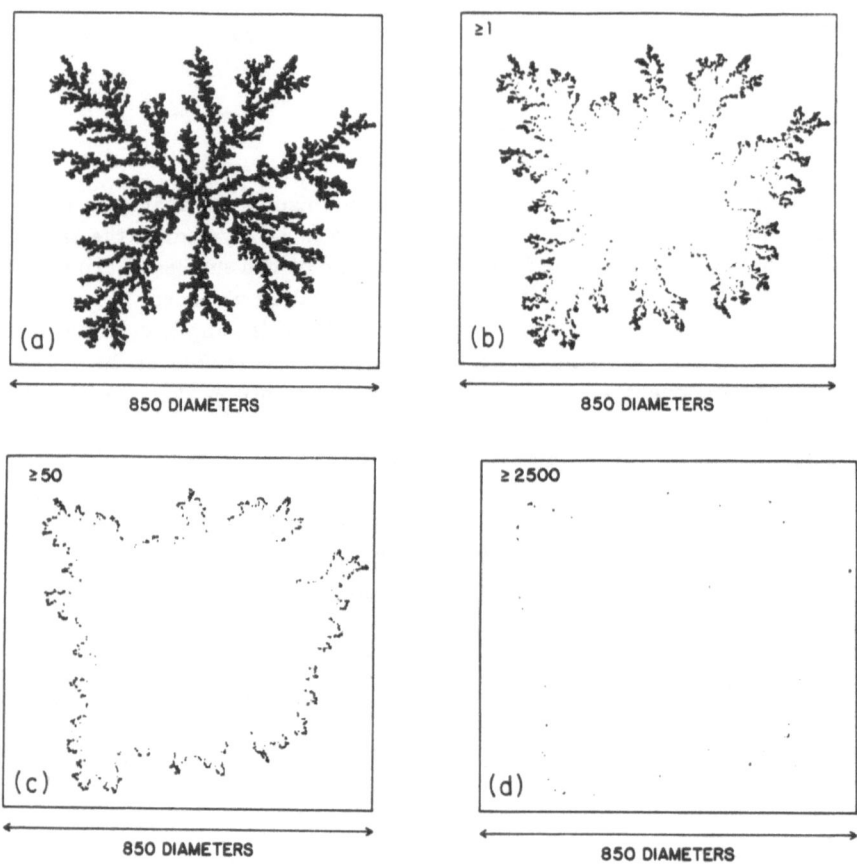

Fig. 5: This figure illustrates the harmonic measure for a 50,000 particle off-lattice 2d DLA aggregate. Figure 3a shows the cluster. Figure 3b shows all 6803 perimeter sites which have been contacted by at least one of 10^6 random walkers (following off-lattice trajectories). Figure 3c shows all of those perimeter sites which have been contacted 50 or more times and Fig. 3d shows those sites which have been contacted 2500 or more times. The maximum number of contacts for any perimeter site was 8197 so that $p_{\max} = 8.2 \times 10^{-3}$. After Meakin et al (1986).

next to grow. Theoretical evidence has been advanced recently to suggest that the numbers p_i form a multifractal set: this set cannot be characterized by a single exponent (as in the case of the DLA aggregate itself) but rather an infinite hierarchy of exponents is required. The physical basis for this fact is that the hottest tips of a DLA aggregate grow much faster than the deep fjords (which hardly grow at all); hence the **rate of change** of the p_i differs greatly when i is a tip perimeter site than when i is a fjord perimeter site.

Although there have been theoretical calculations of the multifractality of DLA (Meakin et al. 1985,1986; Halsey et al. 1986), there had been no experimental tests of these predictions. We have recently carried out the first such tests, and found experimental confirmation of the broad outlines of the theory of multifractals (Nittmann et al. 1987a).

There are many experimental realizations of DLA, and for the present work we will focus upon two-dimensional fractal viscous fingers since it is possible to study the real-time growth using a movie camera and to digitize precisely the observed time development of the DLA fractal. By subtracting two successive "snapshots" we can obtain an accurate estimate of the appropriate normalized growth probability p_i for each perimeter site of the finger (Fig. 3).

We first calculated the distribution function $n(p)$, where $n(p)dp$ is the number of perimeter sites with p_i in the range $[p_i, p_i + dp_i]$. This curve has a long tail extending to the extremely small values of p_i for perimeter sites deep inside fjords. We found good agreement between the experimental $n(p)$ for viscous fingers (Fig. 6a) and the corresponding theoretical $n(p)$ calculated for DLA (Fig. 6b).

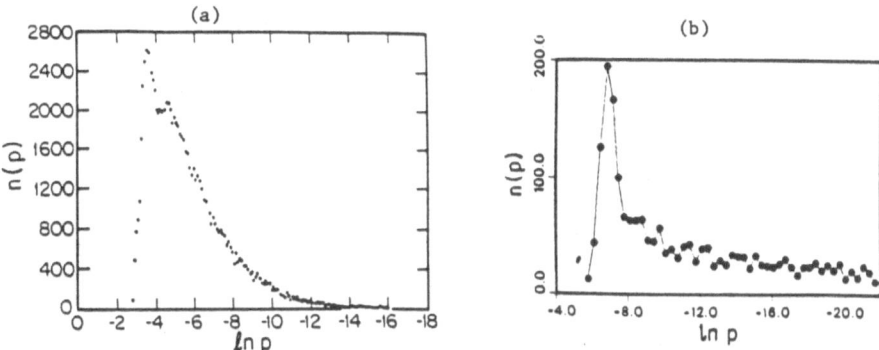

Fig. 6: Comparison between the distribution functions $n(p)$ for simulated (a) and "experimental" (b) viscous fingering patterns. Here $n(p)\delta p$ is the number of perimeter sites with growth probabilities in the range $[p, p + \delta p]$. The simulated patterns and their growth probabilities were obtained using the dielectric breakdown model. The growth probabilities for the experimental patterns were obtained by numerically solving the Laplace equation in the vicinity of a digitized representation of the pattern with absorbing boundary conditions on the sites occupied by the pattern. Similar results were obtained for large α (corresponding to the "tips") by directly subtracting two successive experimental patterns. After Amitrano et al (1986) and Nittmann et al (1987).

We next formed the moments $Z_q = \Sigma(p_i)^q$ which are characterized by the hierarchy of exponents τ_q defined through $Z_q = L^{-\tau_q}$, where L is a characteristic linear dimension. The experimental results (Fig. 7a) show that when q is large, τ_q is linear in q but for q small there is downward curvature in τ_q, showing that the fjords are characterized by different growth rates than the tips. It is conventional to also calculate the Legendre transform with respect to q of τ_q: $-f(\alpha) = \tau(q) - q\alpha$ where $\alpha = d\tau/dq$. Downward curvature in $\tau(q)$ corresponds to upward curvature in $-f(\alpha)$ [Fig. 8a]. The experimental data of Figs. 7a and 8b compare favorably with the theoretical DLA model calculations shown in Figs. 7b and 8b.

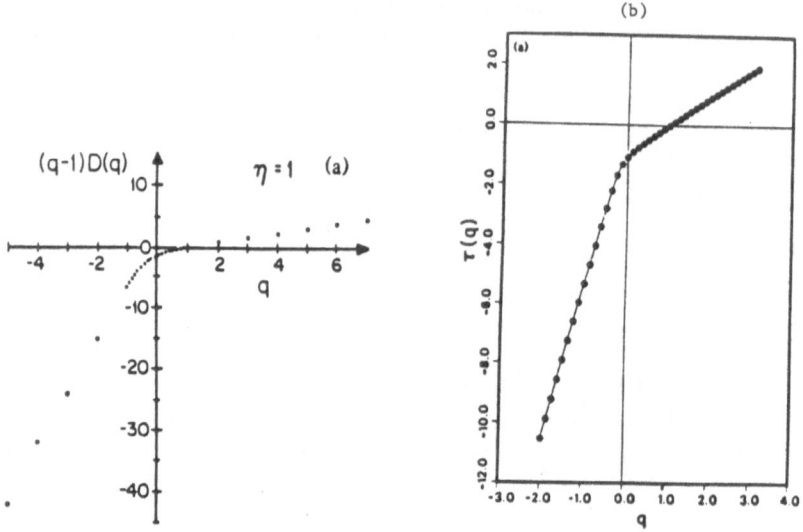

Fig. 7: Comparison of the critical exponents $\tau(q) = (q-1)D(q)$ for the (a) theoretical and (b) "experimental" viscous fingering patterns. In both cases, $\tau(q)$ was obtained numerically (see caption to Fig. 6). After Amitrano et al (1986) and Nittmann et al (1987).

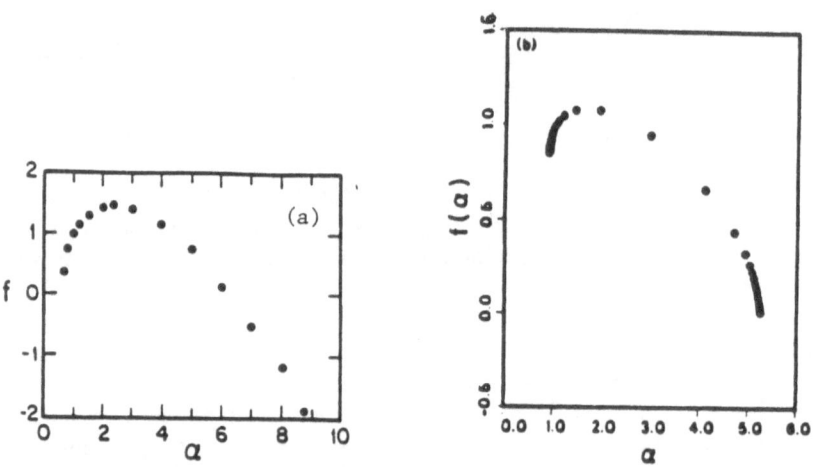

Fig. 8: Comparison between (a) theoretical and (b) "experimental" plots of the function $f(\alpha)$. After Amitrano et al (1986) and Nittmann et al (1987).

"Dendritic Solidification": Variants of the Fluid Mechanical Models

By analogy with the Ising model and its variants, we can modify DLA/DBM to describe other fluid mechanical phenomena. One of the most intriguing of these concerns a variation of the viscous fingering phenomenon in which there is present anisotropy. Ben Jacob et al (1985) imposed this anisotropy from by scratching a lattice of lines on their Hele-Shaw cell. They found patterns that strongly resemble snow crystals! If viscous fingers are described by DLA, then can the Ben Jacob patterns be described by DLA with imposed anisotropy?

Nittmann and Stanley (1986) attempted to answer this question—specifically, they attempted to reproduce the Ben Jacob patterns with suitably modified DLA. A scratch in a Hele-Shaw cell means that the plate spacing b is increased along certain directions, and the permeability coefficient k relating growth velocity to ∇P is proportional to $b^2 (k \propto b^2)$. Hence Nittmann and Stanley calculated DLA patterns for the case in which there was imposed a periodic variation in the k. It is significant that their simulations reproduce snow crystal type patterns, just like the experiments. These simulations relied for their efficacy on the presence of noise reduction.

Noise Reduction

The original DLA and DBM models are prototypes of completely chaotic systems. No discernable pattern emerges. If there is a weak anisotropy, we expect that the resulting pattern reflects this anisotropy. For example, if the simulations are carried out on a lattice, then the presence of the lattice imposes a weak anisotropy (e.g., on a square lattice, it is more likely that particles attach to the westernmost tip if they approach from the west than from the north or south). This weak anisotropy is not visually apparent unless large clusters are grown. However the largest DLA clusters made (Meakin 1986a), with mass about 4 million sites, clearly

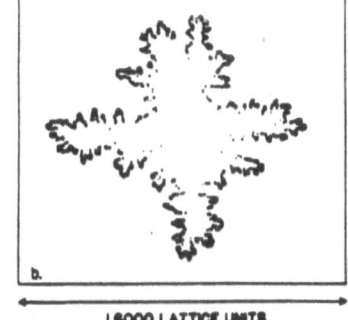

|← 15000 LATTICE UNITS →| |← 16000 LATTICE UNITS →|

Fig. 9: A huge DLA cluster with a mass of 4 million sites grown on a square lattice. Shown is only the last 5% of the growth. In reality, there is structure on all scales less than the width W of the 4 arms. Moreover, W scales with cluster mass as $W \sim (1/M)^{-1/d_f}$, just in the same way as the quantity ξ_L defined in Eq. (1c). The spontaneous appearance of side branches is reminiscent of experimental dendritic growth patterns such as those shown in Fig. 13. After Meakin (1986a).

display the anisotropy (Fig. 9). Unfortunately, no one can afford the computer resources to make such "mega-DLA" clusters each time we wish to model a new phenomenon. Noise reduction is a computational trick that seems to have the property that it speeds up the attainment of this asymptotic limit. In the absence of noise reduction, a perimeter site becomes a cluster site whenever it is chosen (e.g., whenever a random walker lands on that site).

"Noise reduction" means that we associate a counter with each perimeter site; each time that site is chosen, the counter increments by one. The perimeter site becomes a cluster site only after the counter reaches a pre-determined threshold value termed s (Tang 1985, Nittmann and Stanley 1986, Kertész and Vicsek 1986). When $s = 1$, we recover the original noisy DLA. Growth is dominated by the stochastic randomness in the arrival of random walkers. If s is very large, then growth is determined by the actual probability distribution.

For example, suppose we start with a large disc as a seed particle (instead of a single site). The growth probability at all points on the disc surface will be equal, assuming a continuum. By the D'Arcy growth law this disc should evolve in time into a larger disc. On the other hand, for ordinary DLA ($s = 1$), as soon as a random walker touches a single perimeter site on the disc, this site will become part of the cluster and the disc will lose its circular symmetry. The growth probabilities will all be re-calculated, and the perimeter sites close by the one that just grew will have higher growth probabilities. Thus the disc with a single site added to it will be more likely to grow in the direction of that single site. At a later time we will almost certainly not find a cluster with circular symmetry.

Clearly if s is very large, then the initial growth will preserve an almost circular structure. This is because before the first site is added to the circular seed, all the perimeter sites will acquire large numbers in their counters ($s - 1$, $s - 2$, etc.). After the first site is added, these additional perimeter sites will be very close to the threshold for growing while the new perimeter sites that were born when the first cluster site is added will all have counters initialized at zero. A typical cluster grown in this fashion is shown in Fig. 10; actually this cluster is grown on a square lattice with first and second neighbor interactions, not on a continuum. However Meakin et al. (unpublished) have found an almost identical pattern for the continuum case.

At first sight, there is little economy in computational speed, since one needs "s times as many" random walkers to reach a given cluster size. Thus to grow a cluster with merely 4000 sites with $s = 1000$ requires almost as much time as to generate a mega-DLA with 4,000,000 sites and $s = 1$. Fortunately, there is a way around this problem. Instead of using random walkers to solve the Laplace equation (to sample the growth probabilities p_i on each perimeter site), we can directly solve the Laplace equation numerically. This is the approach used when the dielectric breakdown model was first proposed (Fig. 11). Whether one calculates the growth probabilities by sending in random walkers or by solving the Laplace equation is immaterial: the difference between DLA and DBM is the boundary conditions, not the method of calculation.

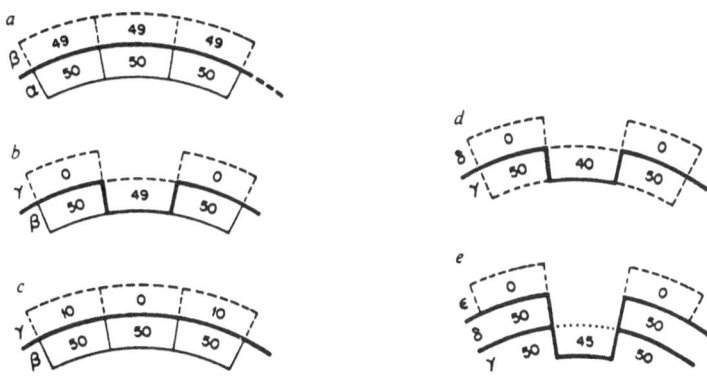

Fig. 10: Schematic illustration of the difference between an outward ('positive') and an inward ('negative') interface fluctuation. A positive fluctuation tends to be damped out rather quickly, as mass quickly attaches to the side of the extra site that is added. On the other hand, a negative fluctuation grows, in the sense that mass accumulates on both sides of the tiny notch. The notch itself has a lower and lower probability of being filled in, as it becomes the end of a longer and longer fjord. This is the underlying mechanism for the tip-splitting phenomenon when no interfacial tension is present. a shows the advancing front (row α) of a cluster with $s = 50$. The heavy line separates the cluster sites (all of which were chosen 50 times) from the perimeter sites (all of which have counters registering less than 50). In a, no fluctuations in the counters of these three sites have occurred yet, and all three perimeter counters register 49. b shows a negative fluctuation, in which the central perimeter site is chosen slightly less frequently than the two on either side; the latter now register 50, and so they become cluster sites in row β. The perimeter site left in the notch between these two new cluster sites grows much less quickly because it is shielded by the two new cluster sites. For the sake of concreteness, let us assume it is chosen 10 times less frequently. Hence by the time the notch site is chosen one more time, the two perimeter sites at the tips have been chosen 10 times (c). The interface is once again smooth (row γ), as it was before, except that the counters on the three perimeter sites differ. After 40 new counts per counter, the situation in d arises. Now we have a notch whose counter lags behind by 10, instead of by 1 as in b. Thus the original fluctuation has been amplified, due to the tremendous shielding of a single notch. Note that no new fluctuations were assumed: the original fluctuation of 1 in the counter number is amplified to 10 solely by electrostatic screening. This amplification of a negative 'notch fluctuation' has the effect that the tiny notch soon becomes the end of a long fjord. To see this, note that e shows the same situation after 50 more counts have been added to each of the two tip counters, and hence (by the 10 : 1 rule) 5 new counts to the notch counter. The tip counters therefore become part of the cluster, but the notch counter has not yet reached 50 and remains a perimeter site. The notch has become an incipient fjord of length 2, and the potential at the end of this fjord is now exceedingly low. Indeed it si quite possible that the counter will never pass from 45 to 50 in the lifetime of the cluster. In our simulations we can see tiny notch fluctuations become the ends of long fjords, and all of the above remarks on the time-dependent dynamics of tip splitting are confirmed quantitatively. After Nittmann and Stanley (1986).

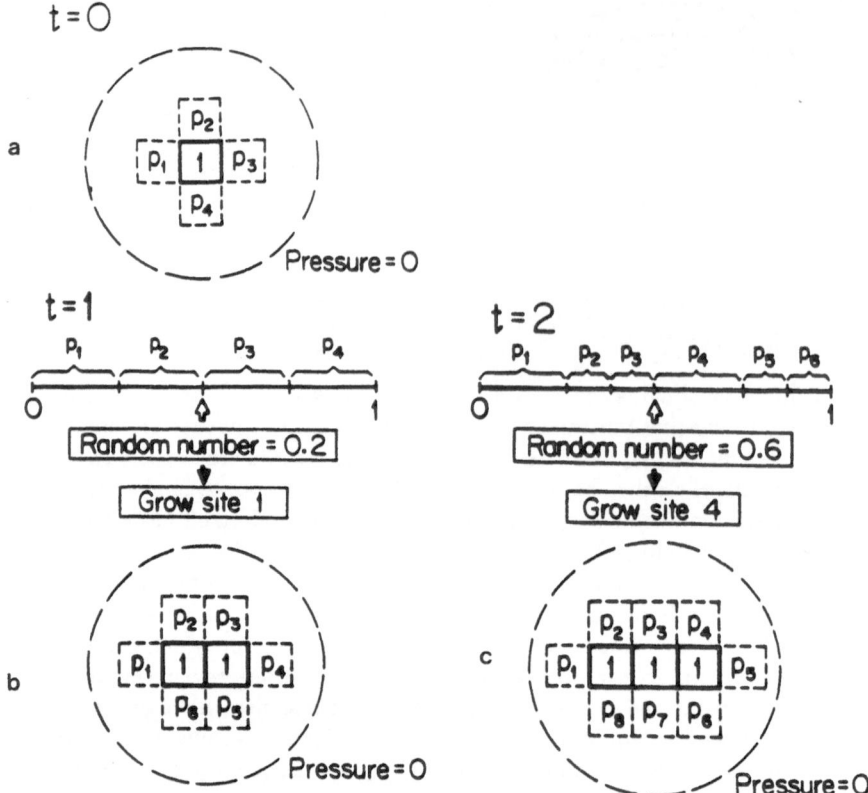

Fig. 11: Schematic illustration of the first steps in the generation of a DLA cluster by solving directly the Laplace equation on a square lattice.

The advantage of the Laplace equation approach when s is large is obvious: one need re-solve the Laplace equation only after a site is actually added to the cluster. In between adding sites, one simply chooses random numbers weighted by the growth probabilities of each perimeter site. This is a relatively rapid procedure for the computer, compared with its counterpart of sending random walkers.

"Snow Crystals"

Of course, real dendritic growth patterns (such as snow crystals) do not occur in an environment with periodic fluctuations in $k(x, y)$. Rather, the *global* asymmetry of the pattern arises from the *local* asymmetry of the constituent water molecules. Can this local asymmetry give rise to global asymmetry? Buka et al (1986) replaced the Ben Jacob experiment (isotropic fluid, anisotropic cell) by the reverse: isotropic cell but anisotropic fluid! To accomplish this, they used a nematic liquid crystal for the high viscosity fluid. Thus the analog of the water molecules in a snow crystal are the rod-shaped anisotropic molecules of a nematic. This experiment shows that the underlying anisotropy can as well be in the fluid as in the environment.

Snow crystal formation is thought to involve mainly the aggregation of tiny ice particles and droplets of supercooled water. To the extent that snow crystals grow by accreting water molecules previously in the vapor or liquid phase, the growth rate is thought to be limited by the diffusion away from the growing snow crystal of the latent heat released by these phase changes. Under conditions of small Peclet number, the diffusion equation describing the space and time dependence of the temperature field $T(\mathbf{r}, t)$ reduces to the Laplace equation. Thus a reasonable starting point is DLA, independent of whether we wish to focus on particle aggregation, heat diffusion, or both.

DLA reflects well the randomness inherent in a wide range of growth processes, including colloidal aggregation, it fails to describe dendritic solidification. While the deterministic models of snow crystals produce patterns that are much too "symmetric," the DLA approach suffers from the opposite problem: DLA patterns are too "noisy." That DLA is too noisy has long been recognized as a defect of this otherwise physically appealing model. Recently, an approach has been proposed (Nittmann and Stanley 1987a) that retains the "good" features of DLA and at the same time produces patterns that resemble real (random) snow crystals.

Firstly, we introduce (Nittmann and Stanley 1987a) controlled amounts of noise reduction of the same sort used previously for both DLA and for DBM. It is believed that noise-reduced DLA is in the same universality class as ordinary DLA—i.e., it has the same fractal dimension d_f, the only difference being an increase in the characteristic local length scale W. One advantage of setting $s > 1$ is that the asymptotic behavior ("mass" $= \infty$) behavior shows up much sooner than if $s = 1$. We do not explicitly introduce anisotropy—the only anisotropy present is the six-fold anisotropy arising from the underlying triangular lattice.

The patterns obtained (Nittmann and Stanley 1987a) have the same general features for all values of s greater than about $s = 100$—the effect of increasing s seems mainly to be that of increasing the width W of the fingers and side branches. The fjords between the 6 main branches contain much empty space. Some snow crystals have such wide "bays" but some do not. A better model would seem to require some tunable parameter that enables the complete range of snow crystal morphologies to be generated. We have found one such parameter, η, that has the desired effect of reducing the difference in the ratio of the growth probabilities between the tips and fjords. Specifically, we relate by the rule $p_i \propto (\nabla\phi)^\eta$ the growth probability p_i (the probability that perimeter site i is the next to grow) to the potential ϕ (e.g., ϕ may be the temperature $T(\mathbf{r})$ at point $\mathbf{r}$, or the probability that a tiny ice particle is at point $\mathbf{r}$). Our model is thus the analog for DLA of the "η model."

We used η to tune the balance between tip growth and fjord growth and found growth patterns that resemble better the wide range of experimentally-observed snow crystal morphologies (Nittmann and Stanley 1987a). To what does the case $\eta \neq 1$ correspond? For $\eta = k$ (k = positive integer), we have a model (Meakin 1986b) in which a site grows only if it is chosen k times in succession ($k = 1$ is pure DLA). It is possible that we have a situation not altogether different from the classic n-vector model of isotropically-interacting n-dimensional classical spins: this model makes physical sense only if n is a positive integer, yet its study for other values of n has led to rich insights—particularly the cases $n = 0$ (the dilute polymer chain limit), $n = \infty$ (the spherical model) and $n = -2$ (the mean field limit). Similarly, the Q-state Potts model makes physical sense only if Q is an integer above 1, yet the cases $Q = 0$ (random resistor network), $Q = 1$ (percolation) and $Q = 3/2$ (a spin glass model) are of great interest.

The fractal dimension d_f is believed independent of the value of the noise reduction parameter s (s renormalizes the cluster mass). We confirmed this belief. However, we found d_f does depend on η. The most reliable estimates were obtained by first calculating estimates of d_f for a sequence of increasing cluster masses, and then extrapolating this sequence to infinite cluster mass. Our values for d_f agreed remarkably well with values we obtained by digitizing photographs of experimentally observed snow crystals. Of course this preliminary study (Nittmann and Stanley 1987a) does not completely "solve" the snow crystal problem:

(i) The initial seed of a snow crystal is almost certainly hexagonal (i.e., quasi-2-dimensional), since this is the local geometry that water molecules take when they form hexagonal ice I_h. Are DBM-type considerations (small growth probability near the center of a plate-like structure) sufficient to explain why a snow crystal remains quasi-2-dimensional as it continues growing? Why does its thickness remain less than its width? It is perhaps appropriate to mention that no adequate explanation has yet been advanced for why a snow crystal remains quasi-2-dimensional throughout its growth, despite the fact that the "assembly plant" is certainly 3-dimensional. Intuition on this subject stems from experience not only from critical phenomena but also from recent theoretical and experimental work on pattern formation, where it was found that even minute amounts of anisotropy are sufficient to stabilize structures of lower effective dimension.

(ii) What are the microscopic mechanisms that give rise to the feature that real snow crystals contain branches (and side branches) which are much more than one molecule thick? Is noise reduction relevant, or is noise reduction merely a "computational trick" that allows one to see the asymptotic form of a DLA cluster using reasonable masses? (E.g., on a square lattice, the same cross-like pattern for a mass of 5,000 sites seen in noise-reduced DLA with a noise-reduction parameter of $s = 500$ is also seen in ordinary "noisy" DLA $(s = 1)$ provided the mass is allowed to increase to roughly 5,000,000 sites! We know that DLA is obtained even if the incoming random walkers have a sticking probability that is less than one. Hence we anticipate that DLA might possibly describe a modest range of phenomena with structural re-arrangement. What is the actual sticking probability for newly arriving water molecules in real snow crystals? Is a value of the sticking probability less than unity sufficient to account for the fact that the arms and sidebranches of real snow crystals have macroscopic thickness.

(iii) Are those real snow crystals which possess relatively compact cores with ramified dressing on their surfaces products of different environments of assembly, or did melting and structural re-arrangement take place after formation? Can one mimic the effect of the changing environments in which a given snow crystal is actually assembled? Do these correspond to varying parameters such as η or γ *in the course of the growth process*? To study this effect, we generated patterns with values of η and γ that change during the growth process–e.g., we might choose $\eta \ll 1$ for an initial fraction f of the growth (thereby creating a hexagonal core), and $\eta = 1$ thereafter (thereby creating a ramified exterior portion).

(iv) Does the presence in the clouds of a wind whose direction and speed varies randomly (both in time and in space, with characteristic time scales and length scales that are microscopic) imply that the actual trajectories of water molecules and water droplets might more resemble those of some extremely "pathological" path than those of a conventional DLA type random walk? We know that the random walk trajectories of DLA correspond exactly to the present electrostatic growth model, the DBM with DLA boundary conditions. What are the trajectories in "real space" corresponding to a choice of the η parameter below unity? One can speculate that a Lévy flight with tunable fractal dimension may be related to the path of a real ice particle buffeted around in a cloud.

(v) How significant, in practice, is the role played by diffusion of latent heat away from the growing aggregate in determining the actual structure of a snow crystal? We know that this phenomenon is of paramount importance in dendritic growth of crystals from a liquid phase How significant is the role played by the capillary length $d_o = \gamma/L$ in vapor phase deposition of water molecules onto a growing snow crystal? (Here L is the latent heat.) An ideal model might encompass *both* the diffusion of heat away from the snow crystal *and* the aggregation of particles toward the snow crystal?

(vi) Are real snow crystals sometimes fractal objects? This intriguing question has been the object of considerable discussion in recent years. Our growth patterns are fractal, for all positive values of η. We found (Nittmann and Stanley 1987a) that the fractal dimension d_f is independent of the value of the noise reduction parameter s (s seems to mainly renormalize the cluster mass), but d_f does depend on η. We also found that these values for d_f agreed well with values we obtained by digitizing the corresponding photographs of experimentally observed snow crystals (Fig. 12).

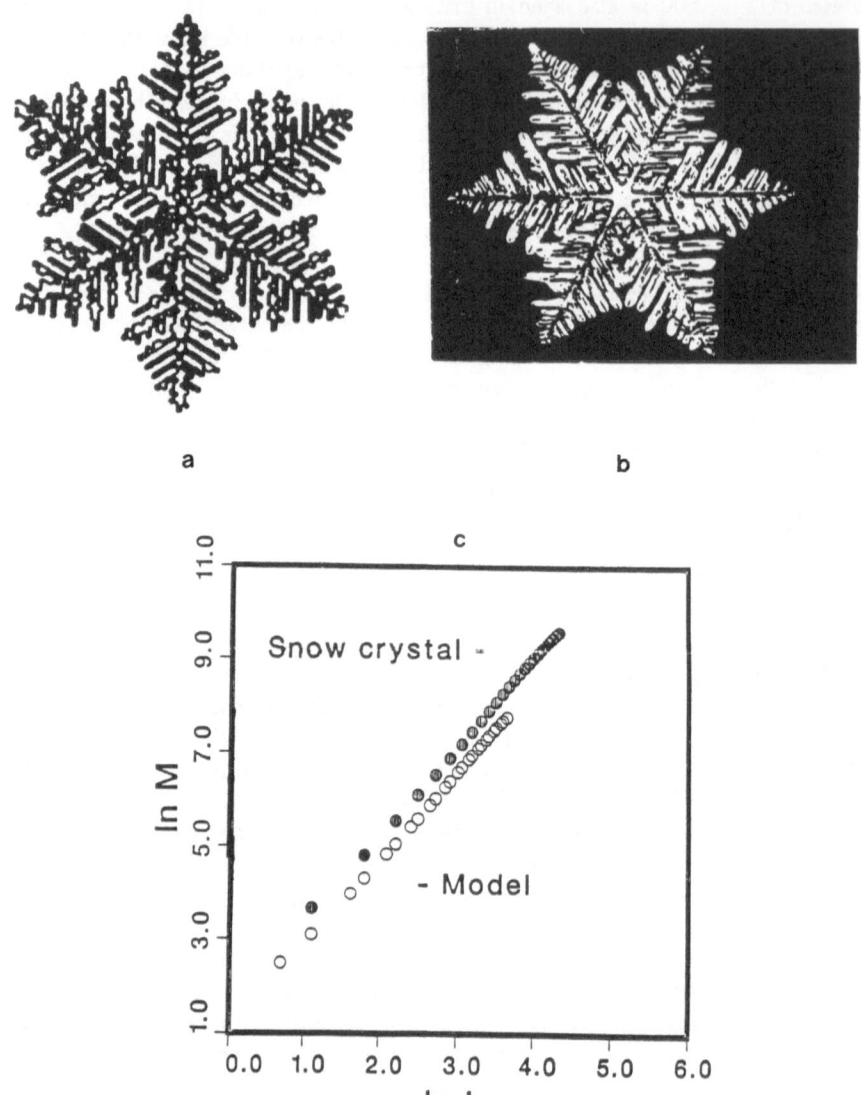

a

b

c

Snow crystal -

- Model

ln M

ln L

Fig. 12: (a) A typical snow crystal from the collection of 2453 photographs assembled in Bentley and Humphreys (1935). Other experimental examples may be found in Nakaya (1954) and LaChapelle (1969). (b) A DLA simulation with noise reduction parameter of $s = 200$ and non-linearity parameter $\eta = 0.5$. (c) comparison between the fractal dimensions of (a) and (b) obtained by plotting the number of pixels inside and $L \times L$ sandbox logarithmically against L. The same slope, $d_f = 1.85 \pm 0.06$, is found for both. The experimental data extend to larger values of L, since the digitzer used to analyze the experimental photograph has 20,000 pixels while the cluster has only 4000 sites. After Nittmann and Stanley (1987a).

Dendritic Growth of NH₄Br

Dendritic crystal growth has been a field of immense recent progress, both experimentally and theoretically. In particular, Dougherty et al (1987) have recently made a detailed analysis of stroboscopic photographs, taken at 20 second intervals, of dendritic crystals of NH_4Br (Fig. 13a). They have found three surprising results: (i) the sidebranches are non-periodic at any distance from the tip, with random variations in both phase and amplitude, (ii) sidebranches on opposite sides of the dendrite are essentially uncorrelated, and (iii) the rms sidebranch amplitude is an exponential function of distance from the tip, with no apparent onset threshold distance. Some of these results are apparently at variance with predictions from recent theories (Saito et al 1987; Ben-Jacob et al 1984; Kessler et al 1984).

How can we understand these new experimental facts? Many existing models reflect the essential physical laws underlying the growth phenomena, but fail to find a tractable mechanism to incorporate the effects of noise on the growth. Growth of a dendrite from solution is controlled by the diffusion of solute towards the growing dendrite. In the limit of small Peclet number, the diffusion equation reduces to the Laplace equation (as mentioned above). The Laplace equation for a moving interface (the growing dendrite) brings to mind the diffusion limited aggregation model (DLA). Growth patterns produced by the various DLA simulation algorithms do *not* resemble dendritic growth patterns: DLA patterns are much too chaotic in appearance. We shall discuss here a related model (Nittmann and Stanley 1987b) whose asymptotic structure does resemble the patterns found experimentally—both in broad qualitative features and in quantitative detail. The picture that emerges is one of Laplacian growth, where noise arises from the fact that there are concentration fluctuations in the vicinity of the growing dendrite (these are estimated to be roughly $\pm 10^5$ NH_4Br molecules per cubic micron).

Our starting point is the observation that minute amounts of anisotropy become magnified as the mass of a cluster increases. In fact, even the weak anisotropy of the underlying lattice structure can become so amplified that clusters of 4,000,000 particles take on a cross-like appearance (cf. Fig. 1 of Ref. 8). A real dendrite has a mass of roughly 10^{16} particles; it is impossible to generate clusters of this size on a computer, since even clusters of size 10^6 require hundreds of hours on the fastest available computers. Fortunately, there is a computational trick—termed *noise reduction*—that speeds the convergence of the pattern toward its asymptotic "infinite mass" limit. The patterns we obtained with noise-reduced DLA resemble Fig. 1 of Dougherty et al (1987), reproduced in Fig. 13a.

A typical result (Nittmann and Stanley 1987b) for a mass of 4000 particles is shown in Fig. 13b. After each 333 particles are added, a contour is drawn:

(i) It is apparent from the "stroboscopic" representation of Fig. 13b that the distance between successive tip positions is a decreasing function of the mass; in fact, we find that $log\,x_{\text{tip}}$ is linear in $log\,M$ with slope 2/3. This result is consistent with the belief that $d_f = 1.5$ for DLA with anisotropy.

(ii) The tip is remarkably parabolic: specifically, when we form $(y_c - y_o)^2$ (where y_c is the contour, and y_o is the centerline of the dendrite) and plot this on linear graph paper as a function of $x - x_{tip}$, we obtain a straight line with an R value of 0.997.

(iii) The sidebranches are non-periodic at any distance from the tip, with random variations in both phase and amplitude. To demonstrate this, we have analyzed our simulations in exactly the same mathematical fashion as Dougherty et al analyzed the experimental dendrite patterns.

An open theoretical question concerns the microscopic origin of the sidebranching phenomenon. One current hypothesis predicts that the sidebranch amplitude would be periodic and the two sides of the dendrite should have correlated sidebranching. Dougherty et al (1987) noted that their experimental data are not consistent with this hypothesis, and we can make similar remarks for the present model. A second hypothesis views sidebranching as a result of the noise arising from concentration fluctuations. To test this hypothesis, Dougherty et al (1987) plot the sidebranch amplitude as a function of $x - x_{tip}$, the distance from the tip. They found that the sidebranch amplitude decreases as the distance variable $x_{tip} - x$ decreases, and shows no sign of a threshold distance below which the amplitude is zero. Moreover, they found that close to the tip the sidebranch am-

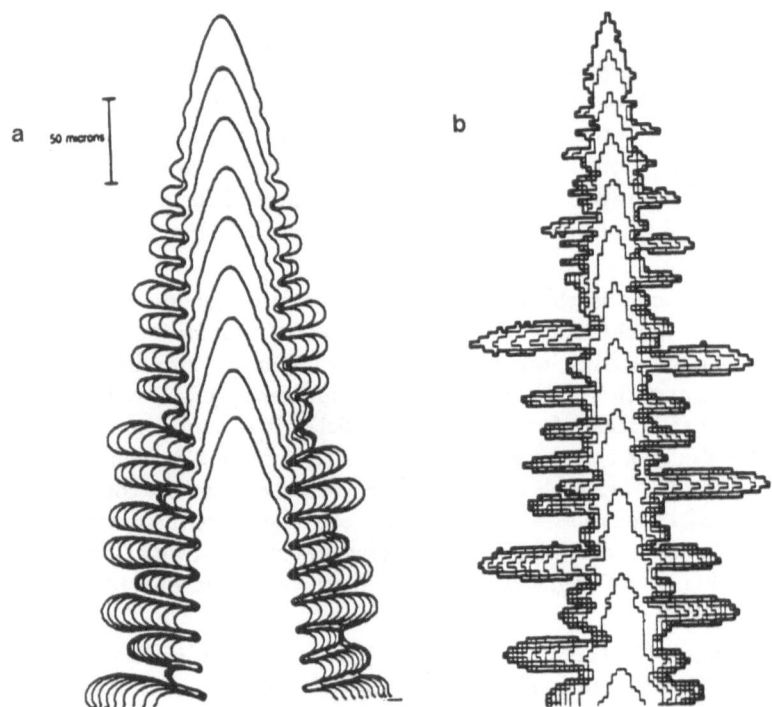

Fig. 13: (a) Experimental pattern of dendritic growth, measured for NH_4Br by Dougherty et al. (1987). (b) DLA simulation with noise reduction parameter $s = 200$ (after Nittmann and Stanley 1987b).

plitude is roughly linear on semi-log paper. If we plot y_c, the amplitude, which should scale roughly as the square root of the area under the peak if the peak maintains its shape as a function of $x - x_{tip}$; we find exactly the same exponential growth of sidebranch amplitude with distance from the tip.

In summary, we have developed a model in which noise reduction is used to tune the effect of noise, and cubic anisotropy is introduced through the use of an underlying square lattice. The resulting patterns obtained strongly resemble the experimental patterns of Dougherty et al (1987), both in their *qualitative* appearance and in the same degree of *quantitative* detail studied experimentally. Sidebranching arises from the fact that an approximately flat interface in the DLA problem grows trees (which resemble "bumps" in the presence of noise reduction); these compete for the incoming flux of random walkers. If one tree gets ahead, it has a further advantage for the next random walker and so gets ahead still more. Thus some sidebranches grow while others do not. The characteristic spacing λ between sidebranches scales with the dendrite mass with the same exponent $2/3$ that characterizes the growth of dendrite length x_{tip}. Moreover, the patterns we obtain are reasonably independent of details of the simulation in that similar patterns are obtained when we vary the surface tension parameter σ over a modest range; we can also alter the boundary conditions of the model with some latitude and even allow for non-linearity in the growth process ($\eta \neq 1$).

The significance of the present findings is that the essential physics embodied in the DLA model—previously used to describe fluid-fluid displacement phenomena ("viscous fingering")— seems sufficient to describe the highly uncorrelated (almost random) dendritic growth patterns recently discovered from the experiments and quantitative analysis of Dougherty et al (1987).

Summary

We have argued that it is worth exploring all the consequences of a straightforward physical model. Our optimism is based on the success of the Ising model and percolation in the past. We must be mindful that substantial variants of the original model may be called for. In our case, e.g., anisotropy must be introduced or else the pattern bears absolutely no resemblance to dendritic growth. Also, noise reduction must be introduced or else the computer time becomes prohibitive.

This modest work perhaps raises more questions than it answers, but it nonetheless might stimulate further investigation of the basic physics of random systems that must be better understood in order to explain experimentally-observed non-symmetric dendritic growth patterns and fluid mechanics patterns. The reader interested in more details than provided here may consult recent books on the subject (Family and Landau 1984; Boccara and Daoud 1985; Pynn and Skjeltorp 1986; Stanley and Ostrowsky 1986; Pietronero and Tosatti 1986; Stanley 1987).

Acknowledgements

I'd like to thank the Organizing Committee for inviting me. The results I report here are largely due to interactions with A. Coniglio, G. Daccord, P. Meakin, E. Touboul, T. A. Witten and, most especially, J. Nittmann. I wish to thank especially Roger Pynn, Tormod Riste, and Gerd Jarrett—this threesome combined forces (and "charms") to make this a truly memorable ASI.

REFERENCES

AMITRANO, C., CONIGLIO, A., and DI LIBERTO, F., 1986, Phys. Rev. Lett., **57**, 1016.

BALL, R., 1986, *Physica*, **140A**, 62

BEN-JACOB, E., GODBEY, R., GOLDENFELD, N. D., KOPLIK, J., LEVINE, H., MUELLER, T., and SANDER, L. M., 1985, *Phys. Rev. Lett.*, **55**, 1315.

BEN-JACOB, E., GOLDENFELD, N., KOTLIER, B. G., and LANGER, J. S., 1984, *Phys. Rev. Lett.*, **53**, 2110.

BENTLEY, W. A., and HUMPHREYS, W. J., 1962, *Snow Crystals* (Dover, NY).

BOCCARA, N. and DAOUD, M. (eds), 1985, *Physics of Finely Divided Matter* [Proceedings of the Winter School, Les Houches, 1985] (Springer Verlag, Heidelberg).

BUKA, A., KERTÉSZ, J., and VICSEK, T., 1986, *Nature*, **323**, 424.

CHANDRASEKHAR, S., 1943, *Rev. Mod. Phys.*, **15**, 1.

CHEN, J. D. and WILKINSON, D., 1985, *Phys. Rev. Lett.*, **55**, 1985.

CONIGLIO, A., 1982, *J. Phys. A*, **15**, 3829.

DACCORD, G., NITTMANN, J., and STANLEY, H. E., 1986, *Phys. Rev. Lett.*, **56**, 336.

DOUGHERTY, A., KAPLAN, P. D., and GOLLUB, J. P., 1987, *Phys. Rev. Lett.* **58**, 1652.

FAMILY, F. and LANDAU, D. P. (eds), 1984, *Kinetics of Aggregation and Gelation* (Elsevier, Amsterdam).

GAWLINSKI, E.T., and STANLEY, H. E., 1981, *J. Phys. A.*, **14**, L291-L299.

GEIGER, A., and STANLEY, H. E., 1982a, *Phys. Rev. Lett.*, **49**, 1749.

GEIGER, A., and STANLEY, H. E., 1982b, *Phys. Rev. Lett.*, **49**, 1895.

HALSEY, T. C., MEAKIN, P., and PROCACCIA, I., 1986, *Phys. Rev. Lett.*, **56**, 854.

ISING, E., 1925, *Ann. Physik*, **31**, 253-258.

KERTÉSZ, J., and VICSEK, T., 1986, *J. Phys. A*, **19**, L257.

KERTÉSZ, J., STAUFFER, D., and CONIGLIO, A., 1985, *Ann. Israel Phys. Soc.* (Adler, Deutscher and Zallen, eds), p. 121-148.

KESSLER, D., KOPLIK, J., and LEVINE, H., 1984, *Phys. Rev. A*, **30**, 3161.

LaCHAPELLE, E. R., 1969, *Field Guide to Snow Crystals* (U. Washington Press, Seattle).

LANGER, J. S., 1980, *Rev. Mod. Phys.*, **52**, 1.

LENZ, W., 1920, *Phys. Z.*, **21**, 613-5.

MÅLØY, K. J., FEDER, J., and JØSSANG, T., 1985, *Phys. Rev. Lett.*, **55**, 2688.

MEAKIN, P., STANLEY, H. E., CONIGLIO, A., and WITTEN, T. A., 1985, *Phys. Rev. A*, **32**, 2364.

MEAKIN, P., CONIGLIO, A., STANLEY, H. E., and WITTEN, T. A., 1986, *Phys. Rev. A*, **34**, 3325-3340.

MEAKIN, P., 1986a, *J. Theor. Biol.*, **118**, 101.

MEAKIN, P., 1986b, *Proc. Israel Conference on Fracture*.

NAKAYA, U., 1954, *Snow Crystals* (Harvard Univ Press, Cambridge).

NIEMEYER, L., PIETRONERO, L., and WEISMANN, H. J., 1984, *Phys. Rev. Lett.*, **52**, 1033.

NITTMANN, J., DACCORD, G., and STANLEY, H. E., *Nature* **314**, 141 (1985).

NITTMANN, J., and STANLEY, H. E., 1986, *Nature* **321**, 663-668.

NITTMANN, J., and STANLEY, H. E., 1987a, *Phys. Rev. Lett.*, (submitted).

NITTMANN, J., and STANLEY, H. E., 1987b (preprint).

NITTMANN, J., STANLEY, H. E., TOUBOUL, E., and DACCORD, G., 1987, *Phys. Rev. Lett.*, **58**, 619.

PIETRONERO, L. and TOSATTI, E. (eds), 1986, *Fractals in Physics* (North Holland, Amsterdam).

PYNN, R. and SKJELTORP, A. (eds.), 1986, *Scaling Phenomena in Disordered Systems* (Plenum, N.Y.).

SAITO, Y., GOLDBECK-WOOD, G., and MÜLLER-KRUMBHAAR, H., 1987, *Phys. Rev. Lett.*, **58**, 1541.

STANLEY, H. E., 1977, *J. Phys. A*, **10**, L211-L220.

STANLEY, H. E., 1987 *Introduction to Fractal Phenomena* (in press).

STANLEY, H. E., and OSTROWSKY, N. (eds), 1986, *On Growth and Form: Fractal and Non-Fractal Patterns in Physics* (Martinus Nijhoff, The Hague).

TANG, C., 1985, *Phys. Rev. A.*, **31**, 1977.

WITTEN, T. A., and SANDER, L. M., 1981, *Phys. Rev. Lett.*, **47**, 1400.

THE GROWTH OF FRACTAL AGGREGATES

Paul Meakin

Central Research and Development Department
E. I. du Pont de Nemours & Company
Wilmington, DE 19898   USA

INTRODUCTION

A large variety of structures with more or less well defined fractal geometries have now been recognized.[1] Many of these structures have come about as a result of nonequilibrium growth or aggregation processes. At present much of what we know about these growth and aggregation processes has come from computer simulations. Almost 20 years ago a simple ballistic aggregation model in which clusters combined with other clusters to generate successively larger and larger clusters was developed by Sutherland.[2,3] This model led to the formation of tenuous structures which exhibited mass-length scaling relationships which would now be recognized as being characteristic of fractal geometry. The mass-length scaling exponents obtained from three dimensional simulations correspond to a fractal dimensionality (D) of about 1.85. More recent larger scale simulations led to values in the range 1.90 to 1.95.[4,5]

About a decade later Forrest and Witten[6] investigated the structure of iron particle aggregates formed by vaporizing an iron coating on thin tungsten wires in a dense ($10^{-3} - 10^{-1}$ atmospheres) helium atmosphere by passing an electrical current pulse through the wire. The aggregates were collected on electron microscope grids and the micrographs were digitized and analyzed. One of these micrographs is shown in Figure 1. A fractal dimensionality of 1.8-1.95 was estimated using five different aggregates. This experimental work motivated the later development of the diffusion limited aggregation (DLA) model by Witten and Sander.[7] In this model particles are added, one at a time, to a growing cluster or aggregate of particles via random walk trajectories. A crude two dimensional version of this model is illustrated in Figure 2. By measuring the radius of gyration (R_g) as a function of the cluster mass (M) and the two point density-density correlation function (C(r)) as a function of distance (r), a fractal dimensionality of about 5/3 was estimated using the relationships

$$R_g = M^\beta, \quad D_\beta = 1/\beta \tag{1}$$

and

$$C(r) \sim r^{-\alpha}, \quad D_\alpha = d-\alpha \tag{2}$$

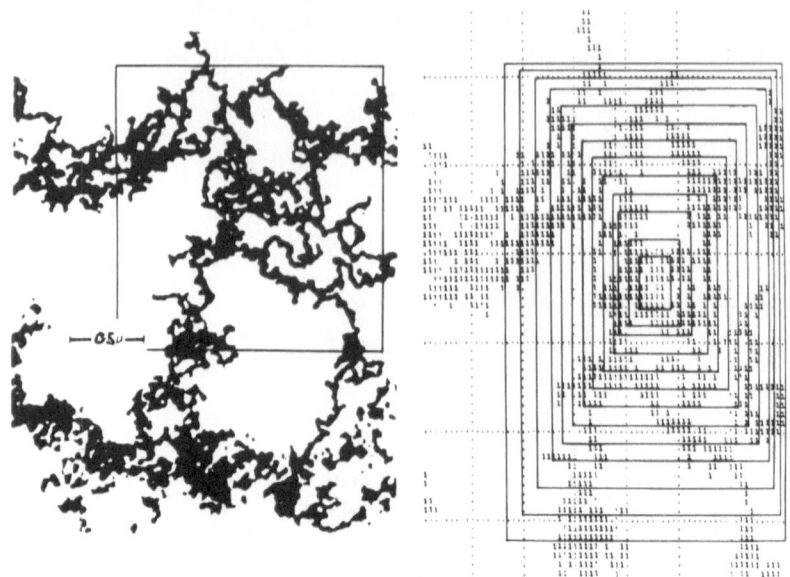

Fig. 1. This figure shows an electron micrograph of an iron particle
aggregate (S. R. Forrest and T. A. Witten, J. Phys. A12:109
(1979)). It represents a projection of a three-dimensional
structure onto a plane. Because the fractal dimensionality
is smaller than 2, essentially all of the structure can be
seen in the projection which has the same fractal
dimensionality as the three-dimensional object. This figure
was provided by T. A. Witten and is reproduced with
permission.

where d is the ordinary Euclidean dimensionality of the embedding space
or lattice. For self similar fractals $D_\alpha = D_\beta = D$ where D is the
"general purpose"[8] fractal dimensionality. Figure 3 shows a 50,000
particle cluster generated using an improved off-lattice version of the
model.[9]

The DLA model does not represent at all well the process of colloidal
aggregation and the three-dimensional DLA model does not generate
structures which resemble that shown in Figure 1. For example $D \simeq 2.5$
for 3d DLA vs. 1.8 - 1.95 in the real aggregates. However, the DLA model
has subsequently been found to be relevant to a large variety of
processes including fluid-fluid displacement in porous media[10,11] and
Hele-Shaw cells,[12,13,14] dielectric breakdown,[15] electrodeposition,[16,17]
the morphology of thin film deposits,[18,19] the dissolution of porous
rock[20] and possibly biological growth processes.[21,22] The development of
the DLA model by Witten and Sander stimulated considerable interest in
non-equilibrium growth and aggregation processes in general. The DLA
model illustrated that simple growth and aggregation models could lead to
valuable insights into important physical (and chemical) processes. This
has motivated much of the subsequent work in this area.

The ballistic cluster-cluster aggregation model of Sutherland[2,3]
leads to structures which resemble that shown in Figure 1. A more
realistic model for most colloidal aggregation processes is the diffusion
limited cluster-cluster aggregation model[23,24] in which particles and
clusters are brought together using random walk rather than ballistic
trajectories. Figure 4 shows a cluster generated using a
three-dimensional off-lattice version of this model in which clusters are

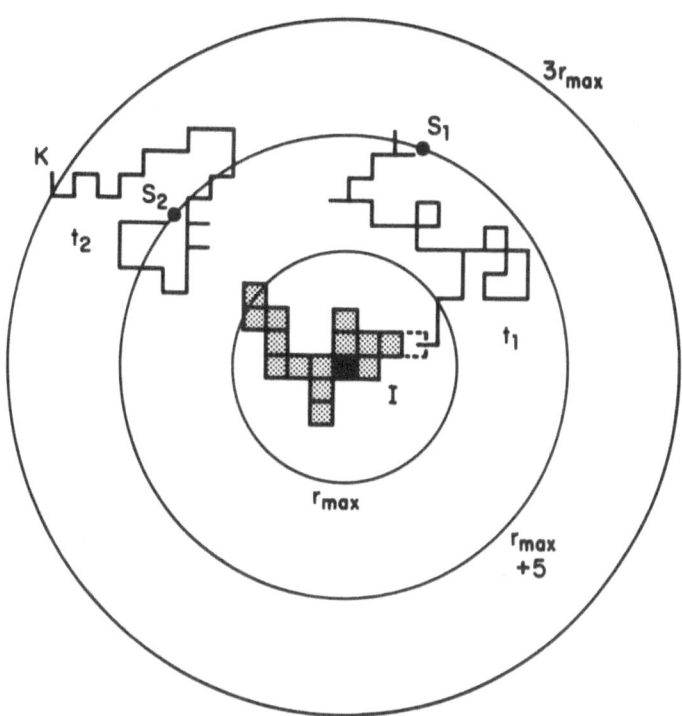

Fig. 2. A simple two-dimensional lattice model for diffusion limited
aggregation. Particles are launched, one at a time, from
randomly selected points (S_1 and S_2) on a circle which
encloses the cluster. They are then moved to the nearest
lattice site and undergo a random walk on the lattice. Two
particle trajectories (t_1 and t_2) are shown. Trajectory t_1
eventually brings the particle into an unoccupied lattice
site on the surface of the cluster (dashed border). At this
point the trajectory is stopped and the unoccupied surface
site will be filled. Trajectory t_2 moves the particle away
from the cluster and when it crosses the "killing" circle
with a radius of $3r_{max}$ (r_{max} is the maximum radius of the
cluster) it is terminated and a new trajectory is started
from a randomly selected position on the launching circle
which, in this case, has a radius of r_{max} + 5 lattice units.
The simulation is started with a single occupied site
(black) and is shown at an early stage where 14 additional
sites have been filled (shaded).

randomly selected from a list of clusters (which initially contains only
single particle clusters) and combined via random walk trajectories
before being returned to the list (which is then shorter by one member).
Models of this type[25] as well as closely related lattice models[26] lead to
clusters with a fractal dimensionality of about 1.78. This value for D is
in good agreement with that found in the experiments of Forrest and
Witten and in more recent experimental work on colloid aggregation under
fast aggregation conditions.[27-30]

Another simple limiting case scenario for colloid aggregation is
represented by the chemically limited aggregation model.[31] In this model

←————————— 1000 DIAMETERS —————————→

Fig. 3. A 50,000 particle cluster generated by an improved
off-lattice version of the two-dimensional DLA models shown
in Figure 2.

rigid clusters are combined irreversibly (as is also the case in the
ballistic and diffusion-limited cluster-cluster aggregation models).
However, in this model it is assumed that very many contacts are required
before sticking occurs and that the system can sample all possible
configurations (or at least a representative sample of them). In a
simplified version of this model[32,33] pairs of <u>particles</u> are picked at
random and brought into contact with each other with a randomly selected
orientation. If a selected paricle is part of a cluster, the associated
particles are carried along with the randomly selected particle. If an
overlap occurs between any pair of particles in the two clusters, the
clusters are returned to the list of clusters and a new pair of particles
is randomly selected. If the two randomly selected particles belong to
the same clusters, a new selection is also made. If, on the other hand,
no overlaps are found, the two clusters are combined, returned to the
list as a single cluster and the list of clusters is shortened by one
member. Figure 5 shows a 3d cluser generated using an off-lattice
version of this model. The ensemble averaged fractal dimensionality
obtained from this model is about 2.09.[34] Values for D close to this
have been found in experimental systems in which slow irreversible
aggregation is occurring.[35-37]

    In most of the early work on the growth of fractal aggregates
interest was initially focussed on their geometric properties. More

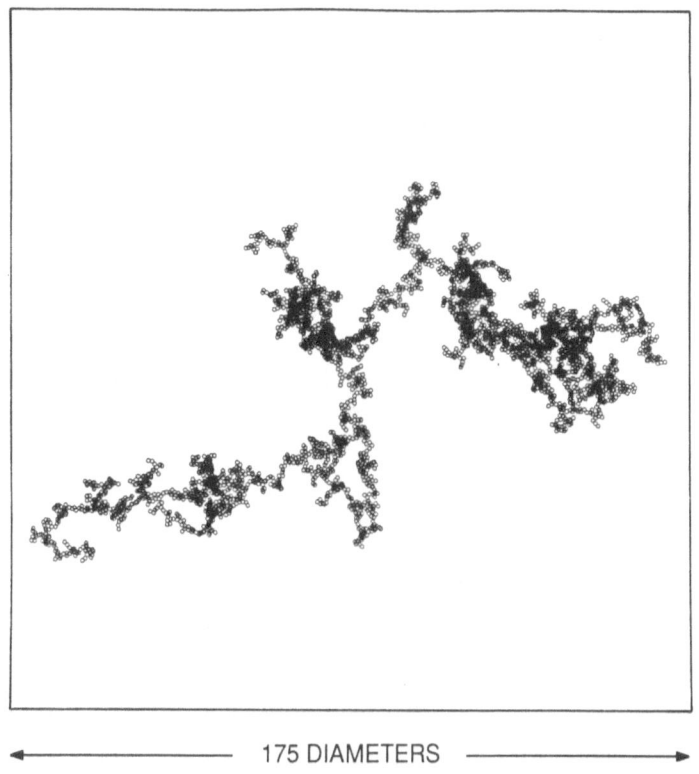

←——————————— 175 DIAMETERS ———————————→

Fig. 4. A cluster of particles generated by a three-dimensional
model of diffusion-limited cluster-cluster aggregation.

recently interest has also developed in the nature of physical and
chemical processes occurring on fractal substrates and on the way in
which fractal structures grow including the time dependence (kinetics) of
the growth process. In these lectures I will be concerned mainly with
this aspect of the growth of fractal aggregates.

The Growth of DLA Clusters

It was realized very early on in the investigation of the DLA model[7]
that one of the most important aspects of the growth process is that
trapping of the random walkers by the extremities of the cluster very
effectively prevents them from penetrating into the interior regions of
the growing aggregate. Figure 6 shows some results obtained from a
simulation in which 10^6 random walkers were used to probe the surface of
a 50,000 particle off-lattice DLA cluster.[38] The "random walkers" are
circles of unit diameter just like those in the cluster. They are
launched from random positions on a circle enclosing the cluster and are
terminated on contact with the cluster or on reaching a distance of
100 r_{max} from the center of the cluster. A record is kept of how many
times each of the particles in the cluster is contacted. Figure 6a shows
the cluster itself and Figure 6b shows all of those particles which have
been contacted after 10^6 total contacts with the cluster. It is apparent
from this figure that very few of the particles have been able to
penetrate very far into the interior of the cluster. The concentration
of the growth probability onto the most outer regions of the cluster is
even more apparent in Figure 6c which shows those particles which have

100 DIAMETERS

Fig. 5. A cluster of 9785 particles generated by a three-dimensional
off-lattice model for chemically limited cluster-cluster
aggregation.

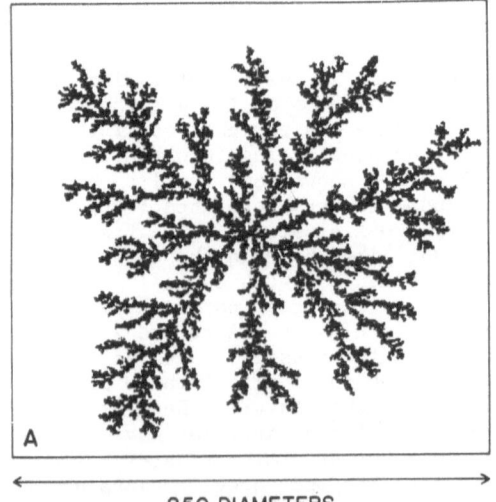

850 DIAMETERS

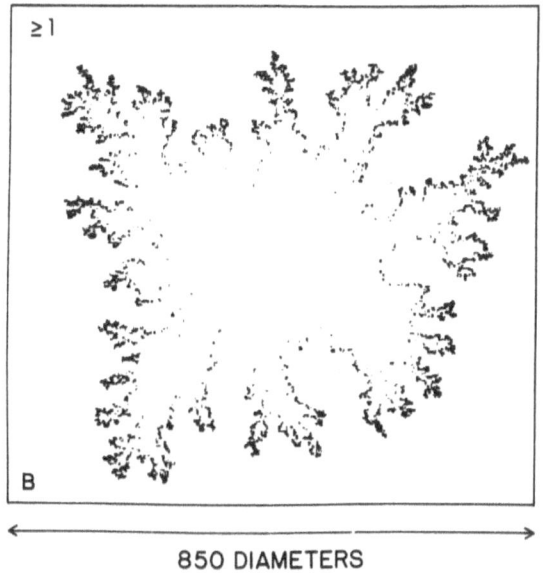

850 DIAMETERS

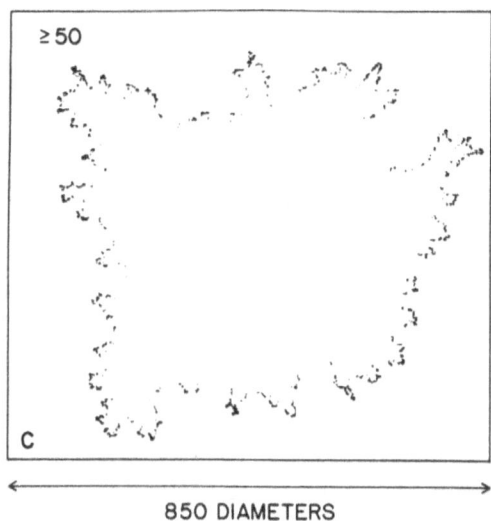

850 DIAMETERS

Fig. 6.   The distribution of contacts for a 50,000 particle
          off-lattice DLA cluster which has been probed using 10^6
          random walkers which are terminated after their first
          contact with the cluster. Figure 6a shows the cluster,
          Figure 6b shows all of the contacted particles and Figure 6c
          shows those sites contacted 50 or more times.

been contacted 50 or more times. Figure 6 provides a partial visualization of the active zone[39] (region where growth is occurring) in the DLA cluster.

It has recently been shown[38,40-44] that the distribution of growth probabilities on the surface of growing DLA clusters (and a variety of other systems) can be described in terms of a fractal measure or multifractal distribution. The fractal measure is characterized by a continuous spectrum of scaling indices and associated fractal dimensionalities.[45-47] This fascinating aspect of the growth of fractal aggregates is not the main topic of these lectures and will be discussed in the lectures by H. E. Stanley.

The implications of Figure 6 for our purpose is that the absorption of diffusing particles by the growing aggregate can be approximated very well by an absorbing circle with a radius of r_a where r_a is the radius of the active zone which grows with increasing cluster mass according to the power law

$$r_a \sim M^\nu \tag{3}$$

The exponent ν has the same value as the exponent β which describe how the radius of gyration, maximum radius and other lengths which characterize how the overall size of the clusters grow with increasing cluster mass (M).[39,48]

Computer simulations indicate that strong screening of the diffusing particles occurs in three as well as two dimensions[49-51] and simple theoretical arguments[52,53] indicate that effective screening will occur for all d.

As originally formulated by Witten and Sander[7] the DLA model specifies an ordered sequence of events but there is no time scale. For spaces or lattices of dimensionality, d, ≥ 3 it would, on average, take a particle an infinite time to find the cluster if its trajectory were not terminated at some surface enclosing the cluster.

One realization of the DLA model is the growth of an absorbing aggregate in a dilute field of diffusing particles. Under these conditions the flux of particles onto the surface of an aggregate in a d dimensional space can be obtained from the steady state diffusion equation

$$1/r^{d-1} \; d/dr \; r^{d-1} \; d/dr \; C(r) = 0 \tag{4}$$

with absorbing boundary conditions at the surface of the aggregate which can be represented by a hypersphere or radius $R \simeq r_a$ (vide supra). From equation (4) it can be shown that the steady state concentration profile which can be used to determine the flux onto the surface is given by

$$C(r) = C_o [1-(R/r)^{d-2}] \tag{5a}$$

for d>3. The concentration profile given in equation (5a) was obtained assuming that C(R) = 0 where R is the radius of an effective hyperspherically symmetric absorbing surface and that $C(r) \rightarrow C_o$ as $r \rightarrow \infty$ where C_o is the initial concentration. From equation (5a) we find that the flux of particles, J, onto the cluster is given by

$$J = dM/dt = \mathcal{D} C_o R^{d-2} \qquad\qquad (5b)$$

where $\mathcal{D}$ is the particle diffusion coefficient. For $d = 2$ it is not possible to obtain a steady state solution to the diffusion equation with an initial concentration boundary condition. (A steady state concentration profile is not formed because two dimensional space is not "big enough" and a depleted region in the vicinity of the absorbing surface continues to grow in size.).

Equation (5) with the power law dependence of M on R ($M \sim R^D$) gives the result[52]

$$R(t) \sim t^{1/(D-d+2)} \qquad\qquad (6)$$

For $d = 3$ both on-lattice and off-lattice computer simulations indicate that D has an effective value of about 2.5[51,53] so that[52,54] $R(t) \sim t^{2/3}$ and $N(t) \sim t^{5/3}$. For $d = 2$ equation (6) would predict $R(t) \sim t^{1/D}$ and $N(t) \sim t$ (In this case $1/D \simeq 0.585$[48].). However, it should be noted that logarithmic corrections are expected in $d = 2$ as a result of the fact that there is no steady state solution to the diffusion equation.

Multi-Particle DLA Simulations

Both two dimensional[55-57] and three dimensional[57] simulations of growth from a "sea" of particles have been carried out. In this model sites are randomly occupied on a lattice until the required concentration or density (ρ) is reached. Particles are then selected randomly and moved by one lattice unit in a randomly selected direction (providing this would not result in multiple occupancy). After each move the nearest neighbors are checked to determine if the particle is at the surface of the growing cluster. If a nearest neighbor is occupied by a site in the cluster, the particle becomes permanently part of the cluster. This model can easily be made time dependent by incrementing the time by 1/N (where N is the number of free particles in the system) after each particle has been selected. Figure 7 shows a DLA cluster growing in this fashion. Clusters grown using the multiparticle DLA model exhibit the fractal structure associated with DLA on short length scales but are uniform ($D = d$) on longer length scales. The correlation length ξ depends on ρ and $\xi \to \infty$ as $\rho \to 0$. It would be reasonable to expect that $\xi \sim \rho^{-1/\alpha}$ where $\alpha = d - D$ (about 0.3). However, 2d computer simulations[55] indicate that $\xi \sim \rho^{-2}$ or $\xi \sim e^{-(\rho 1/2)}$. Since ξ is rather difficult to measure accurately these results do not rule out the simple idea that $\xi \sim \rho^{-1/\alpha}$. Both Voss[55] and Meakin and Deutch[57] find that $M(t) \sim t$ for low densities. Neither of these simulations has been carried out on a large enough scale with enough samples to determine if logarithmic, or other, corrections are present. At higher densities the simulation results differ somewhat. Voss interpreted his results in terms of a power law growth of the mass $M(t) \sim t^{\eta(\rho)}$ where $\eta(\rho)$ is a continuous function of ρ. Meakin and Deutch interpreted their results in terms of a crossover from $\eta = 1$ at short times to $\eta = 2$ at long times. More extensive simulations will be needed to resolve these discrepancies.

In three dimensions Meakin and Deutch found that $M(t) \sim t^{\eta}$ where the exponent η has a value of about 5/3 at low densities. They also found that the cluster radius of gyration (R_g) grows with time according to $R_g \sim t^{2/3}$ in accord with equation (6). At higher densities the effective

Fig. 7.   A simulation of multiparticle DLA with a concentration of
          0.05 particles per lattice site. The stationary cluster
          contains 5,000 sites and the depletion of the mobile
          particles in the vicinity of the cluster is shown. This
          figure was provided by R. F. Voss and is reproduced with
          permission.

exponent $\eta\ (\rho)$ increases and the simulation results are consistent with
either a continuously varying exponent or a crossover[57] from a value of 1
at low densities to 3 at high densities.

Growth From A Laplacian Field

     The DLA model can be used to simulate random growth processes
controlled by a field ϕ obeying Laplace's equation $\nabla^2 \phi = 0$ with absorbing
boundary conditions ($\phi = 0$ on the growing cluster). The close
correspondence between this process and DLA is demonstrated by the
dielectric breakdown model of Niemeyer, Pietronero and Weismann.[59]   This
model is more flexible than the DLA model and is well suited to the
investigation of time dependent problems (but so far not much emphasis
has been placed on this aspect of the model). The main disadvantage of
the dielectric breakdown model is that relatively large amounts of
computer time are required.

     For the cases where the growth probability is linearly related to the
local field or field gradient at the surface of the growing cluster,
equivalent results can be obtained more rapidly and without the
uncertainties associated with the possibility of incomplete relaxation of

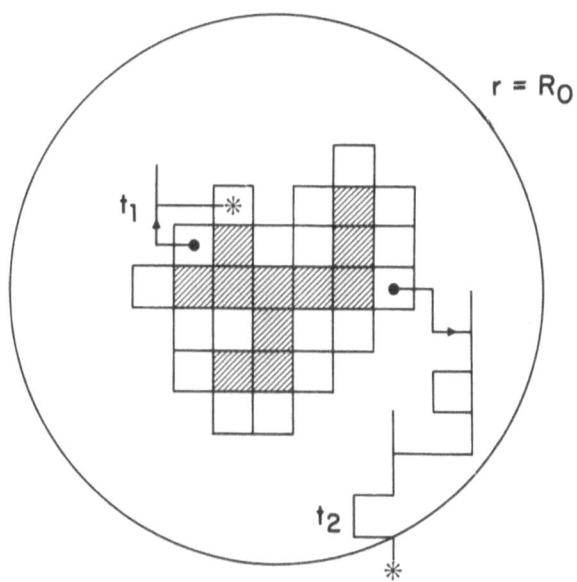

Fig. 8. A model for the displacement of a viscous fluid by a
nonviscous fluid in a porous medium contained in a 2d
Hele-Shaw cell using random walkers to simulate the
Laplacian pressure field in the viscous fluid. Particles
are launched from randomly selected unoccupied surface sites
(empty sites with borders). Figure 8 shows two
representative trajectories. Trajectory t_1 returns to the
surface. In this event a new trajectory is started from a
randomly selected surface site. Trajectory t_2 eventually
reaches the outer boundary. The site from which the
trajectory was started (indicated by a dot) is filled, newly
generated surface sites are identified and an unoccupied
surface site is randomly selected to initiate a new random
walk. Occupied sites (which represent the nonviscous fluid)
are shaded.

the Laplacian field using random walkers to simulate the Laplacian field.
For cases where the growth probability is proportional to ϕ^h (or $\nabla\phi^h$)
random walkers can also be used[60] (particularly if η is an integer) but
the advantages of using random walkers are more dubious.

Figure 8 represents a model for simulating growth controlled by a
Laplacian field. This model was motivated by experiments on fluid-fluid
displacement experiments in a 2d porous medium inside a radial Hele-Shaw
cell.[11] In this case the field obeying Laplace's equation is the
pressure (P) inside the viscous fluid (which is displaced by a nonviscous
fluid) and the boundary conditions are P = Constant (for convenience a
value of 1 can be used since the equations are linear and do not depend
on the magnitude of P) at the fluid-fluid boundary and P = 0 at the outer
boundary of the cell. The simulation is based on the idea that the

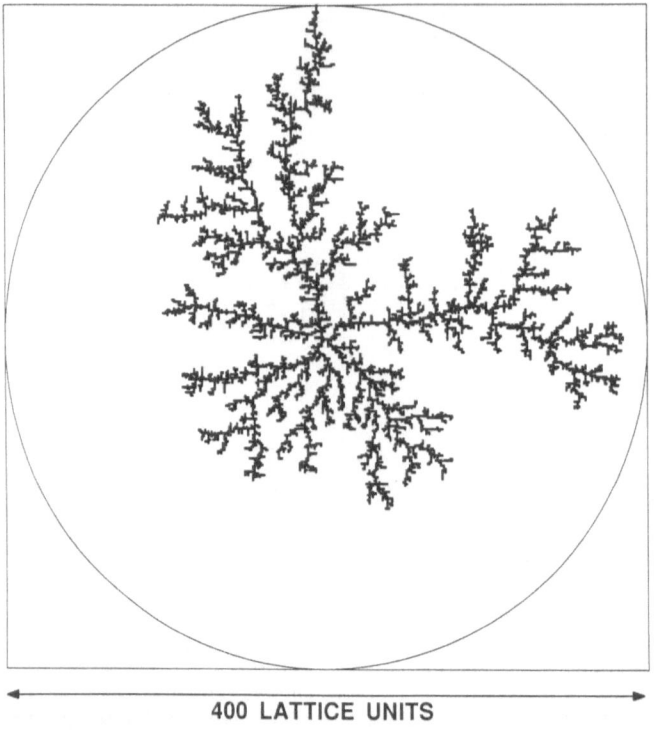

400 LATTICE UNITS

Fig. 9. A cluster grown on a square lattice using the model shown in
Figure 8.

frequency with which a lattice site is visited by a random walker is
proportional to the pressure associated with that site. The boundary
condition P = 0 can be satisfied by terminating any random walker which
reaches the outer boundary. The boundary condition P = 1 on the surface
of the growing cluster can be satisfied by launching random walkers with
equal probabilities from all of the sites on the surface of the growing
cluster. If a random walker returns to a surface site (trajectory t_1 in
Figure 8), it is terminated to preserve the boundary condition P = 1 on
the fluid-fluid interface. Growth occurs when a random walker launched
from the surface reaches the outer boundary (trajectory t_2 in Figure 8).
In this event the site from which the random walker was launched is
occupied. A timescale can be introduced into the model by incrementing
the time by $1/N_s$ every time a new random walk is started at the surface
of the clusters. Here N_s is the total number of surface sites.
Equivalent results could be obtained by launching the random walkers from
the outer boundary and incrementing the time by a constant amount. In
this model a new trajectory would be started if the random walker
returned to the outer boundary or contacted the cluster (in the later
event the newly contacted surface site would be filled).

Figure 9 shows an aggregate grown on a square lattice using the
algorithm outlined above. Because of the influence of the outer boundary
there is an additional instability in the growth process and the
aggregates are much less "symmetric" than those grown using the original

56

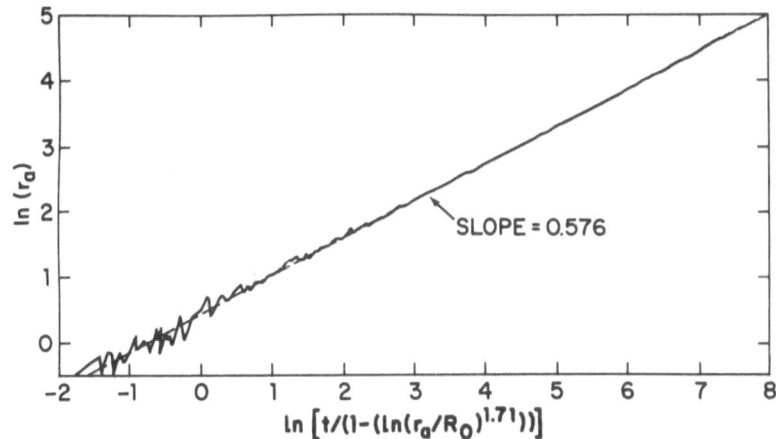

Fig. 10. Figure 10 results from a simulation of a two-dimensional random growth process controlled by a Laplacian field. In this case growth from the center of a circle with a potential of zero at the perimeter and a constant potential on the growing structure was simulated using random walkers. This figure shows the dependence of $\ln(r_a)$ on $\ln[t/(1-1.71 \ln(r_a/R_0))]$ (r_a is the radius of the active zone and R_0 is the radius of the cell). Simple theoretical considerations (outlined in the text) indicate that the slope should be $1/D$ (about 0.585).

DLA model with a distant killing circle. The boundary conditions at the surface of the aggregate correspond to those used in the original DLA model of Witten and Sander.[7] Simulations can be carried out corresponding to the dielectric breakdown[59] boundary conditions by launching and killing random walkers on the aggregate rather than at unoccupied surface sites.[61] In order to model the time dependence of fluid-fluid displacement in a porous medium[11] similar simulations were carried out using a hexagonal instead of a square lattice. For the hexagonal lattice the effects of lattice anisotropy (which are already apparent in Figure 9) are very much reduced.[62]

For a circular interface of radius R the flux onto the interface would be given by

$$J \sim P/\ln(R_o/R) \tag{7}$$

where R_o is the radius of the cell.

Using arguments similar to those employed in the case of growth from a concentration field, we find that

$$dR/dt \sim R^{(1-D)}/\ln(R_o/R) \tag{8}$$

assuming that the effective radius R can be replaced by the radius of the active zone then equation (8) leads to

$$t \sim \int R^{D-1} \ln(R_o/R) dR \tag{9}$$

and the solution to equation (9) is given by

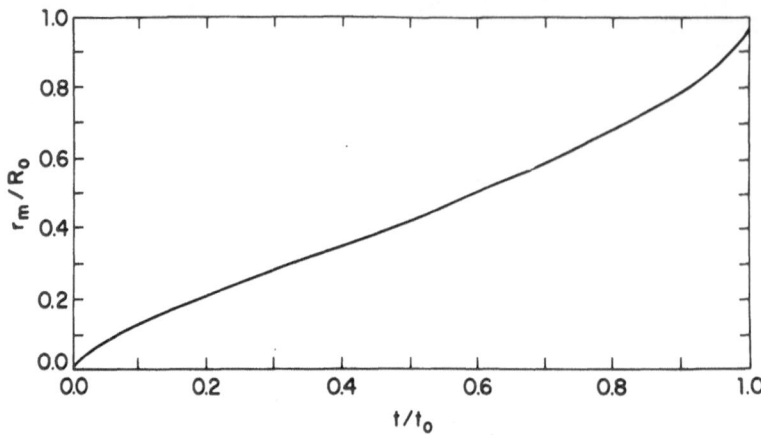

Fig. 11. Dependence of r_m/R_o on t/t_o from the simulations used to
obtain Figure 10. Here r_m is the maximum radius of the
cluster and t_o is the time at which the perimeter is
reached. The results shown here represent the average over
20 simulations.

$$t \sim (r_a/R_o)^D [1-D \ln (r_a/R_o] \qquad\qquad (10)$$

Figure 10 shows the dependence of $\ln(r_a)$ on $\ln[t/(1-D\ln(r_a/R_o))]$. A
least squares fit gives a slope of 0.570 ± 0.003 which is quite close to
the value of $1/D$ (0.585). If $\ln(r_a)$ is plotted against
$\ln[t/(1-2\ln(r_a/R_o))]$ almost identical results are obtained (a slope of
0.568 ± 0.003). These results were obtained from 20 simulations carried
out on a hexagonal lattice with a radius (R_o) of 200 lattice units.
Figure 11 shows the dependence of r_m/R_o on t/t_o obtained from the same
set of simulations. Here r_m is the maximum cluster radius and t_o is the
time at which the edge of the cell is reached. The results shown in
Figure 11 have been compared with experiments in which glycerol is
displaced by air from a two-dimensional porous medium with a constant
pressure difference across the cell.[11] The simulation results were found
to be in good agreement with experiments carried out at a high capillary
number which give fingering patterns which closely resemble DLA.[63] (The
experiments used to investigate fluid-fluid displacement processes in two
dimensional porous media will be discussed in the lecture by Jens Feder.)
In this section we have seen that most aspects of the kinetics of the
growth of DLA can be understood in terms of very simple theoretical ideas
if it can be assumed that the cluster radius is related to mass by a
power law. For the most part computer simulations are in good agreement
with these ideas.

Cluster-Cluster Aggregation

    Three stages in a two-dimensional simulation of cluster-cluster
aggregation on a square lattice at a small but finite concentration are
shown in Figure 12. In this model clusters are selected at random and
moved by one lattice if a random number X uniformly distributed in the
range $0<X<1$ is smaller than $\mathscr{D}(s)/\mathscr{D}_{max}$ where $\mathscr{D}(s)$ is the diffusion
coefficient for clusters of size s and $\mathscr{D}_{max}$ is the maximum diffusion
coefficient for any cluster in the system. After each cluster has been

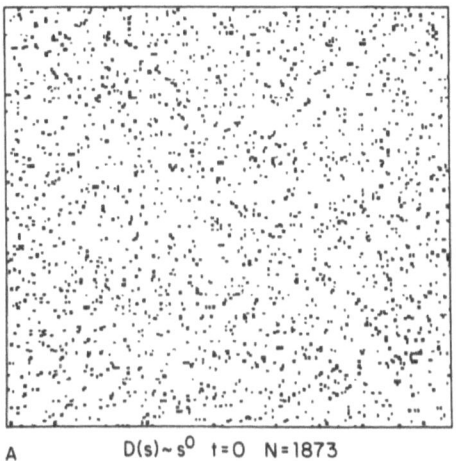

A $D(s) \sim s^0$ $t = 0$ $N = 1873$

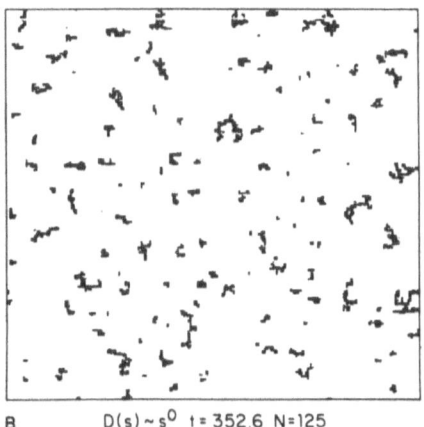

B $D(s) \sim s^0$ $t = 352.6$ $N = 125$

Fig. 12. Three stages in a small scale 2d simulation of
diffusion-limited cluster-cluster aggregation with a mass
independent cluster diffusion coefficient ($D(s) \sim S^0$). The
simulation was carried out on a 200x200 site square lattices
using 2,000 occupied lattice sites to represent 2,000
particles. Figure 12a shows the initial random starting

moved, its perimeter is examined and clusters which move into contact with each other (via nearest neighbor occupancy) are irreversibly combined to form a larger cluster. Simulations of this type can easily be made time dependent by incrementing the time by $1/(N\mathscr{D}_{max})$ where N is the total number of clusters and $\mathscr{D}_{max}$ is the maximum diffusion coefficient.

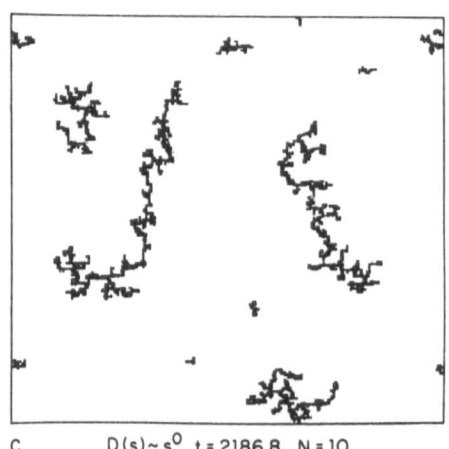

c $D(s) \sim s^0$ t = 2186.8   N = 10

configuration. Because some particles occupy adjacent (nearest neighbor) sites there are 1873 clusters in the system. Figure 12b shows the simulation at the stage where the number of clusters has been reduced to 125 and Figure 12c shows a later stage at which N = 10. The simulation was made time dependent by increasing t by 1/N after each cluster had been moved by one lattice unit.

Under a wide variety of conditions the cluster size distribution evolves into a self-preserving form which is independent of initial conditions. Under these conditions and assuming that there is only one characteristic size, it is natural to postulate the scaling form

$$N_s(t) \sim s^{-\theta} f(s/S(t)) \qquad (11)$$

Here $N_s(t)$ is the number of clusters of size (mass) s at time t and S(t) is the mean cluster size

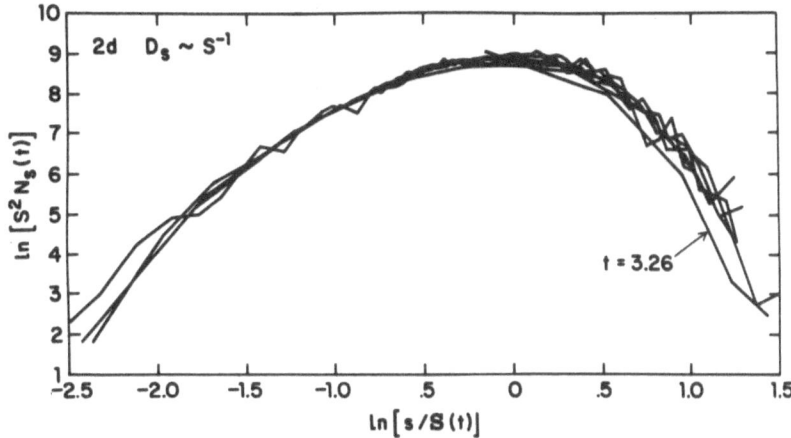

Fig. 13. Scaling of the time dependent cluster size distribution obtained from a 2d simulation of the aggregation of rods assuming that $D(s) \sim s^{-1}$. This figure shows how the size distributions, $N_s(t)$, over a wide range of times (t = 3.26 to t = 31782) can be scaled onto a common curve for all but very short times.

$$S(t) = \sum_{s=1}^{\infty} s^2 N_s(t) / \sum_{s=1}^{\infty} s N_s(t) \qquad (12)$$

at the same time. Conservation of mass requires that $\Theta = 2$ so that we have

$$N_s(t) = s^{-2} f(s/S(t)) \qquad (13)$$

This equation seems to have been derived independently in several different ways (see references 64-69, for example). Figure 13 shows an example of this scaling form for a two-dimensional simulation of the aggregation of rods.[70] This model resembles the diffusion-limited cluster-cluster aggregation model except that sticking is allowed in only the X or Y directions on the lattice.

Many aspects of cluster-cluster aggregation can be understood in terms of the mean field Smoluchowski equation[71]

$$dC_j/dt = - \sum_{i=1}^{\infty} K(i,j)C_i C_j$$

$$+1/2 \sum_{i+k=j} K(i,k)C_i C_k \qquad (14)$$

In equation (14) the first term on the left hand side represents the loss of clusters of size j due to reaction to form larger clusters and the second term represents the formation of clusters of size j by the addition of smaller clusters. In cluster-cluster aggregation simulations on a lattice three (or more) clusters may join to form a single cluster, but this is a rare event which becomes increasingly improbable as the concentration is reduced.

Despite the success of the Smoluchowski equation approach, computer

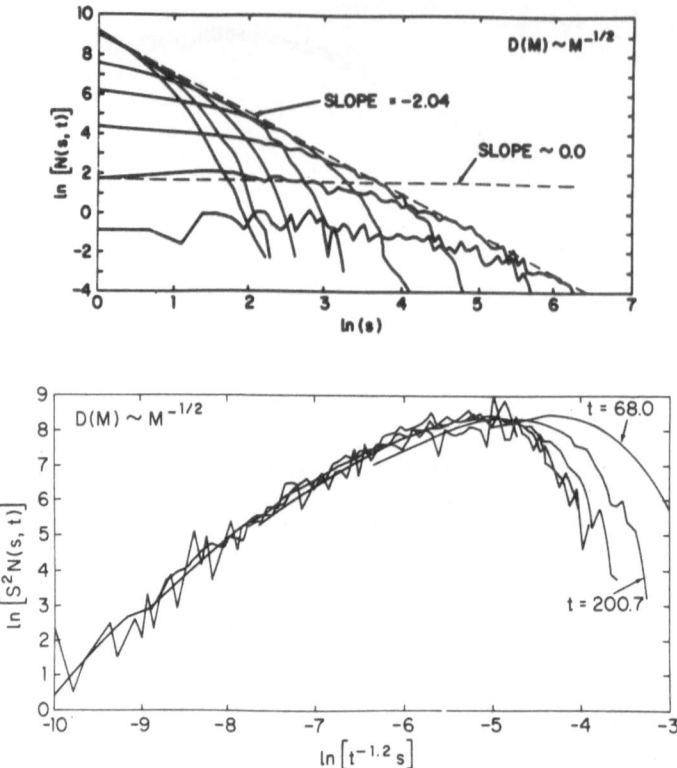

Fig. 14. Some results from a 3d simulation of diffusion-limited cluster-cluster aggregation in which the cluster diffusion coefficients were given by (s) ~ $s^{-1/2}$. These simulations were carried out using particles on 133^3 lattices with periodic boundary conditions. Figure 13a shows the cluster size distributions at several stages during the simulation and Figure 13b shows how these results can be scaled using the scaling function $N_s(t) \sim s^{-2} f(s/t^z)$ with $z \simeq 1.2$. This scaling form is closely related to that given in equation (11) since $S(t) \sim t^z$ with $z \simeq 1.2$ in the scaling regime (at long times but before only a few clusters remain). The results from a number of simulations were averaged to obtain these results.

simulations are still needed for several reasons including (1) The Smoluchowski equations are mean field equations, and it is not always obvious under what circumstances the effects of fluctuations can be neglected. It now seems that d = 2 is the critical dimension (logarithmic corrections are needed) and that the mean field approximation fails for d = 1.[72-75] (2) Analytical solutions are not available for all simple reaction kernels (K(i,j)) However, in most cases scaling arguments can be used to obtain the exponent z which describes the growth of the mean cluster size (S(t)) and decay of the number of

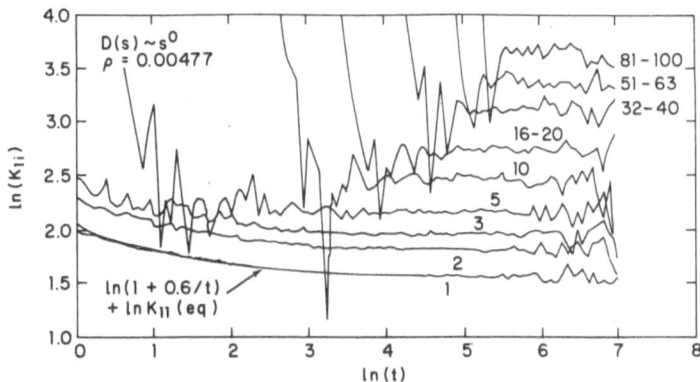

Fig. 15. The effective rate constants (K(1,i)) for the combination of
single particles and clusters containing i particles. These
results were obtained from 483 simulations using 10,000
particles on cubic lattices with 128^3 sites. In these
simulations a mass independent diffusion coefficient was
used.

clusters ($N(t)$) and τ which describes the cluster size distribution
($N_s(t) \sim s^{-\tau}$)).[76,77] (3) It is not always obvious what the correct
reaction kernel is for a particular physical process or model.

Diffusion-Limited Cluster-Cluster Aggregation

To simulate the kinetics of diffusion-limited cluster-cluster
aggregation, it is frequently assumed that the cluster diffusion
coefficient is determined by its mass.[78,79] ($\mathscr{D}(s) \sim s^\gamma$, for example).
For the ease of motion in a dense fluid scaling arguments and
calculations based on Kirkwood Riesman[80] theory[81] indicate that $\mathscr{D}(s) \sim s^{-1/D}$. Figure 14 shows results from a 3d simulation in which it was
assumed that $\mathscr{D}(s) \sim s^{-1/2}$ (not very much different from $\mathscr{D}(s) \sim s^{-1/D}$
since $D \simeq 1.78$). The results shown in Figure 14 and a variety of related
simulations[67-69,81,82] indicate that the scaling form given in equation
13 describes the asymptotic cluster size distribution in 3d
cluster-cluster aggregation. This scaling form also seems to be
applicable for $d = 2$[67-69,82] and $d = 4$[68,89] though these may be
logarithmic corrections for $d = 2$ which have not been detected in the
simulations.

For 3d diffusion-limited cluster-cluster aggregation the mass
dependent diffusion coefficient $\mathscr{D}(s) \sim s^{-1/D}$ leads to the reaction kernel

$$K(i,j) \sim (i^{1/D}+j^{1/D})(i^{-1/D}+j^{-1/D}) \qquad (15)$$

The reaction kernel has been measured directly in 3d simulations by Ziff
et al.[83] and has been found to be consistent with equation (15). Figure
14 shows the measured collision rates (K(1,i)) between single particles
and clusters of size i. Except for short times (where the initial
conditions are important) the elements of the kernel (K(i,j)) are
constant and their magnitudes are in good agreement with equation (15).

Simple Monte Carlo methods can be used to simulate the evolution of
the cluster size distribution governed by a Smoluchowski equation. In
this case geometry is ignored (except for its influence on the reaction

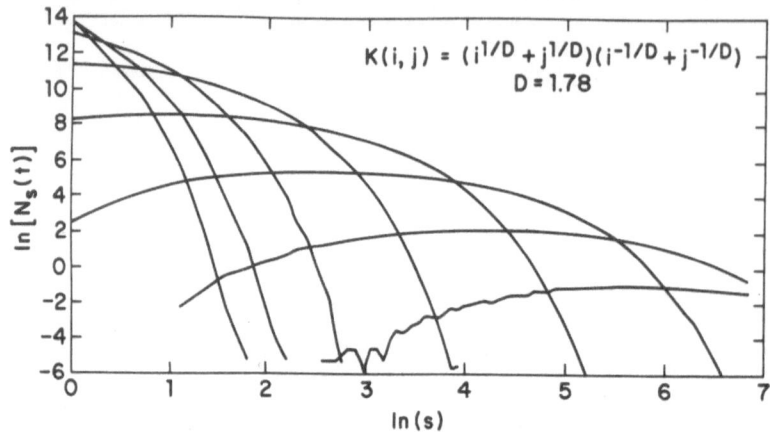

Fig. 16. Monte Carlo simulation of the mean-field Smoluchowski equation with the kernel $K(i,j) = (i^{1/D}+j^{1/D})(i^{-1/D}+j^{-1/D})$ and $D = 1.78$. Each simulation was started with 10^6 particles each with a unit mass and the results from 200 simulations were averaged. In this way the evolution of 2×10^8 particles controlled by the Smoluchowski equation for 3d diffusion-limited cluster-cluster aggregation was simulated. The cluster size distribution is shown at times $t = 4.27 \times 10^{-9}$, 2.14×10^{-8}, 1.07×10^{-7}, 5.37×10^{-7}, 2.69×10^{-6}, 1.35×10^{-5}, 6.76×10^{-5} and 3.39×10^{-5}.

kernel) and it is only necessary to keep track of the size of each cluster in a list. Starting with a long list of objects of unit mass, pairs of objects are selected from the list at random. If a random number X ($0<X<1$) is smaller than $K(i,j)/K_{max}$, the two objects are combined and returned to the list as a single object of mass $i+j$. Otherwise, the two objects of mass i and j are returned to their original places in the list and another selection is made. Here K_{max} is the maximum values of $K(i,j)$ for any pair of objects in the list. Each time a pair of particles is selected (whether or not they are combined), the time is incremented by $1/(N^2 K_{max})$. Figure 16 shows the results of a simulation using the reaction kernel given in equation (16) (with $D = 1.78$). The results shown in this figure are in good agreement with both cluster-cluster aggregation model simulations and experiments on the aggregation of colloidal gold.[37] A different Monte Carlo method for simulating the evolution of cluster size distributions from a Smoluchowski equation has been discussed by Spouge.[84]

Ballistic Cluster-Cluster Aggregation

In order to simulate the kinetics of ballistic cluster-cluster aggregation, it is necessary to make some assumptions about the distribution of cluster velocities. Here we will assume that aggregation is occurring at a very low density in a low density gas. Under a wide range of conditions the cluster mean free path will be much larger than the cluster size and there will be many cluster-gas molecule collisions between each cluster-cluster collision. Under these conditions the reaction kernel can be written down using the kinetic theory of gases[85]

$$K(i,j) \sim \sigma(i,j) \, ((i+j)/ij)^{1/2} \qquad\qquad (16)$$

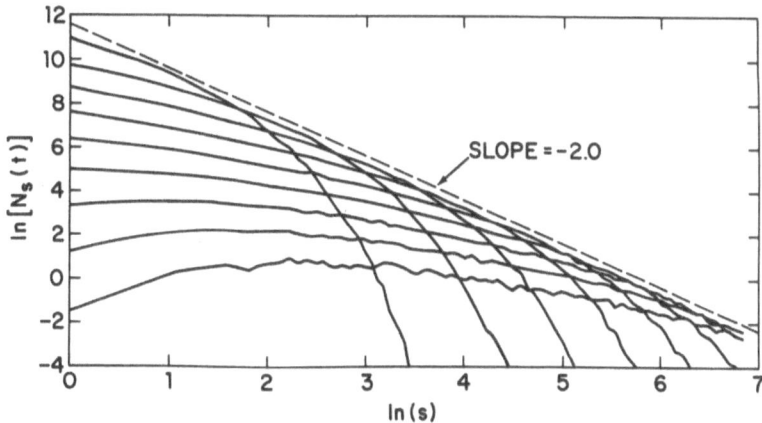

Fig. 17. Results from a simulation of 3d ballistic aggregation in
which the velocity (v) of clusters of different masses were
assumed to be given by the kinetic theory of gases (v ~
$s^{1/2}$). The results shown here were obtained from 32
simulations employing 200,000 identical spherical particles
each. Here the cluster size distributions are shown at the
stages where the system contains (10^5, 5×10^4,
$2.5 \times 10^4 \dots 200,000/2^9$) clusters.

where $\sigma(i,j)$ is the geometric collision cross section for clusters of
size i and j. Conditions such as those assumed to obtain equation (16)
are likely to be met in the aggregation of cosmic dust particles.[86]
Since cluster-cluster aggregates have a fractal dimension of about 1.95
(they are not "transparent" to each other) equation (16) can be
approximated by

$$K(i,j) \sim (i^{1/D} + j^{1/D})^2 \; ((i+j)/ij)^{1/2} \tag{17}$$

Equation (17) gives a good approximation to the true form for K(ij)
except for collisions between very large and very small aggregates (since
D<2 large clusters are transparent to single particles but this effect
becomes important only for very large clusters and by the time very large
clusters have grown, there are no single particles left).

One way of simulating the kinetics of ballistic cluster-cluster
aggregation is to pick pairs of clusters (l and m) randomly from a list
of clusters. If a random number X (0<X<1) is larger than $K'(l,m)/K'_{max}$,
the pair of clusters is returned to the list. Here $K'(l,m)$ is given by

$$K'(l,m) = (r_l^{max} + r_m^{max})^2 \; ((s_l + s_m)/s \; s_m)^{1/2} \tag{18}$$

where r_i^{max} is the maximum radius of the ith cluster and s_i is its mass.
K'_{max} is the maximum value of $K'(l,m)$ for any pair of clusters. If
$X < K'(l,m)/K'_{max}$ one of the two clusters is "fired" at the second cluster
with an impact parameter randomly selected from a circle of radius
($r_l^{max} + r_m^{max}$) and with a random direction and cluster orientations. If
the clusters contact each other, they are joined at their point of first
contact and returned to the list of clusters which is shortened by one
number. If they miss each other, they are returned to their original
places in the list of clusters. In any event after a pair of clusters
has been selected, the time is incremented by $1/(N^2 K'_{max})$. Figure 17
shows the cluster size distributions obtained during a simulation of this
type.[87] It gives results very similar to those obtained from the mean
field Smoluchowski equation with the reaction kernel given in equation (17).

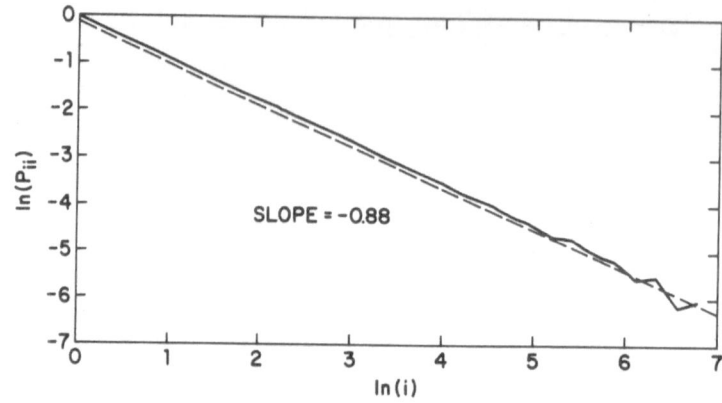

Fig. 18. The dependence of $\ln[\rho(i,i)]$ on $\ln(i)$ obtained from off-lattice simulations of chemically limited cluster-cluster aggregation. $\rho(i,i)$ is the probability that a randomly selected pair of particles in two different randomly oriented clusters of size i can be brought into contact without causing overlap of any other pair of particles.

It should be noted that in our simulations of both diffusion-limited and ballistic cluster-cluster aggregation the effects of cluster rotation have not been fully included. Rotational diffusion does decrease the fractal dimensionality of cluster-cluster aggregates though the effect will be small under realistic conditions.[88] Similarly, the effects of cluster rotation of the kinetics of diffusion-limited cluster-cluster aggregation and ballistic aggregation are also probably quite small. However, they are probably large enough to be seen in very careful experiments. In practice, these effects will, in most cases, be smaller than those caused by other deviations from the idealized models (hydrodynamic interactions, nonrigidity, sticking probabilities, long and short range interactions, etc.)

Reaction-Limited Cluster-Cluster Aggregation

The kinetics of reaction-limited cluster-cluster aggregation[31] has recently been discussed theoretically by Ball et al.[89] In the chemically limited cluster-cluster aggregation model, it is assumed that pairs of clusters are able to sample all possible bonding configurations without bias.[31,89] Ball et al. show that a reaction kernel with the scaling properties $K(i,j) \sim i$ for $j \ll i$ and $K(i,j) \sim i$ for $i \approx j$ (for d = 3) leads to kinetics which are consistent with those found experimentally[35-37] (an exponential growth on the mean cluster size $(S(t))$ and a power law cluster size distribution described by an exponent, τ, of about 1.5). For reaction kernels with the property $K(i,i) \sim i^{\lambda}$ power law growth of $S(t)$ and a "hump" shaped cluster size distribution is expected for $\lambda < 1$ and gellation is expected finite gellation time for $\lambda > 1$.[74,75,90] Ball et al.[89] argue that the system stabilized at $\lambda = 1$ and suggest that if λ grows above 1 the cluster size distribution broadens increasing D and forcing λ down. If λ falls below 1 a more monodisperse cluster size distribution evolves which lowers D and raises the value of λ. The value of the exponent λ has been measured in several 3d simulations. Using a monodisperse (hierarchical) model[31] a value of 1.16 was found while a value of 1.06 ± 0.02 was obtained from a model with a natural evolution of the cluster size distribution.[33]

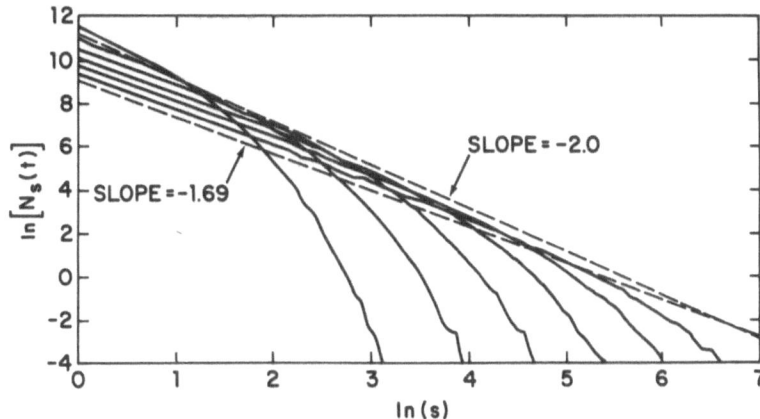

Fig. 19. Cluster size distributions obtained from the same
off-lattice model for chemically-limited aggregation used to
generate the results shown in Figure 18. Here the cluster
size distributions are shown at the stages when N has
reached the values $N_o/(\sqrt{2})^n$ (n = 1-6). N_o is the total
number of particles (200,000 in this case).

Figure 18 shows some results obtained from the 3-dimensional
off-lattice chemically-limited aggregation model described in the
introduction and used to generate Figure 5. These simulations were
carried out using 200,000 particles and were terminated when the maximum
cluster size exceeded 3,000. The results from 123 such simulations were
averaged to obtain Figure 18 which shows the dependence of $\ln[P(i,i)]$ on
$\ln(i)$ where $P(i,i)$ is the probability that two particles in clusters of
size i can be brought into contact without an overlap of any pair of
particles in the two clusters. The slope of -0.88 in Figure 18
corresponds to a value of 1.12 for λ. Figure 19 shows some of the
cluster size distributions obtained from the same simulations.
Simulations such as these can be made fully time dependent by
incrementing the time by a constant amount each time an attempt is made
to add two clusters together (i.e., each time two particles are
selected). The growth of the mean cluster size $(S(t))$ was measured in
this way. The simulation results are consistent with either a power law
growth $(S(t) \sim t^z$ with $z \simeq 4\text{-}5)$ or exponential growth. A somewhat better
fit to the simulation results is obtained with exponential growth in
$S(t)$ $(S(t) \sim e^{kt})$.

The results shown in Figures 18 and 19 do not seem to be in
particularly good agreement with the ideas of Ball et al. which are
briefly outlined above. However, they are not inconsistent with the
behavior proposed by Ball et al. being the asymptotic behavior for
chemically limited cluster-cluster aggregation. Slow crossover to
asymptotic behavior is not unusual and there is evidence from lattice
models closely related to the off-lattice model described above that λ
may be approaching a limiting value of 1.0.

Discussion

One of the main objectives of these lectures has been to illustrate
how Monte Carlo methods can be used to obtain information about the
kinetics of nonequilibrium growth and aggregation processes. Because one
of the most important early applications of Monte Carlo methods was the
simulation of equilibrium systems,[91] a common misconception that Monte
Carlo methods cannot be used to obtain accurate kinetic information has
developed. While it is true that equilibrium properties can be obtained

using Monte Carlo methods which give neither the correct path to equilibrium nor a meaningful time scale, Monte Carlo methods provide a powerful method for exploring the kinetics of both equilibrium and nonequilibrium systems. Some of the main advantages of the Monte Carlo approach include (1) Relative ease of programming and flexibility of Monte Carlo algorithms. (2) Complex boundary conditions can be treated almost as easily as simple boundary conditions. (3) The effects of fluctuation are included automatically. (4) Monte Carlo methods are frequently relatively free from mathematical approximations. The main drawbacks include (1) Monte Carlo methods give results of low precision when compared to most other approaches (large amounts of computer time are often required to get results with reasonably small statistical uncertainties). (2) Monte Carlo methods may give results without very much theoretical insight.

Monte Carlo simulations similar to those described above have been used to investigate the kinetics of a variety of other nonequilibrium growth and aggregation processes including steady state cluster-cluster aggregation[92] (approach its steady state), reversible aggregation,[93] particle coalescence.[94]

Acknowlegement

I would not have been able to contribute effectively to this institute without the help of a large number of colleagues. In particular, these "lecture notes" consist, in large part, of an outline of work carried out on the kinetics of growth and aggregation in collaboration with F. Boger, J. M. Deutch, B. Donn, F. Family, J. Feder, T. Jossang, K. J. Maloy, E. D. McGrady, S. Miyazima, T. Vicsek and R. M. Ziff. I would also like to thank R. F. Voss (IBM) and T. A. Witten (Exxon) for providing Figures 7 and 1 respectively.

References

1. B. B. Mandelbrot, "The Fractal Geometry of Nature", W. H. Freeman and Company, New York (1982).
2. D. N. Sutherland, J. Colloid and Interace Sci. 25:373 (1967).
3. D. N. Sutherland and I. Goodarz-Nia, Chem. Eng. Sci. 26:2071 (1971).
4. P. Meakin, J. Colloid and Interface Sci. 102:505 (1984); P. Meakin and B. Donn, unpublished.
5. R. Jullien, M. Kolb and R. Botet, J. Physique Lett. (Paris) 45:L211 (1984); R. Ball and R. Jullien, J. Physique Lett. (Paris) 45:L1031 (1984).
6. S. R. Forrest and T. A. Witten, J. Phys. A12:109 (1979).
7. T. A. Witten and L. M. Sander, Phys. Rev. Lett. 47:1400 (1981).
8. B. B. Mandelbrot, "Fractals in Physics", Proceedings of the Sixth International Symposium on Fractals in Physics, ICTP, Trieste, Italy, L. Pietronero and E. Tosatti, eds., North Holland Amsterdam (1986).
9. P. Meakin, J. Phys. A18:L661 (1985).
10. L. Patterson, Phys. Rev. Lett. 52:1621 (1984).
11. K. J. Maloy, J. Feder and T. Jossang, Phys. Rev. Lett. 55:2688 (1985).
12. J. Nittman, G. Daccord and H. E. Stanley, Nature 314:141 (1985).
13. L. P. Kadanoff, J. Stat. Phys. 39:267 (1985).
14. H. Van Damme, F. Obrecht, P. Levitz, L. Gatineau and C. Laroche, Nature, 320:731 (1986).
15. L. Niemeyer, L. Pietronero and A. J. Wiesmann, Phys. Rev. Lett. 52:1033 (1984).
16. R. M. Brady and R. C. Ball, Nature 309:225 (1984).
17. H. Honjo, S. Ohta and M. Matsushita, J. Phys. Soc. Japan 55:2487 (1986).
18. W. T. Elam, S. A. Wolf, J. Srague, D. V. Gubser, D. Van Vechten, G. L. Barz, Jr. and P. Meakin, Phys. Rev. Lett. 54:701 (1985).

19. Gy Radnoczi, T. Vicsek, L. M. Sander and D. Grier, preprint (1986).
20. G. Daccord and R. Lenormand, preprint (1986).
21. P. Meakin, J. Theor. Biol. 118:101 (1986).
22. A. A. Tsonis and P. A. Tsonis, Persp. Biol. Med. xx:xxxx (1987).
23. P. Meakin, Phys. Rev. Lett. 51:1119 (1983).
24. M. Kolb, R. Botet and R. Jullien, Phys. Rev. Lett. 51:1123 (1983).
25. P. Meakin, Phys. Lett. 107A:269 (1985).
26. R. Jullien, M. Kolb and R. Botet, J. Physique Lett. 45:L211 (1984).
27. D. A. Weitz and M. Oliveria, Phys. Rev. Lett. 52:1433 (1984).
28. C. Aubert and D. Cannell, Phys. Rev. Lett. 56:738 (1986).
29. J. E. Martin, D. W. Schaefer and A. J. Hurd, Phys. Rev. A33:3540 (1986).
30. G. Bolle, C. Cametti, P. Codastefano and P. Tartaglia, Phys. Rev. A35:837 (1987).
31. R. Jullien and R. Botet, J. Phys. A17:L639 (1984).
32. F. Leyvraz, preprint (1986).
33. W. D. Brown and R. C. Ball, J. Phys. A18:L517 (1985).
34. P. Meakin and F. Family, unpublished.
35. D. W. Schaefer, J. E. Martin, P. Wiltzius and D. S. Cannell, Phys. Rev. Lett. 52:2371 (1984).
36. D. Johnston and G. Benedek, "Kinetics of Aggregation and Gelation", p. 181, F. Family and D. P. Landau, eds., North Holland Amsterdam (1984).
37. D. A. Weitz, J. S. Huang, M. Y. Lin and J. Sung, Phys. Rev. Lett 54:1416 (1985).
38. P. Meakin, A. Coniglio, H. E. Stanley and T. A. Witten, Phys. Rev. A34:3325 (1986).
39. M. Plischke and Z. Racz, Phys. Rev. Lett. 53:415 (1984).
40. T. C. Halsey, P. Meakin and I. Procaccia, Phys. Rev. Lett. 56:854 (1986).
41. P. Meakin, H. E. Stanley, A. Coniglio and T. A. Witten, Phys. Rev. A32:2364 (1985).
42. C. Amitrano, A. Coniglio and F. diLiberto, Phys. Rev. Lett. 57:1016 (1986).
43. L. Pietronero, C. Evertsz and A. P. Siebesma "Stoichastic Processes in Physics and Engineeering". S. Albeverio, Ph. Blanchard, L. Streit and M. Hazewinkel (Proc. BIBOS IV) eds., S. Riedel, Dordrecht (1986).
44. P. Meakin, Phys. Rev. A34:710 (1986); Phys. Rev. Axx:xxxx (1987).
45. B. B. Mandelbrot, J. Fluid Mech. 62:331 (1974).
46. H. G. E. Hentschel and I. Procaccia, Physica 8D:440 (1983).
47. T. C. Halsey, M. H. Jensen, L. P. Kadanoff, I. Procaccia and B. Shraiman, Phys. Rev. A33:1141 (1986).
48. P. Meakin and L. M. Sander, Phys. Rev. Lett. 54:2053 (1985).
49. M. Plischke and Z. Racz.
50. P. Meakin and T. A. Witten, Phys. Rev. A28:2985 (1983).
51. P. Meakin and L. M. Sander, unpublished; P. Meakin, unpublished.
52. J. M. Deutch and P. Meakin, J. Chem. Phys. 78:2093 (1983).
53. P. Meakin, Phys. Rev. A27:604 (1983).
54. T. A. Witten, Proceedings "MACRO '82", p. 88, IUPAC, Amherst, Mass., June (1982).
55. R. F. Voss, Phys. Rev. B30:334 (1984); J. Stat. Phys. 36:861 (1984).
56. R. F. Voss and M. Tomkiewicz, J. Electrochem. Soc. 132:371 (1985).
57. P. Meakin and J. M. Deutch, J. Chem. Phys. 80:2115 (1984).
58. T. A. Witten and L. M. Sander, Phys. Rev. B27:5686 (1983).
59. L. Niemeyer, L. Pietronero and A. J. Wiesmann, Phys. Rev. Lett. 52:1033 (1984).
60. M. Matsushita, K. Honda, H. Toyoki, Y. Hayakawa and H. Kondo, J. Phys. Soc. Japan 55:2618 (1986).
61. L. Pietronero, private communication (1984).
62. P. Meakin, Phys. Rev. A33:3371 (1986).
63. K. J. Maloy, F. Boger, J. Feder, T. Jossang and P. Meakin, preprint.
64. S. K. Friedlander, Smoke, Dust and Haze, Wiley Interscience, New York (1977).

65. A. A. Lushnikov and V. N. Piskunov, Dolklady Phys. Chem. 231:1166 (1976).
66. J. Silk and S. D. White, Astrophys. J. 22:L59 (1978).
67. T. Vicsek and F. Family, Phys. Rev. Lett. 52:1669 (1984).
68. M. Kolb, Phys. Rev. Lett. 53:1653 (1984).
69. R. Botet and R. Jullien, J. Phys. A17:2517 (1984).
70. S. Mijazima, P. Meakin and F. Family, preprint.
71. M. Von Smoluchowski, Z. Phys. 17:585 (1916).
72. D. Toussaint and F. Wilczek, J. Chem. Phys. 78:2642 (1983).
73. P. Meakin and H. E. Stanley, Phys. Rev. A17:L173 (1984).
74. K. Kang and S. Redner, Phys. Rev. Lett. 52:955 (1984).
75. K. Kang and S. Redner, Phys. Rev. A32:435 (1985).
76. P. G. J. Van Dongen and M. H. Ernst, Phys. Rev. Lett. 54:1396 (1985).
77. F. Leyvraz "On Growth and Form: Fractal and Nonfractal Patterns in Physics". H. E. Stanley and N. Ostrowsky, eds., Martinus Nijhoff, Dordrecht (1986), p. 136.
78. P. Meakin, Phys. Rev. B29:2930 (1984).
79. R. Jullien, M. Kolb and R. Botet "Kinetics of Aggregation and Gelation". F. Family and D. P. Landau, eds., Elsevier, North-Holland Amsterdam (1984).
80. J. G. Kirkwood and J. Riseman, J. Chem. Phys. 16:565 (1948).
81. P. Meakin, Z.-Y. Chen and J. M. Deutch, J. Chem. Phys. 82:3786 (1985).
82. P. Meakin, T. Vicsek and F. Family, Phys. Rev. A31:564 (1985).
83. R. M. Ziff, E. D. McGrady and P. Meakin, J. Chem. Phys. 82:5269 (1985).
84. J. L. Spouge, J. Colloid Interface Sci. 107:38 (1985).
85. J. H. Jeans, "The Dynamic Theory of Gases", Dover, New York (1954).
86. B. Donn, private communication.
87. P. Meakin and B. Donn, unpublished.
88. P. Meakin, J. Chem. Phys. 81:4637 (1984).
89. R. C. Ball, D. A. Weitz, T. A. Witten and F. Leyvraz, Phys. Rev. Lett. 58:274 (1987).
90. R. M. Ziff, J. Stat. Phys. 23:241 (1980).
91. H. Metropolis, A. W. Rosenbluth, M. N. Rosenbluth, A. H. Teller, and E. Teller, J. Chem. Phys. 21:1089 (1953).
92. T. Vicsek, P. Meakin and F. Family, Phys. Rev. A32:1122 (1985).
93. P. Meakin, J. Chem. Phys. 83:3645 (1985); P. Meakin, unpublished.
94. K. Kang, S. Redner, P. Meakin and F. Leyvraz, Phys. Rev. A33:1171 (1986).

COLLOIDAL FLOW*

Arne T. Skjeltorp

Institute for Energy Technology
P.O.B. 40, N-2007 Kjeller, Norway

ABSTRACT

A "two-fluid" system consisting of a concentrated "slurry" of polystyrene microspheres dispersed in water as fluid No. 1 and water as fluid
No. 2 is used to study colloidal flow. As the specific weight of polystyrene is about 4% larger than that of water, the "slurry" will flow under
the influence of gravitation. The flow is studied in Hele-Shaw cells
(closely spaced glass plates). This produces various interfacial instabilities which may be related to viscous fingering (vertical/inclined
cells) and "Rayleigh-Bénard"-like convection (horizontal cells).

I. INTRODUCTION

The collective behaviour of the flow of microparticles in liquids is
full of surprises. Examples are enhanced sedimentation in inclined channels ("Boycott" effect[1,2]) and increased sedimentation of small particles
in the presence of larger bouyant particles ("Whitemore" effect[3,4]). The
purpose of this presentation is to report a few novel results using very
uniformly sized polystyrene microspheres to study gravitationally induced
flow between closely spaced plates (Hele-Shaw cell).

The experiments were performed in essentially two different modes:
(i) Colloidal flow along the plates for various inclinations of the cell
(viscous fingering studies[5]), and (ii) flow from the upper plate to the
lower plate in cells placed horizontally. ("Rayleigh-Bénard"-like convection instability[6]).

Some aspects of viscous fingering in traditional binary fluid mixtures
will be reviewed and compared with the present results. The situation (ii)
would entail rather complicated analysis and the results are therefore essentially presented as empirical findings.

II. HELE-SHAW FLOW

Hele-Shaw cells have been widely used to study the patterns of the
interface between two fluids[5]. This experimental realization offers a

---

*See figure D on color page.

clear two-dimensional observation of the interface occupying the entire gap width and therefore also suitable for theoretical modelling. The experiments are traditionally performed in two modes[1] (Fig. 1): In one mode, fingering instability occurs when a fluid of low viscosity (μ_1) displaces another fluid of higher viscosity (μ_2). An interfacial tension between the two fluids (σ) tends to reduce the instabilities. The other mode occurs if a heavier fluid (density ρ_2) is on top of a lighter one (ρ_1). In this case the interface is gravitationally unstable. These two effects can occur simultaneously and in the controlling flow parameters they are interchangeable, as is discussed below.

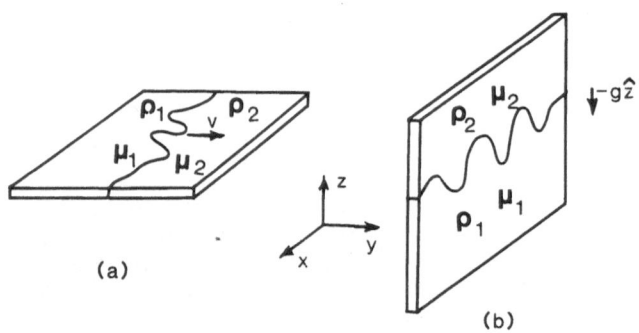

(a)

(b)

Fig. 1   Equivalence of Hele-Shaw cells driven by: (a) pressure and (b) gravitation.

The original work by Saffman and Taylor[8], with air pushing glycerine and water into oil, showed the emergence of a long-time state with only one finger. The ratio of finger width to the channel width was found to be close to 1/2. Subsequent work has shown that the pattern selection of the interface is extremely hard to solve as the interfacial tension between the two fluids approaches zero[5]. It thus appears that the flow becomes chaotic in this limit and may have fractal character.

Recently, Tryggvason and Aref[7] (TA) performed so-called vertex-sheet calculations to map the interfacial instability. They introduced two controlling parameters for the flow:

The dimensionless viscosity contrast expressed as:

$$A = \Delta\mu/2\mu \qquad\qquad\qquad (1)$$

with $\Delta\mu = \mu_2 - \mu_1$ and $\mu = (\mu_1 + \mu_2)/2$ and a dimensionless surface tension

$$B = \sigma b^2 / 12 U^* W^2 \mu. \qquad (2)$$

Here, W is the width and b the thickness of the cell and U^* a characteristic velocity:

$$U^* = |[\Delta\mu V + (1/12) \Delta\varrho g'b^2]/2\mu| \qquad (3)$$

with $\Delta\varrho = \varrho_2 - \varrho_1$ and V throughflow velocity, g' is the effective acceleration due to gravity. This means that when the plates of the Hele-Shaw cell are inclined towards the horizontal with an angle φ, the effect of gravity is confined to the component parallel to the plates and thus

$$g' = g \sin \varphi \qquad (4)$$

with g the gravitational constant.

It has been shown[8] that in the limit of vanishing surface tension, the sign of the term inside the absolute-value sign in Eq. (3) determines the stability. In particular, the interface is unstable when

$$\Delta\mu V + (1/12) \Delta\varrho g'b^2 \geq 0. \qquad (5)$$

This shows that in this formalism the instability due to pressure gradient (first term in Eq. (5)) is interchangeable with instability due to gravitational field (second term in Eq. (5)). In other words, a horizontal Hele-Shaw cell with a throughflow V (e.g. air pushing water) is equivalent to a vertical (or inclined) cell with density stratification as long as boundary effects may be ignored[7].

TA performed extensive simulations varying the parameters A and B in Eqs. (1) and (2). One typical result is shown in Fig. 2. The simulation started by perturbing an initially flat interface by an irregular multi-wavelength disturbance of very small amplitude. As may be seen, the interface evolves in a very complicated fashion with merging of fingers as well as tip splitting. The TA results show that small values of A do not show the Saffman-Taylor picture of one single finger growing.

In another approach, Liang[10] has recently performed random-walk simulations of Hele-Shaw flow using an algorithm related to diffusion-limited aggregation (DLA). His results show a transition from smooth boundary fingers to fractal structures for decreasing values of the interfacial tension.

There has apparently been no report of Hele-Shaw experiments showing fractal structures using Newtonian fluids[5]. However, Nittmann, Daccord and Stanley[11] used a non-Newtonian fluid combined with a Newtonian fluid which produced fractal structures. Recently, Maher[9] used the binary-liquid mixture isobutyric acid plus water and was able to vary the viscosity contrast A (Eq. (1)). His results show that for A $\simeq$ 0 and low B, many fingers are formed and continue to grow whereas if A is large, the number of fingers decrease rapidly with time.

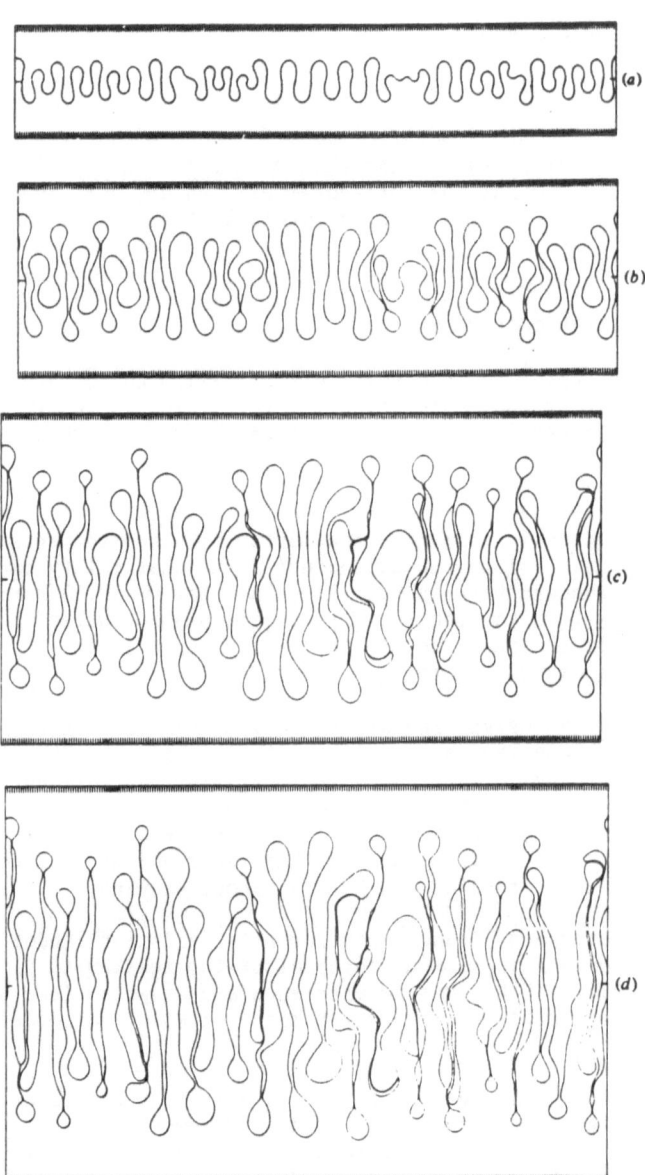

Fig. 2   Simulation by TA (fig. 5 in ref. 7) of the interface evolution
using B = 1.25 10^{-5} and A = 0.0 as discussed in the text.

III. EXPERIMENTAL

The particles used in the present experiments were very uniformly sized polystyrene spheres of diameter d = 1 - 100 μm produced by the Ugelstad swelling technique[2]. (For the results reported here, only 1.9 μm spheres were used.)

The experiments were performed in square (W) Hele-Shaw cells consisting of plane-parallel glass plates with various separations b. The aspect ratio of the cell, Γ = W/2b, was large (typically $\Gamma \sim$ 20 mm/0.2 mm = 100) so the effects of the side walls were minimal. The cells were filled with water dispersed particles of relative concentration n ≃ 10% and stabilized with a small amount of surfactant (0.1 % sodium dodecyl sulphate) to prevent aggregation. The specific weight of the polystyrene spheres is ϱ_p = 1.04 g/cm^3 so they will slowly sediment in water.

The experimental procedure for the viscous fingering experiments is shown in Fig. 3. The cell was kept in a vertical position until the spheres had settled at one end of the cell. The experiment started by turning the cell around making an angle φ with the horizontal plane (φ = 90^0 for vertical position). By varying φ it was thus possible to vary the gravitational field according to Eq. (4).

In the "Rayleigh-Bénard" configuration, the experimental set-up and procedure were quite similar (Fig. 4), but the flow was normal to the cell walls. Here, the cell was kept in a horizontal position until the spheres had settled to a dense slurry at the lower plate. The experiment started by quickly turning the cell around producing a downward flow of the spheres.

The detailed flow patterns and movement of individual spheres were observed in both experiments using a light microscope with video camera attachment and recorder. Besides, a camera with macro-objective was used for direct picture taking of the whole cell.

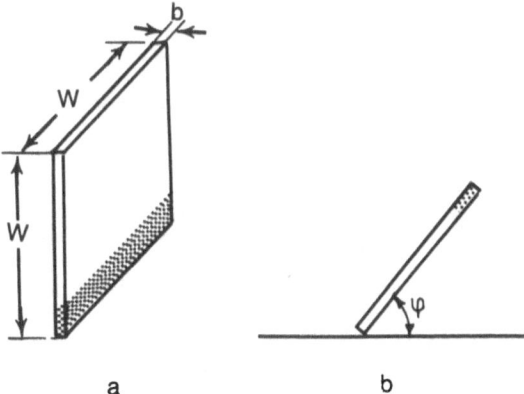

Fig. 3   Hele-Shaw configuration for viscous fingering experiments:
(a) the cell is kept in a vertical position to let the spheres settle at one end; (b) experiment starts by turning the cell around making an angle φ with the horizontal plane.

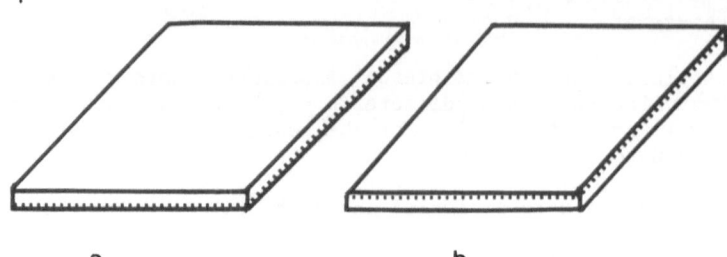

Fig. 4 "Rayleigh-Bénard" configuration for convective flow experiments:
(a) The cell is kept in a horizontal position until the spheres
have settled onto the lower plate; (b) the experiment starts by
turning the cell around 180°.

IV. HELE-SHAW COLLOIDAL FLOW

A. Results

Fig. 5 shows a typical time evolution of the flow of 1.9 μm spheres in
a vertical Hele-Shaw cell (Fig. 3, $\varphi = 90°$, b = 0.1 mm). In Fig. 5a, it is
seen that a relatively regular array of "fingers" has bulged out from the
interface. These perturbations grow in Fig. 5b and already tip splittings
are observed. The interface becomes more and more irregular or chaotic as
time passes. There are some effects from diffusion of the spheres, but
this does not spoil the overall features.

Close inspection of the movements of individual spheres shows a dilute
random arrangement. The main "trunk" and side branches act as support
channels for the flow of spheres to the advancing tips from the upper
"bulk" part. At the same time, there is a backflow of water with few
spheres. Some main fingers grow and take over whereas smaller sidebranches
slow down. During the whole process the fingers show tip splitting.

Figs. 6 and 7 show typical flows for the same cell with inclinations φ
= 75° and 60° to the horizontal plane, respectively. This reduces the
effective gravitational driving force according to Eq. (4). The characteristic features for decreasing φ are as follows: The average separation
between the tips at the initial instability is about the same. The flow
still becomes chaotic (tip splitting) after some time in Fig. 6. Below a
critical angle, the flow becomes rather regular as examplified in Fig. 7
with a constant number of fingers in the long-time limit.

Fig. 8 shows vertical flow ($\varphi = 90°$) for a plate separation of b =
0.05 mm. As may be seen, the flow starts as before, but with shorter
average distances between the fingers than for the b = 0.1 mm case. The
long term evolution is remarkable. After about 12 hrs the flow pattern is
extremely ramified (fractal) both at the advancing tips and also in the
deep "fjords" left behind.

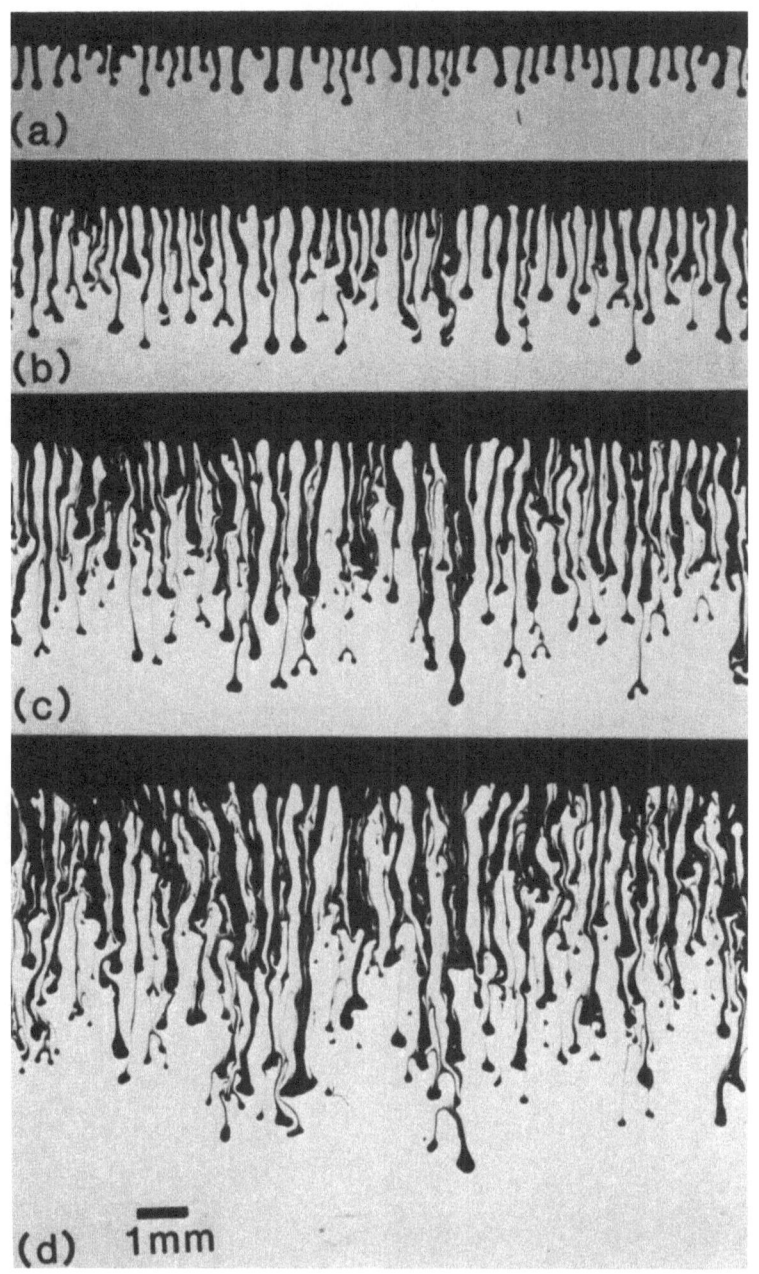

Fig. 5   Gravitational flow of 1.9 μm spheres for a vertical Hele-
Shaw cell (φ = 90°, b = 0.1 mm) for successive times: (a) 30
secs; (b) 60 secs; (c) 90 secs; (d) 120 secs.

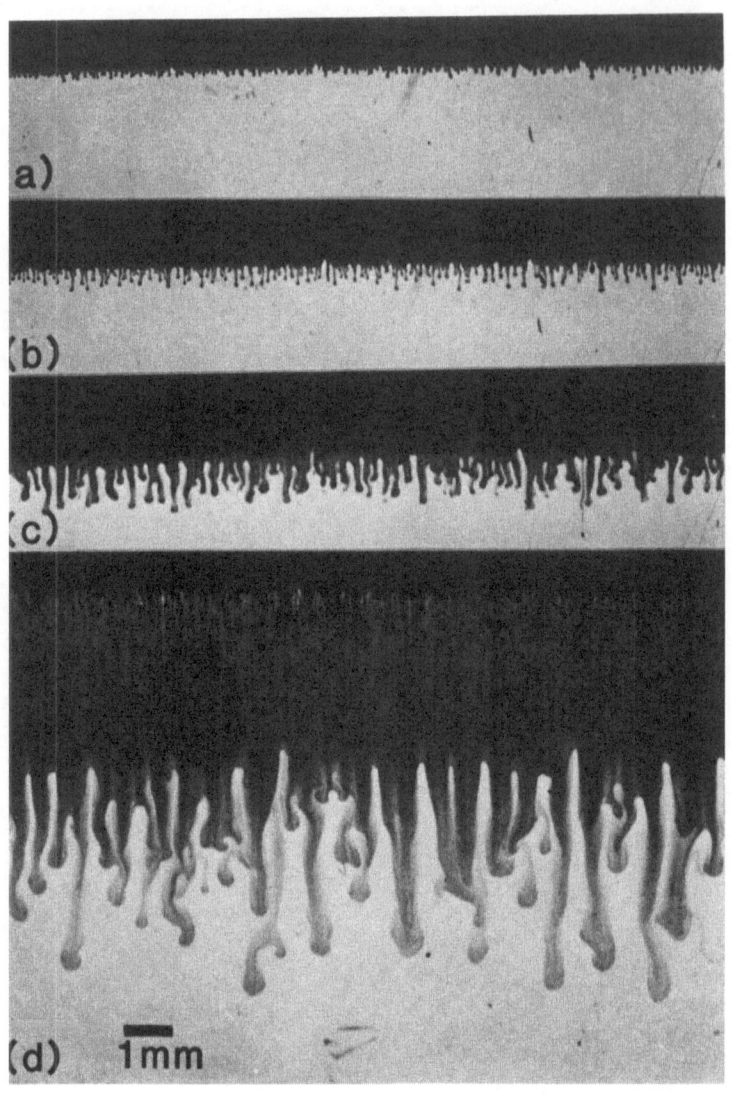

Fig. 6 Flow for an inclined cell ($\varphi = 75^0$, b = 0.1 mm) for successive times: (a) 60 secs; (b) 120 secs; (c) 360 secs; (d) 1800 secs.

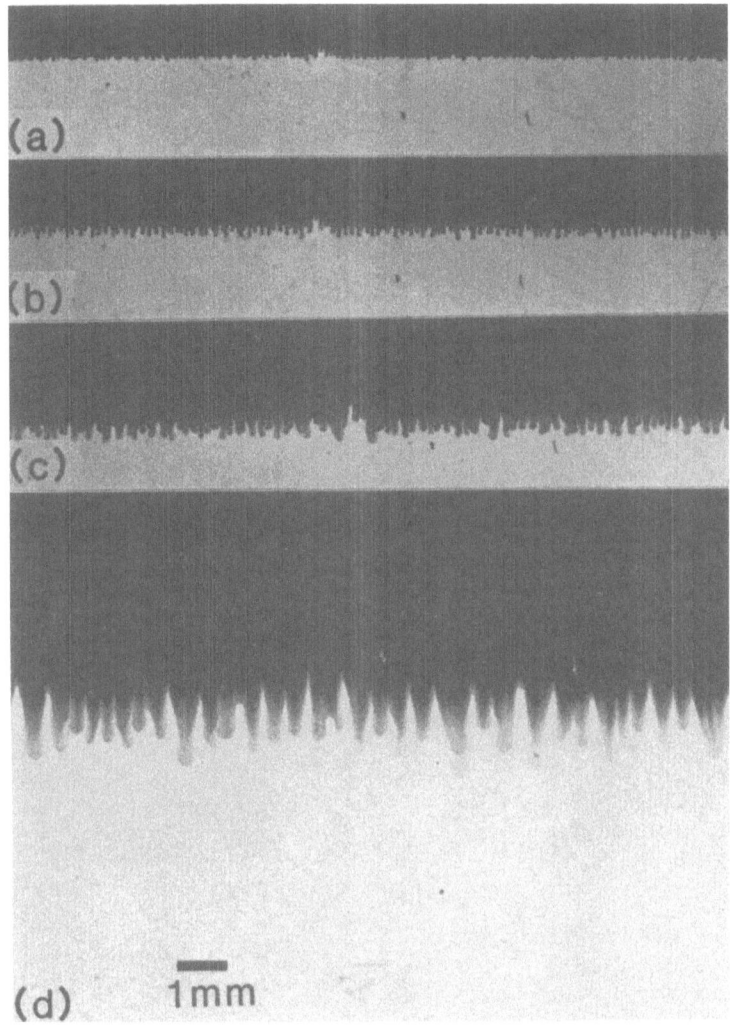

Fig. 7 Flow for an inclined cell ($\varphi = 60^0$, b = 0.1 mm) for successive times: (a) 60 secs; (b) 120 secs; (c) 360 secs; (d) 1800 secs.

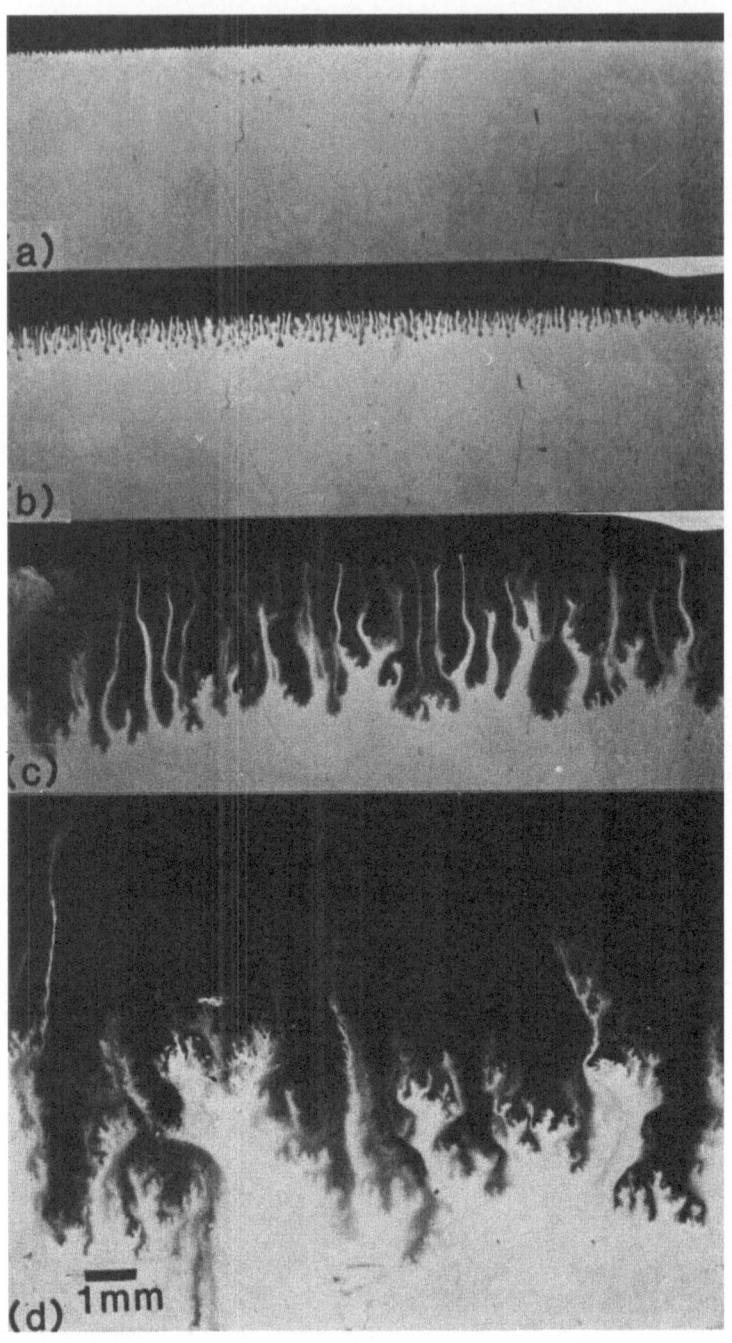

Fig. 8 Flow for a vertical Hele-Shaw cell ($\varphi = 90^u$) with plate separation b = 0.05 mm for successive times: (a) 60 secs; (b) 300 secs; (c) 1 hr; (d) 12 hrs.

B. Discussion

The initial flow patterns shown above (e.g. Fig. 5) have remarkable similarities with TA's simulated pattern as examplified in Fig. 2 except for the asymmetry in the present case. At first sight, the present system does not appear to qualify as a good physical realization of TA's model simulations. First of all, the fluids should be Newtonian and immiscible. Concentrated colloids are usually considered non-Newtonian (viscosity depends on shear rate). However, as noted in the previous section, the sphere concentration in the fingers was low (< 5%) with dilute random particle arrangement. In this limit the dilute slurry thus qualifies as Newtonian[13]. The "liquids" are clearly miscible. However, this does not create a serious problem. The mixing by diffusion is relatively slow compared to the experimental time, but this is a source of noise!

The most interesting feature with the system is clearly the fact that the interfacial tension between the "slurry" and water must be extremely low as it is essentially a water-water interface. In fact, surface tension measurements using a capillary technique showed no difference between the two fluids within experimental error (< 1%). The ratio of the viscosity of the slurry and water is estimated to be $\mu_2/\mu_1 \leq 1.1$ for the concentration in question[14]. The viscosity parameter (Eq. (1)) is thus $A \leq 0.09$. It is therefore argued that the present system may qualify as a good approximation for a Newtonian binary-fluid system with essentially zero interfacial tension ($B \approx 0$) and low viscosity ratio.

Some preliminary results from the present experiments will be discussed in the following. An interesting parameter is the wavelength selection or the average distance λ between the initial fingers, Fig. 9. For $b \geq 0.05$ mm it is found that

$$\lambda \sim b^{3/2}. \tag{6}$$

The crossover to an almost constant λ for $b \leq 0.05$ mm is apparently due to diffusion. A suitable parameter is the Peclet number $P_c = Ub/D_0$ where U is the downward flow velocity and D_0 diffusivity of the spheres. To first order it is found that $P_c \approx 0.3$ for $b = 0.05$ mm. The width of the cell, W, does not appear to influence the λ-relation which is reasonable as $\lambda/W \ll 1$.

Another interesting parameter is the distance h measured from the tip of the longest finger to the initial interface. It is found that h scales with time t as

$$h \sim t^{\alpha}. \tag{7}$$

For not too large values of t (Fig. 5), the fitted value for the exponent is $\alpha = 1.4 \pm 0.2$, Fig. 10. It is interesting to note that for a DLA process, α is related to the fractal dimension D by[10]:

$$D = 1 + 1/\alpha. \tag{8}$$

The value for α found above produces $D = 1.7 \pm 0.2$ which indeed is the DLA-value (two-dimensional case).

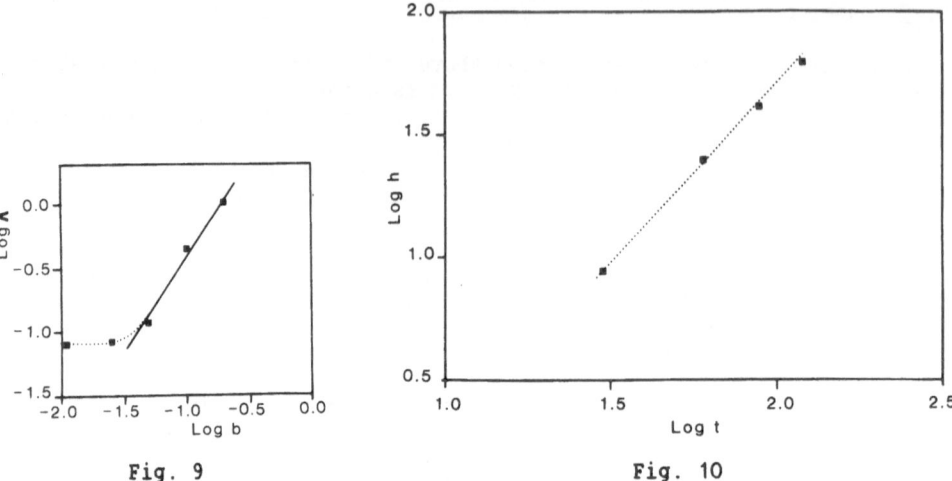

Fig. 9

Fig. 10

Fig. 9 Log-log plot of the initial wavelength selection λ vs plate separation b (in mm). The solid line represents $\lambda \sim b^{3/2}$.

Fig. 10 Log-log plot of distance h of the longest finger (in mm) to the initial interface vs time t (in secs) as discussed in the text. The slope is the fitted exponent $\alpha = 1.4$.

V. "RAYLEIGH-BÉNARD" TYPE FLOW

The primary purpose of this section is to present new observations of colloidal flow from the upper plate to the lower plate in a horizontal Hele-Shaw cell (See Fig. 4). A typical result is illustrated in Fig. 11 for a plate separation b = 0.1 mm and 1.9 μm spheres.

As mentioned earlier (Sec. III), the cell was left in a horisontal position until the spheres had settled at the lower glass plate. The cell was then quickly turned around and the observations displayed in Fig. 11 started. Here, dark and light areas denote regions of high and low sphere concentration, respectively. The emergence of the regular patterns is reminiscent of Rayleigh-Bénard convection driven by heating a fluid from below[5]. A conceptual sketch of the process producing this pheomenon is given in Fig. 12: (a) Initially, fluctuations in the sphere concentration are created as the spheres start to flow downward; (b) the instability is amplified with downward and upward fluid convection (thick arrows) producing high density regions (thin arrows indicate sphere movements); (c) the development of regular arrays; (d) final patterns. Clearly this situation is unstable as there is no upward mass transport of the spheres.

An interesting parameter is the wavelength selection or the average cell size λ_m in the flow pattern. Figs. 13 and 14 thus show the time evolution of flow for other plate separations b = 0.05 and 0.2 mm, respectively. As may be seen, the average cell size decreases with b and in fact

$$\lambda_m \simeq b. \tag{9}$$

82

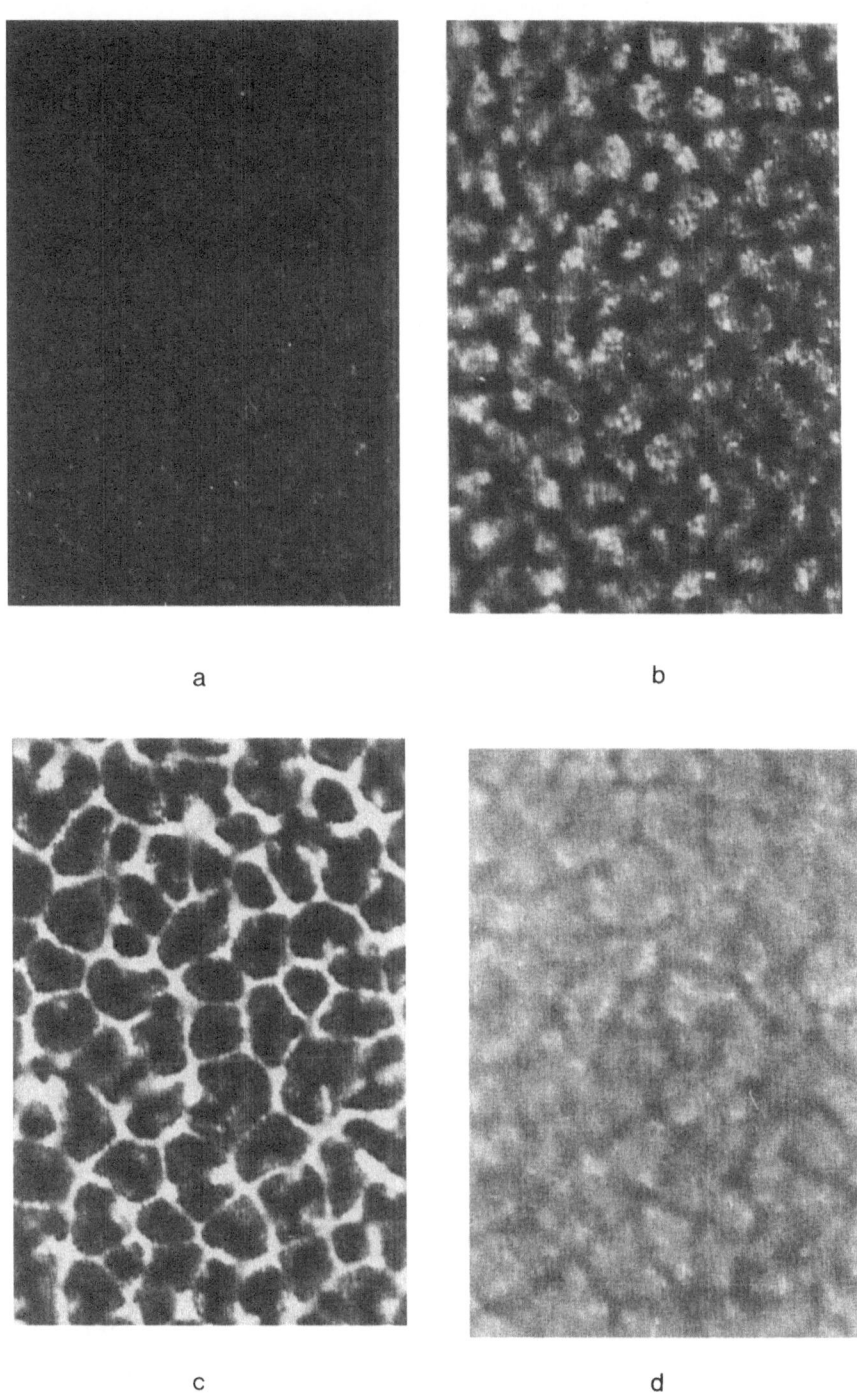

a   b

c   d

0.1mm

Fig. 11   Plane view of concentration patterns developed from the sedimentation of 1.9 µm spheres (b = 0.1 mm), as discussed in the
text for different settling times: (a) 15 secs; (b) 30 secs; (c)
60 secs; (d) 120 secs. Dark areas denote high concentration
regions of spheres (downward stream) whereas light areas denote
upward convection with few spheres.

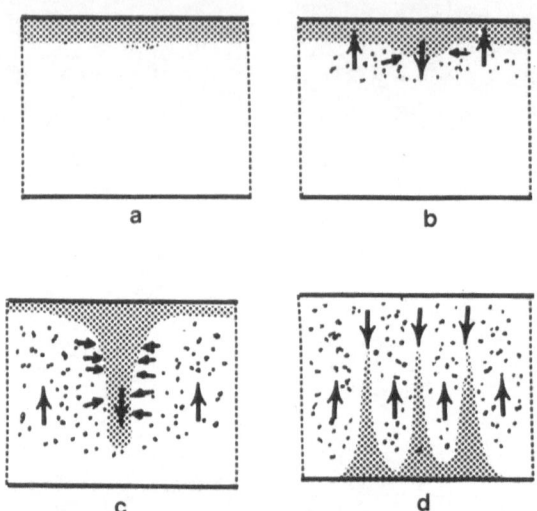

Fig. 12   Schematic of the spontaneous formation of patterns in the
sedimentation of monosized particles: (a) Development of
concentration fluctuations; (b) fingering starts; (c) regular
patterns are formed; (d) final regular patterns. Thick arrows
denote fluid velocity, thin arrows sphere velocity.

It is of interest to give an estimate of the time scale involved in
the growth rate of sphere-accumulation down towards the lower plate. Let
us consider a "blob" of radius r containing an excess concentration Δn of
spheres compared to the average n. This "blob" will sink with a velocity
approximately given by[15]:

$$V_s = A\Delta n v_2 \Delta \varrho g r^2 / \mu_1 \qquad (10)$$

Here A = 2/9 (for Stokes law), $v_2 \simeq 4 \cdot 10^{-12}$ cm^3 is the volume of the
sphere, $\Delta \varrho \simeq 0.04$ g/cm^3 is the difference between the densities of the
spheres and water, $g \simeq 10^3$ cm/sec^2, $\mu_1 = 10^{-3}$ g/cm sec the viscosity, and
$r \leq 0.005$ cm. The approximate sphere-accumulation balance equation is thus

$$(4/3)\pi r^3 (d\Delta n/dt) \approx \pi r^2 n v_2 = [\pi A n v_2 \ \Delta \varrho r^4 g / \mu_1]\Delta n. \qquad (11)$$

The time constant for sphere-accumulation in the sinking region is
thus

$$T_2 \simeq \mu_1 / [A n \ v_2 g \ \Delta \varrho \ r]. \qquad (12)$$

By using $n \simeq 3 \cdot 10^9$ spheres/cm^3 ($\sim$ 10% dispersion) and as an example r =
0.005 cm, $T_2 \simeq 50$ secs. This is a reasonable value compared with the
observed time.

Although the equations and boundary conditions may be formulated for the
present system, they are obviously hard to solve as for the viscous
fingering problem in the limit of zero interfacial tension.

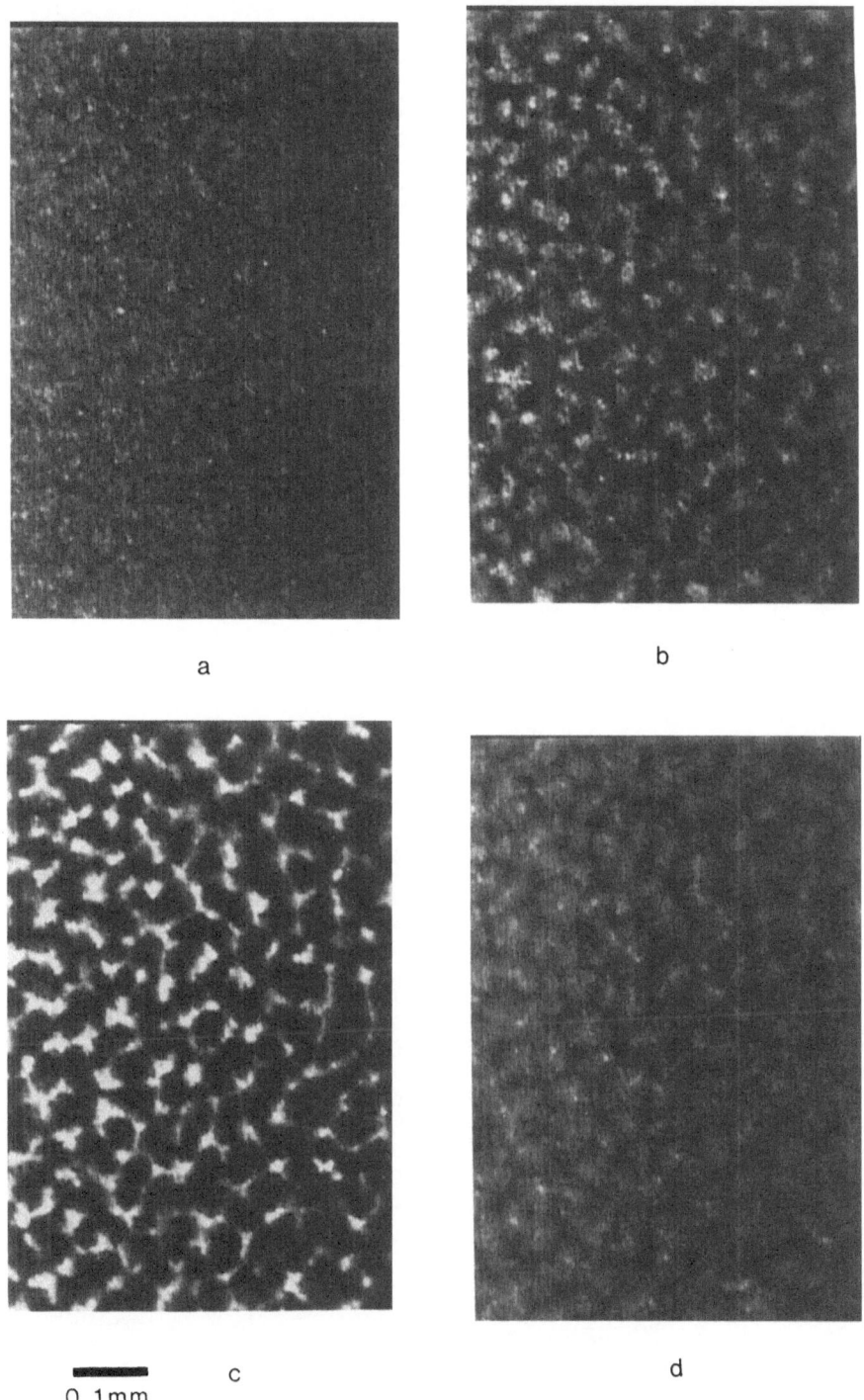

a

b

c

0.1mm

d

Fig. 13   Flow pattern for plate separation b = 0.05 mm as discussed in
the text for different settling times: (a) 15 secs; (b) 30 secs;
(c) 60 secs; (d) 120 secs.

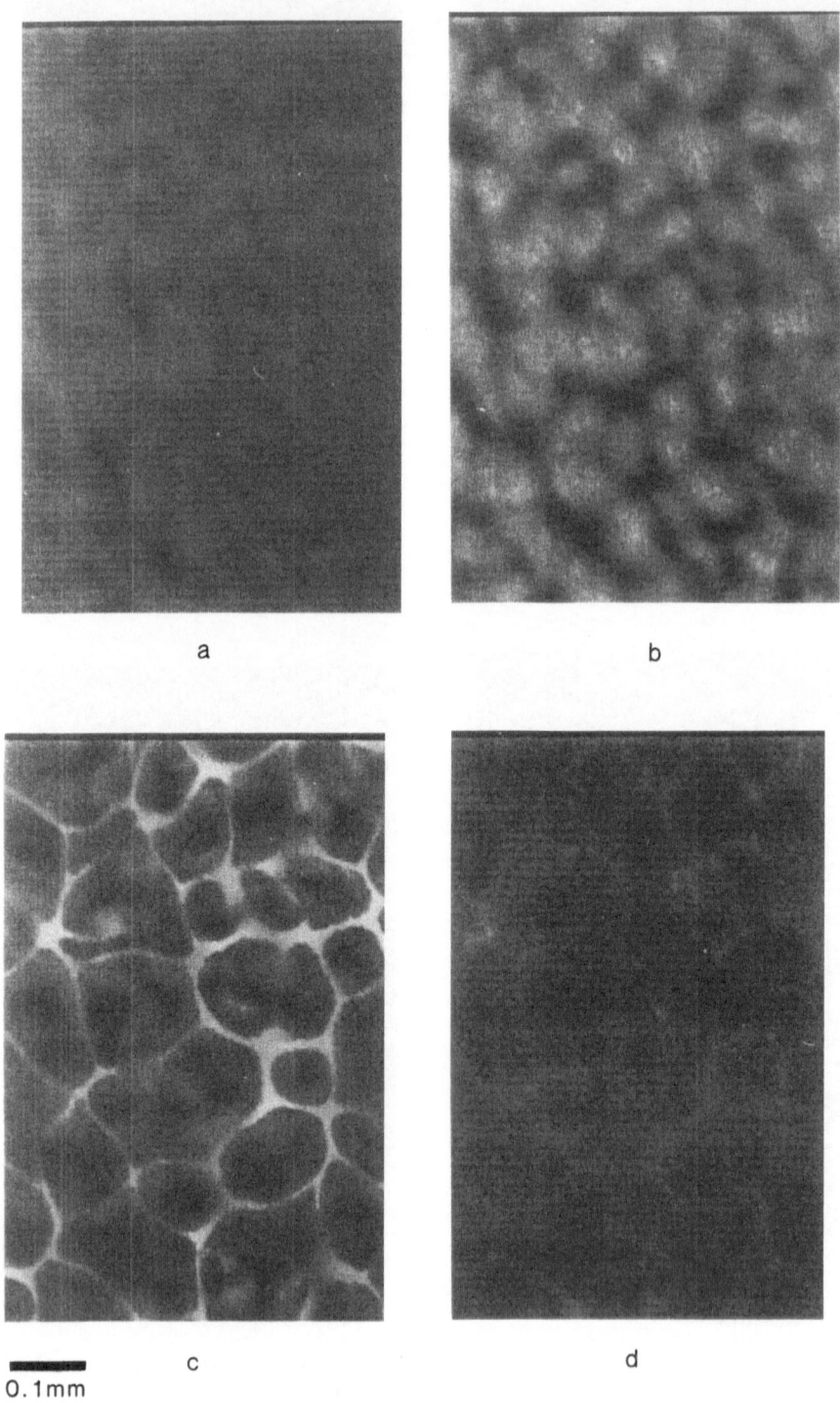

a

b

c

0.1mm

d

Fig. 14   Same legend as in Fig. 13 but for plate separation b = 0.2 mm.

VI. SUMMARY

The aim of this paper has been to serve as an introduction to the use of uniformly sized microparticles to model various phenomena in colloidal flow. On the basis of the results presented here and other published results, it is surprising to find out how little is actually known about such processes even for well-defined systems. It seems likely that with the advent of "taylor-made" microparticles and dispersions, coupled with recent advances in analytical models, old phenomena may be revisited and turned into beautiful fields of the future.

VII. ACKNOWLEDGEMENTS

The research was supported in part by Dyno Particles A/S, the Norwegian Research Council for Science and Humanities (NAVF) and the Nansen Foundation. I also wish to thank John Ugelstad and collaborators at SINTEF for useful advice and for providing monodisperse spheres as well as Geir Helgesen for helpful discussions and programming assistance.

REFERENCES

1. A.E. Boycott, Nature <u>104</u>, 532 (1920).
2. R.H. Davis and A. Acrivos, Ann. Rev. Fluid Mech. <u>17</u>, <u>91</u> (1985).
3. R.L. Whitmore, Brit. J. Appl. Phys. <u>6</u>, 239 (1955).
4. G.K. Batchelor and R.W. Janse von Rensburg, J. Fluid Mech. <u>166</u>, 379 (1986).
5. For a recent review, see D. Bensimon, L.P. Kadanoff, S. Liang, B.I. Shraiman, and C. Tang, Rev. Mod. Phys. <u>58</u>, 977 (1986).
6. M. Velarde and C. Normand, Sci. Am. <u>243</u>, 78 (1980).
7. G. Tryggvason and H. Aref, J. Fluid Mech. <u>136</u>, 1 (1983).
8. P.G. Saffman and G.I. Taylor, Proc. Roy. Soc. Lond. <u>A245</u>, 312 (1958).
9. J.V. Maher, Phys. Rev. Lett. <u>54</u>, 1498 (1985).
10. S. Liang, Phys. Rev. <u>A33</u>, 2663 (1986).
11. J. Nittman, G. Daccord, and H.E. Stanley, Nature <u>314</u>, 141 (1985).
12. J. Ugelstad, P.C. Mork, K.H. Kaggerud, T. Ellingsen, and A. Berge, Adv. Colloid Interface Sci. <u>13</u>, 101 (1980). The spheres are commercially available from Dyno Particles A/S, POB 160, N-2001 Lillestrøm, Norway
13. C.E. Chaffey, Colloidal Polymer Sci. <u>255</u>, 691 (1977).
14. D.J. Jeffrey and A. Acrivos, Amer. Inst. Chem. Eng. J., <u>22</u>, 417 (1976).
15. J.O. Kessler, Contemp. Phys. <u>26</u>, 147 (1985).

THE MICROWORLD OF ELECTRODEPOSITION*

Geir Helgesen

Department of Physics
University of Oslo
Blindern, 0316 Oslo 3, Norway

INTRODUCTION

The diffusion limited aggregation (DLA) model by Witten and Sander[1] has been widely used to decribe ramified pattern formation in processes like particle aggregation, viscous fingering, dielectric breakdown and electrodeposition (ED). The purpose of this presentation[2] is to show a wide range of growth morphologies in ED for silver and tin. The results are similar to earlier ED-experiments[3,4,5], but certain aspects are visualized for the first time.

EXPERIMENTAL

The experimental set-up consisted of two electrodes of silver or tin on a glass microscope slide. The electrodes were either evaporated or painted on the slide. The distance between them was approximately 12 mm, and the width of the slide was 22 mm. An even, thin layer of deionized water was confined between the slide and a cover glass (separation typically 50-100 μm).By turning up the voltage from a voltage supply (0-12 V), the positive electrode started to dissolve, and after some time the dissolved material began to deposit on the negative electrode. Low currents were involved, typically 10-50 μA. The cell was placed under a microscope connected to a video camera, and the growth process was recorded on video tape. The pictures were digitized with a resolution of 512x512 pixels and analysed with an image processing system. The fractal dimension has been estimated using the box counting technique[6].

RESULTS AND DISCUSSION

Typical results from the experiments with tin electrodes are shown in Fig.2 and 3. The deposits were dendrites with a tip growth velocity of 10-50 nm/s. Tip-splitting of the long arms was observed, but the resolution of the video system was not good enough for extensive studies. In the case of compact silver electrodes the deposition process was slower than for tin, and the structures formed were quite different as shown in Fig.4. Silver is almost completely insoluable in pure water, so the concentration of silver ions in the water was extremely low, and the conditions were close to the ideal DLA-case with non-interacting

---

*See figures E and F on color page.

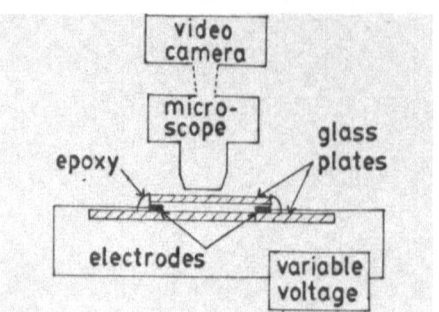

Figure 1: Schematic experimental set-up.

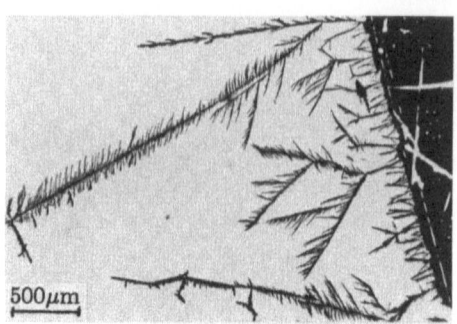

Figure 2: Typical deposits using tin electrodes.

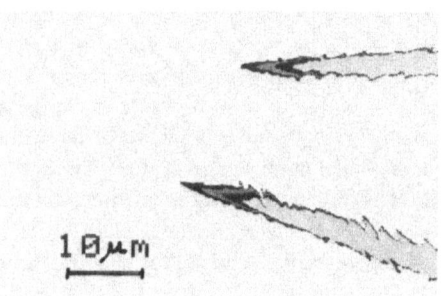

Figure 3: Time evolution of two tin deposit tips at 3 1/2 min. intervals. The sidebranches can not be distinguished here. Tip growth velocity $v_g = 12$nm/s.

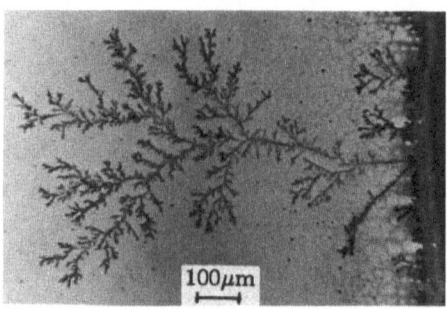

Figure 4: DLA-like growth at very low silver ion concentration. ($D_f \approx 1.6$)

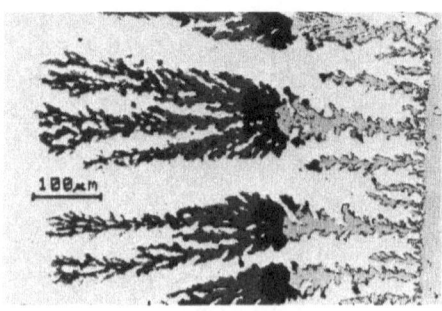

Figure 5: Silver trees formed at higher ion concentration. Pictures taken at 2 min. intervals. ($v_g = 2$-5 μm/s)

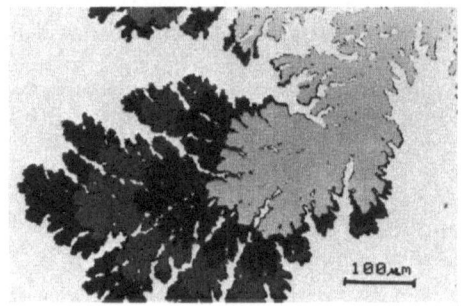

Figure 6: Silver deposit growing at a rate of 6.0 μm/s. ($D_f \approx 1.8$) Time between snapshots is 16 s.

Figure 7: Typical silver deposits at 2 min. intervals.

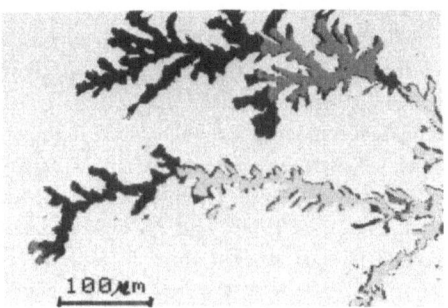

Figure 8: Stringy silver deposit. The time intervals here are 1 minute. ($D_f \approx 1.3$, $v_g = 3.4\mu m/s$)

Figure 9: Silver dendrites. ($D_f \approx 1.4$, $v_g = 2.3\mu m/s$)

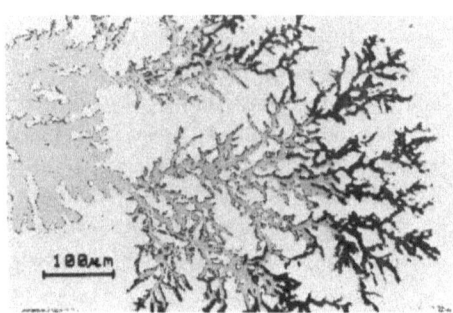

Figure 10: When the leftmost more compact part of this silver deposit apparently did not grow any more, the voltage was increased. The dark branches show the new growth during 45 seconds. ($v_g = 5.1\mu m/s$.)

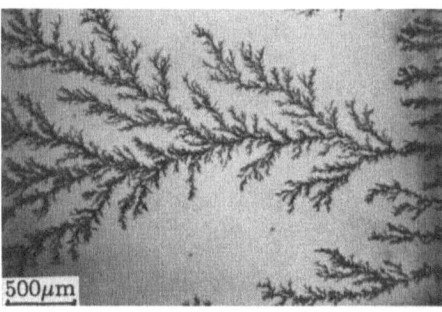

Figure 11: Deposition of silver at low voltage and high ion concentration.

Figure 12: The ion concentration in this case was approximately the same as in Fig.11, but the voltage was higher.

particles moving towards the aggregate. To increase the growth speed, we used porous silver electrodes. With this increased electrode surface, the growth speed was increased by a factor of more than 100, and the deposition was easily seen and recorded on video tape. A few minutes after the voltage was turned on, a cloud of tiny Brownian particles was observed moving from the positive towards the negative electrode before the deposition started. A strong screening of the smallest trees formed is easily seen in Fig.5. After some time the growth gradually changed and more compact deposits with slower growth started to appear (Fig.6 and 7). Other places rather stringy deposits with few sidebranches appeared (Fig.8), and in special cases the deposits formed typical dendrites (Fig.9). The reason for these changes is believed to be local variations in the ion concentration. In earlier ED-experiments[3,4] the ion concentration did not change during deposition. As the growth stopped completely, it could be started again by increasing the applied voltage, but the type of deposits then changed as shown in Fig.10.

In many of these experiments one observed lines of deposited silver particles parallel to the electrodes indicating convection in the system. The distance between these stripes was approximately the same as the distance between the glass plates. In some cases one could see convection phenomena near the tip of the deposits. This shows that there was a competition between diffusion and convection, and this could be the reason for some of the changes in the morphology. By varying the external voltage, the growth rate increased and the structures changed as shown in Fig.11 and 12. An exchange of the polarity of the electrodes during an experiment started deposit growth on the other electrode, but the previously formed aggregates apparently did not dissolve. In one of the experiments an AC-field was used to vibrate the deposits, and it is possible that this technique may be used to find the elastic properties of these aggregates. The calculated fractal dimensions given in the figure captions are in agreement with recent computer simulations on ED[7].

ACKNOWLEDGEMENTS

I would like to thank A.T. Skjeltorp for many stimulating discussions, and Dyno Particles A/S for partial support.

REFERENCES

1. T. Witten and L. Sander, Phys.Rev.Lett. **47** 1400 (1981)

2. A special video has been produced to visualize directly the microscopic growth dynamics.

3. Y. Sawada, A. Dougherty and J.P. Gollub, Phys.Rev.Lett. **56** 1260 (1986)

4. D. Grier, E. Ben-Jacob, R. Clarke and L.M. Sander, Phys.Rev.Lett. **56** 1264 (1986)

5. J.H. Kaufman, A.I. Nazzal, O.R. Melroy and A. Kapitulnik, Phys.Rev. B **35** 1881 (1987)

6. See e.g. R.F. Voss in *Scaling Phenomena in Disordered Systems* eds. R. Pynn and A.T. Skjeltorp (Plenum,1985)

7. O. Zhong-can, Y. Gang and H. Bai-lin, Phys.Rev.Lett. **57** 3203 (1986)

DLA WITH SHORT-RANGE INTERACTIONS

M. A. Novotny

IBM Bergen Scientific Center
Allégaten 36
N-5000 Bergen, Norway

I. INTRODUCTION AND METHOD

Recently there has been a great deal of theoretical and experimental interest in growth processes and associated pattern-formation phenomena. Excellent reviews are provided in other contributions to these proceedings[1] and in Ref. 2. Progess has been made in understanding widely different phenomena, including viscous fingering in Hele-Shaw cells,[3,4] fluid-fluid displacement in porous media,[5,6] electric breakdown,[7] electrodeposition,[8] the formation of snowflakes,[9] and the irreversible kinetic aggregation of gold colloids[10] and uniform polystyrene spheres.[11]

There has been considerable interest in modifying the model for DLA (Diffusion Limited Aggregation) introduced by Witten and Sander.[12] In the traditional DLA model, one starts with a single seed particle, and a single incoming random walker (undergoing Brownian motion) is released from infinity and always sticks to the grown cluster when it makes contact. In an effort to incorporate effects of non-zero surface tension, a number of modifications have been made to make the DLA model better describe fluid-fluid displacement in porous media. One modification was to introduce a sticking probability (the probability that the incoming walker sticks to the cluster) which is less than unity[13] and may depend on either a diverging length scale[14] or the local environment.[15,16] A method of including the effects of surface tension based on the Green's function for the underlying Laplace equation has also been introduced.[17].

Although the modifications described above can change the cluster from a DLA-type fractal to a compact object, it does not allow one to obtain different types of fractals, such as those with a finite density of closed loops — as are seen in the aggregation of gold colloids.[10] I will introduce a different modification of the DLA algorithm. Assume that the aggregation is being done on a square lattice (all simulations in this paper are performed on a square lattice). The incoming random walker is at a certain point in the lattice, and has four nearest-neighbor sites that it can walk to, Fig. 1(A). If neighboring site i has a positive number p_i associated with it, then let the probability that the walker moves to site i be given by $p_i/(p_1 + p_2 + p_3 + p_4)$. The numbers p_i are all unity far from the cluster. Near the cluster, the numbers p_i are given by assuming an interaction between the incoming walker and the particles on the cluster. In particular, consider the energy barrier that the incoming walker experiences when it is near the cluster — this energy barrier enters into the probability that the walker will move in a particular direction as a Boltzmann weight. For nearest neighbor pair-wise interactions, let p be the Boltzmann weight associated the the interaction between the incoming walker and a single nearest neighbor particle on the aggregate. Then if a given square has n nearest neighbor aggregate particles, the number associated with the square is p^n. This is illustrated in Fig. 1(B). It is of course possible to consider extensions to the nearest neighbor pair-wise interactions. In this paper both pair-wise nearest-neighbor interactions and three-body interactions (between the incoming walker and two aggregate particles) will be considered. The four different three-body interactions that are used are shown in Fig. 1(C).

Fig. 1. (A) The shaded square is the location of the random walker, which walks to adjacent square i with (unrenormalized) probability p_i. (B) A representative cluster showing the values of p_i for pair-wise nearest-neighbor interactions. (C) The four different three-body interactions used to grow the clusters in Fig. 3 are shown. The filled squares are particles on the aggregate, and the unfilled square is where S_3 is used to calculate p_i.

Fig. 2. The first 15000 particles of representative aggregates are shown for pair-wise nearest-neighbor interactions of strength S_2. (A) $S_2 = 10$ (B) $S_2 = 0.1$

II. DATA AND ANALYSIS

Due to the size limitation of this manuscript, it will not be possible to present a detailed analysis of the data — this will be done in a future publication.[18] Instead, a description of the various aggregates that can be grown using DLA with short-range interactions will be provided, and a sketch of the analysis will be given. Representative aggregates grown with the algorithm described in the preceding section are shown in Fig. 2 for isotropic pair-wise nearest-neighbor interactions of strength S_2. (In this case $S_2 = p$.) When S_2 is greater than unity (an attractive interaction), the cluster becomes slightly more stringlike than normal DLA (which has $S_2 = 1$), but the effect is not very dramatic. However, when $S_2 < 1$ (a repulsive interaction) the cluster changes locally to have a finite finger width that increases as S_2 decreases. This can be understood physically since for a repulsive interaction, the incoming random walker will have a larger probability of penetrating and hence sticking deep in a fjord. The radius of gyration as a function of the mass of the aggregate, $R_g(M)$, averaged over 100 clusters grown on a 1000×1000 lattice was calculated. The fractal dimension D_f,

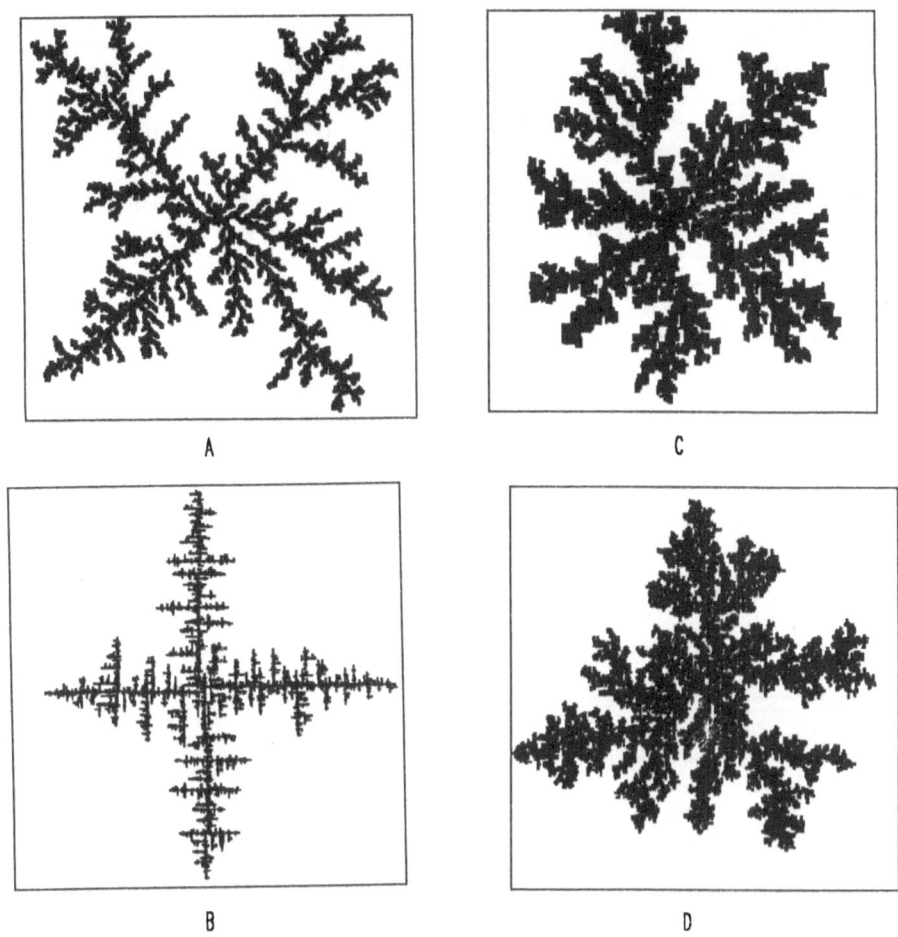

Fig. 3. *Representative aggregates grown with $S_2 = 0.1$ and $S_3 = 10^4$ are shown for the four different three-body interactions shown in Fig. 1(C). The interactions are in the same position in this figure and in Fig. 1(C). The first 15000 particles are plotted in each case.*

defined by $R_g = A M^{\frac{1}{D_f}}$ did not change as S_2 was varied — only the amplitude A changed with S_2. The amplitude varied from $A = 0.42$ for $S_2 = 0.1$ to $A = 0.53$ when $S_2 = 100$. (For $S_2 < 0.01$ the clusters grown on the 1000×1000 lattice were too small to penetrate the asymptotic region — as seen by the curvature in the log-log plot of R_g versus M.) With the aid of histograms of the mass as a function of angle (measured with the origin at the seed particle), it was also found that anisotropy due to lattice effects were most pronounced when $S_2 > 1$, *i.e.* growth along the axis is more probable. For $S_2 < 0.01$ the lattice effects are not seen in the histograms — although they may be seen for aggregates grown on larger lattices.

When both pair-wise nearest-neighbor interactions of strength S_2 and three-body interactions of strength S_3 were included, more dramatic changes in the aggregates were observed. In particular, when $S_2 < 1$ and $S_3 > 1$, the fractal dimension could be varied, and a large number of loops could at times be observed. The structure of the fractal depended not only on the values of S_2 and S_3, but also on the details of the three-body interaction. Figure 3 shows representative samples grown with $S_2 = 0.1$ and $S_3 = 10^4$ for the four different three-body interactions illustrated in Fig. 1(C). By averaging over 100 aggregates grown on 1000×1000 lattices, the fractal dimension was determined. For the interactions shown in Fig. 3(A) through 3(D) (but with $S_2 = 0.01$) the fractal dimensions were $D_f = 1.66, 1.60, 1.84$, and 1.96 respectively. In addition, the effects of lattice anistopy are very different for the four different three-body interactions used. In particular, Fig. 3(A) shows that the growth proceeds predominantly along the diagonal directions, which is also seen using histograms of M versus angle.

III. DISCUSSION AND CONCLUSIONS

The effect of changing the rules for DLA to include short-range interactions have been shown. When only pair-wise nearest-neighbor interactions are included, the clusters become more stinglike for attractive interactions ($p > 1$) and develop a finite finger width for repulsive interactions ($p < 1$). However even though the local structure changes, the fractal dimension associated with the radius of gyration does not change. When pair-wise repulsive interactions and three-body attractive interactions are included, the fractal dimension, the local structure, and the effects of the lattice can all change dramatically. In particular, depending on the nature of the three-body interaction, clusters which contain a large number of loops can be grown. Such growth laws may be relevant to the aggregation of gold colloids[10] where in three dimensions fractals with many loops and with a fractal dimension close to unity are observed. In fact, microscopically one expects that the force between an incoming gold particle and the aggregate will depend on the local environment — and the force may be very anisotropic. Hence the growth law containing both S_2 and S_3 may be of relevance to such experiments. (Of course, off-lattice simulations would be required for a detailed analysis. Such simulations are currently in progress.[19]) When both repulsive S_2 and S_3 are present, the model includes some of the behavior seen in fluid-fluid displacement in porous media with finite surface tension.[5,6] By analyzing the Green's functions of the Laplace equation, one might expect that different continuum limits would be obtained depending of whether the range of the interaction goes to zero or remains finite as the lattice size approaches zero. If this is true, than the modified DLA algorithm may be applied to fluid-fluid displacement with the interaction strength and distance related to the surface tension.

Acknowledgements

The author thanks P. Meakin for many useful discussions. Useful conversations with J. Feder, T. Jøssang, D. P. Landau, K. K. Mon, and K. Måløy are gratefully acknowledged.

REFERENCES

[1] P. Meakin, *Growth of Fractal Aggregates*, in these proceedings.
[2] *Kinetics of Aggregation and Gelation*, edited by F. Family and D. P. Landau
   (North-Holland, Amsterdam, 1984).
   On Growth and Form, edited by H. E. Stanley and N. Ostrowsky
   (Martinus Nejhoff, Boston, 1986).
[3] C. Tang, Phys. Rev. A **31** 1977 (1985).
[4] J. Nittmann, G. Daccord, and H. E. Stanley, Nature **314** 141 (1985).
[5] L. Paterson, Physical Review Letters **52** 1621 (1984).
[6] K. J. Måløy, J. Feder, and T. Jøssang, Phys. Rev. Lett. **55** 2688 (1985).
[7] Y. Sawada, S. Ohta, M. Yamazaki, and H. Honjo, Phys. Rev. A **26** 3557 (1982).
[8] R. M. Brady and R. C. Ball, Nature **309** 255 (1984).
[9] S. Miyazima and T. Tanaka, J. Phys. Soc. Japan **56** 441 (1987).
[10] D. A. Weitz, J. S. Huang, M. T. Lin, and J. Sung, Phys. Rev. Lett. **54** 1416 (1985).
[11] A. T. Skjeltorp, *Colloidal Flow* and *Aggregation and Crystal Growth*,
   in these proceedings.
[12] T. Witten and L. Sander, Phys. Rev. Lett. **47** 1400 (1981).
[13] T. A. Witten and L. M. Sander, Phys. Rev. B **27** 5686 (1983).
[14] K. K. Mon, Physical Review A **34** 4469 (1986).
[15] T. Vicsek, Phys. Rev. Lett. **53** 2281 (1984).
[16] J. R. Banavar, M. Kohmoto, and J. Roberts, Physical Review A **33** 2065 (1986).
[17] L. P. Kadanoff, J. Stat. Phys. **39** 267 (1985).
[18] M. A. Novotny, to be published.
[19] M. A. Novotny and K. Kaski, unpublished.

EFFECT OF ROTATIONAL DIFFUSION ON QUASI-ELASTIC LIGHT SCATTERING FROM FRACTAL CLUSTERS

H.M. Lindsay[*], R. Klein[†], D.A. Weitz[*], M.Y. Lin[*] and P. Meakin[‡]

[*] Exxon Research and Eng, Rt 22E, Annandale NJ 0880.1
[†] Dept of Physics, U. of Konstanz, Konstanz, Germany
[‡] E.I. DuPont Co., Willmington DE 19899

INTRODUCTION

The aggregation of colloids is of substantial interest both fundamentally and practically. The level of interest has risen in recent years with the observation that colloidal aggregates are often well characterized as scale-invariant, or fractal, objects, providing a quantitative description of the structure of these random, irregular clusters.[1-4] The consequences of the scale invariance on light scattering from the clusters has been widely exploited. In static light scattering, the dependence of the scattering intensity on the scattering wave vector q allows a convenient way of determining the fractal dimension of the clusters, while Quasi-Elastic Light Scattering (QELS) has proved useful in monitoring the kinetics of the aggregation process. The combination of the scale-invariant structures of the aggregates and the power-law distributions which often occur leads to elegant scaling behavior of the dynamic light scattering.[5,6] For the large clusters ($qR_g \gtrsim 1$) often found in aggregation, rotational diffusion can play an important role in determining the decay of the autocorrelation of the scattered light measured in QELS. While scaling arguments have been used to account for the contribution of rotational diffusion, it is nonetheless important to determine this effect more quantitatively.

In this paper we develop a general technique for the explicit evaluation of the contributions of rotational diffusion to the decay of the autocorrelation function in QELS, given a known cluster structure. We apply this technique to clusters generated by computer simulations of diffusion-limited cluster-cluster aggregation (DLCA).[7] In determining these rotational effects, it is natural to decompose the structure of the aggregates into a set of spherical harmonics. This provides a more complete description of the structure of a random aggregate than the spherically averaged correlation function probed by static light scattering, and may, in fact, be of more general interest and importance.

QELS

In a QELS experiment we measure fluctuations in the light scattered by a sample. The total intensity of laser light scattered by an ensemble of particles will depend upon their positions, since both the intensity and the phase of the light scattered from the each particle is important. For homodyne scattering from Brownian spheres, the auto-correlation function of the scattered light is just $S(q,t) = Ae^{-2\Gamma t} + B$, where $\Gamma = q^2 D$, $D = k_b T/6\pi\eta R$, and B is the baseline. Hence the autocorrelation function can be used to measure the radius of non-interacting spheres in a fluid.

For non-spherical particles, more than the translational diffusion constant D is needed to describe the light scattering. If we rotate a non-spherical particle, the intensity of light it scatters will fluctuate, leading to an additional decay of the measured autocorrelation function, with the amount of this contribution depending on the exact size and shape of the scatterers.

One process which produces large, non-spherical particles is colloidal aggregation. Their structure is particularly interesting because it is highly random and is irregular on all length scales. In this paper we examine the effects of rotational diffusion on light scattering from gold colloids aggregated under diffusion-limited conditions. We choose this system to study due to the wealth of information already determined by a variety of means for this type of aggregation.[1,2,4,8] The clusters we study are made of gold spheres 150 Å in diameter, and have a fractal dimension $d_f \sim 1.8$. Figure 1 shows the light scattered from a simulated DLCA cluster at several wavevectors as a function of the orientation of the cluster. A scatterer which is small compared to 1/q will act as a point scatterer and show no rotational effects; rotation can play a role only when $qR \gtrsim 1$. These results show that the orientation of these clusters plays an important role in their scattering, and hence that rotational diffusion may have an important role in QELS that must be considered.

CALCULATIONS

To calculate the QELS autocorrelation function, we make several assumptions. We assume that the clusters are dilute and non-interacting, conditions met by our experimental system. We assume that the scattering from the clusters is of the Rayleigh-Debye-Gans form; calculations by Chen et. al.[9] show the validity of this approach for our fractal clusters. We take each cluster as being made up of identical optically isotropic point scatterers of polarizability α_i. Each cluster is characterized by a single hydrodynamic radius R_H, with translational diffusion coefficient D and rotational diffusion coefficient θ following the Stokes relations $D = kT/6\pi\eta R_H$ and $\theta = kT/8\pi\eta R_H^3$. We also assume that translation and rotation are uncoupled. These assumptions are made in recognition that while the aggregates are asymmetrical, they are not

Figure 1.

Light intensity scattered from a single cluster as a function of the angular orientation of the cluster. For the top curve $qR_g = .1$, for the middle curve $qR_g = 1$, and for the bottom curve $qR_g = 10$.

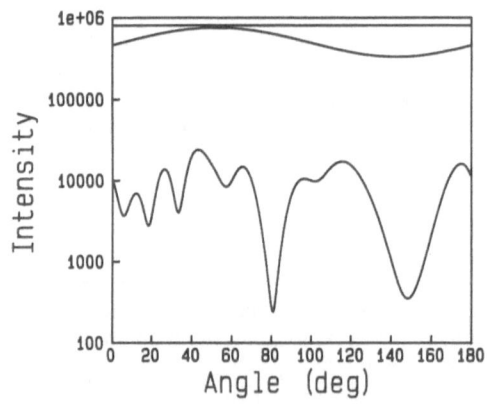

excessively so, with axial ratios generally less than 3, making the use of a single hydrodynamic radius reasonable.

With these assumptions, we can write the autocorrelation function from an ensemble of clusters as

$$S(q,t) = \alpha_i^2 \sum_{\alpha,\beta} \sum_{i,j=1}^{n_\alpha, n_\beta} < e^{i\mathbf{q} \cdot (\mathbf{r}_i^\alpha(t) - \mathbf{r}_j^\alpha(0))} > \qquad (1)$$

where α_i is the polarizability of a particle in the cluster, $r_i^\alpha(t)$ is the position of particle i in cluster α at time t, and n_α is the number of particles in the αth cluster.

We can express the position of a particle in terms of the position of the center of mass of its cluster $R_\alpha(t)$ and its displacement from that center of mass $b_i^\alpha(t)$:

$$\mathbf{r}_i^\alpha(t) = \mathbf{R}_\alpha(t) + \mathbf{b}_i^\alpha(t) \qquad (2)$$

Since translations and rotations are assumed to be independent, and particles in different clusters are uncorrelated

$$S(q,t) = \alpha_i^2 \sum_\alpha <e^{i\mathbf{q} \cdot (\mathbf{R}_\alpha(t) - \mathbf{R}_\alpha(0))} > \sum_{i,j=1}^{n_\alpha} <e^{i\mathbf{q} \cdot (\mathbf{b}_i^\alpha(t) - \mathbf{b}_i^\alpha(0))}> \qquad (3)$$

where the first term gives the contribution due to translational diffusion, and is just $\exp(-q^2 D_\alpha t)$. The second term gives the contribution of rotational diffusion to the autocorrelation function.

To calculate the rotational terms, we expand the scattered light into its multipole contributions, using the spherical harmonics as a basis set

$$S(q,t) = \alpha_i^2 \sum_\alpha e^{-q^2 D_\alpha t} \sum_{\ell=0}^\infty A_\ell^\alpha(q) \, e^{-\ell(\ell+1)\theta_\alpha(t)} \qquad (4)$$

where

$$A_\ell^\alpha(q) = \sum_{M=-\ell}^\ell \left| \sum_{i=1}^{n_\alpha} j_1(qb_i^\alpha) \, Y_{\ell M}(\Omega_i) \right|^2 \qquad (5)$$

We emphasize that this formalism is just a generalization of the treatment for scattering from rods to arbitrarily shaped objects.[10] It requires that we know the positions of the particles in the clusters. Since we cannot determine these positions in our experimental aggregates, we use the coordinates of clusters generated by computer simulations of DLCA, which produce clusters with nearly the same fractal dimension as measured in experiment.

As seen in Eq. 4, the multipole expansion terms $A_\ell(q)$ play a central role in determining the contribution of rotational diffusion to the autocorrelation function, as they determine the intensity of each term in the total decay. To investigate this role, we show in figure 2 the $A_\ell(q)$ multipole terms for a group of clusters, each of which contains 900 particles, and has a radius of gyration of ~20 particle diameters. We see that at small wavevectors, the $\ell=0$ term corresponding to isotropic scattering is dominant, as expected. This merely reflects the fact that when $qR_g \ll 1$, the clusters scatter as points, and

Figure 2.

Amplitude of multipole moments $A_\ell(q)$ as a function of scattering vector for a group of 900 particle DLCA clusters. Terms from $\ell=0$ to 7 are shown, along with the sum of these terms. The q values are in units of inverse particle diameters.

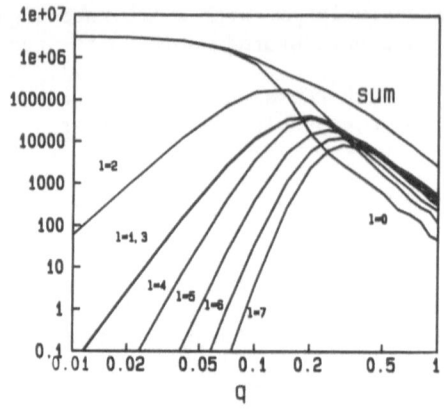

rotational effects are not discerned. The higher order terms become important when $qR_g \gtrsim 1$, and the internal structure of the clusters can be resolved. As ℓ increases, the peak in $A_\ell(q)$ moves to higher wavevectors while the magnitude of that peak decreases.

The relative intensities of the multipole terms are not the only factor which will affect the autocorrelation function of the scattered light; the time-scales of the different mechanisms must also be considered. The translational relaxation time τ_T is q dependent, and scales as R_H/q^2, while the rotational time τ_R is independent of q, and scales as R_H^3. Rotational effects can significant only when $\tau_R \lesssim \tau_T$. In an experimental system, this condition will often be met. For example, if $q=10^5$ cm^{-1} and $R_H=1000$ Å, the time scales are comparable, with $\tau_T=5$ ms, while $\tau_R=7$ ms.

To calculate $S(q,t)$ for an experimental system, we must sum the multipole contributions from a distribution of the clusters. For DLCA, the distribution has been found to be flat, with a cutoff at high mass. In figure 3 we show the calculated value of Γ, the first cumulant of the decay, divided by q^2 to remove the trivial translational diffusion dependence. These values are calculated with a flat distribution of 200 clusters up to a maximum cluster size of 1000 particles, with R_H determined from R_g, using $\beta=R_H/R_g=0.8$. We see that at large q, Γ/q^2 increases substantially due to rotational effects. At the highest q vectors measured here, we see that rotational diffusion is just as important as translational diffusion, as the high-q value of Γ/q^2 is twice the low-q value. The experimental data is also shown in figure 3, and is in good agreement with our calculations.

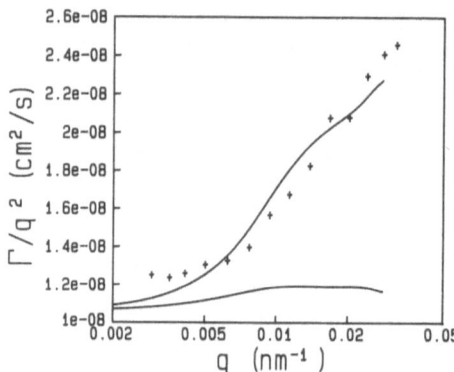

Figure 3.

Γ/q^2 for a distribution of DLCA clusters. The upper curve shows calculated results for a flat distribution of cluster of up to 1000 particles, the lower curve results due to translational diffusion only. The points are experimental values.

CONCLUSIONS

We have developed a general technique to calculate rotational contributions to QELS from irregular particles, which has been used to study diffusion-limited cluster-cluster aggregates, and can in the future be used to study other particles, such as reaction-limited aggregates. This multipole expansion technique provides a way to quantitatively characterize the asymmetry of a cluster, allowing the characterization of the structure of an aggregate beyond just the determination of its fractal dimension. It is of importance here, but may also be of more general significance. We find that rotational diffusion can contribute significantly to the decay of the autocorrelation function $S(q,t)$, in some cases causing as much or more decay than translational diffusion. Thus any quantitative interpretation of QELS data must account for the effects of rotational diffusion when $qR_g \gtrsim 1$. We are currently investigating the validity of using a scaling approach to do so. These calculated rotational effects are consistent with the q dependence of Γ/q^2 which is measured experimentally in gold aggregates.

REFERENCES

1.   P. Dimon, S. K. Sinha, D. A. Weitz, C. R. Safinya, G. S. Smith, W. A. Varady and H. M. Lindsay, Phys. Rev. Lett. **57**, 595(1986).
2.   D. A. Weitz, M. Y. Lin, J. S. Huang, T. A. Witten, S. K. Sinha, J. S. Gethner and R. C. Ball, in *Scaling Phenomena in Disordered Systems*, ed. R. Pynn and A. Skjeltorp (Plenum, New York:1985).
3.   D. W. Schaefer, J. E. Martin, P. Wiltzius and D. Cannell, Phys. Rev. Lett. **52**, 2371(1984).
4.   D. A. Weitz, J. S. Huang, M. Y. Lin and J. Sung, Phys. Rev. Lett. **54**, 1416(1985).
5.   J. E. Martin, this volume.
6.   J. E. Martin and F. Leyvraz, Phys. Rev. **A34**, 2346(1986).
7.   P. Meakin, Phys. Rev. Lett. **51**, 1119(1983).
8.   D. A. Weitz and M. Y. Lin, Phys. Rev. Lett. **57**, 2037(1986).
9.   Z. Chen, P. Sheng, D. A. Weitz, H. M. Lindsay, M. Y. Lin and P. Meakin, to be published.
10. B. J. Berne and R. Pecora, *Dynamic Light Scattering* (Wiley, New York:1976).

THEORY OF DIELECTRIC BREAKDOWN IN METAL-LOADED DIELECTRICS

Paul. D. Beale

Department of Physics
University of Colorado at Boulder
Boulder, CO   80309-0390   USA

Phillip M. Duxbury

Department of Physics and Astronomy
Michigan State University
East Lansing, MI 48824   USA

We have examined a very simple model of dielectric breakdown in random mixtures of metal and dielectric.[1,2,3] We expect this analysis to be relevant for an class of materials which are composed of a random mixture of metallic particles embedded in a dielectric matrix. An example is solid fuel rocket propellant[4] which is a mixture of microscopic aluminum particles (the fuel) in a dielectric matrix composed of oxidizer and rubber binder. The model we analyze is a percolation model in which the bonds of a d-dimensional lattice with lattice spacing a are occupied by conductors with probability p and by capacitors with probability 1-p. The probability p is chosen to be less than the percolation threshold p_c so that no conducting path traverses the entire system. The breakdown process is modeled by assuming that the capacitors can withstand a maximum voltage drop of 1 volt. The entire lattice has a size of L lattice spacings. A macroscopic voltage is applied across the lattice. This voltage is raised until the voltage drop across one of the capacitors exceeds 1 volt. This macroscopic voltage is called V_1, the initial breakdown voltage. The capacitor which fails is replaced by a conducting element. The process of failing one of the capacitors is repeated until a conducting path is formed across the sample. The maximum value of the applied voltage during this procedure is called the complete breakdown voltage and is denoted V_b.

The macroscopic initial and complete breakdown electric fields are given by $E_1 = V_1/La$ and $E_b = V_b/La$ respectively. One would ordinarily expect these fields to be intensive variables. However, in this problem, E_1 is a logarithmically decreasing function of the system size. The initial breakdown field as a function of conductor fraction p and system size L is shown in figure 1. The breakdown field is a monotonic function of p which vanishes at the percolation threshold. Systems in which the metal fraction is close to the percolation threshold are therefore extremely susceptible to dielectric breakdown.

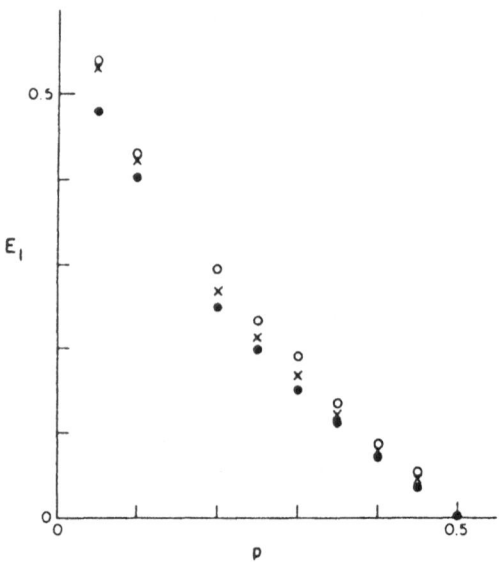

Figure 1. The initial breakdown field, E_1, as a function of metal fraction, p, and system size L. The open circles, crosses and filled circles are L=50, 70 and 100 respectively.

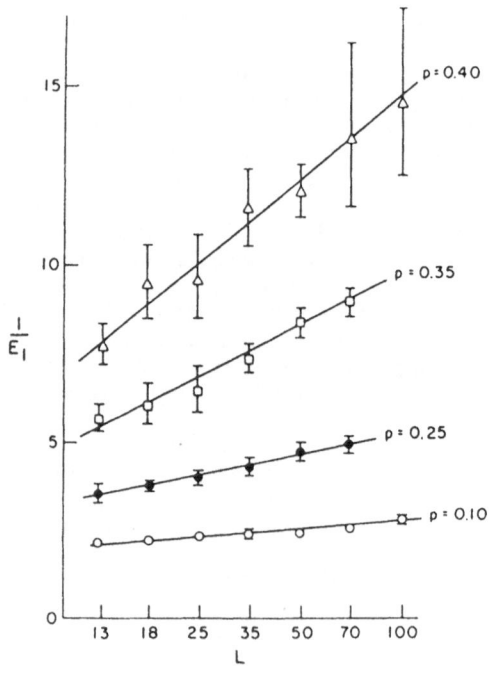

Figure 2. The inverse of the initial breakdown electric field plotted vs. the logorithm of the system size. The good linear fit demonstrates the average breakdown field is given by equation (1).

Note in figure 1 that the breakdown field is a slowly decreasing function of system size for all p. This size dependence is accurately fit by the functional form[3]

$$E_1 \approx \frac{1}{A(p) + B(p) \log(L)}$$ (1)

as is shown in figure 2. The coefficient $B(p)$ (the slope of the line in figure 2) is a singular function of p as $p \to p_c$. A simple scaling argument[3] indicates the $B(p)$ scales as $B(p) \sim (p_c - p)^{-\nu}$ close to the percolation threshold. The exponent ν is the correlation length exponent for the percolation problem.

The logorithmic size dependence (1) can be derived by making an observation concerning the breakdown process. The initial breakdown begins near the tip of the largest percolation cluster which is oriented along the applied electric field. In order to determine the distribution function of breakdown field strengths we need to determine the distribution function of the largest cluster in samples of volume L^d. Clearly the largest cluster in a sample of size L will, on average, be larger than the largest cluster in a system of smaller size L'. An argument based on the statistics of extremes[3] gives the following form for the breakdown field distribution function $F_L(E)$.

$$F_L(E) \approx 1 - \exp\left[- c\, L^d \exp\left(\frac{-k}{E} \right) \right]$$ (2)

This function is the probability that a sample of linear size L will fail at applied electric field less than E. The coeffiecients c and k depend on the metal fraction p. It is then easy to see that $A(p)$ and $B(p)$ in equation (1) are given approximately by

$$A(p) \approx \frac{\ln(c) - \ln(\ln(2))}{k}$$ (3)

and

$$B(p) \approx \frac{d}{k} .$$ (4)

The derivation[3] of the functional form (2) is based on three assumptions. First, the breakdown field is determined by the linear size of the largest cluster. Second, the cluster size distribution function is an exponentially decaying function of cluster size (this is appropriate for p far from p_c). Third, a cluster of linear dimension l enhances the local electric field by of order l.

If the cluster size distribution is a power law decaying function of cluster size (when p is close to p_c so that L is less than the percolation correlation length) then the breakdown field distribution function is

$$F_L(E) \approx 1 - \exp\left[- c\, L^d\, E^m \right] .$$ (5)

This form (5) is called the Weibull distribution.[5]

Figure 3 shows that the form (2) is to appropriate form for the breakdown distribution function for p far from p_c. The data for the distribution of breakdown fields on two different lattice sizes is plotted so that if (2) correctly describes the data then the distributions for the two different lattice sizes will data-collapse onto a single straight line. Plotting this same data in a manner to test the Weibull form results in a very poor fit. However, the Weibull distribution should fit the breakdown data very close to the percolation threshold.

The behavior of the dielectric breakdown field in three-dimensional models is the same as for the two-dimensional results given above. The average breakdown field tends monotonically to zero as p approaches the percolation threshold and the initial breakdown field has a logorithmic size dependence given by equation (1).

Finally, we observe that the complete breakdown field E_b appears to have the same average value as the initial breakdown field. The average value of the complete breakdown field is always found well within the uncertainty range of the initial breakdown field. This has been tested on a number of different lattice sizes and metal fractions in two dimensions. In other words, a field sufficient to cause a single microscopic failure is also sufficient to cause a cascade of microscopic failures resulting in the formation of a macroscopic conducting path across the system. It is along this path that the electrostatic discharge will take place.

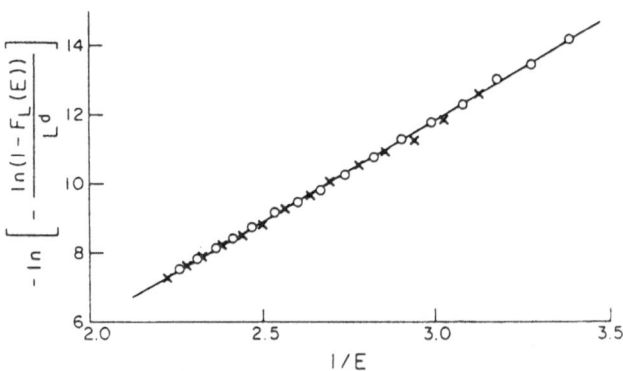

Figure 3. The initial breakdown field distribution function plotted in a manner to test the functional form (2). The data is for two different lattices L=70 (crosses) and L=100 (circles) with metal fraction p=0.100 . If (2) is valid then the data for these two lattice sizes should collapse onto a single straight line. The distribution function was determined using 1000 samples for the L=70 lattice and 744 samples for the L=100 lattice.

REFERENCES

1. D. Bowman, private communication and preprint (1987);
    H. Takayasu, Phys. Rev. Lett. 54, 1099 (1985)
2. L. de Archangelis, S. Redner and H.J. Hermann, J. de Phys. Lett.
    46, L585 (1985);
3. P.M. Duxbury, P.D. Beale and P.L. Leath, Phys. Rev. Lett. 57, 1052
    (1986);
    P.M. Duxbury, P.L. Leath and P.D. Beale, Phys. Rev. B, in press
    (1987);
    P.D. Beale and P.M. Duxbury, submitted to Phys. Rev. B. (1987).
4. R. Kent and R. Rat, J. Electrostatics 17, 299 (1985).
5. W. Weibull, "Fatigue Testing and Analysis of Results," (Pergamon,
    New York, 1962).

FRACTAL INTERPRETATION OF DIELECTRIC RESPONSE

G.A. Niklasson, K. Brantervik and I.A. Serbinov*

Physics Department, Chalmers University of Technology
S-412 96 Gothenburg, Sweden

INTRODUCTION

In this paper we put forward a fractal interpretation of the dielectric properties of disordered materials. In materials having charge carriers of low mobility, the dielectric permittivity $\varepsilon(\omega) = \varepsilon_1(\omega) + i\varepsilon_2(\omega)$ often displays an anomalous low frequency dispersion[1] (ALFD),

$$\varepsilon_1(\omega) - \varepsilon_\infty \sim \varepsilon_2(\omega) \sim \omega^{-p} \quad , \qquad \omega < \omega_c$$
$$\varepsilon_1(\omega) - \varepsilon_\infty \sim \varepsilon_2(\omega) \sim \omega^{n-1} \quad , \qquad \omega > \omega_c \qquad (1)$$

Here ε_∞ is the high frequency dielectric constant, ω_c is a crossover frequency, and the exponents p and n can have values between zero and unity. Below we relate the values of these exponents to the fractal structure of the material and to fractal time processes. The theory is used to interpret experiments on Co-Al$_2$O$_3$ composites, pyrolyzed polyimide and iron oxide films.

THEORY

We model the conduction process by a random walk of the charge carriers on a structure consisting of localized states or conducting inclusions. The dielectric permittivity is related to the frequency dependent diffusion coefficient $D(\omega)$ by

$$\varepsilon(\omega) = i\sigma(\omega)/\omega = iNe^2 D(\omega)/\omega kT, \qquad (2)$$

where $\sigma(\omega)$ is the complex electrical conductivity, N is the charge carrier density, e is the electron charge, k is Boltzmann's constant and T is the temperature. Now[2]

$$D(\omega) = -\frac{1}{6}\omega^2 \int_0^\infty dt\, e^{-i\omega t} \langle r^2(t)\rangle, \qquad (3)$$

---

* Permanent address: Institute of Radio Engineering and Electronics, Academy of Sciences, Marx Avenue 18, Moscow, USSR.

where $\langle r^2(t) \rangle$ is the mean square displacement of the random walker, which can be related to the probability that the walker returns to the origin at time t, $P_0(t)$, by[3,4]

$$\langle r^2(t) \rangle \sim (P_0(t))^{-2/D} \quad , \tag{4}$$

where D is the dimensionality of the structure. We can now use these expressions to distinguish different cases.

(a) A regular lattice gives a dc conductivity behaviour with $\varepsilon_1 \sim$ constant and $\varepsilon_2(\omega) \sim \omega^{-1}$

(b) A regular arrangement of clusters gives an ordinary diffusion behaviour with n = 0.5 at $\omega > \omega_c$, which crosses over to a dc conductivity at lower frequency. The crossover frequency here corresponds to the cluster size.

(c) A random walk on a fractal structure is characterized by the random walk dimension D_W which is defined by

$$\langle r^2(t) \rangle \sim t^{2/D_W} \quad . \tag{5}$$

In this case we cannot obtain the dielectric response directly from eqs. (2) and (3), but we must also scale the total admittance in the diffusion volume.[5] We now obtain $n = 1 - D_f/D_W$, where D_f is the fractal dimension, at $\omega > \omega_c$. A dc conductivity is again predicted below ω_c, which corresponds to the correlation length of the structure.

(d) A fractal time process is realized by a random walker which waits for sometime between each step, if[6] the waiting time distribution $\Psi(t)$ is of the form

$$\Psi(t) \sim t^{-1-D_t} \quad , \tag{6}$$

where D_t is the fractal dimension of the process. Such behaviour can be a result of multiple trapping in an exponential trap distribution, hopping in a random chain, trap-controlled hopping or hopping with an exponential distribution of activation energies. Eq. (6) gives rise to a power-law behaviour of $\varepsilon(\omega)$ at low frequencies. The exponent takes the values $p = 2D_t/(1+D_t)$ in one dimension[3] and $p = D_t$ in higher dimensions.[7]

(e) If a fractal time process is present on a fractal structure the power-law exponents combine multiplicatively,[8] at least to a good approximation. Thus we have $p = D_t$ at $\omega < \omega_c$ and $1-n = D_t D_f/D_w$ at $\omega > \omega_c$. This means that

$$(1-n)/p = D_f/D_w. \tag{7}$$

EXPERIMENTS

The dielectric properties were measured by applying a sine wave from a function generator over the sample and measuring the amplitude and phase of the current through the sample. This method was used for frequencies between 10^{-4} Hz and 4kHz. At higher frequencies we instead used a conventional capacitance bridge.

Co-Al_2O_3 composite films were produced by dual electron-beam evaporation. Low concentrations of cobalt enter directly as ions in the amorphous Al_2O_3 matrix. At cobalt volume fractions, f, in excess of 0.07 also Co particles start to appear. For $f \leq 0.10$ the dielectric permittivity displays an ALFD with $p < 1$. An example is shown in fig. 1. We interpret the data as being due to a fractal time process on a fractal structure, i.e. an "infinite" percolation cluster consisting of localized states and metal particles. The fractal time process is probably connected with the

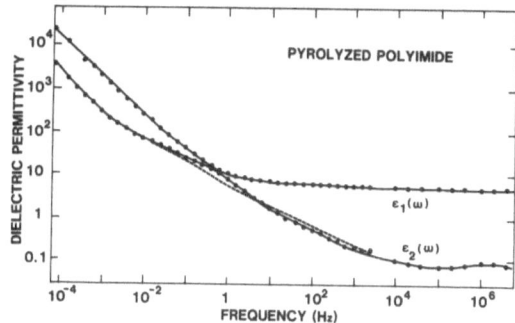

Fig. 1. Dielectric permittivity as a function of frequency for a Co-Al$_2$O$_3$
film with f = 0.1. Dots denote experimental values while full
lines are drawn as a guide for the eye. Dashed curves denote
$\varepsilon_1(\omega) - \varepsilon_\infty$.

presence of cobalt ions in the oxide. Preliminary measurements suggest
that p is independent of temperature. This means that the data cannot be
interpreted in terms of multiple trapping; instead we suggest that trap-
controlled hopping is present in our samples. From an average over five
films we found that $(1-n)/p \sim 0.68 \pm 0.06$, which is, within our error bars,
in good agreement with theoretical predictions by Alexander and Orbach[4].

Polyimide foils (Kapton) were pyrolyzed in vacuum at temperatures
between 400°C and 450°C for various times. This yields a partially carbonized film with conductivity close to the insulator-conductor transition for
this material. The heat treatment was quantified by the variable $\Delta E =
kT\ell n(t/t_0)$ where t is the time and $t_0 \sim 10^{-13}$ s. Fairly insulating films
such as that in fig. 2 show $p \sim 0.9$ and $n \sim 0.5$. We think that conduction
takes place by tunneling or hopping between carbonized inclusions arranged
in linear chains.

Fe$_2$O$_3$ films produced by chemical vapour deposition were partially
converted to conducting Fe$_3$O$_4$ by heat treatment in vacuum. Films close to
the insulator-conductor transition show a dc conductivity at low frequencies
and $n \sim 0.5$ at $\omega > \omega_c$, as seen in fig. 3. This indicates that the conducting phase is arranged in a regular clustered pattern.

CONCLUSION

We have established a connection between the dielectric response of
disordered materials and the concepts of fractal structures and fractal
time processes. Dielectric measurements can give important information on
the structure and the electrical conduction mechanisms in disordered materials.

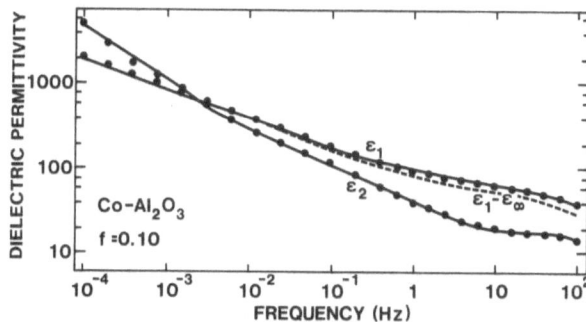

Fig. 2. Dielectric permittivity as a function of frequency for a pyrolyzed
polyimide film with $\Delta E = 2.29$ eV. Symbols are as in fig. 1.

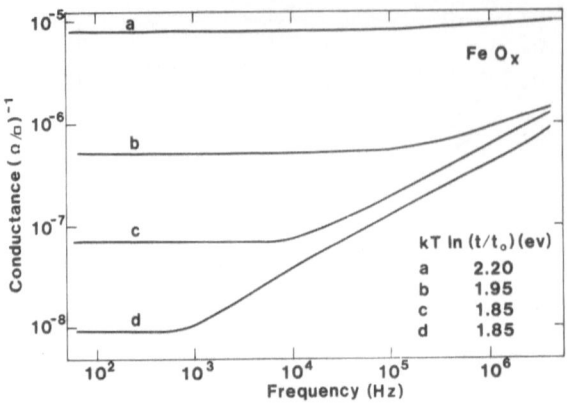

Fig. 3. Conductance as a function of frequency for iron oxide films with compositions close to the insulator-conductor transition. Values of ΔE are given in the inset.

ACKNOWLEDGEMENT

This work was financially supported by grants from the National Swedish Board of Technical Development and from ASEA Research and Development, Västerås, Sweden.

REFERENCES

1. A. K. Jonscher, Low-frequency dispersion in carrier-dominated dielectrics, Phil. Mag. B 38:587 (1978).
2. H. Scher and M. Lax, Stochastic transport in a disordered solid, Phys. Rev. B 7:4491 (1973).
3. S. Alexander, J. Bernasconi, W. R. Schneider and R. Orbach, Excitation dynamics in random one-dimensional systems, Rev. Mod. Phys. 53:175 (1981).
4. S. Alexander and R. Orbach, Density of states on fractals: "fractons", J. Phys. Lett. 43:L-625 (1982).
5. S. H. Liu, Fractals and their applications in condensed matter physics, Solid State Phys. 39:207 (1986).
6. M. F. Schlesinger and B.D. Hughes, Analogs of renormalization group transformations in random processes, Physica A 109:597 (1981).
7. S. Alexander, Anomalous transport properties for random-hopping and random-trapping models, Phys. Rev. B 23:2951 (1981).
8. A. Blumen, J. Klafter, B. S. White and G. Zumofen, Continuous-time random walk on fractals, Phys. Rev. Lett. 53:1301 (1984).

DYNAMICS AND STRUCTURE
OF VISCOUS FINGERS IN POROUS MEDIA

Knut Jørgen Måløy, Finn Boger
Jens Feder and Torstein Jøssang
Institute of Physics, University of Oslo
Box 1048, Blindern, 0316 Oslo 3, Norway

1 Introduction

The displacement of a high viscosity fluid by a low viscosity fluid in a porous medium is a process of both scientific and practical importance. It has recently been shown by Chen and Wilkinson[1] and by Måløy et al.[2] that viscous fingering in a random porous medium at high capillary numbers, $Ca >> 10^{-4}$, generates structures with a fractal[3] geometry. This fractal structure closely resembles that obtained from the diffusion limited aggregation (DLA) model of Witten and Sander[4]. Similar structures have also been obtained by fluid-fluid displacement in radial Hele-Shaw cells using non-Newtonian viscous fluids[5]. The relationship between fluid-fluid displacement in porous media and DLA was first discussed by Paterson[6] and a more detailed analysis has been presented by Kadanoff[7].

A wide variety of non-equilibrium processes such as dielectric breakdown[8], random dendritic growth[11,12], electrodeposition[9,10], and the dissolution of porous materials[13] have also been shown to lead to the formation of structures similar to those associated with the DLA model. Here we describe a study of viscous fingering of Newtonian fluids in a porous medium at high capillary numbers ($Ca = u\mu/\sigma \simeq 0.15$). Here u is the mean velocity of the longest finger, μ is the viscosity of the viscous liquid, and σ is the surface tension. Under these conditions growth occurs mainly at the relatively unscreened (most exposed) tips of the fingers and can be described in terms of an active zone[14] which leaves behind a 'frozen' structure that does not evolve further with increasing time. Since the 'frozen' interior, which forms the majority of the cluster, is formed by the advance of the active zone, it is reasonable to expect that a more complete description of the dynamics of the active zone might lead to a better understanding of the viscous finger structure.

The original DLA model of Witten and Sander specifies a time ordered sequence of events but is not explicitly time dependent. While the time dependent aspects of DLA and closely related processes have been explored theoretically[15,16] and by

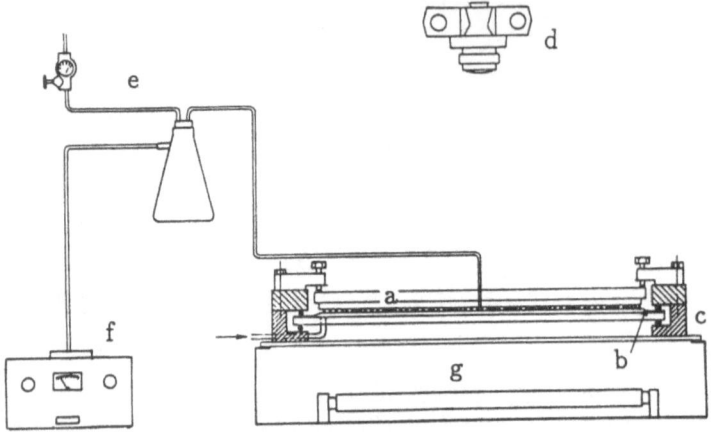

Figure 1: Experimental setup. a: porous model with glass spheres, b: plastic film, c: external aluminium ring, d: camera, f: pressure gauge, e: pressure reduction system, g: light board.

means of computer simulations[17,18], this aspect of DLA has only recently been explored experimentally[2].

Here we describe the results of experiments carried out to explore the dynamics of fluid-fluid displacement in a two-dimensional porous medium. The results of these experiments are compared with simulations on a modified DLA model[41,42] which allows the time dependence of the growth process to be investigated in the zero concentration limit. (These simulations are described by Meakin in these proceedings).

We consider in some detail the dynamics of the active (growth) zone, and introduce a new *fractal measure*[3] on the growing finger structure based upon the observed growth dynamics. We analyze the measure in terms of a '*multifractal*' $f(\alpha)$ curve.

In the last part of this paper we describe a new way of analyzing the static finger structure, by moments of the observed structures. These moments are characterized by scaling exponents d_q, which contain more information about the finger structures than the fractal dimension alone. We apply this analysis also to finger structures observed at intermediate and low capillary numbers.

2   Experimental Methods

The experimental setup is shown in Fig. 1. The porous model consisted of $1\,mm$ diameter glass spheres sandwiched between a stiff plexiglass disk and a thin plastic sheet held in contact with the spheres by compressed air. The model was made by coating a plexiglass disc ($6\,mm$ thick and $40\,cm$ in diameter) with a $0.1\,mm$ layer of transparent epoxy, and spreading a layer of $1\,mm$ glass spheres onto the disk. The epoxy was allowed to partially harden before spreading the glass spheres so that the spheres would stick on contact and not move later. After the epoxy layer hardened, the excess of glass spheres was removed leaving a monolayer. The disc with the monolayer of spheres was placed, face down, onto a plastic film. This assembly was supported from above and below by $10\,mm$ thick glass discs and clamped in an aluminum ring as illustrated in Fig. 1. We connected the space between the plastic film and the lower glass plate to a compressed air supply in order to force the film into contact with all of the glass spheres. This air pressure was higher than the pressure of the fluid injected into the porous model and was held constant during the experiment. By filling the model before increasing the pressure on the plastic film, it is easier to fill pores with high viscosity fluids without trapping air bubbles.

In a typical experiment air injected at the center displaces glycerol filling the pore space of the model. The displaced glycerol was accumulated at the ambient pressure, p_0, in a channel at the rim of the circular model. Air at a constant pressure p_s, was introduced into the porous layer through a $1\,mm$ hole at the center of the plexiglass disc. A stable pressure of the air displacing the glycerol was obtained using a Mortanair B11-M3 pressure regulator and a Norgren low pressure regulator connected with an external reservoir of about $5\,l$. The pressure of the air during the injection process was measured with a Texas Instrument Fused Quartz Pressure Gauge, and the pressure fluctuations at the center were typically less than 0.5%.

The resulting finger structure was photographed with a Nikon F3 camera controlled from an IBM PC. Uniform lighting of the transparent model was provided from below. A typical time between each picture was $0.6\,s$. The pictures were enlarged using an Agfa Gevaert reprodoline 716 transparent film. Fig. 2 shows a typical viscous finger pattern obtained in this way. Note that the picture in Fig. 2 was obtained by superimposing a negative picture of the filled model with a positive picture of the finger structure. We thus subtracted out all the features that remain constant such as the fixed spheres and any unevenness in the illumination. In effect we obtain the difference between the two pictures.

In order to identify the active growth zone at a given time, and to filter out noise from the picture, we subtracted a picture taken at the previous time. The earlier picture was subtracted by superimposing the negative of the earlier picture below the positive of the last picture. In this way we obtained the growth zone shown in Fig. 3, for the fingers structure shown in Fig. 2.

The viscous fingers were analyzed by digitizing the pictures. We used several systems having different resolutions for this analysis. A RCA TC2055CX video camera and a Tecmar Video van Gogh interface in an IBM PC, gave a resolution of 256×256 pixels. A JVC video camera and a Data Translation Video board in an IBM PC, gave a resolution of 512×480 pixels. A Data Copy scanning digitizing camera interfaced to an Apollo DN3000 workstation gave a maximum resolution of 4000×4000 pixels. For this system we limited our analysis to only 2000×2000 pixels since this is enough to resolve the pore structure of our model.

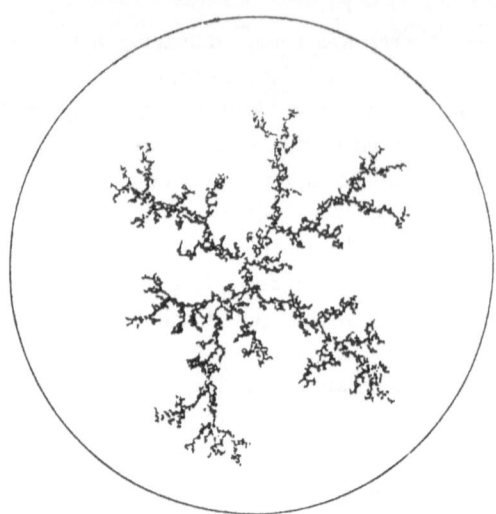

Figure 2: Fractal viscous fingering with a fractal dimension $D = 1.64 \pm 0.04$ in a 2–dimensional porous medium. Air was injected in glycerol with a pressure difference $(p_s - p_0)$ of 20.2 mm Hg and a capillary number Ca of 0.15. The structure was observed at a time $t = .8\,t_0$, where the breakthrough time is $t_0 = 28.6$ sec.

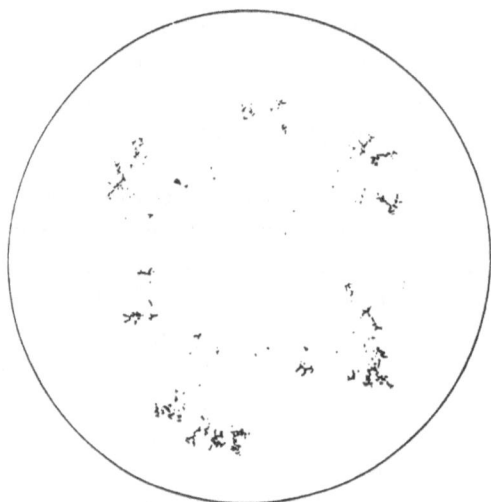

Figure 3: Active growth zone of a viscous fingering structure shown in Fig. 2. The time between the two pictures used to construct the growth zone was $2.8\,s$, which corresponds to a relative time increment of $\Delta t / t_0 = 0.10$.

3   Viscous Fingers in Hele Shaw cells

Saffman and Taylor[20] developed a theory of viscous fingering in a Newtonian liquid between two parallel glass plates separated by a gap of a. They considered the one- dimensional case of channel flow where a fluid of viscosity μ_1 displaces another fluid having viscosity μ_2. The flow was described by Darcy's law

$$\mathbf{U} = \frac{k}{\mu} \nabla p \,, \tag{1}$$

for each of the two fluids separately. Here $\mathbf{U}$ is the fluid flux and $k = a^2/12$, the permeability. They concluded that the interface between the two fluids is unstable whenever the viscosity of the driving fluid is lower than the viscosity of the fluid being displaced. In their linear stability analysis they found that the interface is unstable with respect to pertubations having a wavelength $\lambda > \lambda_c$ and grow at a maximum rate at a wavelength λ_m given by:

$$\lambda_m = \sqrt{3}\lambda_c, \quad \text{with} \quad \lambda_c = \frac{\pi b}{\sqrt{Ca}}, \quad \text{and} \quad Ca = \frac{U\mu}{\sigma} \,. \tag{2}$$

These results are valid for the case where the viscosity of the driving fluid can be ignored.

Paterson[21] solved the same problem for a 2D circular Hele Shaw cell and found that when the circumference increases beyond a critical value $\lambda_c/a = Ca^{-1/2}$, it becomes unstable and splits into fingers with a width of λ_c. We find, however, that the structure of viscous fingers in a porous medium is fractal and is therefore qualitatively different from the structures in a ordinary Hele Shaw, cell which is not fractal at the same capillary numbers.

For an incompressible fluid and a porous medium with homogeneous porosity Darcy's law (1) and $\nabla \cdot \mathbf{U} = 0$ gives the Laplace equation

$$\nabla^2 p = 0 \,. \tag{3}$$

Here we are concerned with a gas injected with a constant pressure p_s and with constant pressure p_0 at the rim. The ordinary Hele Shaw cell and a Hele Shaw cell with a porous medium will satisfy the same differential Eq. (1) and (3). However the boundary conditions are clearly not the same. In the case of a porous medium a new length scale, the size of the pore b, must be taken into account. In a porous medium, the direction of propagation and the width of the viscous fingers depend of the pore size and the geometry of the porous matrix. In an ordinary Hele Shaw cell the finger width is controlled by the capillary length λ_c. Note that Darcy's equation (1) only applies in an average sense. The permeability k is defined only when one considers the porous medium on length scales much larger than typical pore dimensions. The dynamics of displacement fronts which advance on the pore scale are, therefore, not described by the Darcy equation alone—the random pore geometry influences the displacement dynamics.

Nittmann, Daccord and Stanley[5] have studied viscous fingering of water into a solution of polymers in water. The high viscosity solution is non-Newtonian having a viscosity that depends on the shear rate. Also the interfacial tension between the two miscible fluids is small, and the displacement takes place at very high effective capillary numbers. Using a radial Hele Shaw cell they found that their structures could be characterized by a fractal dimension $D = 1.7$ consistent with our result and the DLA simulations.

We note that there are differences between the structure they observed and the finger structure observed in our random porous media and in the DLA simulations. (i) In their experiments the "spokes" of the radial viscous fingers are straight, relative to the sinuous "spokes" of the DLA simulations and our experiments[25]. (ii) There is more "foliage" growing on the DLA "spokes" and the spokes in our experiments compared to the relative barren impression of the radial viscous finger spokes of Daccord et. al.[25]. (iii) If one makes a histogram of tip splitting angle, measured at each of the finger bifurcation one find that the radial viscous fingers of Daccord et al. is narrower than for DLA simulations and our experiments[25].

In a continuum description of fluid-fluid displacement the fingers will grow in the regions where the pressure gradient at the interface is largest. In the diffusion limited aggregation model an additional component is introduced by the stochastic nature of the random walk. In a random porous medium the pressure gradients across the pore necks will fluctuate along the interface because of the capillary pressure fluctuations and the viscous drag fluctuations at the pore level. This pressure fluctuation is important and should not be neglected even at high capillary numbers. To understand the dynamics of viscous fingers at high capillary numbers it is therefore important to consider both the global pressure given by Darcy's law and the pressure fluctuations at the pore level.

4 Viscous Fingers in Porous Media

The fractal Dimension

The observed viscous fingering structures were analyzed by digitizing pictures such as the one shown in Fig. 2. Black pixels in the digitized pictures correspond to pore space where air has displaced glycerol. The number $N(r)$ of black pixels was measured as a function of distance, r, from a point near the center of injection. Fractal structures exhibit scaling and we expect $N(r)$ to have the form [2,3]

$$N(r) = N_0(r/R_g)^D f(r/R_g) . \tag{4}$$

Here D is the fractal dimension of the structure and R_g is its radius of gyration. N_0 is the total number of black pixels, which correspond to pore space from which glycerol has been displaced, i.e. the mass of the displaced glycerol. The scaling power-law Eq. (4) is expected only for the range $a < r < R_0$, where a is a typical pore dimension, and R_0 is the radius of the model. We therefore must include the crossover function $f(x)$ in Eq. (4). The crossover function is constant in the range

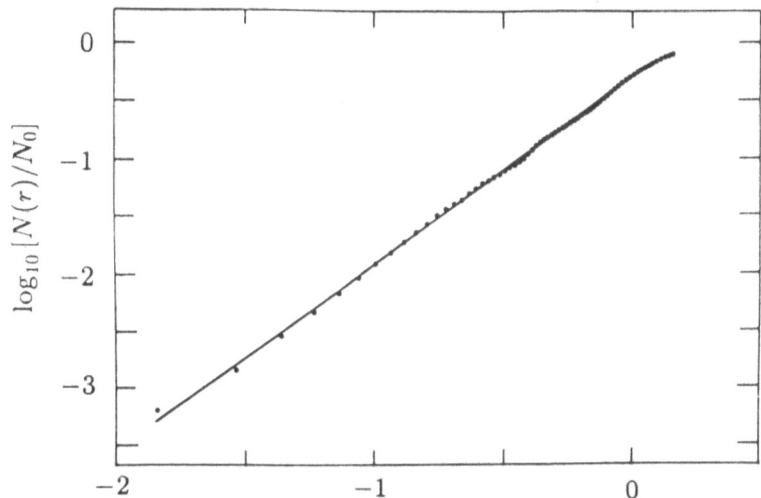

Figure 4: The normalized finger structure area $N(r)/N_0$ as function of the reduced radius r/R_g. In this experiment air displaces glycerol at $Ca = 0.15$, in a two-dimensional porous model. The solid line obtained from a fit to the experimental data has a slope of 1.64 ± 0.04.

$a/R_g < x < 1$, and approaches to x^{-D} for $x > 1$, so that $N(r) \to N_0$ for $r >> R_g$. In Fig. 4 we plot $\log[N(r)/N_0]$ as function of $\log(r/R_g)$ for the structure shown in Fig. 2. By fitting Eq. (4) in the range $2a < r < R_g$, we find a fractal dimension $D = 1.64 \pm 0.04$. This value is the mean value of D obtained by choosing different centers (on the cluster) inside an area with a radius of about 8 pixels from the center of injection. The uncertainty given expresses the range of D values obtained within this area. This value for the fractal dimension differs somewhat from that obtained from simulation results[22] ($D \simeq 1.71$). However, the shape of the viscous fingers shown in Fig. 2 bear a striking resemblance to the shapes generated by two dimensional DLA models. We believe that a DLA model describes the essential physics of our experiments.

The longest finger and the radius of the active zone scale as R_g

The average radius of the active zone is defined by

$$R_a = \frac{1}{n} \sum l_i , \qquad (5)$$

where l_i is the distance from the center of injection to i-th black pixel in the growth zone shown in Fig. 3, and n is the total number of such pixels. We have measured the length, R_m, of the longest finger, the radius, R_a, of the active growth zone, and the radius of gyration, R_g, of the total cluster as function of time t, for different experiments with the same pressure difference, $(p_s - p_0)$ equal to $20.2mm\,Hg$.

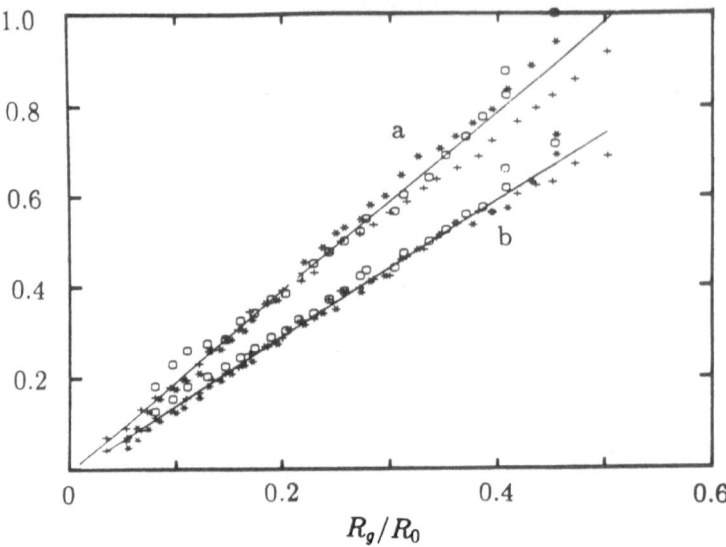

Figure 5: (a): The longest finger R_m/R_0 as function of the radius of gyration R_g/R_0. (b): The average radius R_a/R_0 of the active zone as a function of the radius of gyration R_g/R_0. These results were obtained with air displacing glycerol, in three different experiments as indicated by different symbols in the figure. The pressure difference $(p_s - p_0)$ was $20.2\,mm\,Hg$.

If the average radius of the growth zone, is plotted as function of the radius of gyration, R_g, of the viscous finger structure of the last picture, we obtain the result shown in Fig. 5. In the same figure we have also plotted the length R_m of the longest finger as function of the radius of gyration of the total cluster for the same experiments. Linear functions shown as a solid lines in Fig. 5 fit the data points well.   Within the accuracy of our experiment the average radius of the growth zone, the longest finger and the radius of gyration of the total cluster have the same functional form.

$$
\begin{aligned}
R_a &= c_1 R_g \sim N_0^{1/D} \\
R_m &= c_2 R_g \sim N_0^{1/D}
\end{aligned}
\tag{6}
$$

The values of the constants c_1 and c_2 from fitting to linear functions are $c_1 = 1.49 \pm 0.02$ and $c_2 = 1.98 \pm 0.03$. This result is consistent with the DLA simulation of Plischke and Rácz, [14] who found the same exponent $1/D$ for both R_a and R_g as a function of N_0.

The Width of the Active Growth Zone

The radius of gyration of the growth zone, $R_{a,g}$, is defined by

$$
R_{a,g} = \sqrt{\frac{1}{n} \sum l_i^2} \, ,
\tag{7}
$$

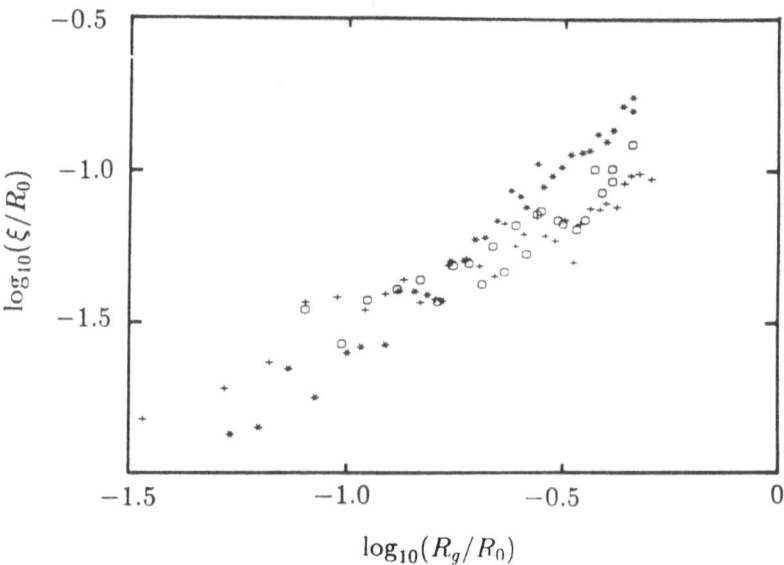

Figure 6: The width ≤ the active zone ξ/R_0 as a function of the gyration radius R_g/R_0 obtained from the three experiments used to obtain Fig. 8. The pressure difference was $(p_s - p_0)$ of $20.2\,mm\,Hg$, with air displacing glycerol

and the width, ξ, of the growth zone is defined by

$$\xi = \sqrt{R_{a,g}^2 - R_a^2} \, . \tag{8}$$

The relationship between the width of the active zone (the screening width) and the radius of gyration has been a subject of considerable interest[14,22,23,24], and relates to the exponent κ in the scaling relation:

$$\xi \sim R_g^\kappa \, . \tag{9}$$

Plischke and Rácz[24] argue that, since DLA clusters scale differently in the radial and tangential directions, the width of the active zone should scale differently from the radius of gyration of the total cluster. They suppose that the screening depth is proportional to the average distance between the branches. From their simulation[14] of clusters containing 10 000 particles they found $\kappa = 0.83$.

Meakin and Sander[22] found that when the size of the cluster increased the exponent κ increased. By using 1 000 off-lattice clusters each containing 50 000 particles they found that the effective value of κ increased with increasing cluster size. For clusters containing 50 000 particles they found that $\kappa = 0.93$ and suggested that the limiting value for $(N_0 \rightarrow \infty)$ should be 1.0. The asymptotic value for the exponent κ is still an unresolved question.

In Fig. 6 we have plotted ξ as function of R_g for three different experiments in a double logarithmic plot. There is too much noise in our data to resolve the question as to the value of κ. If we assume a power law behavior as given in Eq. (9), we find that the exponent κ lies between 0.8 and 1.2.

5 The Dynamics of Finger Growth

The Time Dependence of Longest finger R_m

We have measured the length ,R_m, of the longest finger as function of time t, for different experiments with the same pressure difference $(p_s - p_0)$ equal to $20.2 mm Hg$ The result of these measurements are shown in Fig. 7, as function of t/t_0, where t_0 is the 'break through' time when the longest finger reaches the rim of the model.

As a frame of reference for our discussion of the *dynamics* of fractal finger growth let us first consider the growth of a bubble injected at the center of a circular Hele Shaw cell. For a circular bubble of radius r, at a fixed pressure p_s, expanding into the viscous fluid, equations (1) and (2) give the relation[2]

$$t/t_0 = (r/R_0)^2 \left(1 - ln(r/R_0)^2\right) , \qquad (10)$$

when the pressure is fixed at p_0 at the circular rim at radius R_0 of the cell. The time t_0 is the 'break-through' time when the bubble reaches the rim. This gives the dashed curve in Fig. 7, and it describes the data well. This indicates that the longest fingers control the potential flow and generate a 'Faraday cage' with a radius almost equal to R_m screening the internal structure of shorter fingers.

Clearly Eq. (10) should be modified for fractal growth[2,41]. Matsushita et al.[9] studied the fractal structure of 2–dimensional zinc metal leaves grown by electrode-position and obtained $D = 1.66 \pm 0.03$. They proposed that the 'effective' radius of the structure should grow as $r \sim t^{1/D}$. This proposal suggests that the exponent 2 in Eq. (10) be replaced by D in order to account for the fact that less of the fluid is displaced by fractals than by a bubble. This replacement gives the thin fully drawn curve in Fig. 7. We expect this replacement to be valid for the 'effective' radius. We find, however, that the exponent 2 fits the results for the dynamics of the *longest* finger better than the exponent 1.64, see Fig. 7. In Fig. 7 we also show the results of recent DLA simulations[41], of the dynamics of the longest finger. We see that these simulations fit the observations rather well.

As discussed in the next section it may be more appropriate to consider other characteristic length scales for the growing structure than the longest finger, which after all, represents an extreme in a particular realization of what really is a stochastic process.

The Time Dependence of R_a and R_g

We have also measured the time dependence of the radius, R_a, of the active zone and of the radius of gyration. The result of these measurements together with, R_m, are shown in Fig. 8 as function of t/t_0 where t_0 is the 'break-through' time when the

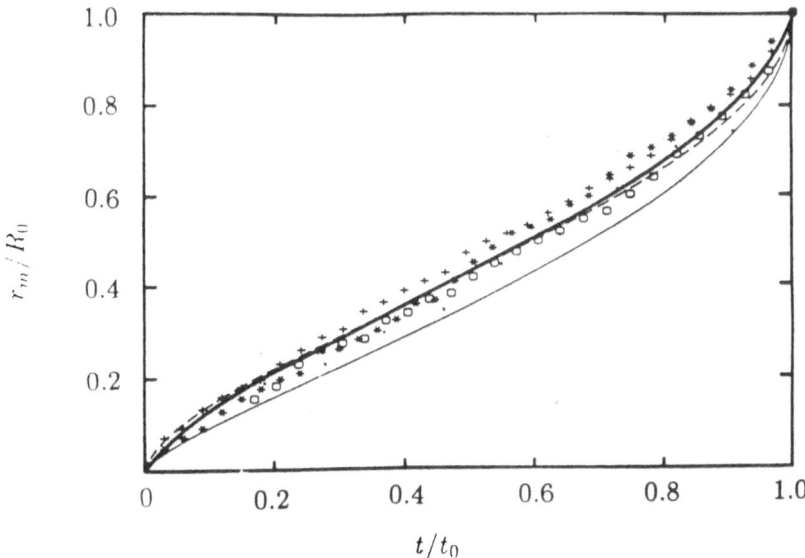

Figure 7: The length r_m/R_0 of the longest finger as function of time t/t_0, where t_0 is the time when the longest finger reaches the rim of the model. Four experiments on a 2–dimensional system of $1\,mm$ glass spheres. The dashed curve is Eq. (10), the thin fully drawn curve represent the same equation but with exponent 1.64. The thick fully drawn curve is the result of DLA simulations[41].

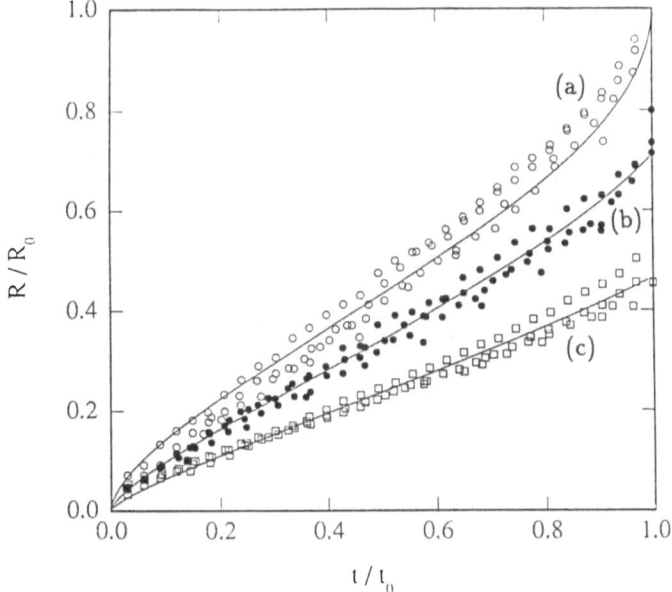

Figure 8: Radius R/R_0 as function of time t/t_0. (a): $R = R_m$ = length of longest finger. (b): $R = R_a$ = average radius of active zone. (c): $R = R_g$ = radius of gyration.

Table 1: Parameters of the fit of Eq. (11) to the observation of fingering dynamics.

$t/t_d = (r/R_0)^d \left(1 - d\ln(r/R_0)\right)$		
r	d	t_d/t_0
R_m — longest finger	2.00 ± 0.04	1
	1.91 ± 0.04	0.96 ± 0.01
R_a — radius of active zone	1.74 ± 0.02	1.13 ± 0.01
R_g — radius og gyration	1.61 ± 0.02	1.54 ± 0.02

longest finger reaches the rim of the model.   In order to get a feel for what is to be understood by the term 'effective radius' for the purpose of the fingering dynamics, we replace the exponent 2 in Eq. (10), with a free exponent d and the break-through time, t_0, with a free parameter, t_d, and analyze our experimental results using the phenomenological relation[1]

$$t/t_d = (r/R_0)^d \left(1 - d\ln(r/R_0)\right) . \tag{11}$$

Eq. (11) was fitted using a nonlinear least squares fit to the length of the longest finger, R_m, the average radius of the active zone, R_a, and the radius of gyration of the total cluster, R_g, as function of time. The results of these fits are shown in Fig. 8.

The parameters of the fit are shown in Table 1. The results in Table 1 show that the data of the gyration radius fit very well to the Eq. (11) with an exponent d very close to the fractal dimension, $D = 1.64$, of these structures. This indicates the the 'effective radius' should be identified with the radius of gyration. The fit for R_m, has been made in two ways. Insisting on the observed break-through time gave $d = 2$. If we take t_d to be a free parameter, also in this case, we find $t_d = 0.96 \pm 0,01$, which is compatible with our time resolution. Therfore $d = 1.91$ is an equally acceptable value for d.

We have also compared our results with Meakin's simulations[42] of DLA fingering dynamics. The simulations were carried out on a hexagonal lattice with the outer boundary having a radius of 200 lattice units. The results shown in Fig. 7 are the average from 20 simulations. (The details of these simulations are described by Meakin in these proceedings).

Meakin[42] analyzes his dynamic DLA simulations using the following form:

$$t \sim (r/R_0)^d \left(1 - \hat{d}\ln(r/R_0)\right) , \tag{12}$$

with $\hat{d}$ fixed at 1.71 — the fractal dimension of DLA clusters. Fig. 9 shows $\ln(r/R_0)$, with $r = R_m$, R_a, and R_g as function of $\ln\left((t/t_0)/(1 - \hat{d}\ln(r/R_0))\right)$. Least squares fits of straight lines to the experimental results gave the values for d given in Tab. 2.

---

[1] A discussion of this relation can be found in the paper by Meakin in these proceedings.

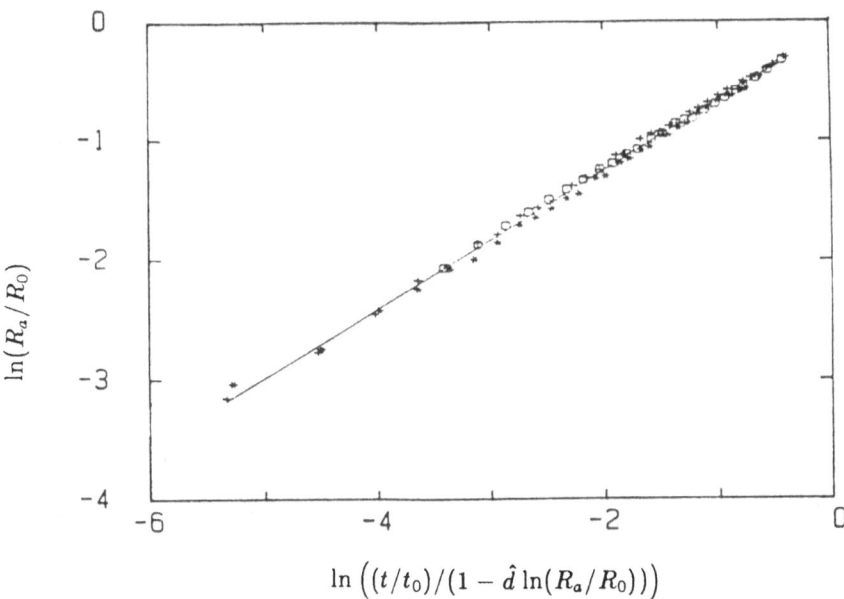

Figure 9: The dependence of $\ln(R_a/R_0)$ on $\ln\big((t/t_0)/(1 - \hat{d}\ln(l/R_0))\big)$, where $\hat{d} = 1.71$. A least square fit to these data is shown as a solid line in the figure with a slope of 0.576 ± 0.003.

Table 2: The results of linear fits of Eq. (12) to the observation of fingering dynamics. The numbers in parenthes are the corresponding d values found by Meakin in his fits to the results of DLA simulations.[42] From the covariance matrix of the fit we estimate the standard deviations in the parameters of to be ± 0.02.

$t/t_d = (r/R_0)^d \left(1 - \hat{d}\ln(r/R_0)\right)$			
$\hat{d}$	d		
	$r = R_m$	$r = R_a$	$r = R_g$
1.64	1.78 (1.78)	1.73	1.69
1.71	1.79 (1.80)	1.73 (1.74)	1.70 (1.73)
2.00	1.84 (1.83)	1.76 (1.76)	1.72

As shown by the results in Tab. 2, we find that the d values obtained by fits to the experimental results, were rather insensitive to the $\hat{d}$ values chosen for the logarithmic term. Our values compare very well with results obtained in Meakin's simulations[42]. We find the good agreement between the simulations and the experiments very satisfying.

6 Fractal Measures of the Active Zone

The growing finger structure develops in the active zone. In our experimental observations of the active zone, we measure the 'mass' m_i of the growth 'islands' shown in Fig. 3. The islands are numbered in an arbitrary way by the index $i = 1, 2, \ldots, N_I$, where N_I is the number of sites at which we observe growth. The time between pictures used to determine the set $\{m_i\}$ was typically $0.6\,s$ and was held constant for each experiment. If instead of subtracting two subsequent pictures we subtracted the last picture at break through, we found that the number of islands increased by less than 10% for all the pictures.

Let the total mass of the islands be $m_0 = \sum_i^{N_I} m_i$, and introduce the normalized mass μ:

$$\mu_i = \frac{m_i}{\sum m_i} = \frac{m_i}{m_0} \, . \tag{13}$$

The set $\mathcal{M} = \{\mu_i\}$, characterizes the *observed* growth of the structure. We consider $\mathcal{M}$ to be a new measure that is easy to measure experimentally and which gives further understanding of the growth process. This measure is to be distinguished from the *harmonic measure* discussed earlier in the context of DLA.[28–30].

Any observed viscous finger structure in our models is a realization of a stochastic process. At any instant the growing finger structure may invade any pore on the perimeter of the structure. We label the perimeter sites with the index k. The dynamics of the finger growth is then in principle controlled by the set $\mathcal{H} = \{p_k\}$, of probabilities that the k-th pore on the perimeter is invaded next. The set $\mathcal{H}$ of growth probabilities is directly related to the *harmonic measure*, that has been discussed earlier in the context of DLA.[29–30]. The measure $\mathcal{H}$, changes as soon as a pore is invaded since the perimeter of the structure changes. The actual growth at any site changes the growth probabilities at all the other sites. The measure $\mathcal{M}$, expresses the integrated effect of the sequence of pore invasion processes and therefore $\mathcal{M}$ is only indirectly related to the harmonic measure.

Although fractal measures were discussed by Mandelbrot[31] already in 1974, it was only recently that new developments[32] triggered a widespread interest in the subject. For DLA clusters Meakin et al.[28], Halsey et al.[29] and Amitrano et al.[30], have considered the growth probabilities in the growth zone in view of a multifractal description (fractal measures). Similar descriptions have been used in both experimental and theoretical works in other areas of physics.[30 – 40]

Recently Nittmann et al.[37] analyzed the moments $Z(q) = \sum p_k^q$ of the growth probabilities p_k determined by analyzing a displacement experiments of a non New-

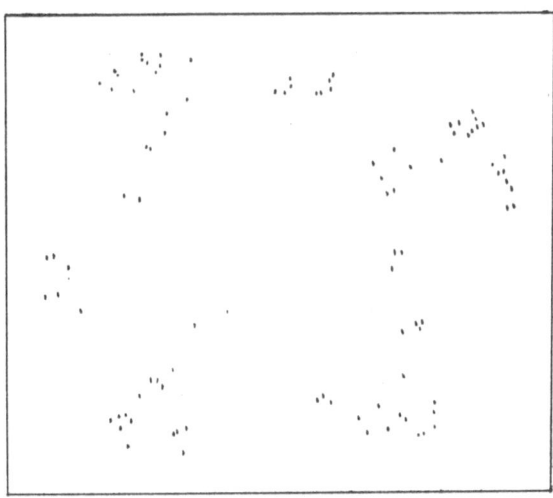

Figure 10: The sites of the growing interface

tonian fluids in a Hele Shaw cell. They analyze the observed fingering structures by calculating the set of growth probabilities $\mathcal{H}$, in the same way Amitrano et al., analyzed their clusters generated by DLA simulations. They state that their result for $Z(q) = \sum p_k^q$ and the $f(\alpha)$ curves are in agreement with the results for DLA simulations obtained by Amitrano et al.

In the following sections we analyze the fractal measure $\mathcal{M}$ obtained in our measurements.

The Fractal Set of Growth Sites

We first consider the set of points $\mathcal{N}$ at which we have observed growth (see Fig. 10). The number of points in this set that have $\mu_i > 0$ is N_I, and we find that N_I increases with the size of the viscous fingering structure. For a fractal structure we expect N_I to be given by:

$$N_I = a \left(\frac{R_g}{\delta} \right)^{D_I} . \tag{14}$$

Here D_I is the dimension of the growing interface, δ is the pixel size at which the structure is analyzed and R_g is the radius of gyration. We could equally well have used the mean radius R_a of the growth zone instead of the gyration radius R_g, because these lengths are proportional within the experimental uncertainties. Eq. (14) gives an increase in N_I with increasing size of the growing finger structure, and a decrease in N_I, as the pixel size δ is increased, decreasing the resolution at which the set of points $\mathcal{N}$ is analyzed.

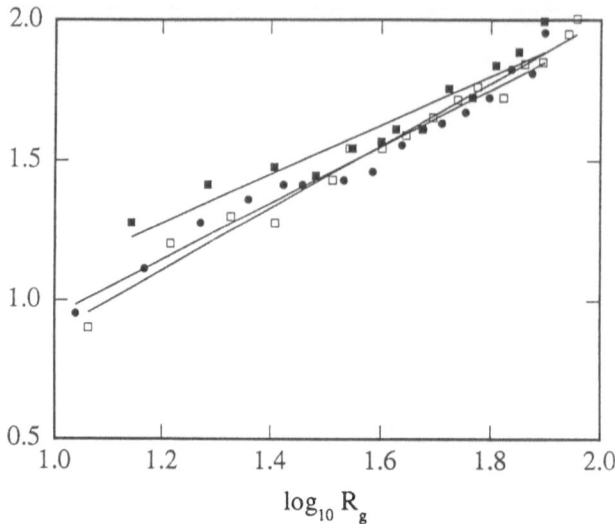

Figure 11: The number growth sites as function of the radius of gyration, for three experiments where air displaces glycerol. The straight lines represent fits of Eq. (14) to the observed values. The parameters of the fit are : • – $a = 0.9 \pm 0.2$, $D_I = 1.01 \pm 0.05$. – $a = 1.7 \pm 0.6$, $D_I = 0.87 \pm 0.08$. – $a = 0.6 \pm 0.1$, $D_I = 1.11 \pm 0.04$. The fractal dimension of the growing interface is $D_I = 1.0 \pm 0.1$.

We have counted N_I for three sequences of pictures of our experiments. In Fig. 11 we have plotted N_I as a function of the corresponding radius of gyration R_g on a log-log plot. From the fits in Fig. 11 we find that the set of points $\mathcal{N}$ is fractal with a dimension

$$D_I = 1.0 \pm 0.1 , \qquad (15)$$

and with an amplitude $a = 1.1 \pm 0.5$.

We have also analyzed sets of points such as the one shown in Fig. 10, at a given radius of gyration, by changing the box size δ. Fig. 12 shows an example, where we determine the 'box counting' dimension of the set $\mathcal{N}$, from the log-log plot of $N_I(\delta)$ versus δ. Fitting Eq. (14) to the observed points for intermediate box sizes we find an estimate for the interface fractal dimension $D_I = 1.0 \pm 0.2$. Note that $N_I(\delta \rightarrow 0) = constant$, corresponding to dimension 0, as it should be for any finite set of points. The crossover to $D = 0$, is simply due to the finite resolution of our observations.

Meakin and Witten[38] have studied the growing interface in DLA simulations and found that N_I increases as $N_0^{0.603 \pm 0.02}$, where N_0 is the cluster mass. This result implies a dimension of $D_I = 1.03 \pm 0.03$, since $N_0 \sim R_g^D$, where $D = 1.71$, is the fractal dimension of DLA clusters. Meakin and Witten also determined the number of particles N_I touching an initial cluster of size N_0 after k particles had been added to the cluster. They found that the number N_I converged to a limit when $k \rightarrow \infty$.

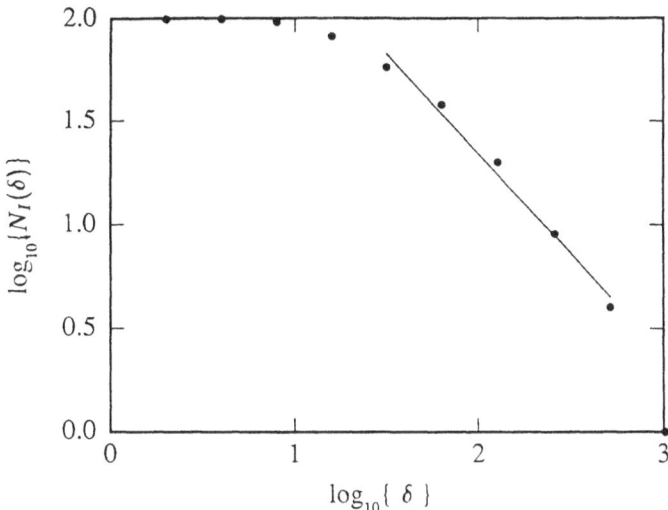

Figure 12: The number $N_I(\delta)$ of 'boxes' of size δ, needed to cover the observed set of growth sites as function of the size δ of the boxes. The box counting dimension is $D_I \sim 1.0$.

We have obtained a satisfying agreement between experiment and simulation and we conclude that both experiments and simulations give $D_I = 1$ as the fractal dimension of the growing interface.

6.1 The $f(\alpha)$ Curve

So far we have only discussed the set of points at which $\mu_i > 0$. We may specify a *subset* $\mathcal{N}_\mu$, consisting of all the growth sites for which $\mu \leq \mu_i \leq \mu + \Delta\mu$. If we specify the measure in a scale independent way we may find that such subsets are fractal sets. We therefore choose to specify subsets of the growth sites by the scaling relation

$$\mu = \left(\frac{\delta}{R_g}\right)^\alpha .$$

(16)

This relation is simply a definition of α

$$\alpha = \frac{\ln \mu}{\ln(\delta/R_g)} .$$

(17)

We could, of course, equally well choose R_a, instead of R_g, in the definition of α. This would only correspond to a shift in α that is irrelevant for large clusters.

Let us choose α to be in the range α to $\alpha + \Delta\alpha$. From Eq. (16) we then find the corresponding range of μ_i for a finger structure having a radius of gyration R_g, observed at a resolution δ. The set of growth sites that have islands that give μ_i in the specified range form a set of points $\mathcal{N}_\alpha$. The set of all growth sites may then be written as the union of such sets:

$$\mathcal{N} = \bigcup_\alpha \mathcal{N}_\alpha . \tag{18}$$

If $\mathcal{N}_\alpha$ is a fractal set then we expect the number of points in the set $N_\alpha(\delta, R_g)$, to satisfy a scaling relation similar to Eq. (14)

$$N_\alpha(\delta, R_g) = \Delta\alpha\, \rho_\alpha(\delta, R_g) = \Delta\alpha\, b_\alpha \left(\frac{R_g}{\delta}\right)^{f(\alpha)} . \tag{19}$$

The number of points in the set is simply proportional to the range $\Delta\alpha$, so we have introduced the density, ρ_α, that is independent of this range. At this point let us again stress that the finite sets of points, which we necessarily consider when we discuss experimental results, only represent samples of the fractal sets $\mathcal{N}_\alpha$, which are defined only in the asymptotic limit of infinite systems or infinite resolution.

Eq. (19) may in principle be used to determine the fractal dimension $f(\alpha)$ of the set that 'supports' the values of the measure specified by α, in the same way we determined the fractal dimension of the growth sites in the previous section. Unfortunately, we find that our models are too small to allow this direct approach to be used. We may nevertheless get an estimate for the $f(\alpha)$ curve using the measured values of μ_i. First we note that we may find the maximum value of $f(\alpha)$, by using the fact that the total number of growth sites is given by

$$N_I = \int d\alpha\, \rho(\alpha) . \tag{20}$$

From Eq. (14) and Eq. (19) we then find that

$$a\left(\frac{R_g}{\delta}\right)^{D_I} = \int d\alpha\, b_\alpha \left(\frac{R_g}{\delta}\right)^{f(\alpha)} . \tag{21}$$

This relation is valid for a large range in R_g and δ only if the integrand has a sharp maximum at some value α_0. If this is the case, we may evaluate the integral by the method of steepest decent and we find that

$$f(\alpha_0) = D_I . \tag{22}$$

The amplitude a depends on the functional form of b_α and $f(\alpha)$, and cannot be determined in a general way. We also note that since the sets $\mathcal{N}_\alpha$ are subsets of the set of growth sites we have the relation

$$0 \le f(\alpha) \le D_I , \tag{23}$$

consistent with Eq. (22).

In order to obtain the '$f(\alpha)$ – curve' we use the observed values of $\{\mu_i\}$ to make a histogram of $\rho(\alpha)$, and plot

$$f(\alpha) = \frac{\ln(\rho(\alpha)) - \ln b_0}{\ln(R_g/\delta)} , \tag{24}$$

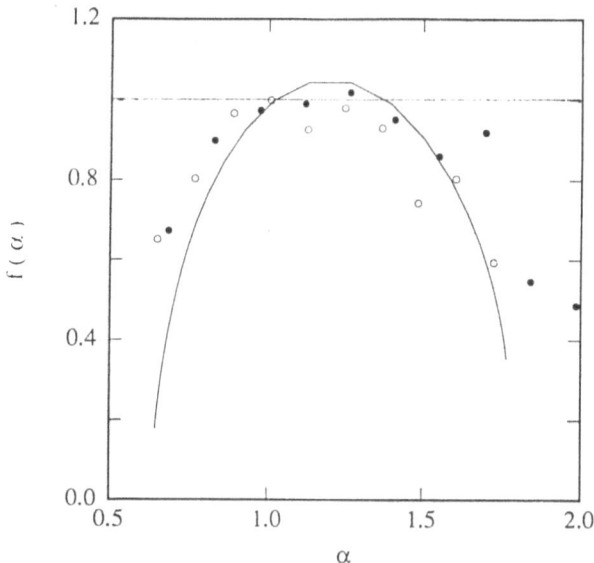

Figure 13: The $f(\alpha)$ curve for the growth zone of fractal viscous fingering in porous models. Filled and open circles correspond to structures having $R_g/\delta \sim 65$ and 42 respectively. The curve is the $f(\alpha)$ obtained from $\tau(q)$ as discussed in the next section.

as a function of α given by Eq. (17). The parameter b_0 represents the scale independent part of the integral in Eq. (21), and is chosen so that the maximum value of $f(\alpha)$ is D_I, which occurs for $b_0 \sim 1.4$. The result of this analysis for the three independent experiments discussed so far in this paper is shown in Fig. 13. Note that b_α could in principle depend on α, and by using a fixed value, valid for $\alpha = \alpha_0$, we obtain an *effective* exponent $f(\alpha)$.

The Moments of the Measure

The fractal measure defined on the growth sites may also be analyzed by considering the moments of the observed measure μ_i, defined by

$$Z_q = \sum_{i=1}^{N_I} \mu_i^q = M(q) \left(\frac{R_g}{\delta}\right)^{\tau(q)}. \tag{25}$$

For $q = 0$, we find that $Z(q = 0, R_g)$ equals to the total number of islands N_I in the growth zone. Wee see from Eq. (14) and Eq. (25), that $\tau(q = 0) = D_I$, and $M(q = 0) = a$.

We determined $Z(q, R_g)$ with q in the range of -10 to 10 for thirteen different growth zones from the same experiments discussed so far. We then determine an *effective* value for the exponent $\tau(q)$ by using $M(0)$ instead of $M(q)$ in Eq. (25) The dependence of $\tau(q) = \log\{Z(q, R_g)/M(q = 0)\}/\log(R_g/\delta)$ on q is shown in Fig. 14.

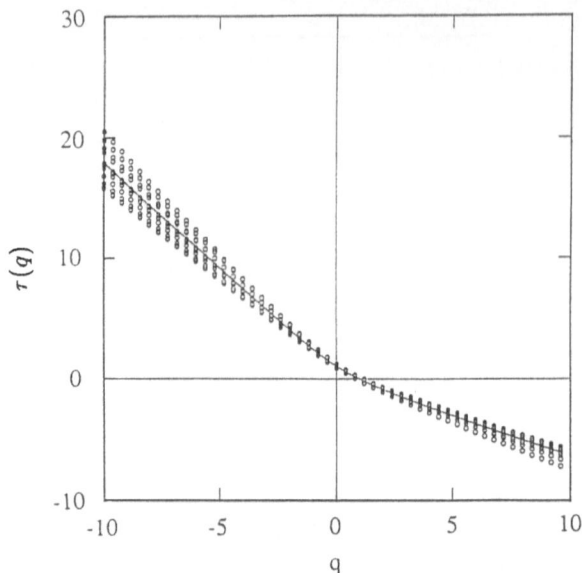

Figure 14: The dependence of the effective $\tau(q)$ on q. Data have been taken
from thirteen different realizations of the growth zones. The curve
represents the average of the experimental results.

For large positive q this curve clearly must be linear in q because the largest
island with $\mu_i \sim \mu_{max}$ will dominate the sum in Eq. (25). Therefore we have that
$\tau(q) \sim q \log(\mu_{max}) + const$, in this region. For large negative q the smallest islands
$\mu_i = \mu_{min}$ will dominate the sum and $\tau(q) \sim q \log(\mu_{min}) + const$. For large negative
q the slope of the curve is limited to the resolution of the digitizing equipment. In
this experiments we used a camera with an resolution of 500×500 pixels.

We may express the moments Z_q in terms of α and $\rho(\alpha)$ introduced in the
previous section as follows:

$$Z_q = \int d\alpha\, \rho(\alpha) \mu_\alpha^q \,. \tag{26}$$

Combining this equation with Eq. (16), (19) and (25), we find that

$$M(q) \left(\frac{R_g}{\delta}\right)^{\tau(q)} = \int d\alpha'\, b_{\alpha'} \left(\frac{R_g}{\delta}\right)^{f(\alpha')-q\alpha'} \,. \tag{27}$$

This equation can be valid for a large range of R_g and of δ only if the integrand has
a sharp maximum at some value of $\alpha' = \alpha$. It follows therefore that

$$\tau(q) = f(\alpha) - q\alpha \,. \tag{28}$$

We note that for $q = 0$, we have $\tau(0) = f(\alpha_0) = D_I$. Since $f(\alpha)$ is not a function of
q, it follows that we may determine α from taking the derivative of Eq. (28) with
respect to q, keeping α fixed:

$$\alpha = -\frac{d\tau(q)}{dq} \,. \tag{29}$$

130

The pair of equations (28) and (29), in effect constitute a Legendre transformation from the independent variables τ and q to the independent variables f and α. Such relations have been discussed extensively in the literature.[28–37] Using this transformation we converted the average $\tau(q)$ curve in Fig. 14 into the $f(\alpha)$ curve shown in Fig. 13. The maximum point of this curve is close to the value 1.0 consistent with the fractal dimension of the point set $\mathcal{N}$. This curve is convex as it should be,[36] and the slope at the point where $f(\alpha) = \alpha$ is very close to $q = 1$.[36]

Large values α represent small μ_i. In our experiments we find it very difficult do determine very small values of μ because of the finite experimental resolution. This limits the accuracy of our results for large values of α. We emphasize that the same set of observations $\{\mu_i\}$, for three different experiments and thirteen different observations of the growth zone, have been used both in the present $\tau(q)$ analysis and in the direct analysis of the previous section. The scatter of points found in the direct analysis gives a better representation of the limited experimental resolution than the smooth averaged curve determined from $\tau(q)$. Note that in both cases we used the observations of the scaling properties of N_I to fix unknown amplitudes. This in effect, sets the maximum of $f(\alpha)$ to one.

Amitrano et al. have analyzed DLA simulations and estimated the harmonic measure $\mathcal{H}$, and presented the results as a $f(\alpha)$ curve.[30] Nittmann et al.[37] have estimated the growth probabilities on the interface of viscous fingers in a non Newtonian liquid, using the methods of Amitrano et al.. We will emphasize that the measure, $\mathcal{H}$, they used is different from the measure $\mathcal{M}$ representing the observed growth. The perimeter of DLA clusters is proportional to the cluster mass, and therefore one expects that $f(\alpha)$ has a maximum value of 1.71, at an α value corresponding to $q = 0$. The simulation by Amitrano et al. have a value of $f_{max} \sim 1.5$ which is close to the fractal dimension 1.7 of the DLA aggregates. The result of Nittmann et al. and Amitrano et al. are only consistent for low values of α. The experiments by Nittmann et al. have maximum of f somewhat above 1.

There is a striking difference between the multifractal $f(\alpha)$ curve calculated from the measure $\mathcal{M}$ and the $f(\alpha)$ curve measured by Nittmann et al. from the growth probabilities. They measured extremely small growth probabilities, $p_i \sim 3 \cdot 10^{-10}$! The fact that our curve drops abruptly already for $\alpha < 2$, may be due to finite resolution in our experiments. We would like to point out, however, that in a porous medium the displacement takes place at the pore level and capillary effects in the screened regions where the growth probabilities in the DLA model is very small, may very well introduce a cut-off into the problem so that only sites with $p_i > p_c$ will contribute to the observed μ_i. We believe that these effects give rise to the characteristic and narrow $f(\alpha)$ curve observed in our experiments.

7 The Structures of Viscous Fingers

The structures observed in a displacement experiment in a porous medium depend of the pressure of the injected fluid. When the injection pressure is reduced

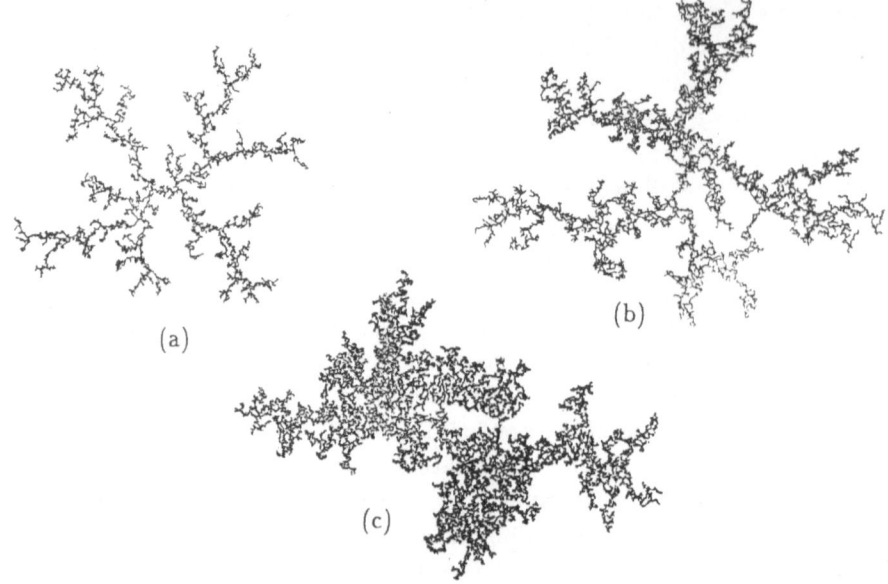

Figure 15: Viscous fingers obtained at different injection pressure. a – 20.2 mm Hg. b – 9.3 mm Hg. c: – 6.9 mm Hg.

the capillary forces become more important and we observe a crossover to more compact structures as shown in Fig. 15.

If, instead of constant pressure, we used constant injection rates, the structures observed at low injection rates can be described by an invasion percolation model. The invasion percolation process was first studied by Wilkinson and Willemsen.[43] Lenormand and Zarcone[44] determined a fractal dimension $D = 1.82$ for a situation where air displaces glycerol from one side to the other in an etched network. A displacement experiment in the invasion percolation regime is shown in Fig. 16, and we see that there are trapped regions of all sizes indicating that the structure is statistic self similar. In the DLA-like regime the structures are clearly not statistically self-similar. In this case the cluster has the center of injection as a natural center and the scaling relation (4) is only valid when the distance r is measured from this center. On the other hand, when the injection pressure is decreased, the structures get more compact. In the constant velocity case, the structure crosses over from the DLA regime to the statistically self-similar invasion-percolation regime.[45]

The fractal dimension alone is not sufficient to characterize structures that are not statistic self-similar. In order to extract more quantitative information about these structures, we have calculated the moments

$$N(q, \delta) = \sum n_i^q , \qquad (30)$$

where n_i is the mass inside boxes of size δ, and the sum is over all boxes containing any black pixels. These moments are related to similar moments used by Voss[39] and Meakin and Havlin.[40] We make the following scaling assumption

$$N(q, \delta) \sim B(q) \delta^{-\bar{f}(q)} \qquad (31)$$

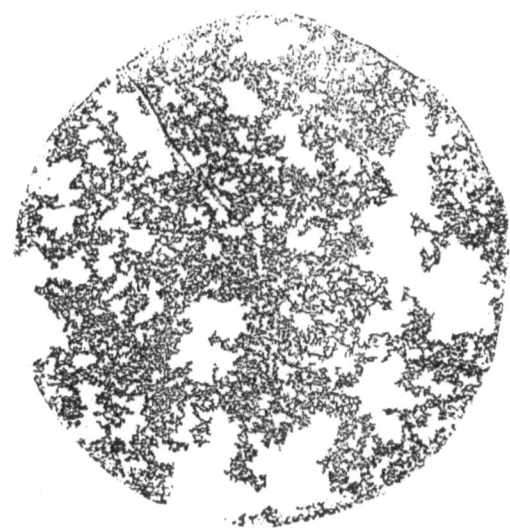

Figure 16: Invasion percolation cluster, obtained in a very low constant injection rate (1 ml/h) experiment.

By fitting the measured $N(q, \delta)$ as a function of δ we determine $\hat{\tau}$. An example for $q = 5.2$ for the invasion percolation structure shown in Fig. 16, is shown in Fig. 17 The results of th $\hat{\tau}(q)$ curve for cluster (a) in Fig. 15 is shown in Fig. 18. For

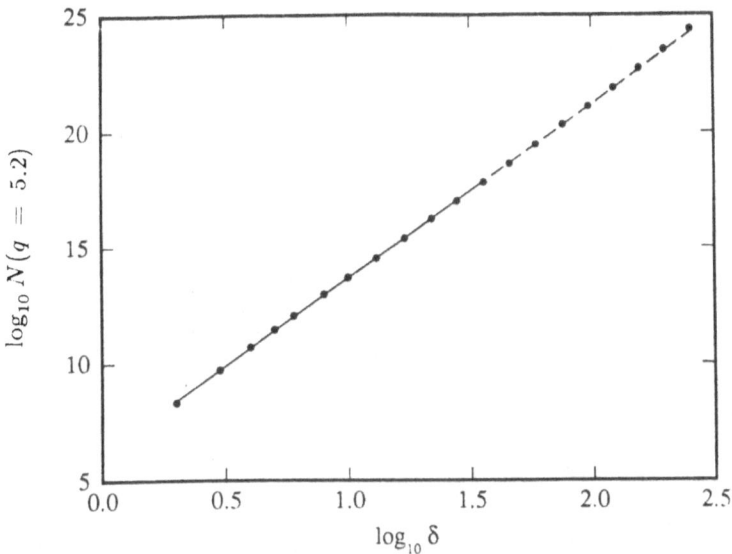

Figure 17: The moment $N(q = 5.2)$ as function of box size δ, for the percolation structure in Fig. 16. The straigth line represents a fit to the data with $\hat{\tau} = -7.53$.

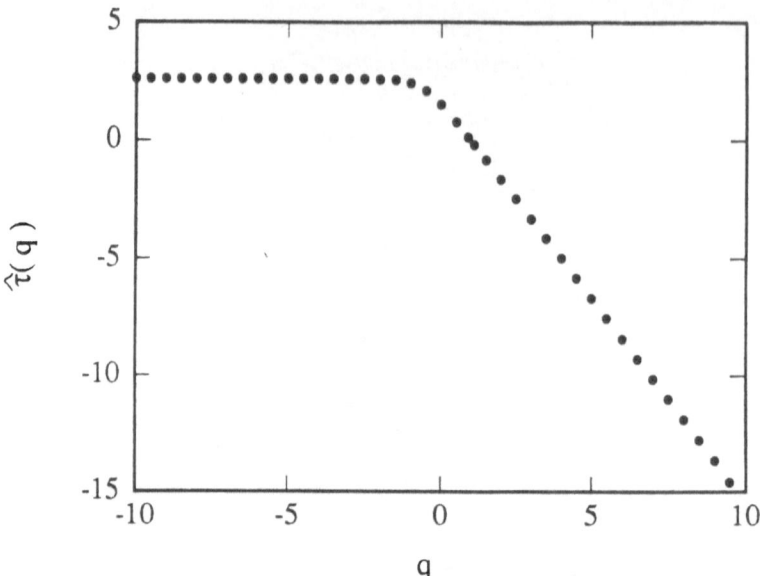

Figure 18: The dependence of $\hat{\tau}(q)$ on q for structure (a) in Fig. 15.

Table 3: The various dimensions determined for the stuctures a,b and c in Fig. 15, and d from Fig. 16.

	(a)	(b)	(c)	(d)
D	1.62	1.67		1.81
$\hat{\tau}(q = -10)$	2.6	2.7	2.9	3.0
D_b	1.51	1.55	1.71	1.73

large negative q only the boxes with $n_i = 1$ will give considerable contributions to the moments $N(q, \delta)$. This explains the constant behavior of $\hat{\tau}(q)$ at large negative q values. At large positive q values $\hat{\tau}(q)$ is nearly linear with q. At $q = 0$ we get $N(q = 0, \delta) = N \sim \delta^{-D_b}$ where $D_b = 1.51$ is the box counting dimension of the cluster. In Fig. 19 we have plotted $d(q) = \hat{\tau}(q)/(q - 1)$ as function of q for the structures shown in Fig. 15 and Fig. 16.

For large negative values $\hat{\tau}(q)$ is constant and $d(q) \sim 1/(q-1)$. For $q = 0$ we find the box counting dimension of the structures. At positive q the slope of this curve is lower for the compact structures (c) and (d) than for the more open structures (a) and (b) in the high pressure regime. This slope gives an indication of 'the degree of self-similarity' of the stuctures, since we have $d(q) = D$, for a strictly selfsimilar structures with fractal dimension of D .

In table 3 is shown the result of D_b, $\hat{\tau}(q = -10)$ and the fractal dimension D, determined from $N(r)$.

For structure (b) it was impossible to determine a cluster dimension D accurately

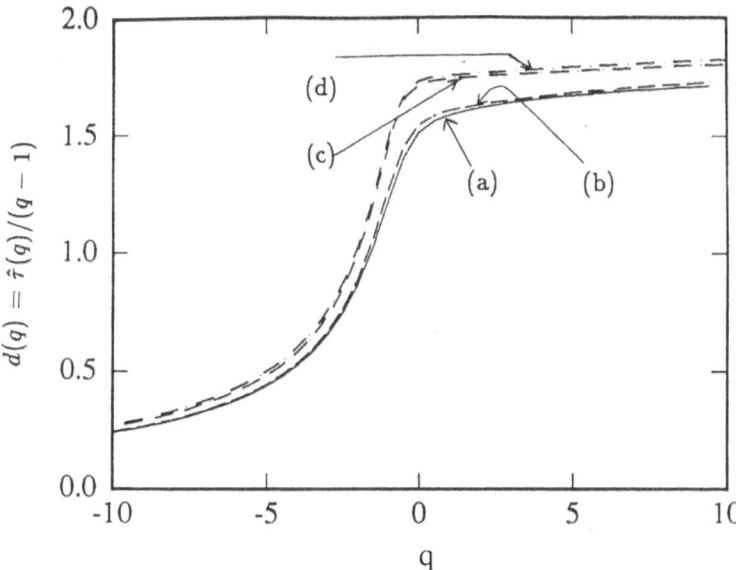

Figure 19: The dependence of $d(q) = \hat{\tau}(q)/(q-1)$ on q, for the structures shown in Fig. 15 a, b, c and for Fig. 16 d.

because the mass $\log N(r)$ was not linear with respect to $\log r$. From table 3 we see that generally the box counting dimension gives a lower value than the fractal dimension D estimated using equation (4).

We find that the set of exponents $d(q)$, gives a useful characterization of the observed finger structures, and allows a quantitative distinction of DLA type and invasion-percolation type structures.

8 Discussion

The dynamic properties of viscous fingers with fractal dimension 1.64 ± 0.04 in the DLA regime have been studied. The growing cluster is screened by the longest fingers. We have measured the longest finger as function of time and found that the data fit very well to DLA simulations. Our results have also been compared with the analytical expression for a growing bubble, Eq. (10).

Both the mean radius R_a of the growth zone, and the length of the longest finger R_m are proportional to the gyration radius R_g of the total cluster. R_m, R_a and R_g therefore follow the same scaling law $R_m \sim R_a \sim R_g \sim N_0^{1/D}$.

We have introduced a new measure M that characterizes the observed growth dynamics. We have studied the structure of the growth zone and find that it is a fractal set with a fractal dimension $D_I = 1$, consistent with the simulations by Witten and Meakin.[38]. We have also determined the spectrum of fractal dimensions $f(\alpha)$ that characterize subsets of the growth zone specified by the measure M with scaling exponents α. The maximum value of $f(\alpha)$ is D_I. The resulting $f(\alpha)$ curve has a characteristic parabolic shape. In contrast the harmonic measure estimated by Amitrano et al.[30] for DLA clusters, has a maximum of $f(\alpha)$ of about 1.5, close to the expected 1.71 for DLA. For non Newtonian fingering in Hele Shaw cells Nittmann et al.[37], somewhat above 1. For low values of α we find that our $f(\alpha)$ and the results obtained from estimates of the harmonic measure coinside, as might be expected. For larger values of α our results show a sharp drop in f for $\alpha < 2$. We interpret this effect to be a result of the effective cut-off in the growth probabilities caused by surface tension effects in the porous medium, and finite resolution of the digitizing equipment.

We have also studied the cross-over that occur in the observed structures when the displacement rate is reduced. The cross-over from the DLA regime to the invasion percolation regime   leads to quantiative changes of a set of scaling exponents $d(q)$ for the observed structures.

Acknowledgment

We thank Amnon Aharony for stimulating and useful discussions essential to the development of our ideas. We had many interesting discussions with Paul Meakin, particularly regarding the dynamics of DLA. This research has been supported by VISTA, a research cooperation between the Norwegian Academy of Science and Letters and Den Norske Stats Oljeselskap a/s (STATOIL). Two of us (K.J.M, and F.B.) wish to acknowledge the receipt of fellowships from STATOIL. We wish to thank Geir Holm for special technique of photographic work, essential to this project.

9 References

1. J. D. Chen and D. Wilkinson, Phys. Rev. Lett. **55**, 1892 (1985).

2. K. J. Måløy, J. Feder, and T. Jøssang. Phys. Rev. Lett. **55**, 2688 (1985).

3. B. B. Mandelbrot *"The Fractal Geometry of Nature"*,
   (W. H. Freeman, San Fransisco 1982)

4. T. A. Witten and L. M. Sander, Phys. Rev. Lett. **47** 1400 (1981).

5. J. Nittmann, G. Daccord and H. E. Stanley, Nature **314**, 141 (1985). G. Daccord, J. Nittmann, H. E. Stanley. In On Growth and Form (H. E. Stanley, N. Ostrowsky editors, Martinus Nijkoff, Boston 1986) p. 203.

6. L. Paterson, Phys. Rev. Lett. **52**, 1621 (1984).

7. L. P. Kadanoff, J. Stat. Phys. **39**, 267 (1985).

8. L. Niemeyer, L. Pietronero and A. J. Wiesmann, Phys. Rev. Lett. **52**, 1033 (1984).

9. M. Matsushita, M. Sano, Y. Hayakawa, H. Honjo, and Y. Sawada, Phys. Rev. Lett. **53**, 286 (1984).

10. R. M. Brady and R. C. Ball, Nature **309**, 225 (1984).

11. W. T. Elam, S. A. Wolf, J. Sprague, D. V. Gubser, D. Van Vechten, G. L. Barz Jr. and P. Meakin, Phys. Rev. Lett. **54** 701 (1985).

12. H. Honjo, S. Ohta and M. Matsushita, J. Phys. Soc. Japan **55**, 2487 (1986).

13. G. Daccord, Phys. Rev. Lett.**58**, 479 (1987);
    G. Daccord and R. Lenormand, Nature **325**, 41 (1987).

14. M. Plischke and Z. Rácz, Phys. Rev. Lett. **53**, 415 (1984).

15. J. M. Deutch and P. Meakin, J. Chem. Phys. **78**, 2093 (1983).

16. H. G. E. Hentschel, J. M. Deutch and P. Meakin,
    J. Chem. Phys. **81**, 2496 (1984).

17. P. Meakin and J. M. Deutch, J. Chem. Phys. **80**, 2115 (1984).

18. R. F. Voss, Phys. Rev. B **30**, 334 (1984).

19. P. Meakin, Phys. Rev. A **33**, 3371 (1986).

20. P. G. Saffman and G. Taylor, Proc. Soc. London, Ser. A **245**, 312 (1958).

21. L. Paterson, J. Fluid Mech. **113**, 513 (1981).

22. P. Meakin and L. M. Sander, Phys. Rev. Lett. **54**, 2053 (1985).

23. P. Meakin, Phys. Rev. A **32**, 453 (1985).

24. M. Plischke and Z. Rácz, Phys. Rev. Lett. **54**, 2054 (1985).

25. G. Daccord, J. Nittmann, and H. E. Stanley, Phys Rev Lett. **56**, 336 (1986).

26. M. Plischke and Z. Rácz. Phys. Rev. A. **31**, 985 (1984).

27. P. Meakin, J. Phys. A, **18**, L661 (1985).

28. P. Meakin, A. Coniglio, H. E. Stanley and T. A Witten, Phys. Rev. A. **34**, 3325 (1986)

29. T. C. Halsey, P. Meakin and I. Procaccia, Phys. Rev. Lett, **56**, 854 (1986).

30. C. Amitrano, A. Coniglio F. di Liberto, Phys. Rev. Lett, **57**, 1016 (1986).

31. B. B. Mandelbrot, J. Fluid Mech. **62**, 331 (1974).

32. H. G. E. Hentschel and I. Procaccia, Physica **8D**, 835 (1983)

33. L. Di Archangelis, S. Redner and A. Coniglio, Phys. Rev. B **31**, 4725 (1985).

34. A. Renyi, Probability theory, North Holland, Amsterdam (1970).

35. R. Renzi, G Paladin, G. Parisi and A. Vulpiano, J. Phys. A **17**, 3521 (1984).

36. T. C. Halsey, M. H. Jensen, L. P. Kadanoff, I. Procaccia and B. I. Shraiman, Phys. Rev. A **33**, 1141 (1986).

37. J. Nittmann, H. E. Stanley, E. Touboul and G. Daccord Phys. Rev. Lett **58**, 619 (1987).

38. P. Meakin and T.A Witten Jr. Phys. Rev. A **28**, 2985 (1983).

39. R. F. Voss Physica. Schripta. **T13**, 27, (1986)

40. P. Meakin S. Havlin. Preprint.

41. K. J. Måløy, F. Boger, J. Feder, T. Jøssang and P. Meakin, Phys Rev A, to be published.

42. P. Meakin, these proceedings.

43. D. Wilkinson and J. F. Willemsen, J. Phys. A: Match. Gen. **16**, 3365 (1983).

44. R. Lenormand and C. Zarcone, Phys. Rev. Lett **54**, 2226 (1985).

45. R. Lenormand, invited lecture, Statphys 16, Boston 1986, p. 114-123.

VISCOUS FINGERING INSTABILITIES IN POROUS MEDIA

J.P. Stokes[1], D.A. Weitz,[1] R.C. Ball,[2] and A.P. Kushnick[1]

[1] Exxon Research and Eng, Rt. 22E, Annandale, NJ 08801
[2] Cavendish Lab, Madingly Road, Cambridge CD3 OHE, UK

We study patterns formed by the viscous fingering instability in a
porous media. When the displacing fluid preferentially wets the medium,
the finger width is much larger than the pore size and, when normalized by
the square root of the permeability, is found to scale with capillary
number as $Ca^{-1/2}$. While traditional theories based on Hele-Shaw geometry
give this dependence for the most unstable wavelength, they are unable to
explain the magnitude of the finger. We consider here the effect of a
velocity dependent capillary pressure in addition to the more conventional
static term, and suggest that it may control the scaling of the finger
width on Ca. We demonstrate the existence of this dynamic capillary
pressure, which offers new insight into the basic physics of the motion of
a fluid interface in porous media.

When a fluid of low viscosity displaces a fluid of high viscosity in
a porous media, a fingering instability can occur, resulting in a rich
variety of interfacial patterns.[1] These patterns have been studied
experimentally using both model porous media[2-4] and the simpler geometry
of a Hele-Shaw cell[5,6], which consists of two parallel, closely spaced
glass plates. In both cases the flow is described by Darcy's law, with
the velocity determined by the gradient of the pressure. There has been
considerable progress recently, both experimentally and theoretically, in
our understanding of the behavior of the instability in the Hele-Shaw
geometry.[1] Less effort has been directed toward our understanding of
viscous fingering in porous media. Nevertheless, significant advances
have been made with the recognition that since the pressure field in the
viscous fluid satisfies Laplace's equation,[7] if the interfacial tension is
ignored, the viscous fingering instability is analogous to the problem of
diffusion-limited aggregation. Thus the patterns which result can often
be described as fractals.

In a previous paper[4] we described results of viscous fingering in
glass bead packs. We found that the patterns formed depended critically
on a number of parameters, and that neither the traditional Hele-Shaw cell
approach, nor the DLA machinery, could adequately describe the behavior
observed. In particular, the wetting properties of the porous media play
a critical role on determining the finger width or the lowest length scale
of the patterns. In the non-wetting case, in which the porous media is
preferentially wet by the displacing fluid, the finger width is of order a
bead size independent of other parameters, and so the lowest length scale

is determined by the characteristic length of the porous medium itself. By contrast, in the wetting case, where the displacing fluid preferentially wets the beads, the finger width was found to be many bead diameters wide and depended on a number of other factors. Thus there is a new characteristic length scale in the problem. In this paper we will summarize our results in the wetting case and suggest a new physical mechanism that can account for this new length scale and correctly predict the scaling behavior of the finger width with capillary number.

The experiments on viscous fingering were done in a modified linear Hele-Shaw geometry where the space between the two glass plates was filled with beads of uniform size. The equation describing the flow of fluid in a porous media is Darcy's Law, $U = -(\kappa/\mu)\nabla P$, which relates the far field or average velocity, U, to the gradient in the pressure, P, through the fluid viscosity, μ and the permeability κ, which is the conductance to flow, and for uniform spherical beads scales as the square of the bead radius. The average velocity can be related to the volumetric injection rate, Q, by $U = Q/Ap$, where A is the cross-sectional area and p is the porosity. While Darcy's Law describes the pressure drop due to viscous forces, a pressure drop also exists at the interface due to the surface tension between the two fluids, γ. The capillary number $Ca = \mu U/\gamma$ is a measure of the relative importance of these two forces.

Typical experimental results for wetting displacement are shown in the top two pictures in Figure 1 for two different bead sizes. The light regions indicate the fingers formed by the invading water displacing the more viscous oil. We find that it is possible to define a finger width, w, which depends only on the capillary number and permeability. As the capillary number decreases w increases. By contrast, as the permeability increases w increases. However, in all cases, w is greater than a bead size.

Figure 2 is a summary of the finger widths for a wide variety of conditions for wetting displacement. We have normalized the finger width by the square root of the permeability to make it dimensionless. Data taken with three different bead sizes with κ varying by two orders of magniture are plotted as a function of the tip capillary number, where we use the velocity at the tip of the growing finger, U_{tip}, as opposed

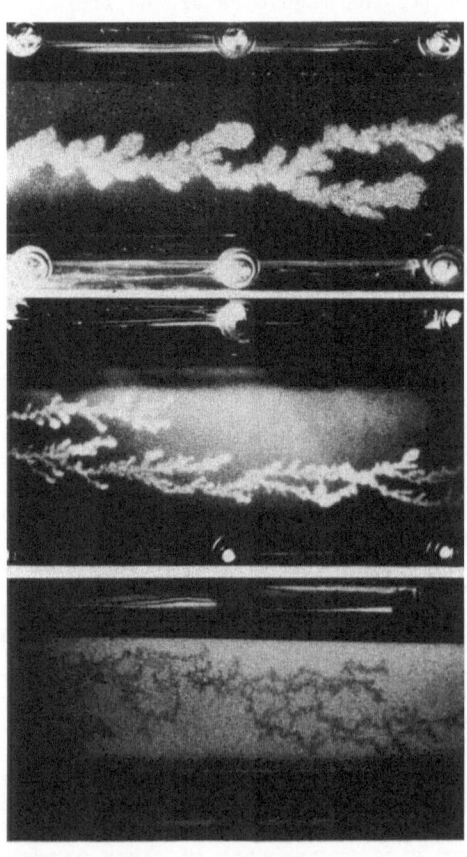

Fig. 1.

Top: Wetting displacement with water (light color) displacing oil (dark) $\kappa = 2 \times 10^{-6}$ cm^2, Ca$= 6.9 \times 10^{-4}$. Middle: Wetting displacement with $\kappa = 2 \times 10^{-7}$ cm^2 and Ca$= 3.8 \times 10^{-4}$. Bottom: Non-wetting displacement with oil (dark) displacing water and glycerol (light), $\kappa = 2 \times 10^{-6}$ cm^2 and Ca$= 2.3 \times 10^{-4}$.

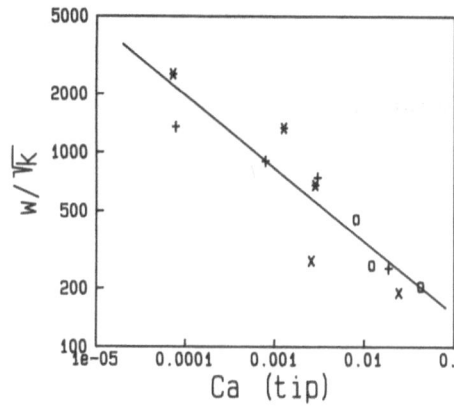

Fig. 2.

Normalized finger width $(w/\kappa^{1/2})$ vs Ca_{tip} for wetting displacement. The data points refer to $*:\kappa=2\mathrm{x}10^{-7}$ cm^2; x: $\kappa=2\mathrm{x}10^{-6}$ cm^2; o: $\kappa=2\mathrm{x}10^{-5}$ cm^2; all with Ca_{tip} varied by changing U_{tip}. +: $\kappa=2\mathrm{x}10^{-6}$ cm^2 with Ca_{tip} varied by changing γ.

to the far field velocity for determining Ca. We independently measured the dependence of w on U_{tip}, μ, and γ. We have varied U_{tip} by two orders of magnitude, μ by a factor of 5, and γ by a factor of 5. Within experimental error, the data all fall on a straight line. This verifies that Ca is a control parameter and also that the finger width scales as the pore size rather than any dimension of the cell. From a least squares fit to the data, we measure the slope to be -0.5±0.1.

In contrast, the behavior for the non-wetting displacement is markedly different, as shown in the bottom picture of Figure 1. Here the finger width is comparable to the characteristic length scale of the pore space and is roughly independent of Ca. We note that this non-wetting case corresponds closely to the behavior observed in two-dimensional micro-model experiments,[8-10] where the fluid displacement is well described by invasion percolation[11] at low Ca and by analogy to diffusion limited aggregation at high Ca.[7] The patterns formed in our cell are consistent with this picture and for the remainder of this paper we will concentrate instead on the wetting case.

The first problem that must be addressed for the wetting case is the origin of the new length scale in the patterns. The form of the observed scaling behavior might be explained on physical grounds, by analogy to the Hele-Shaw cell. We assume that the instability occurs when the viscous and capillary pressure drops are comparable on the length scale of the finger width. This would result in w scaling as $(\kappa/Ca)^{1/2}$. Similar results are obtained from a linear stability analysis for the fastest growing wavelength.[2] However, this analysis assumes that the relevant interfacial tension between the two fluids is determined by the **macroscopic** curvature at the finger tip. Since the interface must locally satisfy the correct boundary conditions and so have the appropriate contact angle with the pore walls, it is quite rough on the scale of a pore size. Thus, the interfacial tension is given by γ/r_{th}, where r_{th} is a characteristic pore size. This is the same at all points on the interface, independent of the macroscopic curvature. Therefore, it is not clear to what extent the gross curvature of the finger tip can determine the interfacial tension. Furthermore, this analysis also drastically underestimates the characteristic length scale and can be made to agree only by replacement of the interfacial surface tension by an empirically determined effective surface tension which depends on the wettability of the medium. Indeed, the correct order of magnitude of the interfacial tension is actually given by γ/r_{th} rather than γ/w.

Clearly the capillary pressure at the interface can not account for the measurements. Instead, we postulate that a capillary force which depends on velocity stabilizes the growing front. If the capillary "pull" is lowest at the highest velocity, this would tend to "hold back" or reduce the fingering at the tip. Thus the dynamic capillary pressure could behave in a fashion analogous to a macroscopic surface tension, but could be of the same order of magnitude as the local interfacial tension on the bead size. We assume that the dynamic capillary pressure has a power law dependence on velocity, through the capillary number, and therefore write the capillary pressure as $P_{cap} = \gamma/r_{th}(-1+KCa^x)$, with K a constant.

To quantify the effects of this dynamic capillary pressure on the finger width, we make a simple dimensional argument, similar to that made above. The viscous pressure drop over the width of the growing finger is $\mu U w/\kappa$. We equate this to the difference in the dynamic capillary pressure across the same width. We neglect the static capillary pressure because it is constant everywhere on the interface and cannot, therefore, stabilize the fingering. Solving for the finger width gives,

$$w = \frac{\kappa}{r_{th}} \; K \; Ca^{x-1}$$

Since $\kappa \sim r_{th}^2$, this form agrees with our measurements for the dependence of w on $\sqrt{\kappa}$. Further, if $x \sim 0.5$, the observed scaling with Ca is recovered. However, the fundamental physical origin of this scaling behavior is completely different from the traditional view, as the new length scale arises from the effects of a dynamic capillary pressure.

To test this conjecture, we have performed experiments on thin tubes filled with glass beads to look for a velocity dependent term in the capillary pressure. The experiment consists of displacing a non-wetting fluid with a wetting fluid where the viscosities of the two fluids are matched to eliminate any fingering instability and reduce the viscous pressures. The expected pressure drop across the tube will be

$$\Delta P = \frac{\gamma \cos\theta}{r_{th}} + \frac{\mu U}{\kappa} L + K' U^x$$

where the first term on the right is the usual static capillary pressure, the second term is the viscous drop, and the final term reflects the nonlinear, velocity-dependent capillary pressure. By measuring the

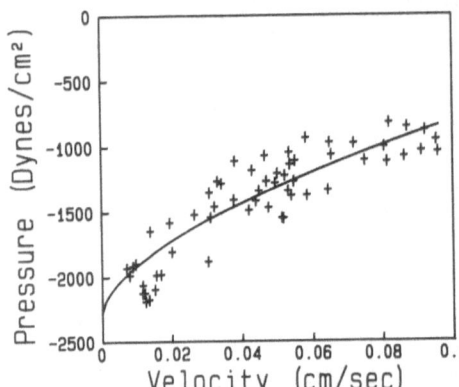

Fig. 3.

Pressure vs velocity for an interface between water and decane in 500μm beads in a 3 mm diameter tube. The pressure due to the viscous drop has been subtracted off. The solid curve represents the calculated capillary pressure including a dynamic term that scales as $U^{.5}$.

pressure-velocity profile with and without the interface in the bead pack, we are able to determine the pressure drop due to the interface. Figure 3 shows the results for a 3 mm diameter tube filled with $500 \mu m$ beads. The viscous term has been subtracted off, leading to the noise in the data. For comparison, the solid line is a calculation using the above equation, with $x=0.5$ and $\gamma/r_{rh}=2400$ dyne/cm^2. and as can be seen, can account for the data reasonably well.

While the results of these experiments are only preliminary, they clearly show that a velocity dependent capillary pressure does indeed exist. The data are consistent with an exponent of ~0.5 which could account for the scaling behavior of the finger width. Furthermore, the data allow us to determine the constant K, from which we can predict the absolute finger width. The predicted finger width agrees remarkably well with the measured width for beads of the same size, given the crudeness of our model.

The pressure-velocity measurements of the motion of an interface in a porous medium are quite analogous to voltage-current measurements of a non-linear resistor, and the data obtained possess a similar richness in behavior. One can view the motion of an interface through a porous media as an example of nonlinear collective transport in a disordered media. There have been numerous studies done on related systems, such as the motion of a charge density wave[12] or the interface between magnetic domains in a disordered magnetic system.[13] These systems have much in common: a pinning force for motion of the interface, jerky motion of the interface when this pinning force is exceeded, and non-linear motion in the applied force. We are presently investigating the motion of the fluid-fluid interface in a porous media as an example of this type of behavior. Preliminary results show that there is pinning for a region of pressure, and nonlinear motion above this threshold. We are also investigating the dynamics of the motion of the interface by measuring the noise in the flow above the pinning pressure. Hopefully, these studies will lead both to a better picture of the flow of an interface in a porous medium as well as an understanding of the origin of the dynamic term in the capillary pressure.

REFERENCES

1. For a review, see G.M. Homsy, Ann. Rev. Fluid Mech. **19**, 271 (1987).
2. R.L. Chuoke, P. Van Meurs and C. Van der Poel, Tran. AIME **216** 188 (1959).
3. E. Peters and D. Flock, Soc. Pet. Eng. J. **21**, 249 (1981).
4. J.P. Stokes, D.A. Weitz, J.P. Gollub, A. Dougherty, M.O. Robbins, P.M. Chaikin and H.M. Lindsay, Phys. Rev. Lett. **57**, 1718 (1986).
5. P. Tabeling and A. Libchaber, Phys. Rev. **A33**, 794 (1986).
6. C.W. Park and G.M. Homsy, J. Fluid Mech., **139**, 291 (1984).
7. L. Patterson, Phys. Rev. Lett. **52**, 1621 (1984).
8. K. Maloy, J. Feder and T. Jossang, Phys. Rev. Lett. **55**, 2688 (1985).
9. R. Lenormand and C. Zarcone, Phys. Rev. Lett. **54**, 2226 (1985).
10. J.D. Chen and D. Wilkinson, Phys. Rev. Lett. **55**, 1892 (1985).
11. D. Wilkinson and J.F. Willemsen, J. Phys. **A16**, 3365 (1983).
12. See "Charge Density Waves in Solids," ed. Gy. Hutiray and J. Solyom, Lecture Notes in Physics Vol 217, (1985).
13. G. Bruinsma and G. Aeppli, Phys. Rev. Lett. **52**, 1547 (1984).

MULTIFRACTALS *

H. Eugene Stanley

Center for Polymer Studies and Department of Physics
Boston University, Boston, MA 02215 USA

The neologism 'multifractal phenomena' describes the concept that each region of an object can be characterized by a different fractal dimension. Multifractal scaling provides a quantitative description of a rich range of heterogeneous phenomena.

In recent years, a wide range of complex structures of interest to physicists and chemists have been quantitatively characterized using the idea of a **fractal** dimension d_f: an effective dimension that corresponds in a unique fashion to the geometrical shape under study, and often is not an integer.[1-7] The key to this progress is the recognition that many random structures obey a symmetry as striking as that obeyed by regular structures. This "scale symmetry" has the implication that objects look the same on many different scales of observation. Stated precisely, if the length scale L of our observation is increased by a factor λ, the "mass" of the object in question increases by a factor λ^{d_f}:

$$M(\lambda L) = \lambda^{d_f} M(L), \tag{1a}$$

with $\lambda^{d_f} \leq \lambda^d \ldots$ i.e., $d_f \leq d$. Equation (1a) is equivalent to the statement that[7]

$$M(L) \sim L^{d_f}. \tag{1b}$$

Patterns arising from phenomena as diverse as dielectric breakdown,[8,9] viscous fingering,[10-26] electrochemical deposition,[27,28] dendritic solidification,[29,30] and chemical dissolution ("acidizing of rock")[31] are characterized by the same numerical value of d_f. Corresponding analogies in the basic physics underlying these phenomena can be elucidated.

Of course, functional equations like (1a) cannot hold for all values of the parameter λ. For example, if λ is larger than the size of the object, or if L itself is comparable to the size of subunits comprising the object, then (1a) will certainly fail. Thus when the concepts of fractal geometry are applied to the structure of

---

* A more extended account of this work will be submitted for publication as an invited "minireview" to Nature, under the authorship HES and Paul Meakin

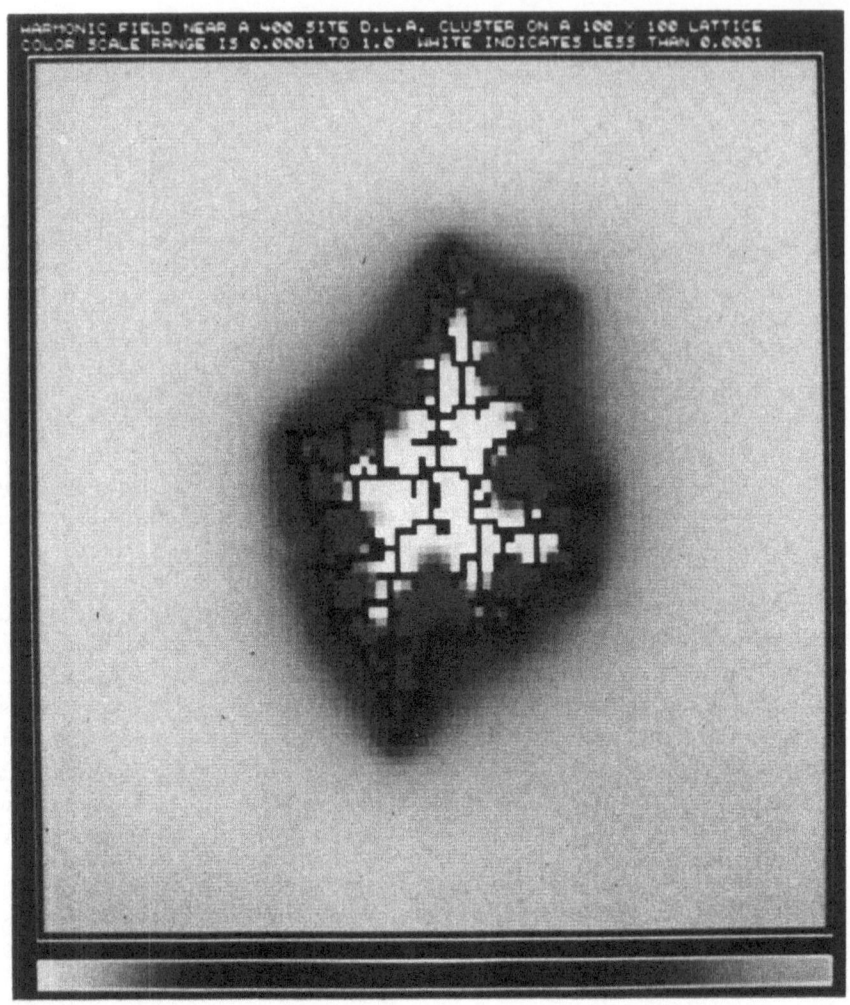

A multifractal set that has been extensively studied recently. Shown in black is a 400-site DLA cluster on a 100x100 square lattice. Think of this cluster as a digitized representation of a non-viscous fluid (e.g., water) being forced under pressure into a viscous fluid (e.g., oil). then one can numerically obtain the pressure at every pixel in the box. The numerical values of this potential are color coded, from orange (pressure zero at the edge of the box) to white (pressure one in the non-viscous fluid). The time-development of this pattern is governed by the growth probabilities at each site on the "water-oil" interface, which are in turn given by the gradient of the pressure at each point on the interface; these are evidently large near the tips and extremely small in the "fjords". Multifractal theory is a tool to describe this distribution of growth probabilities. (H.E. Stanley)

objects like proteins[32-34] that comprised of a relatively small number of subunits, attention must be paid to the range over which fractal phenomena are observed. Scattering experiments determine the intensity $I(q)$ of radiation scattered through a wave vector q, and so provide information concerning structure on the length scale q^{-1}. In fact, a log-log plot of $I(q)$ against q is linear over the range of length scales for which the scattering object is fractal; the slope of this straight-line portion is simply $-d_f$. For example, Ref. 35 presents data[35] from both light and neutron scattering, extending over several decades in q, for a large aggregate of 70Å Ludox[R] spheres.

Our purpose here is to provide an elementary introduction to the subject of **multifractals**, an extremely useful generalization of the concept of fractal geometry.

Multifractals arise naturally when one seeks to define a "measure" on an object.[36-59] For example, consider the diffusion limited aggregation model[60-61] (DLA), which has been successful in describing a variety of aggregation patterns as well as in describing a wealth of other growth phenomena (including those cited in the first paragraph) where the physics is controlled by the Laplace equation. In this case, an important measure is the set of growth probabilities $\{p_i\}$, where the index i runs over all the M_{TOT} perimeter sites of the DLA cluster. Here p_i is the probability that perimeter site i is the next to grow. Thus specification of the set $\{p_i\}$ completely describes the statistical growth of the object.

It is convenient to "bin" the numbers $\{p_i\}$, introducing a distribution function $N(p)$ such that $N(p)dp$ gives the number of perimeter sites whose growth probability is in the range $[p, p + dp]$. We will see that $\log p$ is the natural variable in this problem, which motivates a logarithmic binning [such as $\frac{1}{2} < p < 1, \frac{1}{4} \leq p < \frac{1}{2}, \frac{1}{8} < p < \frac{1}{4},\ldots$]. Thus we define $n(p)$ to be the number of sites for which $\log p$ is in the range $[\log p, \log p + \Delta \log p]$. In the limit of small Δp, $n(p) = pN(p)$.

Like all distribution functions, $n(p)$ is completely characterized by its moments

$$Z_q \equiv \sum_{bin} n(p)p^q. \tag{2a}$$

The Z_q are not known exactly for DLA, but they can be calculated exactly in a mean field limit[62] in which one assumes that the growth occurs at a subset M_u of the "unscreened" perimeter sites. Although not qualitatively correct, this assumption is **plausible** in light of experimental data showing that indeed, most of the growth occurs in a tiny annulus near the periphery of the cluster (Fig. 1). Thus the mean-field *Ansatz* is then the statement that

$$p_i = \bar{p} \qquad [i = 1, 2, \ldots, M_u]$$
$$= 0 \qquad \text{otherwise.} \tag{2b}$$

Hence the qth moment is given by

$$Z_q = [(\bar{p})^q M_u]. \tag{3}$$

Now $\bar{p}$ is fixed by the normalization condition

$$\sum_{i=1}^{M_{TOT}} p_i = \sum_{i=1}^{M_u} \bar{p} = 1. \tag{4}$$

Fig. 1a

Fig. 1: (a) A large DLA cluster of 50,000 particles grown off-lattice. The particles are color coded to signify their time of addition to the cluster. The fact that in this particular cluster, no green particles touch the white or violet particles indicates that the growth probability p_i in the deeply screened interior of this cluster is extremely small. The multifractal approach described here shows that these numbers p_i form a multifractal set. (b) The analog of part a. except that instead of a single seed particle there is an entire row of 300 seed particles. Random walkers are released from the top of this figure and form the hierarchy of treelike structures shown. The same color coding is used as before, and again we find no green particles in this particular picture that occupy the perimeter sites of either white or violet particles. The geometry of part (a) is relevant to a wide variety of aggregation and breakdown phenomena, ranging from colloidal flocs on the one hand to radial viscous fingers on the other. The geometry of part (b) is relevant to dendritic growth, being a statistical realization of the classic Mullins-Sekerka instability. (c) Pattern obtained from the screened growth model, which is a variation of the Eden model in which each perimeter site i is weighted with a factor that depends on the distance between site i and all of the cluster sites. Its major advantage is that the fractal dimension d_f is known exactly in terms of the parameter ϵ that arises in the weight function, so it can be used to generate clusters with a continuously tunable d_f ($d_f = 1.75$ for the case shown here).

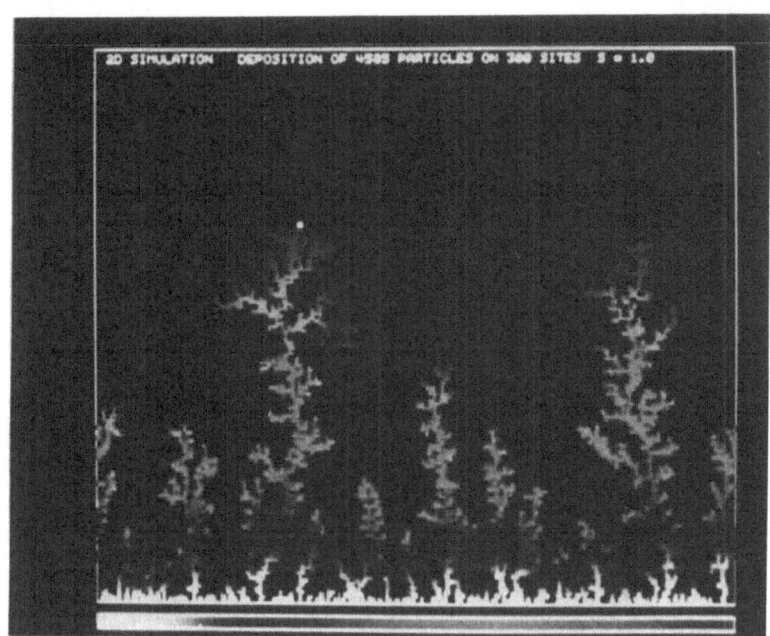

Fig. 1b

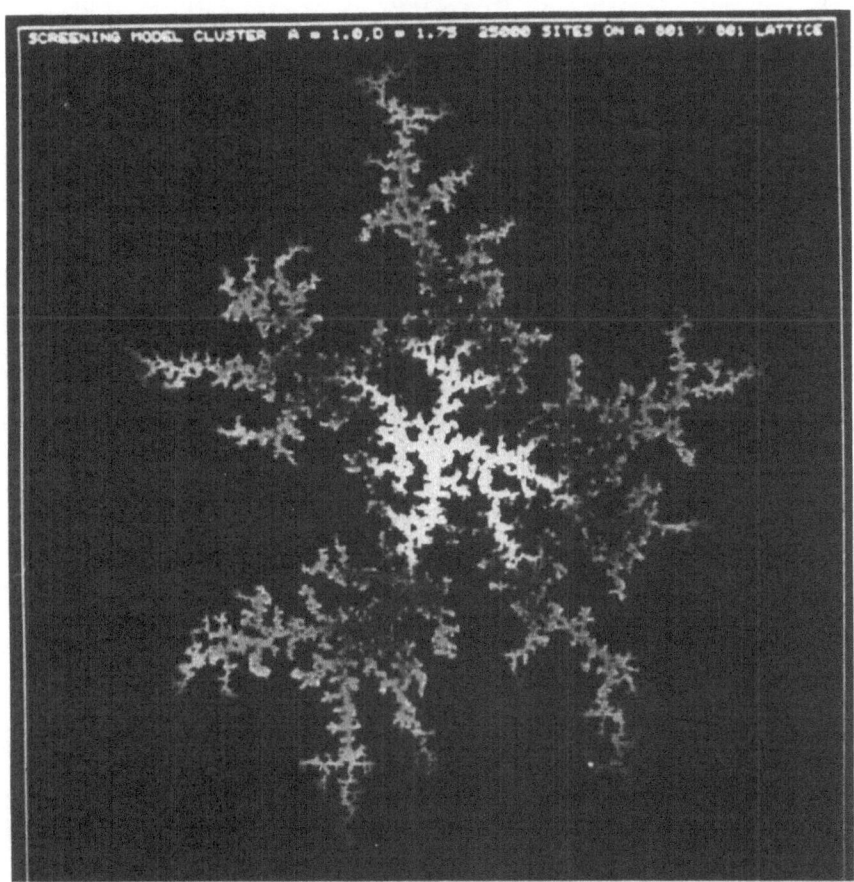

Fig. 1c

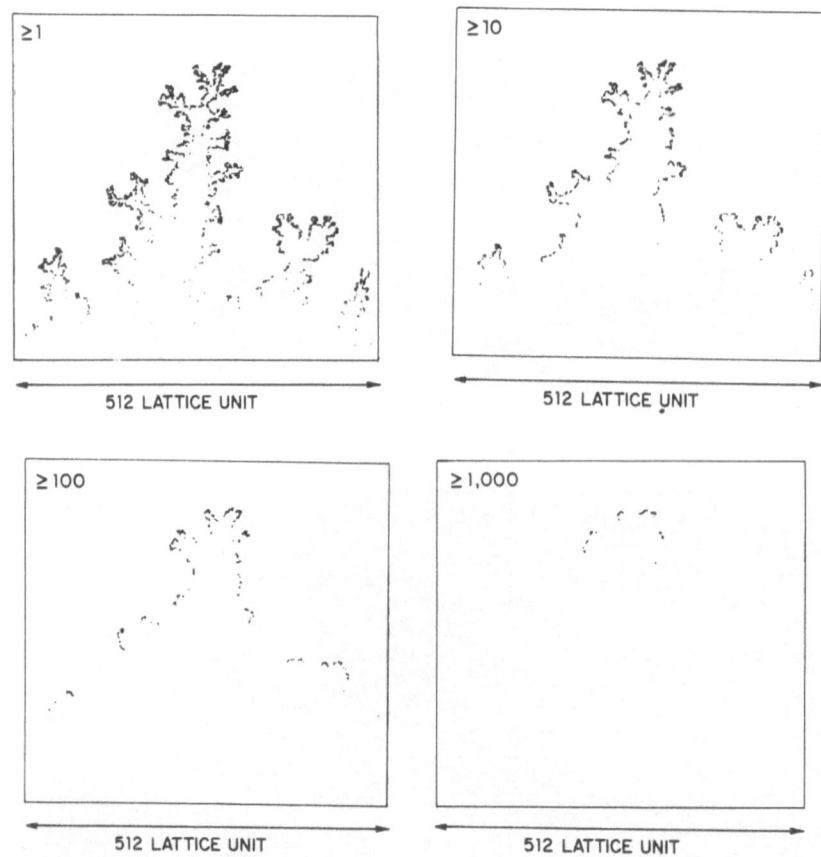

Fig. 2: This figure shows the results of a simulation in which random walkers were used to estimate the harmonic measure (or growth probability measure) for DLA on a strip 512-latticed units wide using periodic boundary conditions. The upper left corner shows all of the sites which were contacted, the upper right corner shows those sites which were contacted ten or more times, the lower left corner shows those sites which were contacted 100 or more times, and the lower right corner shows those sites which were contacted at 1000 times. A new random walk is started after each random walker has contacted the aggregate for the first time. After P. Meakin (preprint). See also Ref. 5.

From (4) it follows that

$$\bar{p} = 1/M_u. \tag{5}$$

Substituting (5) into (3), we find that in the mean field limit the scaling of $Z_q^{-1/(q-1)}$ is independent of q, with

$$Z_q = M_u^{-(q-1)} \sim L^{-(q-1)d_u}. \tag{6}$$

By making specific assumptions on the spatial distribution of the M_u unscreened perimeter sites, the "surface" exponent d_u can be expressed in terms of the "bulk" exponent d_f. This expression, $d_u = [d_f - 1] + [(d - d_f)/2]$, suggests a simple physical interpretation. The first term is the surface exponent one would find if there were no "penetration" of the fractal by the diffusion field (the random walkers). The second term represents a rough incorporation of the fact that for any finite value of q there is some penetration of the fractal. In the $q = \infty$ limit, we expect to find $d_u = d_f - 1$, since only the most exposed tips of DLA are contacted and hence only the first term contributes. This prediction is borne out by detailed numerical calculations, as well as theoretical arguments.[63]

For q sufficiently large, we would surely expect the mean field result (6) to fail, since only the most exposed tips of the DLA perimeter are favored in the moment sum (2a). In fact, for large values of q, (6) fails quite seriously. Here we replace (6) by

$$Z_q \sim L^{-(q-1)D(q)}, \tag{7}$$

where $d_u = D(q)$ is now q-dependent.

For general q, the sum in (2a) is dominated by some value $p = p^*$, where in general

$$p^* = p^*(q) \tag{8}$$

depends on the value of q. Thus $p*$ is the value of p that maximizes the summand $n(p)p^q = \exp[\ln n(p) + q \ln p]$. It is the solution of the equation

$$\frac{d \ln n(p)}{d \ln p} = -q. \tag{9}$$

Let us define the new exponents $\alpha(q)$ and $f(q)$

$$p^*(q) \sim A(q)L^{-\alpha(q)}, \tag{10a}$$

and

$$n(p^*) \sim B(q)L^{f(q)}. \tag{10b}$$

Then

$$(q-1)D(q) = q\alpha(q) - f(q). \tag{11}$$

From (9) it follows that

$$\frac{d}{dq}\left[(q-1)D(q)\right] = \alpha(q). \tag{12}$$

Thus if we know $D(q)$, then $\alpha(q)$ follows from (12) and, finally, $f(q)$ from (11).

We note that the normalization condition $\sum_{p*} n(p*)p* = 1$ implies

$$\sum_q L^{f(q)-\alpha(q)} = 1.$$

Hence

$$f(q) \leq \alpha(q), \tag{13}$$

with the equal sign holding for $q = 1$.

Note also that $p^*(q) \to p_{\max}$ as $q \to \infty$, so if we set $\alpha_{\max} = \alpha(\infty)$,

$$p_{\max} \sim L^{-\alpha(\infty)}. \tag{14}$$

From (14) and (9), we find that in the asymptotic large-L limit,

$$x \equiv \frac{\ln p*}{\ln p_{\max}} = \frac{\alpha(q)}{\alpha(\infty)}, \tag{15}$$

plus terms of order $(1/\ln L)$. Inverting (15), we have

$$q = q(x). \tag{16}$$

Substituting in (10), we finally obtain

$$n(p) = C(x)L^{\phi(x)}, \tag{17}$$

where $x = \ln p/\ln p_{\max}$, $C(x) = B[q(x)]$, and $\phi(x) = f[q(x)]$. Note that we have replaced $p*$ by p since $p*$ can be considered to be an independent variable. That is, (15) is a functional equation valid for all values of $p*$ and hence for all values of p.

The theoretical arguments presented above can be put to direct tests. To accomplish this goal, it is necessary to calculate the set of numbers $\{p_i\}$. This calculation is perhaps most easily done by allowing a large number N of random walkers to touch the cluster. A counter at each perimeter site i records the number of touches T_i experienced (Fig.1). Then p_i is obtained by the normalization relation $p_i = T_i/N$. The results of such numerical experiments are also displayed in the color-coded figure on the cover of this issue.

Equation (17) gives the scaling behavior for the hit distribution function $n(p)$. The actual exponent in (11) depends on the ratio $\ln p/\ln p_{\max}$, which reduces to the usual scaling with only one scaling exponent when $\alpha(q)$ is independent of q. The scaling prediction (17) can be then tested directly. Figure 2 shows the results obtained from a non-equilibrium growth model (the screened growth model) which is similar to DLA.

Thus far, we have described the basic multifractal phenomenon for two single systems —DLA— and one screened growth model. In fact, analogous phenomena are found for a wide variety of systems. They were first found by Mandelbrot in his pioneering studies of fluid turbulence,[36-37] and by Grassberger, Hentschel and Procaccia in the course of analyzing non-linear dynamical systems.[38-40] Recently it has been demonstrated[65] that experimental data on the onset of turbulence can

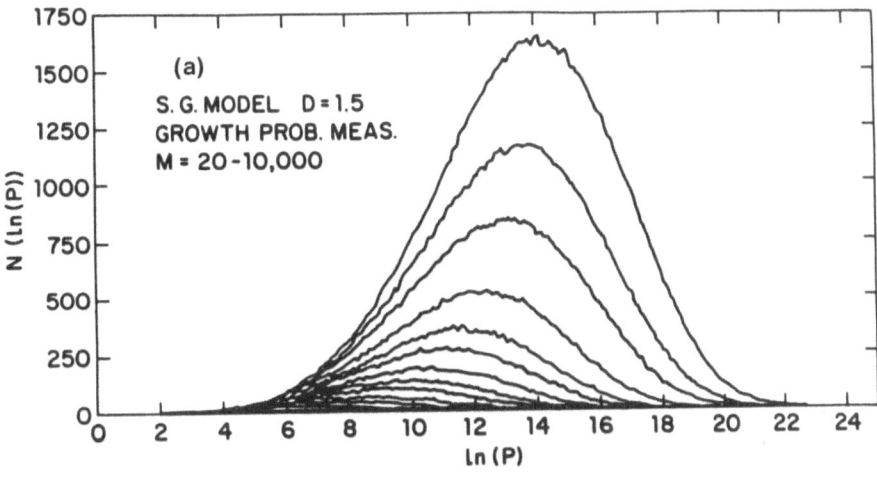

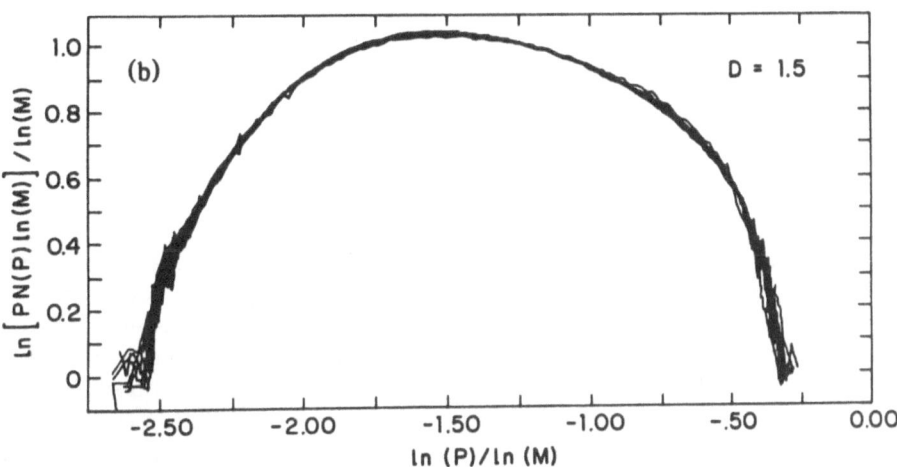

Fig. 3: The probability distribution $n(p)$ for 7 cluster masses in the range $1000 <$ $M < 10000$ both (a) **before**, and (b) **after** scaling. These results[54] were obtained from the screened growth model[78-79] ($d_f = 1.75$ for the case shown here). After Ref. 54.

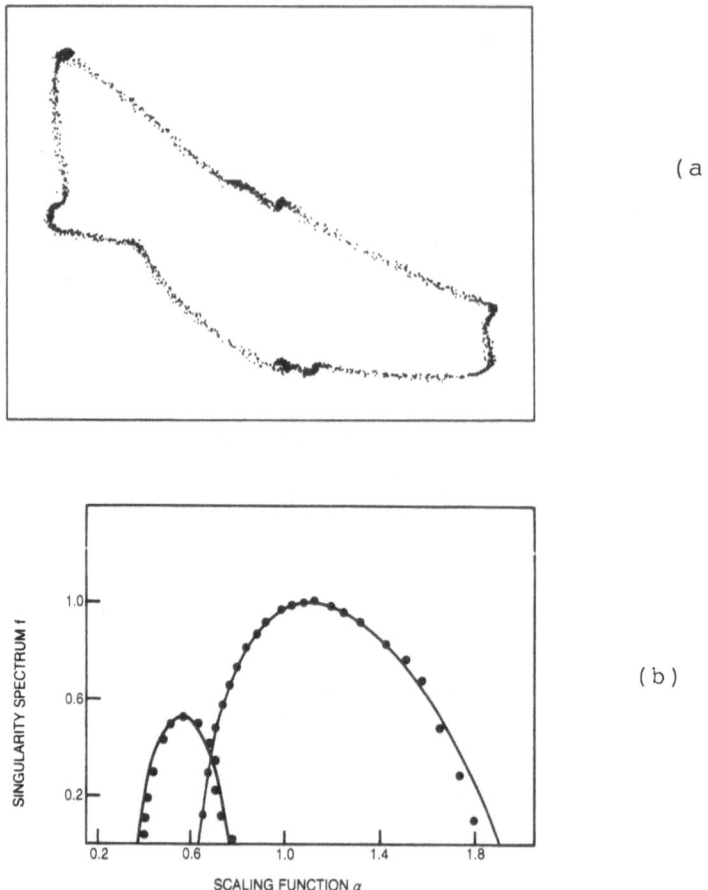

(a)

(b)

Fig. 4: Another example of multifractal phenomena: experiments germane to the onset of turbulence. Shown in (a) is the "critical orbit"—the trajectory of a dynamical system in phase space as the system becomes turbulent—determined from temperature measurements of a cell filled with liquiid Hg that is made to undergo convective and electromagnetic oscillations simultaneously. Above a threshold value of the temperature gradient, the convective rolls in the cell oscillate in the following fashion: an imaginary line between the rolls and parallel to their axes moves from side to side. One applies a dc magnetic field parallel to the axes of the convective rolls, and a vertical ac current is applied to the center of the Hg cell. The system trajectory traces out a torus in phase space, when the two frequencies are incommensurate. The system approaches a turbulent state as the ac current is increased in amplitude, and the torus becomes distorted. This figure shows the intersection of this distorted torus with a plane. The fractal dimension α_i changes with the reference point i. The distribution of values of α_i can be interpreted as the function $f(\alpha)$, which is shown in (b). After Ref. 65.

be analyzed in a fashion suggested by multifractal theory (Fig. 3). Similar scaling is also found for the current distribution in a random resistor network.[44−47] Here, the growth probability p_i for perimeter site i is replaced by the current I_i across bond i in the backbone of a percolation cluster. Various multifractal sets have been mapped[66] onto the thermodynamics of one-dimensional spin models. The depletion of a diffusion substance near an absorbing polymer has been studied; it is found that the scaling with distance r of each moment of the Laplace field is governed by an independent exponent.[67] Several authors have studied the multifractal proper-ties of random walks.[68] In particular, Evertsz and Lyklema[68] have shown that **the** fractal dimension $d_f = 1/\nu$ is a member of a continuous set of scaling exponents; consideration of the entire hierarchy of scaling exponents provides a more complete description of random walks than was possible previously. The main idea is to characterize an infinite walk with an exponent α that measures how fast its total probability decays to zero with increasing mass of the walk. The analog of the function $f(\alpha)$ discussed above is the growth rate $z(\alpha)$ for the subset of walks with decay rate α. The analog of $\tau(q)$ is the Legendre transform of $z(\alpha)$. A log-normal distribution has been found for the first passage time in percolation.[69] Some un-derstanding of the conditions under which such a log-normal distribution will occur has also developed recently.[70−71] We should point out that there exists one rigorous result which in spite of its simplicity is highly non-trivial [N. G. Makarov, Soviet Math. Dokl. **31**, 117 (1985); see also P. Grassberger, unpublished preprint (1984)]: $D(1) = \ldots = 1$. Its generalization to more than 2 dimensions is still open.

One does not need to have a fractal structure in order to find multifractal phenomena. For example, consider a charged needle—a non-fractal object of di-mension one! Calculate the electric field E_i at every point i on the surface. The set $\{E_i\}$ of electric field values is formally analogous to the set $\{p_i\}$ of DLA growth probabilities, and indeed one finds that the $\{E_i\}$ also form a multifractal set.[51]

The question that naturally arises is "Under what conditions can one expect to find multifractal phenomena?" As the example of the needle suggests, the key ingredient is that one must define a measure on an object such that this measure has a different fractal dimension in different regions of the object. Thus when the length of a needle is doubled, the electric field near its tip changes by a factor which is **dif-ferent** from the factor by which the electric field near the center changes. Similarly, the growth probability near the tips of a DLA structure changes by a **different** factor when the mass of DLA is doubled than the growth probability deep in the fjords. This is because the screening in the deep fjords increases dramatically with increas-ing cluster size. Similarly, for the random resistor network the singly-connected "red" bonds carry all the current, so the current in these bonds is independent of cluster size. However the current in the large multiply-connected "blobs" decreases dramatically with cluster size: the size of the largest blobs increases by a factor $2^{1.62}$ each time the cluster mass is doubled and hence the current in each bond of a blob decreases dramatically.

We have seen that multifractal theory permits the characterization of complex phenomena in a fully **quantitative** fashion. Just as completely random phenomena in nature adopt shapes that are fractal, phenomena with spatial correlations may be described by multifractals. For example, randomly porous media are traditionally modeled by the random resistor network of percolation theory: the resistance of each element corresponds to the permeability of each pixel in a suitable digitization of the porous medium. Although this much-studied model captures much of the essential physics, it neglects the phenomenon of spatial correlation which in turn leads to both short and long range heterogeneities in the porous medium. For this reason, it has recently been proposed that atmospheric turbulence and porous media be modelled by a multifractal lattice.[72-76] This is obtained by the procedure shown in Fig. 4. Transport in such a lattice can be anomalously slow, just as it is in the random resistor network model. However the exponent d_w describing the anomaly can be continuously tuned; in fact the slowing down can under suitable conditions become large without limit. A similar model was solved analytically in one dimension[77] with the result $d_w = \tau(1) + 1$. Models of this type have also been used to represent the distribution of energy dissipation in turbulent fluids.

(a)

(b)

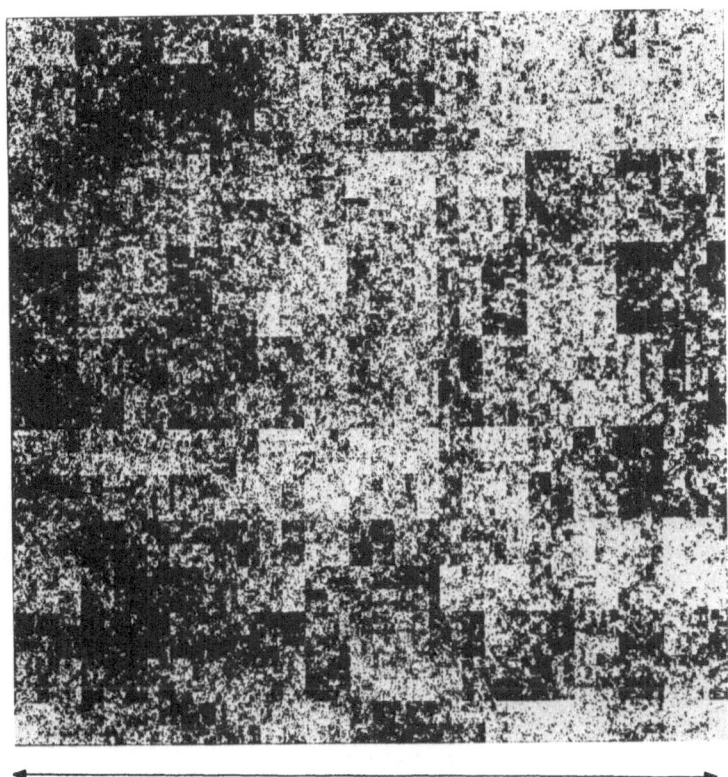

512 LATTICE UNITS

$P_1 = 0.5$ $P_2 = 0.5$ $P_3 = 1.0$ $P_4 = 1.0$

Fig. 5: A multifractal lattice. (a) Generator, (b) Second stage of construction, (c) Results after 8 generations. Lattice shown contains 2^8 pixels along each edge, and shading is proportional to the value of π_i for pixel i. Here π_i is the product of the 8 probabilities assigned to that pixel as a results of the 8 generations. After Ref. 73.

In summary, we have provided a simple introduction to multifractal phenomena. By concentrating on a single example, DLA, we have been able to illustrate the basics of multifractal theory. We have seen that this theory agrees with experiments on radial viscous fingers. We have briefly described a range of other systems that display multifractal phenomena. Finally, we have discussed the possibility that actual structures in nature are multifractal objects.

Our own work on multifractals is mainly in collaboration with G. Daccord, T. C. Halsey, J. Nittmann, I. Procaccia, E. Touboul and, especially, A. Coniglio and T.A.Witten; we owe much to them for our understanding of this subject. We also benefitted from helpful interactions with C. Amitrano, L. de Arcangelis, F. di Liberto, L. Pietronero, and S. Redner. Comments on the first draft of this mini-review are gratefully appreciated from A. Bunde, P. Grassberger, and S. Havlin. I wish to thank especially Roger Pynn, Tormod Riste, and Gerd Jarrett—this threesome combined forces (and "charms") to make this a truly memorable ASI.

REFERENCES

1. Mandelbrot, B. B., *The Fractal Geometry of Nature* (W. H. Freeman & Co., San Francisco, 1982), and references therein.

2. Family, F. and Landau, D. P. (eds), *Kinetics of Aggregation and Gelation* (Elsevier, Amsterdam, 1984).

3. Stanley, H. E., and Ostrowsky, N. (eds) *On Growth and Form: Fractal and Non-Fractal Patterns in Physics* (Martinus Nijhoff, The Hague, 1985).

4. Boccara, N. and Daoud, M. (eds), *Physics of Finely Divided Matter* [Proceedings of the Winter School, Les Houches, 1985] (Springer Verlag, Heidelberg, 1985).

5. Pynn, R. and Skjeltorp, A. (eds.) *Scaling Phenomena in Disordered Systems* (Plenum, N.Y., 1986).

6. Pietronero, L. and Tosatti, E. (eds), *Fractals in Physics* (North Holland, Amsterdam, 1986).

7. Stanley, H. E., *Introduction to Fractal Phenomena* (in press).

8. Niemeyer, L., Pietronero, L. and Wiesmann, H. J., *Phys. Rev. Lett.* **52**, 1033 (1984).

9. Pietronero, L. and Wiesmann, H. J., *J. Stat. Phys.* **36**, 909 (1984).

10. Paterson, L., *Phys. Rev. Lett.* **52**, 1621 (1984).

11. Nittmann, J., Daccord, G., and Stanley, H. E., *Nature* **314**, 141 (1985).

12. Daccord, G., Nittmann, J., and Stanley, H. E., *Phys. Rev. Lett.* **56**, 336 (1986).

13. Van Damme, H., Obrecht, F., Levitz, P., Gatineau, L., and Laroche, C., *Nature* **320**, 731 (1986).

14. Chen, J. D. and Wilkinson, D., *Phys. Rev. Lett.* **55**, 1985 (1985).

15. Måløy, K. J., Feder, J., and Jøssang, T., *Phys. Rev. Lett.* **55**, 2688 (1985).

16. Stokes, J. P., Weitz, D. A., Gollub, J. P., Dougherty, A., Robbins, M. O., Chaikin, P. M., and Lindsay, H. M., *Phys. Rev. Lett.* **57**, 1718 (1986).

17. Tang, C., *Phys. Rev. A* **31**, 1977 (1985),

18. Sherwood, J. D. and Nittmann, J., *J. Physique* **47**, 15 (1986).

19. Buka, A., Kertèsz, J., and Vicsek, T., *Nature* **323**, 424 (1986).

20. Lenormand, R. and Zarcone, C. *Phys. Chem. Hydrodyn.* **6**, 497-506 (1985).

21. Ben-Jacob, E. *et al. Phys. Rev. Lett.* **55**, 1315-1318 (1985).

22. DeGregoria, A. J. and Schwartz, L. W. *J. Fluid Mech.* **164**, 383-400 (1986).

23. Ben-Jacob, E., Deutscher, G., Garik, P., Goldenfeld, N.D. and Lereah, Y. *Phys. Rev. Lett.* **57**, 1903 (1986).

24. Maher, J. *Phys. Rev. Lett.* **54**, 1498 (1985).

25. Couder, Y., Cardoso ,O., Dupuy, D., Tavernier, P. and Thom, W. *Europhys. Lett.* **2**, 437- (1986).

26. Måløy, K. J., Boger, F., Feder, J., Jøssang, T., and Meakin, P., *Phys. Rev.* **35**, xxx (1987).

27. Sawada, T. *Physica A* **140**, 134-141 (1986).

28. Brady, R. M. and Ball, R. C. *Nature* **309**, 225 (1984).

29. Matsushita, M., Sano, M., Hakayawa, Y., Honjo, H. and Sawada, Y. *Phys. Rev. Lett.* **53**, 286 (1984).

30. Nittmann, J. N., and Stanley, H. E. *Nature* **321**, 663-668 (1986).

31. Daccord, G. *Phys. Rev. Lett.* **58**, 479-482 (1987); Daccord, G. and Lenormand, R. *Nature* **325**, 41-43 (1987).

32. Stapleton, H. J., Allen, J. P., Flynn, C. P., Stinson, D. G. and Kurtz, S. R. *Phys. Rev. Lett* **45**, 1456-1459 (1980).

33. Hellman, J.S., Coniglio, A, and Tsallis,C. *Phys. Rev. Lett.* **53**, 1195-1198 (1984).

34. Herrmann, H.J. *Phys. Rev. Lett.* **56**, 2432 (1986).

35. Schaefer, D. W., Martin, J. E., Wiltzius, P. and Cannell, D. S. *Phys. Rev. Lett.* **52**, 2371 (1984).

36. Mandelbrot, B. B. *J. Fluid Mech.* **62**, 331 (1974).

37. Mandelbrot, B. B. in *Proceedings of 15th IUPAP Conference on Statistical Physics* (eds Cabib, E., Kuper, C. G. and Reiss, I.) (Adam Hilger, Bristol, 1978).

38. Grassberger, P. *Physics Letters A* **97**, 227 (1983).

39. Hentschel, H. G. E. and Procaccia, I. *Physica* **8D**, 435 (1983).

40. Frisch, U. and Parisi, G. in *Turbulence and Predictability in Geophysical Fluid Dynamics and Climate Dynamics*, Proceedings of the International School of Physics "Enrico Fermi," Course LXXXVIII (eds Ghil, M., Benzi, R. and Parisi G.) (North-Holland, Amsterdam, 1983).

41. Grassberger, P. and Procaccia, I. *Physica* **13D**, 34 (1984); Grassberger, P. *Phys. Lett.* **107A**, 101 (1985) (here the definition of $D(q)$ via "partition functions" was introduced, among other things); Badii, R. and Politi, A. *J. Stat. Phys.* **40**, 725 (1985).

42. Benzi, R., Paladin, G., Parisi, G. and Vulpiani, A. *J. Phys. A* **17**, 3521 (1984); Ibid. **18**, 2157 (1985).

43. Meakin, P., Stanley, H. E., Coniglio A. and Witten, T. A. *Phys. Rev. A* **32**, 2364 (1985).

44. Rammal, R., Tannous, C., Breton, P. and Tremblay, A. M. S. *Phys. Rev. Lett.* **54**, 1718 (1985).

45. Rammal, R., Tannous, C. and Tremblay, A. M. S. *Phys. Rev. A* **31**, 2662 (1985).

46. Rammal, R. *J. Phys. (Paris)* **46**, L129 (1985).

47. Rammal, R. *Phys. Rev. Lett.* **55**, 1428 (1985).

48. de Arcangelis, L., Redner, S. and Coniglio, A. *Phys. Rev. B* **31**, 4725 (1985).

49. Coniglio, A. in *On Growth and Form: Fractal and Non-Fractal Patterns in Physics* (eds Stanley, H. E. and Ostrowsky, N.) (Martinus Nijhoff, Dordrecht, 1985), p. 101.

50. Halsey, T. C., Meakin, P. and Procaccia, I. *Phys. Rev. Lett.* **56**, 854 (1986).

51. Meakin, P., Coniglio, A., Stanley, H. E. and Witten, T. A. *Phys. Rev. A* **34**, 3325 (1986).

52. de Arcangelis, L., Redner, S. and Coniglio, A. *Phys. Rev. B* **34**, 4656-4673 (1986).

53. Coniglio, A., in *Fractals in Physics* (eds Pietronero, L. and Tosatti, E.) (North Holland, Amsterdam, 1986).

54. Meakin, P. *Phys. Rev. A* **34**, 710 (1986).

55. Amitrano, C., Coniglio, A. and di Liberto, F. *Phys. Rev. Lett.* **57**, 1016 (1987).

56. Castellani, C. and Peliti, L. *J. Phys. A* **19**, L429 (1986); Ioffe, L. B., Sagdeev, I. R., and Vinokur, V. *J. Phys. C* **18**, L641 (1985).

57. Meir, Y., Blumenfeld, R., Aharony, A. and Harris, A. B. *Phys. Rev. B* **34**, 3424 (1986); Blumenfeld, R., Meir, Y., Harris, A. B. and Aharony, A. *J. Phys. A* **19**, L791 (1986); Blumenfeld, R., and Meir, Y., Aharony, A., and Harris, A. B. *Phys. Rev. B* **35**, 3524-3535 (1987); Gefen, Y., Shih, W. H., Laibowitz, R. B., and Viggiano, J. M. *Phys. Rev. Lett.* 3097 (1986); Aharony, A., *Phys. Rev. Lett.* comments; Y. Park, A. B. Harris, and T. C. Lubensky, *Phys. Rev.* **35**, xxx (1986).

58. Rammal, R. and Tremblay, A.-M. S. *Phys.Rev. Lett.* **58**, 415-418 (1987); Fourcade, B., Breton, P. and Tremblay, A.-M. S., "On the Infinite Set of Exponents of Fractal Objects" preprint.

59. Halsey, T. C., Jensen, M. H., Kadanoff, L. P., Procaccia I. and Shraiman, B. *Phys. Rev. A* **33** 1141 (1986).

60. Witten, T. A. and Sander, L. M. *Phys. Rev. Lett.* **47**, 1400 (1981).

61. Witten, T. A. and Sander, L. M. *Phys. Rev. B* **27**, 5686 (1983).

62. Coniglio, A. and Stanley, H. E. *Phys. Rev. Lett.* **52**, 1068 (1984).

63. Leyvraz, F. *J. Phys. A* **18**, L941 (1985).

64. Nittmann, J., Stanley, H. E., Touboul, E. and Daccord, G. *Phys. Rev. Lett.* **58**, 619 (1987).

65. Jensen, M. H., Kadanoff, L. P., Libchaber, A., Procaccia, I. and Stavans, J. *Phys. Rev. Lett.* **55**, 2798 (1985).

66. Katzen, D. and Procaccia, I., preprint.

67. Cates, M. E. and Witten T. A. *Phys. Rev. Lett.* **56**, 2497 (1986); Cates, M. E. and Witten T. A., Phys. Rev. A. **35**, 1809-1824 (1987)

68. Evertsz, C and Lyklema, J.W. Phys. Rev. Lett. **58**, 397-400 (1987). Lyklema, J.W. , Evertsz, C.J.G, and Pietronero, L. *Europhys. Lett.* **2**, 77 (1986); Lyklema,J.W. and Evertsz, C *J. Phys. A* **19**, L895 (1986); Peliti, L. and Pietronero, L. (preprint).

69. Trus, B., Havlin, S. and Stauffer, D., preprint; see also Roman, E., Bunde, A., and Havlin, S., preprint.

70. Pietronero, L. and Siebesma, A. *Phys. Rev. Lett.* **57**, 1098 (1986).

71. Coniglio, A. *Physica A* **140**, 51-61 (1986).

72. Frisch, U., Sulem, P. and Nelkin, M. *J. Fluid Mech.* **87**, 719 (1978).

73. Meakin, P. *Phys. Rev. A* **36**, xxx (1987).

74. Meakin, P. *J. Phys. A* **19**, xxx (1987).

75. Lovejoy, S. and Schertzer, D. *Bull. Am. Meteor. Soc.* **67**, 221 (1986).

76. Schertzer, D. and Lovejoy, S. In IUTAM symposium on turbulence and chaotic phenomena in fluids, Kyoto, Japan, pp. 141-144 (1983).

77. H. Waissmann and S. Havlin, preprint

78. P. Meakin, *Phys. Rev. B* **28**, 6718- (1983).

79. P. Meakin, F. Leyvraz, and H.E. Stanley, *Phys. Rev. A* **31**, 1195 (1985).

MULTIFRACTALITY ON PERCOLATION CLUSTERS

Amnon Aharony

School of Physics and Astronomy
Tel Aviv University, Tel Aviv 69978, Israel

1. INTRODUCTION

Following the introduction of fractals into statistical physics some six years ago,[1] fractals became a growing active discipline in physics research. It has become clear that, at least over some range of length scales, many physical structures exhibit fractal geometry.[2] Research has then divided into two main directions: First, attempts have been made to understand the physical mechanisms which govern the growth of structures (e.g. aggregates) into their particular fractal shapes.[3,4,5] As recently commented by Kadanoff,[6] there remain many open questions to be studied in this direction. In the second direction, the fractal geometry is taken as given, and the physical properties of the structure are then studied.[7] Different physical properties turn out to be determined by subsets of sites (or bonds, or particles) on the structure, each having its own fractal nature. At the present time several infinite sets of independent fractal dimensionalities, or critical exponents, have been identified and studied. This paper aims to review this plenitude of exponents. Although the paper does not specifically discuss dynamics, many of the concepts introduced here are also used to describe dynamic phenomena, e.g. random walks, waves and breakdown phenomena. This lecture is meant to serve as an introduction to those subjects.

To be more specific, this paper will be restricted to the most widely studied system, i.e. that of the infinite incipient cluster at the percolation threshold. However, everything described here equally applies to other fractal systems, e.g. growing aggregates. We start in Sec. 2 with a short review of percolation clusters and of the fractal dimensionalities of their total masses, including corrections, and of their various perimeters. The remainder of the paper concerns two-terminal properties. Sec. 3 lists some of these, e.g. resistance, backbone mass, singly connected bonds, random and self-avoiding walks, etc. The following two sections then present two infinite sets of exponents, i.e. those describing resistance fluctuations (noise) or current moments (Sec. 4) and those describing non-linear resistor networks (Sec. 5). Both sets contain many of the previously listed physically relevant dimensionalities. Sec. 6 concludes the paper, with some open questions.

2. FRACTAL DIMENSIONALITIES AT PERCOLATION

2.1.Percolation Model

The percolation model is widely used to describe alloys of materials with different properties, e.g. metal-insulator, rock-pores, magnet-non-magnet, etc.[8,9] Bonds (or sites) on a lattice are occupied (with probability p) or empty (with probability 1-p) at random.

Connected bonds (sites) form clusters. On an infinite sample, all clusters remain finite below the percolation threshold, p_c, with exponentially few clusters of linear size larger than

$$\xi \sim |p-p_c|^{-\nu}.$$ For $p>p_c$, an infinite cluster exists with finite probability per site, $P_\infty \sim (p-p_c)^\beta$.

As defined above, the percolation model is purely geometric and static. However, in many cases percoaltion occurs as a time dependent phenomenon. In the first historical realization of percolation thoery, in the context of polymer gelation,[10] bonds are added at random, and the concentration p increases linearly with time. Alternatively, a forest fire starts at the origin, and spreads outwards, hitting nearest neighboring trees (sites) at a constant unit speed (each neighbor burns with probability p). In this case, the connected cluster grows until all its neighbors are found not to burn.[11] The time of growth is equal to the shortest path (through cluster sites) from the origin to the growing site. This is also called the "chemical" or "topological" distance.[12]

A closely related dynamic process is that of "invasion percolation":[13] a cluster grows into a sample through selection of paths of least resistance. In dimensions larger than 2 (but not equal to 2), the geometry of the cluster is the same as that of static percolation clusters.

2.2 Fractal Dimensionality of Cluster Mass

At $p=p_c$, $P_\infty=0$. However, finite samples of large linear size L have a finite probability per site, of order $L^{-\beta/\nu}$, to be on a spanning cluster. Alternatively, if one picks a site on such a cluster, and draws a box of linear size L around it, the average number of sites on the same cluster within that box scales as

$$M(L) \sim L^D, \tag{2.1}$$

with the fractal dimensionality $D=d-\beta/\nu$, where d is the Euclidean dimensionality.[14]

In fact, Eq. (2.1) is only asymptotic, for $L \to \infty$. At finite L, one expects correction terms, of the form

$$M(L) = AL^D + A_1 L^{D_1} + A_2 L^{D_2} + \dots , \tag{2.2}$$

with $D>D_1>D_2> \dots$. The exponents $(D_i-D)/\nu$ may be related to the correction to scaling exponents which appear due to irrelevant fields and non-linear scaling fields in critical phenomena.[15] As such, the infinite set of independent exponents D_i is understood, and can be explicitly calculated within specific approximations. It is my strong belief, that each D_i is also interpretable as the fractal dimensionality of some subset on the cluster.[15] This is somewhat supported by the fact that plots of $D_{eff}=d\ell n\, M/d\ell nL \simeq D + a_1 L^{D_1-D} + \dots$ at d=2 exhibit a linear dependence on 1/L, identifying D_1 with (D - 1), i.e. the dimensionality of a random linear cut through the cluster.[16] The geometrical interpretation of other D_i's remains to be studied.

2.3 Perimeters

Another family of dimensionalities concerns the perimeters of the cluster. If a walker is placed on an external edge of a two-dimensional cluster on a square lattice, and instructed to walk around the cluster (e.g. by moving to the right whenever possible, otherwise straight, otherwise left), it will count the sites on the "hull" of the cluster.[17] The average number of sites on the hull of a cluster of linear size L scales as L^{D_h}, with $D_h \simeq 1.75$. The hull is directly relevant to the motion of charged particles on equipotential lines at strong magnetic fields.[18]

Alternatively, one can place the walker on an unoccupied site neighboring the above end point, and instruct it to move to the right on a nearest-neighbor unoccupied site which also neighbors a site on the cluster. The number of sites on this "external perimeter" walk turns out to be much more ramified; it scales as L^{D_e}, with[16] $D_e \simeq 1.35$. This reflects the existence of many inpenetrable "fjords", on all length scales.

One can now define a sequence of external perimeters, with varying connectivities. A priori, this sequence will define a sequence of fractal dimensionalities, $D_h \geq D_e \ldots \geq 1$. However, a recent paper by Coniglio et al.[19] may imply that all the higher order perimeters have the same dimensionality, D_e. Ref. 19 identifies the special self-avoiding walk (SAW) characterizing the "hull" with the theta point of polymers (which separates compact collapsed polymers, with $D=d$, from ramified ones, with the usual SAW dimension $D \simeq 4/3$). Since the external perimeter SAW is restricted by more self-repulsion interaction than the hull, it probably falls into the universality class of the usual SAW's, and $D_e \simeq 4/3$. Higher order perimeters will then also fall into the same class, if indeed only two such classes exist.[19] Preliminary computer simulations[20] seem to confirm this prediction.

3. TWO-TERMINAL PHYSICAL PROPERTIES

3.1 Resistance, Backbone and Singly Connected Bonds

We now turn to properties which are defined with respect to two special sites on the infinite percolation cluster (called "terminals"), at a Euclidean distance L from each other. If each bond represents a unit resistance, then the average resistance measured between such terminals (for $L \gg 1$) is expected to scale as

$$R(L) \sim L^{\tilde{\zeta}_R} \ . \tag{3.1}$$

If we put a current I into one terminal and take it out through the other, then only a subset of the bonds on the cluster will have finite currents. This subset is called the "backbone", and its mass scales as

$$M_B(L) \sim L^{D_B} \ , \tag{3.2}$$

where D_B, the fractal dimension of the backbone, clearly obeys

$$D \geq D_B \geq \tilde{\zeta}_R \ . \tag{3.3}$$

Recently there has been much interest in the distribution of values of the currents in the backbone bonds.[21-23] Clearly, the largest current, I, occurs only in "singly connected" bonds, which separate the cluster into two regimes (each connected to only one of the terminals). The number of these singly connected bonds was proven[24] to scale as

$$M_{SC}(L) \sim L^{1/\nu} \ , \tag{3.4}$$

and clearly $\tilde{\zeta}_R \geq 1/\nu$. The backbone can be divided into singly connected bonds and blobs.[25] The size distribution of the blobs was recently studied by Herrmann and Stanley.[26]

The exponent $\tilde{\zeta}_R$ describes many "linear" physical problems, whose equations of motion can be mapped onto the resistor network Kirchhoff equations. In particular, the Einstein relation between the conductivity and the diffusion coefficient leads to a relation between $\tilde{\zeta}_R$ and the exponent describing the time dependence of random walks. This subject is discussed elsewhere in these proceedings.

3.2 Self-Avoiding Walks

Consider a self-avoiding walk (SAW) which connects the two terminals. Clearly, this walk must go through all the singly-connected bonds. However, it usually has several distinct routes through each "blob". The shortest and the longest SAW's are expected to scale as

$$M_{min}(L) \sim L^{\tilde{\varsigma}min} \quad ,$$

$$M_{max}(L) \sim L^{\tilde{\varsigma}max} \quad . \tag{3.5}$$

The average SAW scales as

$$M_{SAW}(L) \sim L^{1/\nu_{SAW}} \quad , \tag{3.6}$$

and one has

$$D_B \geq \tilde{\varsigma}_{max} \geq 1/\nu_{SAW} \geq \tilde{\varsigma}_{min} \geq \tilde{\varsigma}_R . \tag{3.7}$$

M_{min} is related to the chemical or topological distance, mentioned above.[12] The difference $\left[\tilde{\varsigma}_{max} - \tilde{\varsigma}_{min}\right]$ gives some measure of the asymmetry of the cluster.

4. NOISE, CURRENT DISTRIBUTION AND MULTIFRACTALS

4.1 Resistance Fluctuations and Current Distribution

Eq. (3.1) assumed that all the bonds had exactly the same basic resistance, r_0. If the $M_B(L)$ resistors are different from each other, then R(L) will be a function of all of these resistances, $R=R(r_1, r_2, ... , r_{M_B})$. If the resistances r_i fluctuate in a narrow range around r_0, $r_i=r_0+\delta r_i$, then we can express the cumulants of the total resistance in terms of those of the distribution of each r_i,[27]

$$\Gamma_k = \langle \delta R^k \rangle_c = \sum_i \left[\frac{\partial R}{\partial r_i}\right]^k \langle \delta r^k \rangle_c . \tag{4.1}$$

From the expression for the total power in the network, $RI^2 = \Sigma r_i I_i{}^2$, one finds that

$$R = \sum_i r_i i_i{}^2 \quad , \tag{4.2}$$

where $i_i = I_i/I$ is the fraction of the total current going through the i'th bond. Thus, $(\partial R/\partial r_i) = i_i{}^2$ and

$$\Gamma_k \sim \sum_i i_i^{2k} \quad . \qquad\qquad (4.3)$$

The behavior of Γ_k thus yields information on the distribution of currents in the network.[21]

Like all other quantities at $p=p_c$, each Γ_k scales with a distinct exponent $\tilde{\psi}(k)$,

$$\Gamma_k \sim L^{\tilde{\psi}(k)} \quad . \qquad\qquad (4.4)$$

The infinite set of exponents $\tilde{\psi}(k)$ has been the subject of much recent study.[21-23,27]

When $k \to \infty$, only the singly connected bonds (with $i_i \equiv 1$) survive in Eq. (4.3), hence $\tilde{\psi}(\infty) = 1/\nu$. When $k \to 0$, all the backbone bonds yield $i_i^0 \equiv 1$, hence $\tilde{\psi}(0) = D_B$. Trivially, $\tilde{\psi}(1) \equiv \tilde{\zeta}_R$. For $k=2$, Γ_2 represents the behavior of the second cumulant of the resistance, associated with the noise in the system.[27] Since the function $\tilde{\psi}(k)$ relates all of these properties, it has been very useful to study its functional properties. Indeed, many of the inequalities mentioned above result from the monotonicity of $\tilde{\psi}(k)$, which follows directly from that of i_i^{2k} ($i_i \leq 1$). Similarly, $\tilde{\psi}(k)$ has been shown to be a convex function, and its values have been estimated using various techniques.[23]

Much interest has recently concentrated on mixtures of superconducting (concentration p) and normal bonds.[28,22] As done above, one can construct moments of the currents on the normal bonds, and expect a multifractal distribution. At p_c, the largest currents will be on the "singly disconnected" bonds, whose fractal dimension is also equal to $1/\nu$.

De Arcangelis et al.[22] also consider the distribution of voltage drops on bonds which neighbor a given superconducting cluster, of linear size L. Again, this distribution is multifractal. Since there is no voltage drop inside screened "fjords", the fractal dimension of the perimeter bonds which have non-zero voltage drops coincides (in 2 dimensions) with D_e.

4.2 Multifractals

The monotonic decrease in $\tilde{\psi}(k)$ indicates that as k increases, Γ_k contains smaller subsets of the bonds in the network (going from the whole backbone at $k=0$ to the singly connected bonds as $k \to \infty$). Although the subsets of bonds which determine $\tilde{\psi}(k)$ in the two extreme limits ($k=0, \infty$) are easily identified, it is less easy to identify those corresponding to intermediate values. To some extent, this has been attempted using the notion of multifractals.[29] In this approach, one first defines a set of "probabilities", $p_i = r_i i_i^2 / R(L)$, which denote the fraction of the total power which applies to the i'th resistor. One then considers moments of the p_i's,

$$\Gamma_k = \sum_i p_i^k \sim \sum_i (i_i^2/R(L))^k \sim \Gamma_k/R^k \sim L^{\tilde{\psi}(k)-k\tilde{\psi}(1)} \quad . \qquad\qquad (4.5)$$

This can be rewritten as

$$\Gamma_k = \sum_p n(p) \, p^k \quad , \qquad\qquad (4.6)$$

where n(p) is the number of bonds which have $p_i=p$, and one can use steepest descent to find the value of p which dominates this sum, p_k^*. Both p_k^* and $n(p_k^*)$ are then expected to scale as powers of L,

$$p_k^* \sim L^{-\alpha(k)} \, , \, n(p_k^*) \sim L^{f(k)} \, . \tag{4.7}$$

Since $\partial \ell n(p)/\partial \ell np = -k$ at $p=p_k^*$, we have $\partial f/\partial \alpha = \underline{k}$. Using $\tilde{\psi}(k)-k\tilde{\psi}(1)=f(k)-k\alpha(k)$, we thus find

$$\alpha(k) = \tilde{\psi}(1) - \frac{d}{dk} \tilde{\psi}(k) \, ,$$

$$f(k) = \tilde{\psi}(k) - k \frac{d}{dk} \tilde{\psi}(k) \, . \tag{4.8}$$

Since n(p) counts bonds, f(k) may be interpreted as the fractal dimensionality of the subset of bonds which dominates Γ_k. Indeed, we recover $f(\infty) = \tilde{\psi}(\infty) = 1/\nu$ and $f(0) = \tilde{\psi}(0) = D_B$.

Before concluding this section it is worth mentioning that the multifractal description may be problematic for k<0. The negative moments are dominated by the small currents, and are thus very sensitive to rare configurations which may have specific bonds with very small currents. The smallest currents usually occur on rare ladder-like parts of the cluster. In some cases, these currents decrease exponentially with L, and not as the power law (4.7). Thus, the multifractal description may break down.[23]

Finally, two warnings. On simple symmetric hierarchical models,[21,22] the distribution of the currents is log-binomial. One is therefore tempted to approximate it as log-normal. If this was true then moments like $\sum_i (\log i_i)^k$ would have a constant gap exponent, and

scale as the k'th power of $\sum_i \ell n \, i_i$. This is not the case, since the distribution deviates strongly from log-normal, especially for the small currents.

Another temptation is to use the behavior of the positive moments, $\Gamma_k(k=0,1,2, \ldots)$, to reconstruct the whole current distribution, via an inverse Laplace transform.[30] This inversion is possible only if the Γ_k's are known exactly. As shown in Ref. 23, there exist two different approximants for $\tilde{\psi}(k)$, which almost coincide for k≥0, but deviate strongly for k<0. One of these, which we find more reliable, diverges to infinity at a finite negative k (reflecting the problem with small currents, mentioned above). No numerical estimates of $\tilde{\psi}(k)$ for k>0 will ever yield a reliable distribution of the currents.

5. NON-LINEAR RESISTORS

5.1 Non-Linear Resistors and Geometry

Ohm's law, V=RI, may be generalized to non-linear resistors, if each resistor obeys the voltage-current relation[31-33]

$$V = r|I|^\alpha \text{sign } I \, . \tag{5.1}$$

The form (5.1) is then preserved for the voltage between the two terminals, and therefore, at p_c,

$$R_\alpha(L) \sim L^{\tilde{\zeta}(\alpha)} \quad . \tag{5.2}$$

It turns out that the new infinite set of exponents, $\tilde{\zeta}(\alpha)$, contains many of the specific exponents discussed above. In particular, one has[33]

$$\tilde{\zeta}(\infty) = 1/\nu,$$

$$\tilde{\zeta}(1) = \tilde{\zeta}_R,$$

$$\tilde{\zeta}(0^+) = \tilde{\zeta}_{min},$$

$$\tilde{\zeta}(0^-) = \tilde{\zeta}_{max},$$

$$\tilde{\zeta}(-1) = D_B. \tag{5.3}$$

For $\alpha < 0$, the generalized non-linear Kirchhoff equations have more than one solution, corresponding to locally stable minima of the generalized power, $\sum_i r_i |I_i|^{\alpha+1}/(\alpha+1)$. In the limit $\alpha \to -\infty$, the lowest minimum yields

$$\lim \tilde{\zeta}(\alpha)/|\alpha| = z, \tag{5.4}$$

where $N_{max} \sim L^z$ denotes the largest number of bonds which are cut by a single self avoiding surface which breaks the cluster into two domains, each connected to one terminal.[33] Such surfaces may be relevant to energy barriers for Ising spin model dynamics.

It should be noted that one can also discuss moments of the non-linear currents, and generalize Eqs. (4.3)-(4.4) into $\sum |i_i|^{(\alpha+1)k} \sim L^{\tilde{\psi}(\alpha,k)}$. Various theorems[33] relate the infinite sets $\tilde{\zeta}(\alpha)$ and $\tilde{\psi}(k)$ to $\tilde{\psi}(\alpha,k)$. Although the functions $\tilde{\zeta}(\alpha)$ and $\tilde{\psi}[(\alpha+1)/2]$ have the same values at $\alpha=0,1,\infty$, they are quite distinct for other values of α (particularly $\alpha=0^\pm$, $\alpha<0$). They thus represent two distinct infinite sets of exponents, which characterize the physics on the percolating cluster.

5.2 Crossover from Linear to Non-Linear Resistance

Usually, non-linearity arises only at large currents, and Eq. (5.1) is replaced by[34]

$$V = r I + a I^\alpha . \tag{5.5}$$

It is now no longer true that the voltage between the terminals will also have the exact form (5.5). However, the total power has the form

$$P = \frac{1}{2} \sum_i r |I_i|^2 + \frac{1}{\alpha+1} \sum_i a |I_i|^{\alpha+1}. \tag{5.6}$$

At very small a, one can show[35] that $(\partial P/\partial a)_{a=0} = \sum_i |I_i^0|^{\alpha+1}/(\alpha+1)$, where I_i^0 are the solutions of the linear problem. Thus,

$$P = \frac{1}{2} R(L) I^2 + [a/(\alpha+1)] \Gamma_{(\alpha+1)/2} |I|^{\alpha+1} , \qquad (5.7)$$

with $\Gamma_{(\alpha+1)/2} \sim L^{\tilde{\psi}[(\alpha+1)/2]}$. The linear resistance will thus show deviations from the current independent value $R(L)$ for currents of order

$$I_x(L) \sim L^{\left[\tilde{\zeta}_R - \tilde{\psi}[(\alpha+1)/2]\right]/(\alpha-1)} . \qquad (5.8)$$

Since $\tilde{\psi}(k)$ is monotonically decreasing and convex, the exponent in (5.8) is positive and monotonically decreasing with α. Thus, even if individual resistors become non-linear at low currents, this crossover current increases with the size of the system.

We thus see, that the multifractal moment $\Gamma_{(\alpha+1)/2}$ turns out to be very useful in discussing crossover from one behavior to another. Similar effects are expected in other crossover situations.

5.3 Alternative Nonlinear Effects

Gefen et al.[34] observed a non-linear growth of the current in a metal-insulator mixture, which could not be explained by the mechanism described above. As an alternative, they proposed a dynamic breakdown mechanism, in which nonconducting bonds become conducting at sufficiently high voltages. To study this model, one needs to know the voltage drops over nonconducting bonds. Fixing the voltage between the two teminals, V_{ext}, these voltage drops will have a distribution of values between zero and V_{ext}, which is most probably also multifractal. Assuming that the maximal voltage drop scales as $\Delta V_{max} \sim V_{ext} L^{-x}$, one finds that the first breakdown occured, one must solve Kirchhoff's equations again, to find the new voltages. The new overall resistance decreases, and therefore the new current will increase beyond its "linear" value.

6. CONCLUSION

Multifractals appear whenever one places a measure on a fractal structure. Here we used the distribution of currents as a measure. One expects similar phenomena for other distributions, e.g. for the self avoiding walks described in Sec. 3.2 or for the voltages on the non-conducting bonds, discussed in Sec. 5.3. In other lectures at this meeting, multifractals concern the distribution of growth probabilities on the perimeter of growing aggregates, which is also a distribution of potential drops (in the context of the dielectric breakdown).

Not all infinite sets of different exponents fall into the framework of multifractals. The correction terms, Eq. (2.2), and the non-linear resistors, Eq. (5.2), represent examples of infinite sets which have other origins. Can all these sets be related to each other? Is there a simple basis, with a few parameters, that contains all of them? These questions remain for future studies.

ACKNOWLEDGEMENTS

Much of the work described here has been the result of collaborations with R. Blumenfeld, T. Grossman, A. B. Harris and Y. Meir. This research was supported in parts by grants from the U.S.-Israel Binational Science Foundation (BSF) and the Israel Academy of Sciences.

REFERENCES

1. Gefen, Y., Mandelbrot, B. B. and Aharony, A., Phys. Rev. Lett. 45, 855 (1980).
2. Mandelbrot, B. B., The Fractal Geometry of Nature, (Freeman, New York, 1983).
3. Family, F. and Landau, E. P., editors, Kinetics of Aggregation and Gelation, (North-Holland, Amsterdam, 1984).
4. Stanley, H. E. and Ostrowsky, N., editors, On Growth and Form: Fractal and Non-Fractal Aspects, (M. Nijhoff-Kleuver, Boston, 1986).
5. Pietronero, L. and Tosatti, E., editors, Fractals in Physics, (North Holland, Amsterdam, 1986).
6. Kadanoff, L. P., "Fractals: Where's the Physics", Phys. Today, Feb. 1986, p. 6.
7. Aharony, A., "Fractals in Physics", Europhys. News 17, 41 (1986).
8. Stauffer, D., Introduction to Percolation Theory, (Taylor and Francis, London, 1985).
9. Aharony, A., "Percolation", in Directions in Condensed Matter Physics, edited by Grinstein, G. and Mazenko, G. (World Scientific, Singapore, 1986).
10. Flory, P. J., J. Am. Chem. Soc. 63, 3083, 3091, 3096, (1941); Stauffer, D., Coniglio, A. and Adam, M., Adv. Polymer Sci., 44, 103 (1982).
11. Alexandrowicz, Z., Phys. Lett. 80A, 284 (1980).
12. Havlin, S. and Nossal, N., J. Phys. A17, L427 (1984).
13. Wilkinson, D. and Willemsen, J. F., J. Phys. A16, 3365 (1983).
14. Kapitulnik, A., Aharony, A., Deutscher, G. and Stauffer, D., J. Phys. A16, L269 (1983).
15. Aharony, A., Gefen, Y., Kapitulnik, A. and Murat, M., Phys. Rev. B31, 4721 (1985).
16. Grossman, T. and Aharony, A., J. Phys. A19, L745 (1986).
17. Voss, R. F., J. Phys. A17, L373 (1984).
18. Azbel, M. Y. and Entin-Wohlman, O., Phys. Rev. B32, 562 (1985).
19. Coniglio, A., Jan, N., Majid, I. and Stanley, H. E., "Exact Conformation of a Polymer Chain at the θ-Point", BU preprint (1984).
20. Grossman, T. and Aharony, A., to be published.
21. de Arcangelis, L., Redner, S. and Coniglio, A., Phys. Rev. B31, 4725 (1985).
22. de Arcangelis, L., Redner, S. and Coniglio, A., Phys. Rev. B34, 4656 (1986).
23. Blumenfeld, R., Meir, Y., Aharony, A. and Harris, A. B., "Resistance Fluctuations in Randomly Diluted Networks", Phys. Rev. B35, 3514 (1987).
24. Coniglio, A., Phys. Rev. Lett. 46, 250 (1981).
25. Stanley, H. E., J. Phys. A10, L211 (1977).
26. Herrmann, H. J. and Stanley, H. E., Phys. Rev. Lett. 53, 1121 (1984).
27. Rammal, R., Tannous, C. and Tremblay, A. M. S., Phys. Rev. A31, 2662 (1985).
28. Wright, D. C., Kantor, Y. and Bergman, D. J., Phys. Rev. B33, 396 (1986).
29. e.g. Halsey, T. C., Jensen, M.H., Kadanoff, L. P., Procaccia, I. and Shraiman, B. I., Phys. Rev. A33, 1141 (1986).
30. Fourcade, B., Breton, P. and Tremblay, A.-M. S., "On the Infinite Set of Exponents of Fractal Objects", Sherbrooke Preprint (1986).
31. Kenkel, S. W. and Straley, J. P., Phys. Rev. Lett. 49, 767 (1982).
32. Blumenfeld, R. and Aharony, A., J. Phys. A18, L443 (1985).
33. Blumenfeld, R., Meir, Y., Harris, A. B. and Aharony, A., J. Phys. A19, L791 (1986).
34. Gefen, Y., Shih, W. H., Laibowitz, R. B. and Viggiano, J. M., Phys. Rev. Lett. 57, 3097 (1986).
35. Aharony, A., "Crossover from Linear to Non-Linear Resistance Near Percolation", Phys. Rev. Lett. (in press).

MULTIFRACTALS: FORMALISM AND EXPERIMENTS

Mogens H. Jensen

Nordita

2100 Copenhagen Denmark

Abstract

We review briefly the formalism for studying multifractal scaling properties. The scaling structure is conveniently described by means of an f-α spectrum. For the onset of chaos via quasiperiodicity and period doubling we obtain universal spectra. These spectra are compared with spectra obtained from a forced Rayleigh-Benard experiment and very good agreement is found between theory and experiment. Finally, we show that the experimental spectra can be inverted and give information about the underlying dynamical process.

Introduction and formalism

In the last few years it has become clear that most fractals in nature are multifractals. These are structures interwoven of infinitely many fractals. The nature of such structure was first discussed by Mandelbrot[1] and has recently been intensively studied by Parisi and co-workers,[2] Halsey et al,[3,4] and many others.[5-13]

To characterize a multifractal one dimension is not sufficient. Instead one needs a continuum of dimensions, the Renyi[14] dimensions D_q, which were introduced in the context of dynamical systems by Hentschel and Procaccia.[15] A similar characterization is obtained by the spectrum scaling indices, the $f(\alpha)$ spectrum, introduced by Halsey et al.[4] The D_q-function and the $f(\alpha)$ spectrum are related by a Legendre transformation. In the following we shall briefly outline this formalism, apply it on the attractors obtained at the onset of chaos and compare the theoretical spectra with spectra obtained from experiments on a forced Rayleigh-Benard system.[6,16]

First of all we need a probability density, or a measure, on a fractal set (i.e. a fractal measure). We partition the set into N pieces of sizes l_i and denote the probality of the i'th piece as p_i. Then a partition sum is constructed as[4]

$$\Gamma(q,\tau) = \sum_{i=1}^{N} \frac{p_i^q}{l_i^\tau} \quad .$$

(1)

As $l = \max_i l_i \to 0$, three possibilities may occur. If $\tau > \tau(q)$ then

$\Gamma(q,\tau)$ diverges to infinity; if $\tau<\tau(q)$ then $\Gamma(q,\tau)$ converges to zero. At $\tau=\tau(q)$ the partition sum will tend to a constant defining the function $\tau(q)$. $\tau(q)$ is related to the generalized dimensions as $\tau(q) = (q-1)D_q$.

Next we assume that the density in the i'th piece scale like a power law as the scale tends towards zero[3]

$$p_i \sim l^{\alpha_i} \quad . \tag{2}$$

defining the scaling index α_1. The same power law might be found at the j'th piece $p_j \sim l^{\alpha_j}$. Actually, we find this exponent on a subset of the whole set of a dimension $f(\alpha_1)$.[2,3,4] So this set with dimension $f(\alpha_1)$ is one the of the interwoven subfractals. Similarly, we can find another value of the scaling exponent, α_2 say, on a subset of dimension $f(\alpha_2)$ and so on. In general α assumes values over an interval and a continuous $f(\alpha)$ spectrum is defined on this interval. We now transform the partition sum Eq.(1) into an integral in α. As a typical length scale we use $l = \max_i l_i$. The number of times α assumes a value in the interval $[\alpha', \alpha'+d\alpha']$ is then[4]

$$d\alpha'\rho(\alpha')l^{-f(\alpha')} \quad . \tag{3}$$

Inserting this and the scaling ansatz Eq.(2) into Eq.(1) we obtain

$$\Gamma(q,\tau) = l^{-\tau}\int d\alpha'\rho(\alpha')l^{q\alpha'-f(\alpha')} \quad . \tag{4}$$

In the limit $l\to 0$, the minimal value of the exponent to l under the integral will dominate so we perform a saddle-point approximation

$$\frac{d}{d\alpha'}[q\alpha' - f(\alpha')]|_{\alpha'=\alpha(q)} = 0 \quad . \tag{5}$$

This leads to the following Legendre transformation which we use to obtain the $f(\alpha)$ spectrum:[4]

$$\frac{d}{dq}\tau(q) = \alpha \quad , \quad \tau(q) = \alpha q - f \tag{6}$$

$$\frac{df}{d\alpha} = q \quad , \quad \frac{d^2 f}{d\alpha^2} < 0 \quad .$$

We observe that the $f(\alpha)$ spectrum is convex and that the slope in each point is q. As $q\to\infty$ the largest p_i (i.e. the most concentrated part of the multifractal) dominate the partition sum Eq.(1). This corresponds to the place where the $f(\alpha)$ curve vanishes with infinite slope which is at its left-most part for the minimum α value. As $q\to-\infty$ the smallest p_i dominate (i.e. the least concentrated part) and at the corresponding α-value the right-most part of the $f(\alpha)$ curve vanishes with (negative) infinite slope.

The theoretical spectra

Let us consider the simplest multi-fractal we can imagine, namely a two-scale Cantor set as shown in Fig.1.

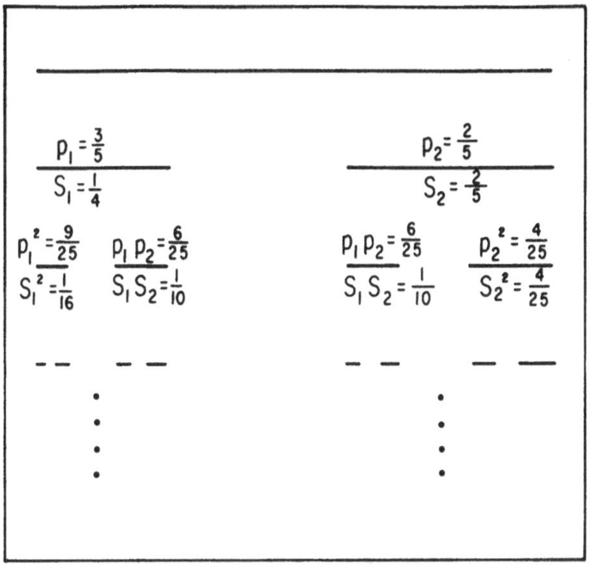

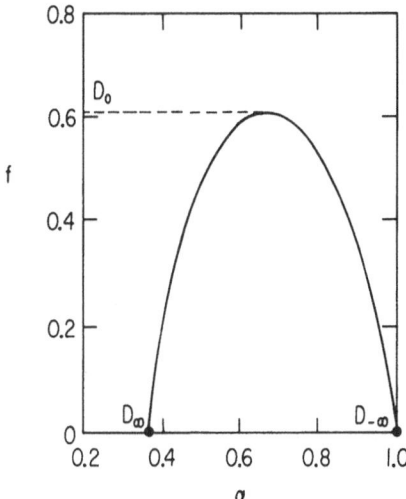

Fig.1 The construction of a two-scale Cantor set.

Fig.2 The $f(\alpha)$ spectrum for the two-scale Cantor set in
Fig.1.

On the second level the partition function for this set is very simple

$$\Gamma = (\frac{p_1^q}{l_1^\tau} + \frac{p_2^q}{l_2^\tau}) \qquad (7)$$

and will on higher levels follow a binomial expansion.[4] We calculate $\tau(q)$ and obtain the $f-\alpha$ spectrum shown in Fig.2. The spectrum has the expected form, the maximun point is the dimension D_0, etc. So the formalism 'works' on this simple example and let us therefore immediately go to a more realistic example, namely the period dou- bling attractor obtained at the accumulation point.[17] This attractor exhibits large variations in density and possesses infinite number of scaling numbers as encoded in Feigenbaums scaling-function, σ.[18] In this case we simply construct the partition function by setting $p_i = 2^{-n}$, where n is the doubling index and take as the l_i the dis- tances from the i'th point to its nearest neighbor. The correspond- ing spectrum is shown as the curve in Fig.3.

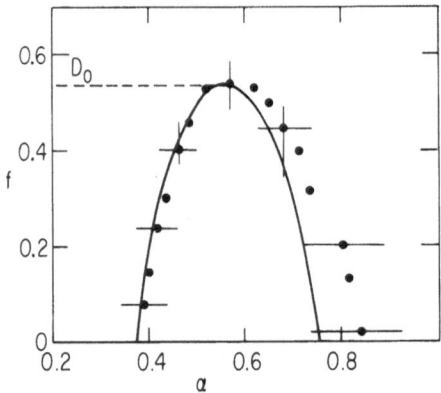

Fig.3 The $f(\alpha)$ spectrum for period-doubling (Ref.4). The dots represent the experimental measurement by Gla- zier et al (Ref.16).

This spectrum is universal and can actually be calculated to any accuracy using an exact renormalization group calculation.[19,20] Since the spectrum is universal, the experimentalists should measure it. The dots in Fig.3 are the experimental measurements by Glazier et al.[16]

Another well-known transition to chaos is the transition via quasiperiodicity.[21] Such a transition is quite easy to establish. We consider a system with a specific internal freqeuncy ω_{int}. The system is then perturbed by an external perburbation with a specific frequency ω_{ext} (an ac-current, an oscillation, etc.). When the ration $\frac{\omega_{int}}{\omega_{ext}}$ is close to a rational number, the oscillations resonate and lock to each other. These resonances form what is know as Arnold tongues in the phase digram.[22,23] Alternatively one can try and tune the frequencies such that $\frac{\omega_{int}}{\omega_{ext}}$ is an irrational number and one can drive the system into a chaotic state keeping the system at a specific irrational ratio. We shall consider the number $w* = \frac{\sqrt{5}-1}{2}$, the golden mean.

Theoretically, such transitions can be studied by maps of the circle onto itself[21]

$$\theta_{n+1} = \theta_n + \Omega - \frac{K}{2\pi}\sin(2\pi\theta_n) \quad . \tag{8}$$

The ratio between the frequencies is the winding number

$$W = \lim_{n \to \infty}\frac{\theta_n - \theta_0}{n} \quad . \tag{9}$$

The Arnold tongues will appear in the (Ω, K)-plane. We tune the parameters in Eq.(8) to golden mean winding number and consider the critical point where chaos sets in ($K=1$). The attractor found from the map Eq.(8) at this critical point is shown in Fig.4.

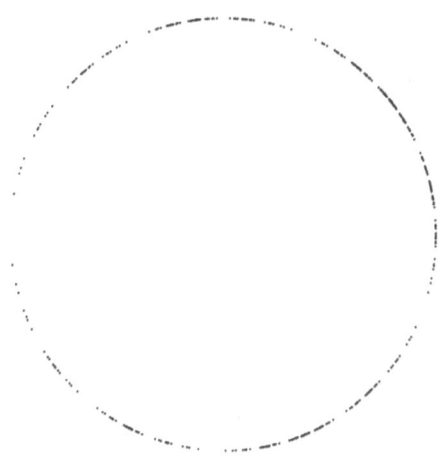

Fig.4 The orbit with golden mean winding number of the circle map Eq.(8).

Clearly there is a large variation in the density of points along the circle. How can we quantify this variation in density ? Our formalism answers this type of question. We calculate the partition function; in this case we also set all p_i's equal to $p_n = \frac{1}{F_n}$, where F_n is a Fibonacci number and as the l_i we choose the closest neighbor distances of the first F_n iterates along the circle. The result of this calculation is shown as the curve in Fig.5.

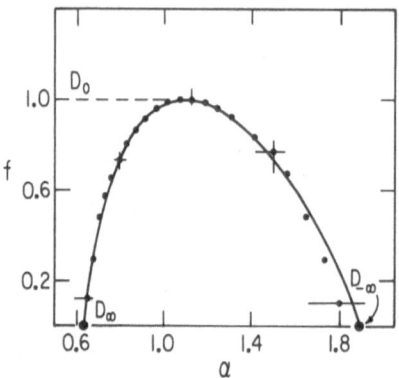

Fig.5 The $f(\alpha)$ spectrum for the golden mean attractor shown in Fig.4. The dots correspond to the experimental measurements by Jensen et al (Ref.6).

This curve is universal for transitions to chaos via quasiperiodicity with golden mean rotation. So it should be compared with experiments.

Rayleigh–Benard experiments

The theoretical predictions outlined here have been compared to data from the beautiful Rayleigh-Benard experiment developed by Libchaber and co-workers.[24,25] The experiment has been described in details by Stavans et al.[25] A small box filled with mercury is heated from below. At a critical temperature difference between top and bottom of the cell, convection with two rolls sets in. A magnetic field is applied along the rolls. At a higher temperature difference an oscillatory instability, travelling along the rolls, sets in. This instability exhibits the internal frequency ω_{int}. The external oscillation is an ac (pulsed) current perpendicular to the box with a specific frequency ω_{ext}. The two frequencies couple. The Arnold tongues of the experiment is shown in Ref.25. The experiment is tuned to the golden mean ratio between the frequencies and is pushed towards the chaotic regime. Just when chaos sets in (as detected in the Fourier spectrum) the temperature on the bottom of the cell is measured in units of the ac-frequency and the attractor shown in Fig.6 is constructed.

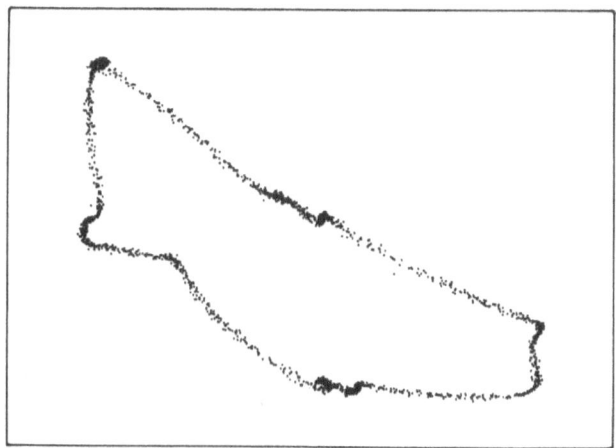

Fig.6 The experimental 'golden mean' attractor from the forced Rayleigh-Benard experiment (Ref.6). Plotted is the temperature at the bottom of the cell measured in units of the ac-period, T_{n+1} versus T_n.

Again we observe a variation in density along the attractor. How should we quantify this variation ? The answer is to calculate the $f-\alpha$ spectrum. The dots in Fig.5 represent the $f-\alpha$ calculation of this attractor. The error bars are largest on the rightmost part (i.e. for negative q) because of low statistics. Within error the two curves are identical. Thus we can claim that the circle map model and the Rayleigh-Benard experiment belong to the same universality class. To the naked eye the theoretical attractor Fig.4 and the experimental attractor Fig.6 are quite different. However the $f-\alpha$ analysis confirm that they from a scaling point of view behave identically. This universality was recently confirmed in a completely different experiment, a driven photoconductor, by Gwinn and Westervelt.[26] They were able to collect many more data point than in the Rayleigh-Benard experiment and obtained astonishing agreement between theory and experiment.

The forced Rayleigh-Benard experiment was also used to test the universality predictions for the $f-\alpha$ spectrum for period-doubling. Within an Arnold tongue both the circle map model and the experiments exhibit period-doublings on their way to chaos. These cascades take place above the critical line for the onset of chaos via quasiperiodicity.

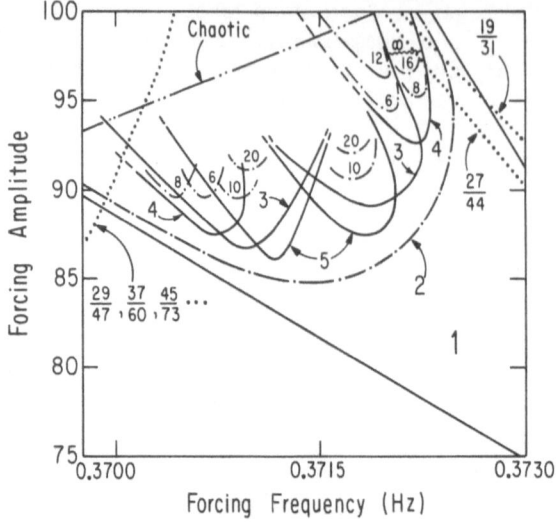

Fig.7 The experimentally found 8/13 tongue (Ref.16). We used the attractor at the accumulation point for the rightmost cascade (marked ∞) to calculate the dots in Fig.3.

Fig.7 shows the 8/13-tongue found in the experiment.[16] Clearly the behavior is quite complicated (see Ref.16 for details). Period doubling cascades are observed both at the left and right sides of the tongue. Between them there are several periodic windows which also period double. We analyzed the attractor at the accumulation point for the right-most cascade. The dots in Fig.3 are the results of this analysis. Again we see that within the error bars the theoretical and the experimental curves are identical. The error bars for negative q are here larger than for the quasiperiodicity experiment. This is caused by a more unstable signal since the experiment here is performed at quite high external amplitude. Still, it is quite remarkable that this complicated fluid dynamical system behaves like the simple circle map.

Inversion of f-α spectra

One can ask the following interesting question: Is it possible, given an f-α spectrum, to uncover the dynamical process which led to this spectrum. Phrased in other words: Is it possible to retrieve dynamical information from 'static' information. This is particularly relevant to experimental situations: if we have some data, but do not know the precise underlying mechanism, is it possible to uncover this mechanism ? The answer to the question is partly yes.

The key to answering this question is provided by the work of Feigenbaum[7] on the thermodynamic formalism.[27,28] We restrict to situations where the probability p_i is identical for each box and follows a power law in an index n of the refinement of the set, $p_i \sim a^{-n}$ (i.e. for period doubling $a=2$, for a golden mean trajectory $a=\frac{\sqrt{5}+1}{2}$). Next each box is labelled by a string $(\varepsilon_n, \ldots, \varepsilon_0)$. Here we restrict to situations where ε_i are binary. The partition function on the n'th level is[29]

$$\Gamma_n = a^{-qn} \sum_{(\varepsilon_n, \ldots, \varepsilon_0)} |\, l(\varepsilon_n, \ldots, \varepsilon_0)\,|^{-\tau} \quad . \tag{10}$$

For large n, $\Gamma_n = \Gamma_{n-1}$ and we obtain the following equation

$$\sum_{\varepsilon_{n+1}, \ldots, \varepsilon_0} |\, l(\varepsilon_{n+1}, \ldots, \varepsilon_0)\,|^{-\tau} = a^q \sum_{\varepsilon_n, \ldots, \varepsilon_0} |\, l(\varepsilon_n, \ldots, \varepsilon_0)\,|^{-\tau} \quad . \tag{11}$$

After inserting the Feigenbaum scaling function[18]

$$\sigma(\varepsilon_{n+1}, \ldots, \varepsilon_0) = \frac{l(\varepsilon_{n+1}, \ldots, \varepsilon_0)}{l(\varepsilon_n, \ldots, \varepsilon_0)} \quad , \tag{12}$$

we obtain

$$\sum_{\substack{\varepsilon_{n+1}, \ldots, \varepsilon_0 \\ \varepsilon_n', \ldots, \varepsilon_1'}} \delta_{\varepsilon_n, \varepsilon_n'} \cdots \delta_{\varepsilon_1, \varepsilon_1'} \sigma^{-\tau}(\varepsilon_{n+1}, \ldots, \varepsilon_0) \,|\, l(\varepsilon_n', \ldots, \varepsilon_1', \varepsilon_0)\,|^{-\tau} = \tag{13}$$

$$a^q \sum_{\varepsilon_n, \ldots, \varepsilon_0} |\, l(\varepsilon_n, \ldots, \varepsilon_0)\,|^{-\tau} \quad ,$$

where summations over dummy variables with appropriate delta functions have been added.[29] Now

$$T_{(\varepsilon_{n+1}, \ldots, \varepsilon_1)(\varepsilon_n', \ldots, \varepsilon_1', \varepsilon_0)} = \sigma^{-\tau}(\varepsilon_{n+1}, \ldots, \varepsilon_0) \delta_{\varepsilon_n, \varepsilon_n'} \cdots \delta_{\varepsilon_1, \varepsilon_1'} \quad , \tag{14}$$

can be interpreted as a transfer matrix of an Ising model whose largest eigenvalue is a^q.[7]

To lowest approximation the transfer matrix is given by

$$T^{(1)} = \begin{bmatrix} \sigma_{00}^{-\tau} & \sigma_{01}^{-\tau} \\ \sigma_{10}^{-\tau} & \sigma_{11}^{-\tau} \end{bmatrix} \quad , \tag{15}$$

with the eigenvalue equation

$$\lambda^2 - (\sigma_{00}^{-\tau} + \sigma_{11}^{-\tau})\lambda + (\sigma_{00}\sigma_{11})^{-\tau} - (\sigma_{01}\sigma_{10})^{-\tau} = 0 \tag{16}$$

$$\lambda_{max} = a^q \quad . \tag{17}$$

Our strategy is now the following. Let us say that we are given an f-α spectrum found from an experiment. We transform this spectrum to a $\tau(q)$ curve. In Eq.(16) there appear three unknown values of the scaling function σ_{00}, σ_{11}, $\sigma_{01}\sigma_{10}$. Also a is an unknown. Thus inserting four coupled values $(\tau_1, q_1), (\tau_2, q_2), (\tau_3, q_3), (\tau_4, q_4)$ enable us to calculate the unknowns.

We applied this strategy[29] on experimental data from the forced Rayleigh-Benard system, both from the experiment on quasiperiodicity[6] (Fig.5) and on period doubling[16] (Fig.3). Table 1 shows our results. We find $a\sim1.618$.... to very good accuracy for the golden mean experiment and $a\sim2.0$ for the period doubling case, thus confirming that an inversion is possible. We also find $\sigma_{11}=0$ in the golden mean case, as is required by the theory (see Ref.29 for details). Finally we find in both cases that σ_{00} and $\sigma_{01}\sigma_{10}$ are related to the universal scaling numbers and tell us about the nature of the underlying dynamical system (quadratic maximum, cubic inflection point etc.). So the experimental $f-\alpha$ spectra contain information about the underlying mechanisms. One can ask if it is possible to extract even further information. This is probably hard[7] since a larger transfer matrix will result in certain degeneracies. However the most essential dynamical information is retrieveable.

Table 1. Typical results of the inversion of the data. Other values of $\tau(q)$ give similar results (Ref.29).

Golden mean 2x2 matrix				Golden mean 4x4 matrix			Period Doubling 2x2 matrix			
Values of q	a	σ_{00}	σ_{11}	Values of q	σ_{000}	$\sigma_{000}\sigma_{010}\sigma_{101}$	Values of q	a	σ_{00}	σ_{11}
0.3,0.6,0.9	1.618	.467	-1.1e-11	0.3,0.6	.456	.261	1.3,1.6,1.6	2.003	.441	.194
0.3,0.9,1.2	1.619	.469	3.8e-82	0.9,1.2	.468	.227	1.3,2.8,4.9	1.999	.379	.176
0.3,0.9,1.5	1.619	.469	3.7e-72	1.2,1.5	.450	.252	1.3,1.9,5.2	2.002	.405	.174
0.6,0.9,1.2	1.619	.471	1.3e-73	1.5,2.1	.441	.269	1.6,2.2,4.9	2.000	.409	.180
0.9,2.1,2.4	1.607	.344	1.1e-54	-.3,-.6	.425	.255	1.9,2.2,4.9	1.998	.400	.179
0.3,-0.3,-0.6	1.652	.457	5.8e-3	-.6,-1.8	.419	.259	1.9,2.5,4.3	1.992	.391	.186
Theory:	1.618	.467	0	Theory:	.467	.281	Theory:	2.000	.399	.159

Acknowledgements

The reviewed work has been performed at the Materials Research Laboratory at the University of Chicago. I thank D. Bensimon, M.J. Feigenbaum, J.A. Glazier, T.C. Halsey, L.P. Kadanoff, A. Libchaber, I. Procaccia, B.I. Shraiman and J. Stavans for the most pleasant collaborations.

References

1   B.B. Mandelbrot, J.Fluid.Mech. , 62:331 (1974).

2   U. Frisch and G. Parisi, "Varanna School LXXXXVIII", M. Ghil, R. Benzi, and G. Parisi, eds., North-Holland, New York (1985), p.84; R. Benzi, G. Paladin, G. Parisi, and A. Vulpiani, J.Phys.A , 17:352 (1984).

3   T.C. Halsey, P. Meakin, and I. Procaccia, Phys.Rev.Lett. , 56:854 (1986).

4   T.C. Halsey, M.H. Jensen, L.P. Kadanoff, I. Procaccia, and B.I. Shraiman, Phys.Rev.A , 33:1141 (1986).

5   L. de Arcangelis, S. Redner, and A. Coniglio, Phys.Rev.B , 31:4725 (1985).

6   M.H. Jensen, L.P. Kadanoff, A. Libchaber, I. Procaccia, and J. Stavans, Phys.Rev.Lett. , 55:2798 (1985).

7   M.J. Feigenbaum, J.Stat.Phys. (to be published).

8   M.E. Cates and T.A. Witten, Phys.Rev.Lett. , 56:2497 (1986).

9   R. Blumenfeld, Y. Meir, A.B. Harris, and A. Aharony, J.Phys.A , 19:L791 (1986).

10   T. Bohr and D. Rand, Physica , 25D:387 (1987).

11   L. Pietronero and A.P. Siebesma, Phys.Rev.Lett. , 57:1098 (1986).

12   R. Badii and A. Politi, Phys.Rev.Lett. , 52:1661 (1984).

13   C. Amitrano, A. Coniglio, and F. di Liberto, Phys.Rev.Lett. , 57:1016 (1986).

14   A. Renyi, "Probability Theory", North-Holland, Amsterdam (1970).

15   H.G.E. Hentschel and I. Procaccia, Physica , 8D:435 (1983); see also P. Grassberger, Phys.Lett. , 107A:101 (1985).

16   J.A. Glazier, M.H. Jensen, A. Libchaber, and J. Stavans, Phys.Rev.A , 34:1621 (1986).

17   M.J. Feigenbaum, J.Stat.Phys. ,19:25 (1978); J.Stat.Phys. , 21:669 (1979).

18   M.J. Feigenbaum, Comm.Math.Phys. , 77:65 (1980).

19   D. Bensimon, M.H. Jensen, and L.P. Kadanoff, Phys.Rev.A , 33:3622 (1986).

20   E. Aurell, Phys.Rev.A , 34:5135 (1986).

21   Scott J. Shenker, Physica , 5D:405 (1982); M.J. Feigenbaum, L.P. Kadanoff, and Scott J. Shenker, Physica , 5D:370 (1982); S. Ostlund, D. Rand, J.P. Sethna, and E.D. Siggia, Physica , 8D:303 (1983).

22   V.I. Arnold, Am.Math.Soc.Trans. , Ser. 2, 46:213 (1965).

23   M.H. Jensen, P. Bak, and T. Bohr, Phys.Rev.Lett. , 50:1637 (1983); Phys.Rev.A , 30:1960 (1984).

24   A. Libchaber, C. Laroche, and S. Fauve, Physica , 7D:73 (1983); J.Phys (Paris) Lett. , 43:L211 (1982).

25   J. Stavans, F. Heslot, and A. Libchaber, Phys.Rev.Lett. , 55:596 (1985).

26   E.G. Gwinn and R.M. Westervelt, Phys.Rev.Lett.

27   D. Ruelle, "Statistical Mechanics, Thermodynamic Formalism", Addison-Wesley, Reading (1978).

28   E.B. Vul, Ya.G. Sinai, and K.M. Khanin, Usp.Mat.Nauk ,39:3 (1984) [Russ.Mat.Surveys ,39:1 (1984)].

29   M.J. Feigenbaum, M.H. Jensen, and I. Procaccia, Phys.Rev.Lett. , 56:1503 (1986).

ON MULTIFRACTALS: THERMODYNAMICS AND CRITICAL EXPONENTS

Preben Alstrøm

Nordisk Institut for Teoretisk Atomfysik, Blegdamsvej 17
DK-2100 Copenhagen Ø, Denmark

ABSTRACT

Using the two-scale Cantor set as a starting point the thermodynamical formalism is discussed, including entropy, free energy, and dimensions. The connection to the $f - \alpha$ spectrum is illustrated, in particular when the probability measure is a Gibb's ensemble. Regarding transitions to chaos as phase transitions critical exponents is defined. The way these enter the thermodynamics is outlined. As a main result the general interpretation of the cross-over scale as the scale where the system change from a fractal to a nonfractal behavior is shown to be ambiguous. As a representative example the global quasi-periodic transition (including noise) described by the sine map is treated.

I. INTRODUCTION

By a multifractal we generally mean a fractal with not one but a whole distribution of characteristic exponents. As a simple example consider first a Cantor fractal obtained by cutting out from every interval a middle interval starting with the unit interval , see fig. 1. In a function terminology take the tent map

$$f_a(x) = \begin{cases} ax & \text{if } x < 1/2 \\ a(1-x) & \text{if } x \geq 1/2 \end{cases} \qquad (a > 2), \qquad (1)$$

and consider the intervals which survive within the unit interval after n iterations, that is

$$A_n = \{x \mid f_a^n(x) \in [0;1]\}. \qquad (2)$$

The Cantor fractal is the intersection of all A_n. The characteristic exponents are $1/n$ times the (natural) logarithm of the derivatives of f_a^n, or if $\Delta_n(i)$ denote the lengths of the intervals at level n,

$$E(i) = -n^{-1}\ln\Delta_n(i) = \ln a. \qquad (3)$$

As we see there is only one characteristic exponent, namely $E = \ln a$.

Now we make the tent map just a little skew, in general write

$$f_{a,b}(x) = \begin{cases} ax & \text{if } x < b/(a+b) \\ b(1-x) & \text{if } x \geq b/(a+b) \end{cases} \qquad (ab > a+b > 0). \quad (4)$$

Then at level n we get for each m between 0 and n (both included) $N(m) = \begin{bmatrix} n \\ m \end{bmatrix}$ intervals of length

$$\Delta_n(m) = a^{-m} b^{m-n}. \qquad (5)$$

In the limit of large n we can use Stirling's formula which for $m = xn$ gives the characteristic exponents to be

$$E(x) = -n^{-1}\ln\Delta(x) = x \ln a + (1-x)\ln b. \qquad (6)$$

We observe that every number between $\ln a$ and $\ln b$ is there, only in the case $a = b$ the distribution collapses.

As the lengths of the intervals $\Delta(x)$ change exponentially with n so do the numbers $N(x)$, the exponents

$$S(x) = -x \ln x - (1-x)\ln(1-x) \qquad (7)$$

express the average gain of information, also called entropy[1] (loosely speaking, information is a measure of knowledge, if I have a hidden ball in one of N boxes you must ask me $\ln N /\ln 2$ questions to find it). In fig. 2 the entropy function $S(E)$ is shown for the two-scale Cantor set evolving from iterations of the skew tent map. Generally, Cantor sets which like this evolve from iterations of maps with different branches are called cookie-cutter sets.

For every characteristic exponent (or energy) E we have (in the limit of large n)

$$N(E) \propto e^{nS(E)} \propto (e^{-nE})^{-S(E)/E} \propto \Delta(E)^{-D(E)}, \qquad (8)$$

with

$$D(E) = S(E)/E. \qquad (9)$$

Hence, the two-scale multifractal is a union of Cantor fractals $A(E)$ with Hausdorff dimension $D(E)$. The Hausdorff dimension D_H of the entire multifractal is the largest of those,

$$D_H = \max_E \{D(E)\}, \qquad (10)$$

or in other words D_H is the value of β for which the β-measure

$$Z_\beta = \sum_E N(E)\Delta(E)^\beta \qquad (11)$$

is finite.

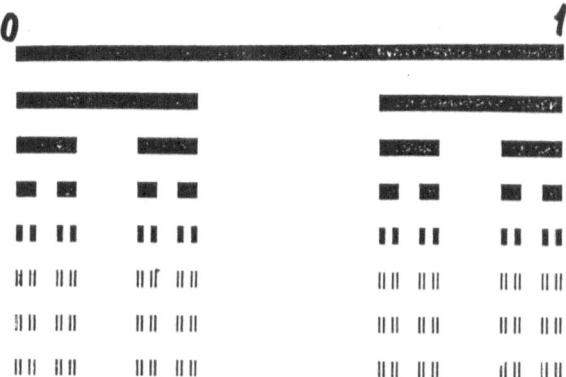

FIG. 1. Construction of a Cantor fractal.

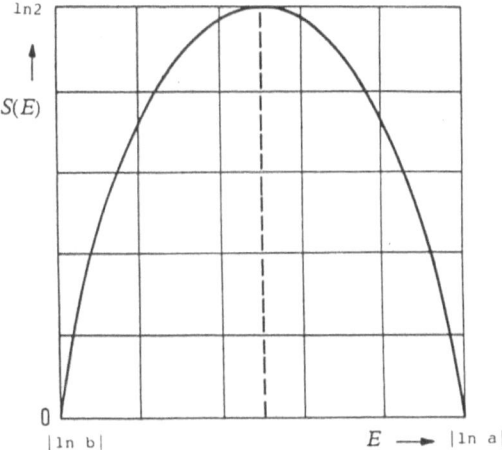

FIG. 2. Entropy function $S(E)$ for the two-scale Cantor set.

II. THERMODYNAMICS

Since this strongly looks like thermodynamics, Z_β is denoted as the partition function (β being the invers temperature). Free energy or pressure might be defined from the logarithm of Z_β, by laziness we just write (in the limit of large n)[2]

$$F(\beta) = -n^{-1}\ln Z_\beta. \tag{12}$$

For the two-scale Cantor set

$$F(\beta) = -\ln(a^{-\beta} + b^{-\beta}). \tag{13}$$

A probability measure $\{p(i)\}$ can be introduced, and if this is only energy dependent, i.e. $\{p(i)\} = \{p(E)\}$ (or simply $p(i) = p(j)$ if $\Delta(i) = \Delta(j)$), a simpel connection $a(E)$ to the $f - a$ formalism[3] exists:

$$f(a(E)) = D(E), \tag{14}$$

and

$$\Gamma(q,\tau) = \sum_i \frac{p(i)^q}{\Delta(i)^\tau} = \sum_E N(E) \frac{p(E)^q}{\Delta(E)^\tau}. \tag{15}$$

In particular, for the Gibb's ensemble[1] $\{p_\beta(E)\}$, where

$$p_\beta(E) = Z_\beta^{-1} \Delta(E)^\beta, \tag{16}$$

we find

$$a(E) = \beta - F(\beta)/E. \tag{17}$$

Note that for $\beta = D_H$, $F(\beta) = 0$, and the $f - a$ curve collapses. Furthermore, if E_β denotes the energy which dominates Z_β we obtain for a,

$$a(E_\beta) = D(E_\beta) = f(a), \tag{18}$$

which means (since $q = f'(a) = 1$) that $D(E_\beta)$ is the information dimension related to the β-measure (in general, if no measure is mentioned explicitely, the natural measure ($\beta = 1$) is understood).

The free energy $F(\beta)$ expresses how fast our β-measure increase ($\beta < D_H$) or decrease ($\beta > D_H$). Introducing

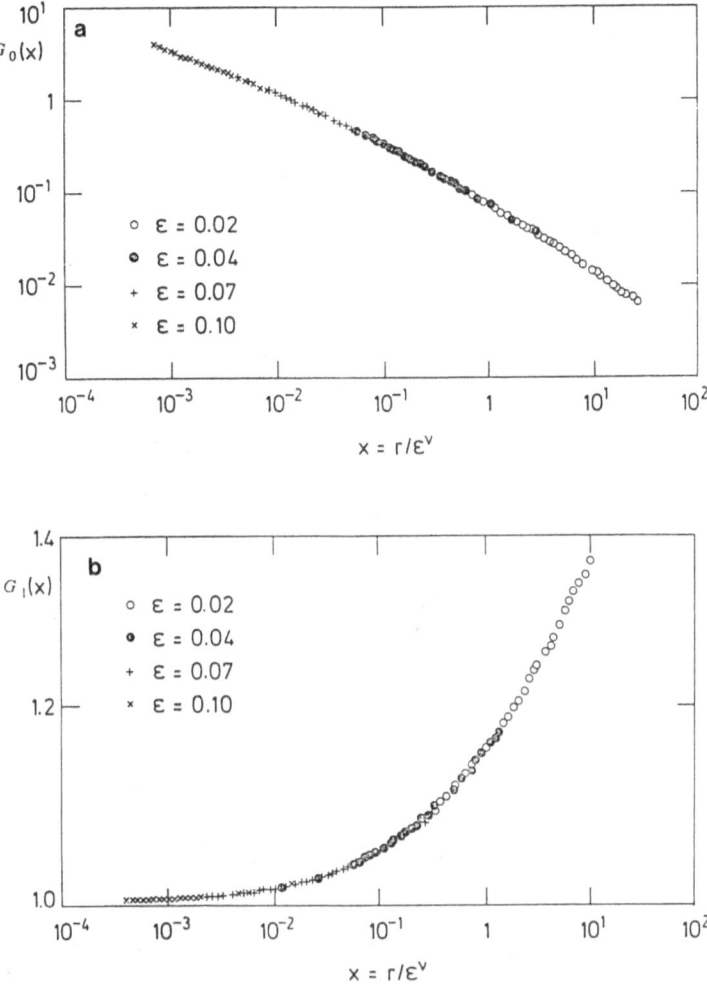

FIG. 3. Scaling functions describing the cross-over for the sin map. (a) For the counting (G_0), $v(0)=2.63$. (b) For the natural measure (G_1), $v(1)=2.35$.

$$\varepsilon = e^{-n} \qquad (19)$$

as the distance from criticality yields

$$Z_\beta(\varepsilon) \propto \varepsilon^{F(\beta)}, \qquad (20)$$

which interpretes the free energy as a critical exponent.

III. CROSS-OVER EXPONENTS

The ideas of the entropy function as well as of the $f - a$ spectrum have great applicability measuring experimental strange and/or fractal sets.[1,3] However, sometimes it is advantageous (or even necessary) to consider the complementary set of intervals, e.g. in phase-locking systems D_{II} and decay exponents at criticality have been found from the size of stability intervals[4] (which constitute the complementary set).

Led by the terminology from phase transitions we study the cross-over: define the β'th moment M_β,

$$M_\beta(r,\varepsilon) = \begin{cases} \sum_{\Delta > r} \Delta(\varepsilon)^\beta & \text{if } \beta \leq D_H \\ \sum_{\text{all } \Delta} \Delta(0)^\beta - \sum_{\Delta > r} \Delta(\varepsilon)^\beta & \text{if } \beta > D_H \end{cases} \cdot (21)$$

(In fact, the value of β for which M_β diverges is the Besicovitch-Taylor dimension, but in general it coincides with D_H). $\{\Delta(\varepsilon)\}$ is now the complementary set of intervals, and ε is the distance to criticality. For cookie-cutter sets like that evolving from the skew tent map $\varepsilon = e^{-n}$, though $\varepsilon = \delta^{-n}$, $\delta = 4.669..$, is preferred for the period-doubling route, and $\varepsilon = \lambda^{-n}$, $\lambda = 1.618..$, is preferred for the quasi-periodic transition at the golden mean.

For a global quasi-periodic transition described by a circle map behavior ε is the minimal slope: as a main example consider the sine map

$$g_\varepsilon(x) = x + \Omega - (1 - \varepsilon)\sin(2\pi x)/2\pi. \qquad (22)$$

$\{\Delta(\varepsilon)\}$ is here the set of stability intervals for the rotation number

$$R_\varepsilon(\Omega) = \lim_{m \to \infty} g_\varepsilon^m(0)/m \qquad (23)$$

$(\{\Omega \mid R_\varepsilon(\Omega) = P/Q\} = \{\Omega \mid g_\varepsilon^Q(x_0) = P \text{ for some } \Omega, x_0\})$.

When $r=0$ the moments follow a power law behavior,

$$M_\beta(\varepsilon) \propto \varepsilon^{F(\beta)} \qquad (24)$$

where $F(D_H) = 0$ (for $\beta = D_H$ logarithmic divergences will in general occur). For $\varepsilon = 0$ the critical behavior is

$$M_\beta(r) \propto r^{\beta - D_H} \qquad (25)$$

($\beta = D_H$ is again a special case).

For the two scale Cantor set the cross-over between the behavior (24) and (25) becomes sharp in the limit of small ε (large n). The cross-over scale is β-dependent, $r_\beta = (1 - a^{-1} - b^{-1}) \varepsilon^{\vee(\beta)}$, where

$$F(\beta) = \vee(\beta)(\beta - D_H). \qquad (26)$$

However, invoking the formalism from critical phenomena,[5] generally smooth scaling functions G_β describing the cross-over is expected,

$$M_\beta(r, \varepsilon) = \varepsilon^{F(\beta)} G_\beta(r/\varepsilon^{\vee(\beta)}). \qquad (27)$$

The cross-over or "correlation length" exponent $\vee(\beta)$ is related to $F(\beta)$ by eq. (26). For the global quasi-periodic transition described by the sine map the scaling functions G_0 for the counting and G_1 for the natural measure are shown in figs. 3a,b.

The convexity of $F(\beta)$ implies that $\vee(\beta)$ decreases with β. For the sine map transition the limiting behavior of $\vee(\beta)$ as $\beta \to \infty$ is determined at the golden mean,

$$\vee(\infty) = y/z = 2.053.., \qquad (28)$$

where $y = 2.164..$ and $z = 1.054..$ are Shenker's exponents.[6]

In the context of phase transitions a conjugate field can be introduced in different ways, e.g. by multiplying by a factor of $e^{-\Delta/h}$ (as in percolation theory[7]) or by adding noise[8] (h being the noise amplitude). Critical exponents α, γ, and δ can be defined, and the scaling as $h \to 0$ examined. Analogous to the $\varepsilon \to 0$ limit cross-over scaling is described by a behavior

$$M_\beta(r, h) \propto h^{\mu(\beta)(\beta - D_H)} H_\beta(r/h^{\mu(\beta)}). \qquad (29)$$

Figs. 4a,b show the two scaling functions H_0 and H_1 obtained when at each iteration a randomly distributed variable in the interval $[-h/2; h/2]$ is added to the sine map. The change in μ is found to be very small, within computational uncertainty $\mu(0)$ and $\mu(1)$ are both found to be about 0.85.

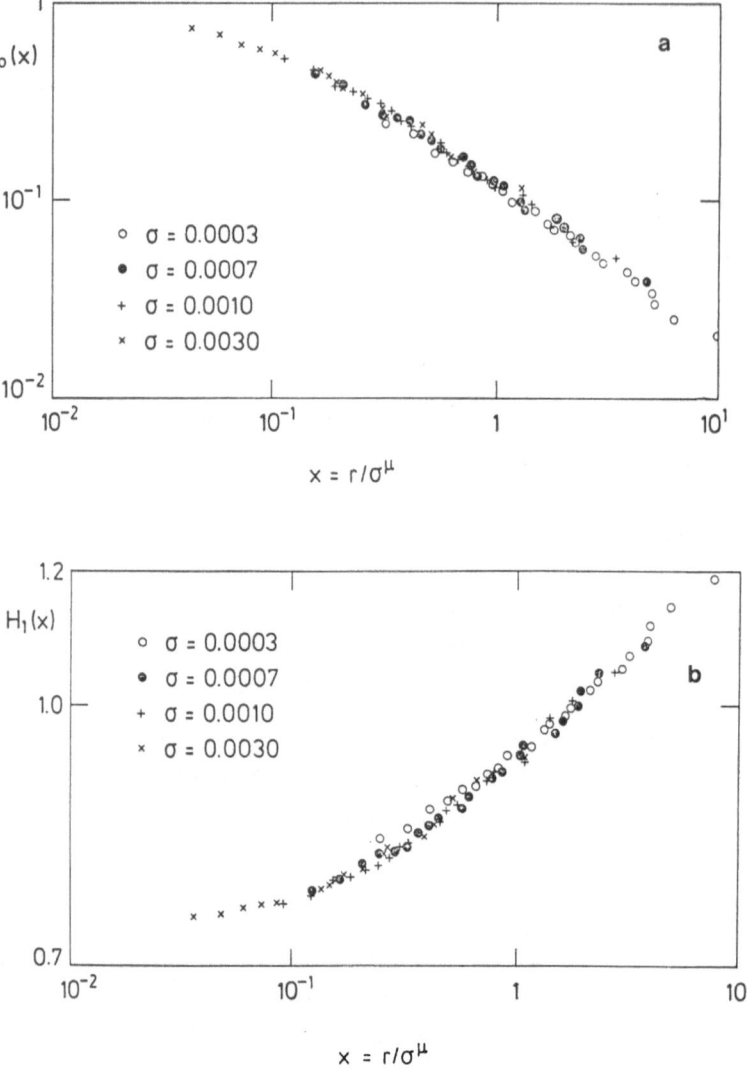

FIG. 4. Scaling functions describing the cross-over for the sin map when noise is added. (a) For the counting (H_0), $\mu(0)=0.86$. (b) For the natural measure (H_1), $\mu(1)=0.83$.

IV. CONCLUSIONS

In conclusion, the formalisms inherited from thermodynamics and phase transitions are of great advantage measuring multifractals. The $f - \alpha$ formalism constitutes a significant link between the Hentschel-Procaccia description[9] (where all covering intervals $\Delta(i)$ are taken equal) and the thermodynamics defined from Gibb's ensemble.

Moreover, the critical exponents as those occurring at transitions to chaos can be interpreted thermodynamically. In particular, this implyes that cross-over exponents in general depend on the moment on which they are defined.

As main examples the skew tent map and the sine map have been treated. However, the presented formalism employed for embedding dimension $d=1$ can be extended to higher dimensions, especially it finds use in percolation.

REFERENCES

1   D. Ruelle, *Thermodynamic Formalism,* in *Encyclopedia of Mathematics and its Applications, Vol. 5* edited by G.-C. Rota (Addison-Wesley, Massachusetts, 1978). T. Bohr and D. Rand, preprint (to appear in Physica D).

2   M. J. Feigenbaum, preprint.

3   T. C. Halsey, M. H. Jensen, L. P. Kadanoff, I. Procaccia, and B. I. Shraiman, Phys. Rev. A **33**, 1141 (1986).

4   P. Alstrøm and M. T. Levinsen, Phys. Rev. B **31**, 2753 (1985); *ibid.* **32**, 1503 (1985). P. Alstrøm, B. Christiansen, P. Hyldgaard, M. T. Levinsen, and D. R. Rasmussen, Phys. Rev. A **34**, 2220 (1986).

5   C. Domb and M. S. Green, *Phase Transitions and Critical Phenomena* (Academic Press, London, New York, 1972-76), see e.g. L. P. Kadanoff, *Scaling, Universality, and Operator Algebras*, in *Vol.* 5a , p. 1.

6   S. J. Shenker, Physica **5D** , 405 (1982).

7   D. Stauffer, Phys. Rep. **54**, 1 (1979); *Introduction to Percolation Theory* (Taylor and Francis, London, 1985).

8   B. A. Huberman and J. Rudnick, Phys. Rev. Lett. **45**, 154 (1980). B. Shraiman, C. E. Wayne, and P. C. Martin, Phys. Rev. Lett. **46**, 935 (1981). M. J. Feigenbaum and B. Hasslacher, Phys. Rev. Lett. **49**, 605 (1982). J. P. Crutchfield, D. Farmer, and B. A. Huberman, Phys. Rep. **92**, 45 (1982).

9   H. G. E. Hentschel and I. Procaccia, Physica **8D** , 435 (1983).

ON THE CHARACTERIZATION OF CHAOTIC

SYSTEMS USING MULTIFRACTALS

Vicent J. Martínez

NORDITA. Blegdamsvej, 17
DK-2100 Copenhagen. Denmark

I. INTRODUCTION

A dynamical system is a set of differential equations describing the time evolution of a physical system which initial
conditions are knowed. The solution of the system is a trajectory in phase space (each point of this trajectory represents
a state of motion of the system). Systems can be conservative,
when they preserve the volume in phase space, and also dissipa-
tive, if that volume is shrinked continuously. However it is
possible to observe chaotic or stochastic motions in both cases,
but it is not clear how to characterize globally the stochasti-
city or how to measure the randomness. In this paper we apply
the multifractal formalism and in particular the $f(\alpha)$ spectrum
of singularities[1] in order to characterize both kinds of systems.

II. THE $f(\alpha)$ SPECTRUM

Let consider N points representing the trajectory. We need
to give a probability measure to that point set, labeling the
points by x_i, the probability p_i will be define in terms of the
number of neighbours within a distance ε. It is physically reasonable[2] that these probabilities may have a power law expres-
sion $p_i = \varepsilon^\alpha$, where α are the scaling indices of the set. Studying
how these values are distributed provides a useful tool to study
the set itself. The number of times α takes a value in $(\alpha, \alpha+d\alpha)$
can be expressed as $n(\alpha)d\alpha = \varepsilon^{-f(\alpha)}d\alpha$, where $f(\alpha)$ is the fractal
dimension of the points in the set which have the same value for
the scaling index α. Defining the partition function,

$$\Gamma(q,\tau,\varepsilon) = \varepsilon^{-\tau} \sum_i (p_i(\varepsilon))^q \qquad (1)$$

The behaviour of Γ in the limit when $\varepsilon \to 0$ and $N \to \infty$ will be,

$$\Gamma(q,\tau) = \begin{cases} \infty & \text{if } \tau > \tau(q) \\ \text{constant} & \text{if } \tau = \tau(q) \\ 0 & \text{if } \tau < \tau(q) \end{cases} \qquad (2)$$

by this way we define the function $\tau(q)$, which is related with
the generalized dimensions[3] D_q by $D_q = \tau(q)/(q-1)$. D_0 is the

Hausdorff dimension, D_1 is the information dimension and D_2 is the correlation dimension.

The variables (q,τ) are related with (α,f) through the Legendre transformation[1], the curve $f(\alpha)$ has a unique maximum, and the value of the spectrum in that point is the Hausdorff or fractal dimension of the set. We can see that α takes values in a particular range $(\alpha_{min}, \alpha_{max})$ where

$$\alpha_{min} = \lim_{q \to \infty} D_q \qquad\qquad \alpha_{max} = \lim_{q \to -\infty} D_q$$

α_{min} is the fractal dimension of the region of the set where the concentration of points is maximum, and α_{max} corresponds to the most rarefied parts.

In the process of calculating $\tau(q)$ there are practical problems because we cannot take the limit when $\varepsilon \to 0$ and $N \to \infty$. The curve $f(\alpha)$ will be shifted in the vertical axis depending on the value we adopt for the constant which appears in (2). Assuming that $\Gamma(q,\tau) = C(\varepsilon,N)$ with C independent of q, we can calculate an effective $\tau(q)$. The choice of this constant can be done fixing the maximum of the $f(\alpha)$ curve, i.e. the Hausdorff dimension D_0 as initial condition. Since $\tau(0) = -D_0$ we see from (1)

$$\Gamma(0,\tau(0)) = \varepsilon^{-\tau(0)} \sum_i (p_i(\varepsilon))^0 = C(\varepsilon,N)$$

But we need of course, to calculate previously the Hausdorff dimension. We will apply an statistically corrected box-counting method, which provides good accuracy even for small data sets. The Hausdorff dimension can be calculated by partitioning the region where the points lie into cells of size λ and counting for each value of λ the number of occuped cells $N(\lambda)$. To avoid the dependence of the particular position of the grid we take $\overline{N}(\lambda) = \langle N(\lambda) \rangle$ where the average is over different "realizations", understanding each of them as the result of the counting process when shifting the grid from one position to another with a random vector $\vec{a}$ with components $|a_i| < \lambda$. D_0 will be the slope of the regression line of the plot $\log(\overline{N}(\lambda))$ vs. $\log(1/\lambda)$.

III. CHARACTERIZING ATTRACTORS OF DISSIPATIVE SYSTEMS

In this section, we apply the previous formalism to one well known attractor, the Hénon map[1], $x_{n+1} = y_n + 1 - ax_n^2$, $y_{n+1} = bx_n$; with a=1.4, b=0.3 and the initial conditions $x_0 = 1$, $y_0 = 0$, (fig. 1). In this kind of systems we can define the probabilities in two different ways:

a.- Spatial probabilities. Generating N points of the attractor, to compute probabilities for each point is an exercise of counting, $p_i = n_i(\varepsilon)/N$ where $n_i(\varepsilon)$ is the number of neighbours at a distance less than ε from the point x_i.

b.- Temporal probabilities. It has been proposed[2] another way to assign a probability measure to the strange set. If the points are generated by time series, we can count the number of time-steps since one point come until another arrives within a distance ε of the first one, say m_i; then the probability at that point should be $p_i \propto 1/m_i$

In both cases, in order to have a true probability measure we must correct the values of p_i, with a normalization condition,

$$\sum_i p_i(\varepsilon) = 1$$

Having these probabilities from two different approaches, we can obtain the $f(\alpha)$ spectrum corresponding to each one, as

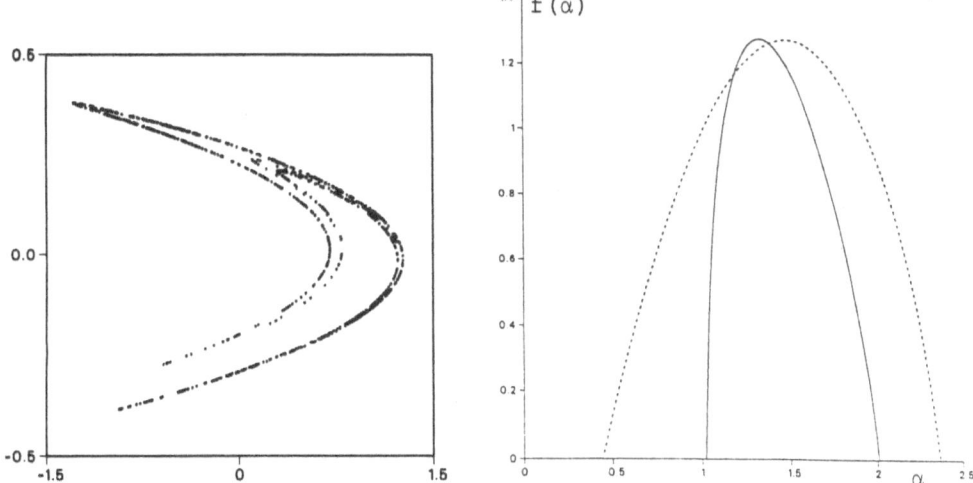

Fig. 1. The Hénon map. On the left we have plotted 1000
points of this map. On the right we can see the
$f(\alpha)$ spectrum of this attractor, calculated with
spatial probabilities: solid line, and temporal
probabilities: dashed line.

it has been explained in the previous section. (For ergodic
systems, we expect that the curve $f(\alpha)$ of the same attractor
will be the same in both cases). We have calculated the $f(\alpha)$
spectrum for the Hénon map, taking only a small number of points
(1000), and using both approaches. As we can see in fig. 1, the
$f(\alpha)$ curve obtained from temporal probabilities, dashed line,
is far to agree with the real spectrum obtained using the spatial
approach. The reason is that to get accurate probability measures
for a system using its dynamics, we need a large ammount of information, i.e. much more points. The mean values of the spectrum
are $D_0 = 1.278$, $\alpha_{max} = 2.006$ and $\alpha_{min} = 1.021$ with errors around 10%.

IV. CHARACTERIZING POINCARE MAPS OF CONSERVATIVE SYSTEMS

The Hénon-Heiles hamiltonian[5] allows us to study how possible is to apply the multifractal formalism in conservative systems; the hamiltonian is

$$H(p_1,p_2,q_1,q_2) = \frac{1}{2}(p_1^2+p_2^2+q_1^2+q_2^2) + q_1^2q_2 - \frac{1}{3}q_2^3$$

Solving numerically the corresponding hamiltonian equations
we can consider the Poincaré sections, plotting in the plane
$q_1 - p_1$ the points for wiich $q_2 = 0$ and $p_2 > 0$, (i.e. we plot the
points of the trajectory when they traverse the surface of
section in a particular sense). The transition to chaos occurs
when the conserved energy is above $E=1/8$. At this point there
are periodic orbits but also some of them look like random
points. We are going to choose two particular orbits with the
same energy but different behaviour[6], they correspond to the
initial conditions,

I.- $p_1(0) = 1/3$, $p_2(0) = 0.129314$, $q_1(0) = 1/4$, $q_2(0) = 1/5$

II.- $p_1(0) = 0.1$, $p_2(0) = 0.467618$, $q_1(0) = 0.1$, $q_2(0) = 0.1$

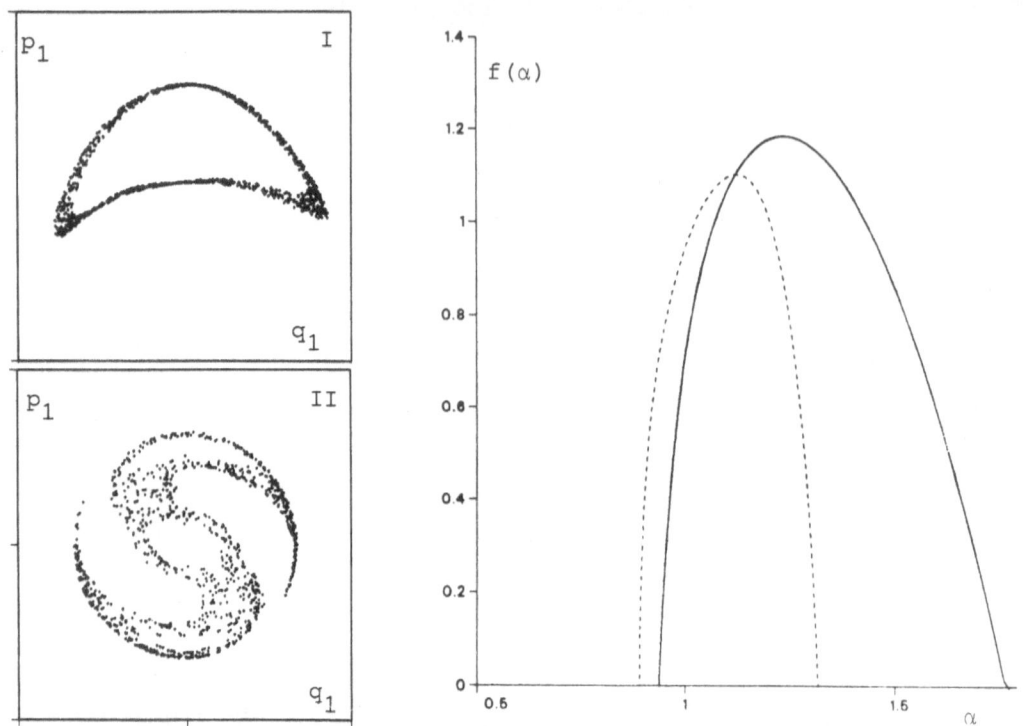

Fig. 2. On the left the Poincaré sections of the orbits
corresponding to the initial conditions I and II.
On the right the f(α) spectrum, dashed line corresponds to orbit I and solid line to orbit II.

The different behaviour of these two orbits is showed in
fig. 2. The diffusion can be observed in both of them, in the
first one looking at the extremes of the orbit, in the second
one the randomness has increased. The mean values of the $f(\alpha)$
are given in table 1. It is remarkable that $D_{-\infty}$ (fractal dimension of the rarefied parts), is much greater in the orbit II
than in I, providing a good tool to measure the diffusion.

Table 1. Mean values of the $f(\alpha)$ spectrum

	D_0	D_∞	$D_{-\infty}$
Orbit I	1.102	0.891	1.317
Orbit II	1.192	0.929	1.795

REFERENCES

1. T.C. Hasley, M.H. Jensen, L.P. Kadanoff, I. Procaccia and
   B.I. Shraiman, Phys. Rev. A33, 1141 (1986)
2. M.H. Jensen, L.P. Kadanoff, A. Libchaber, I. Procaccia and
   J. Stavans, Phys. Rev. Lett. 55, 2798 (1985)
3. H.G.E. Hentschel and I. Procaccia, Physica 8D, 435 (1983)
4. M. Hénon, Commun. Math. Phys. 50, 69 (1976)
5. M. Hénon and C. Heiles, Astron. J. 73, 964 (1969)
6. C.M. Bender and S.A. Orszag, "Advanced Mathematical Methods
   for Scientist and Engineers". McGrawll-Hill, (1978)

VISCOUS FINGERING ON PERCOLATION CLUSTERS

Unni Oxaal[1], Michael Murat[2], Finn Boger[1]
Amnon Aharony[2], Jens Feder[1] and Torstein Jøssang[1]

[1] Institute of Physics, University of Oslo
Box 1048, Blindern, 0316 Oslo 3, Norway

[2] School of Physics and Astronomy
Raymond and Beverly Sackler Faculty of Exact Sciences
Tel Aviv University, Israel

1 Introduction

The physical phenomena that occur when a low-viscosity fluid is forced into a high-viscosity one, inside a porous medium (e.g. water pushing oil in a rock), are clearly of much practical interest. These phenomena became the center of much recent scientific interest when it was realized that under some conditions flow instabilities yield *viscous fingers* (VF) [1,2,3], that are *fractal* [4,5,6]. Fractals are self-similar objects, which look the same at different magnifications [7]. Their number of sites, or mass $M(L)$ within a region of linear scale L behaves as

$$M(L) \sim L^D , \qquad (1)$$

with a usually non-integer fractal dimensionality, D.

The interest in viscous fingers grew even further when it was observed that for two immiscible fluids, with a high capillary number for the displaced one, the fractal shapes of the VF (generated experimentally in two-dimensional Hele-Shaw cells and in porous media) have strong qualitative and some quantitative similarities to those observed in a diverse range of apparently different problems, including *dielectric breakdown* [8], diffusion-limited *electrodeposition* or growth in aqueous solution [9], and computer simulations of *diffusion-limited aggregation* (DLA) [10]. All of these structures have fractal dimensionalities $D = 1.70 \pm 0.05$ in two dimensions. These striking similarities suggested the existence of *universality* among *growth patterns*. This suggestion was put forward by Paterson [11] who showed that two-fluid displacement in porous media can be described by Laplace's equation, with boundary conditions similar to those used in the DLA problem. Since DLA is a process that is simple to simulate numerically (particles are released at the boundary, and perform random walks until they hit (and stick to) the growing aggregate [10]), it was suggested that such simulations may be used to imitate two-fluid flow under various conditions [11,12].

The expectation that flow in Hele–Shaw cells (two parallel glass plates, with a small separation) exhibits the same properties as in real porous media is based on Darcy's law, relating the fluid velocity $\vec{v}$ to the pressure gradient,

$$\vec{v} = -\frac{k}{\mu}\nabla P, \qquad (2)$$

where k is the *average* permeability of the porous medium and μ is the fluid viscosity. For incompressible fluids, $\nabla \cdot \vec{v} = 0$ and therefore,

$$\nabla \cdot (k\nabla P) = 0. \tag{3}$$

If the permeability k is constant over the whole system (as it is in conventional Hele–Shaw cells [1,2,4]), Equation (3) reduces to the Laplace equation, $\nabla^2 P = 0$. If the viscosity of the 'pushing' fluid is negligible, then the pressure in it is uniform, $P = P_0$. Neglecting the surface tension, this is also the pressure in the displaced fluid, at the interface. The pressure on the free boundaries of this fluid is equal to another constant, $P = P_{ext}$. Since the same Laplace equation, with similar boundary conditions, applies to the electrostatic potential in the dielectric breakdown problem [8], and to the equilibrium probability density of random walkers in the DLA problem [10], the analogy between all the problems follows [11].

In *real porous media*, the local permeability k fluctuates randomly in space. Therefore, the arguments connecting DLA and VF must be critically reconsidered. Chen and Wilkinson [13] solved the flow Eq. (2) inside a square-lattice network of tubes with a statistical distribution of radii using a deterministic growth algorithm. Their resulting viscous fingers were similar to DLA clusters, with $D \simeq 1.72$.

Although Chen and Wilkinson [13] introduced a distribution of channel radii, they had no tubes which were *completely* blocked. In real porous media, there is usually a finite concentration, $(1 - p)$, of blocked paths (which may either be completely blocked or be so narrow that no fluid flows in them in practice). As the concentration p decreases, approaching the *percolation threshold* p_c, there appears a correlation length $\xi \sim |p - p_c|^{-\nu}$, with $\nu = 4/3$, such that the percolating network of open pores has a fractal geometry on length scales $(L < \xi)$ [14]. In many cases, physical phenomena exhibit a sharp *crossover* from a fractal behavior $(L < \xi)$ to that of a homogeneous random system $(L > \xi)$ [15]. Although the work of Chen and Wilkinson [13] demonstrated that VF have the same behavior for the $(p = 1)$ or $L > \xi$ case as is found in the uniform Hele–Shaw cells [16] it has not yielded results for the *fractal* regime, $L < \xi$. This regime was recently studied [17], using computer simulations of a stochastic growth model similar to that used for dielectric breakdown [8]. Viscous fingers on the fractal network which occurs at p_c turned out to be very different from those of the homogeneous regime. In particular, their fractal dimension decreased to about 1.3.

As discussed below, the computer simulations involve a variety of approximations and of *ad hoc* procedures. It is therefore not at all clear that they represent the true physical situation. On the other hand, real porous media contain too many uncontrolled parameters, which make it difficult to compare them with a theoretical analysis. In order to study viscous fingers on a controlled fractal medium, we performed both real experiments and computer experiments on the *same* model two dimensional dilute system, at the percolation threshold.

2   Geometry

Allthough the above discussion related to blocked bonds, or tubes, percolation phenomena have the same behaviour for site or bond dilution [15]. For our simulations and experiments we chose a specific realization of a site diluted square lattice, of size 147×147, at the site percolation threshold $p_c = 0.593$, in which the central site belongs to the spanning cluster. Flow was allowed only between occupied sites, and was possible only on the spanning cluster which connects the site at the center of the lattice (chosen to be on this cluster) with the boundaries. Fig. 1 (see colour plate in the front of this book) shows a picture of this cluster, where all the finite clusters were eliminated.

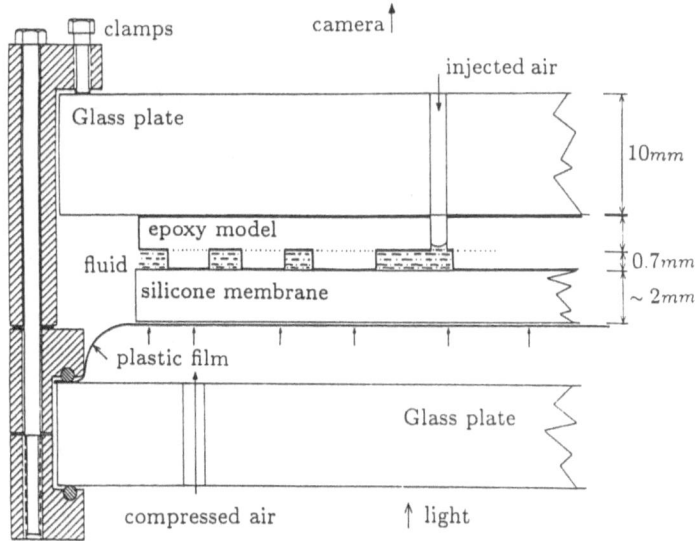

Figure 2: The experimental setup. The model consists of an epoxy cast with the pore pattern on one side that is closed by a silicone rubber sheet. This assembly is supported by glass plates. The epoxy and the silicone rubber sheets are kept in contact by pressurized air as shown. This setup is transparent, lighted from below and photographed from above.

In the basic process to be analyzed, the system is first filled with an incompressible viscous fluid (glycerol dyed with Negrosine, in our experiments), and then a non-viscous fluid (air in our case) is pushed at a fixed rate into the central site, pushing the other fluid out at the outer boundaries. Since the displaced fluid is incompressible, the viscous fingers cannot move into dangling branches of the cluster, which are connected to the rest of the percolating cluster via a single site. Therefore, the flow is limited to the *backbone* of the cluster. The backbone of our cluster is also shown in Fig. 1, in white.

In two dimensions, the percolating cluster at p_c has a fractal dimension of $D = 91/48$ [18], while the backbone has a fractal dimension of $D_{BB} \simeq 1.61$ [19]. In our 147×147 sample, their respective masses were 6261 and 3341 sites.

3   Experimental Setup

We constructed a porous model with a pore structure given by the percolation cluster in Fig. 1, using a method developed by Bonnet and Lenormand [20]:

The model consists of an epoxy resin plate where the pores and connecting channels form a recessed pattern. The epoxy resin plate was made in the following way. A plotter was used to draw the desired pattern on a drafting sheet, and a photographic transparency of the proper size was made from this original. The channels appear in black while the regions where flow will not be permitted are transparent. The film was placed onto a photosensitive nylon plate (BASF Nyloprint WA 175), and this assembly was exposed to ultraviolet radiation. This hardened the unmasked regions of the nylon plate. The masked regions remained soft and were subsequently etched with water, leaving the desired pattern of channels in the nylon plate.

The nylon model is not suitable for the experiments, mainly because the nylon

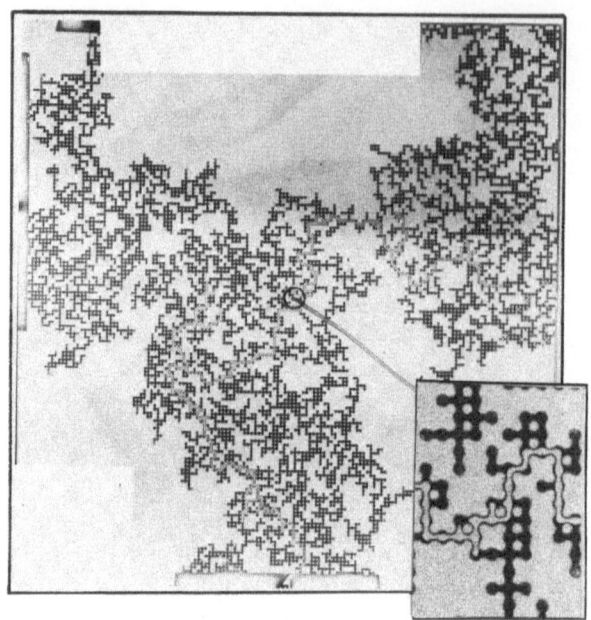

Figure 3: Photograph from a 'fast' displacement experiment at 100 ml/h ∼ 30 pores per second, just after breakthrough. Pores filled with glycerol appear black. Pores where the glycerol was displaced by air appear white. The total number of pores invaded by air is 447. As shown in the insert glycerol was trapped in corners of the pores invaded by air because of capillary effects.

absorbs fluid and swells, distorting the geometry. Therefore we made a silicone rubber mold (RTV 700 and catalyst Beta 4 from General Electric) from the nylon plate, and cast the final epoxy (Araldite XW 396 with catalyst XW 397) model using this mold.

The model is closed by a $2mm$ flexible silicone rubber membrane (Eccosil 2CN with catalyst 50 from Emerson and Cuming) held in contact with the plate by compressed air. This restricts the fluid flow to the channels. Reservoirs were made at the boundaries in order to keep the model open to the flow and the boundary pressure at ambient conditions. The dimensions of the model are 16.5 × 16.5 cm, the pore volume is $\simeq 1 \mu l$, channel depth 0.7 mm, pore diameter 1.1 mm, and bond width 0.7 mm. To ensure uniform wettability the model and membrane are coated by absorbing a monolayer of collagen, from a 2 % solution. Water and glycerol wet models treated in this way in a uniform manner.

The model is filled with glycerol colored black by water soluble Negrosine, before it is closed. The model is supported from both sides by 10 mm glass plates. The lower of the two glass plates is clamped together with a plastic film, and pressurized air is applied to the space between. In this way we force the silicone rubber membrane into contact with all the regions of the model where fluid flow is prohibited. The air pressure was kept constant and sufficiently high to keep the membrane in place during the experiment.

In the experiments the glycerol in the channels was displaced by air, injected through a 0.7 mm hole in the epoxy model. The point of injection was chosen at the cluster site at the geometrical center of the model. The air was injected at constant pump rate, using a piston type pump (Pharmacia P-500). The displacement process was photographed from above using a Nikon F3 camera controlled by an IBM PC. Lighting was provided from below. The experimental setup is shown in Fig. 2.

Initially the network of channels appears black on a transparent background. Pores invaded by air appear white, as shown in Fig. 3. As shown in the insert, some glycerol remains on the walls of the channels due to wetting effects.

4 Experimental Results

The experiments were performed at several flow rates. At low flow rates a typical interval between photographs was 30 seconds while at high flow rates it was typically 1.3 seconds. Basically, we find all the experiments with rates larger than ~ 50 ml/h to have very similar results. Fig. 3 shows a typical picture of such a '*fast*' experiment, for a flow rate of 100 ml/h, 29 seconds after the start of the experiment.

In order to analyze the results quantitatively we measured the coordinates of the invaded pores on the photographs using a Tektronix 495 digitizing tablet controlled by an IBM PC. Fig. 4 (see colour plate) shows the same data as in Fig. 3 , and Fig. 5 (see colour plate) shows the result of a 'slow' experiment, at a flow rate of 1 ml/h ~ 0.33 pore/sec. Both digitized pictures contain data from four different experimental pictures, with the masses of the growing aggregate as indicated.

The fast and slow experiments can be quantitatively distinguished by looking at plots of the mass M of the fingers (i.e. the volume of injected air), as counted on the digitized pictures, as function of time, wich is known from the timing on the camera. As can be seen from Fig. 6, there is a qualitative difference between the functions $M(t)$ for the two kinds of experiments. Fig. 6a shows $M(t)$ for the slow flow. Scince our experiments were performed using a constant pump rate when injecting the air, the linearity of Fig. 6a (with a *measured* slope equal to the pump rate), proves that the motion of the interface in the slow flow does not depend on the pressure gradient in the displaced fluid. This is not consistent with Eq.(2), or with the stochastic DLA (or dielectric breakdown) algorithm. This result led us to apply a new simulation model, as described below in Sec. 6.

At high injection rates, Fig. 6b shows a non-linear dependence of $M(t)$. We attribute this non-linearity to the increase in the pressure gradient in the viscous fluid, as the interface approaches the external boundary of the sample. This is consistent with Eq. (2) and with the stochastic model. An approximate form for the shape of $M(t)$ is obtained as follows: Scince the cluster is at p_c, there exist several independent channels from the origin to the boundary. Each of these channels is built of singly connected sites, through which air must pass, connecting intermediate blobs [21]. The flows in these channels are decoupled from each other. Approximating each of these by a quasi-one-dimensional channel, of total length L (between origin and external boundary), the location x of the intreface should obey $dx/dt \sim \nabla P \sim P_0/(L - x)$, hence $t = Ax - Bx^2$. In our experiments, we found that the radial distance from the origin to the end of the fastest tip is proportional to to the radius of gyration, $r \sim m^{1/D}$, where m is the mass of the finger. Assuming that $x \sim r$, we expect

$$t = t_0 + am^{1/D} - bm^{2/D} .\tag{4}$$

Here t_0 is the time at the start of the experiment. The full curve in Fig. 6b is a fit this expression, fixing $1/D = 0.75$. Note that the last two points are below the fitted line, probably due to boundary effects.

5 Simulations of Fast Flows

As demonstrated by our discussion of Eq.(4), we believe that the flow of the viscous fluid is governed by Eqs.(2) and (3), which could in principle be solved explicitly and deterministically [13]. In the case of the *fast* flow rate, we believe that this will give the same results as the stochastic approach, invented in Ref. [8]

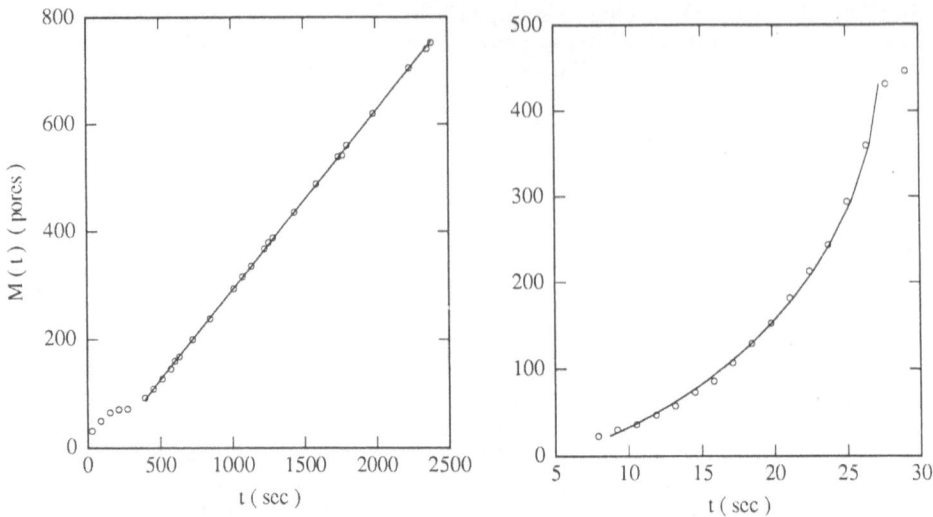

Figure 6: The mass of the fingers as function of time. a – slow flow, 1 ml/h
~ 0.33 pore/sec (full line). b – fast flow, 100 ml/hr $\sim$ 30 pores/sec.
The full line represents the fit of Eq. (4) to the observations. The
parameters in Eq. (4) are: $t_0 = 3.51$, $a = .51$ and $b = 2.9 \cdot 10^{-3}$.

and used in Ref. [17]: At every time step, we first solve a discrete version of the
Laplace equation inside the viscous fluid,

$$P_{\vec{r}} = \frac{1}{4} \sum_{\vec{\delta}} P_{\vec{r}+\vec{\delta}}, \tag{5}$$

where sites $\vec{r} + \vec{\delta}$ are the neighbors of site $\vec{r}$. If a site at $\vec{r} + \vec{\delta}$ is blocked, we set
$P_{\vec{r}+\vec{\delta}} \equiv P_{\vec{r}}$ in Eq. (5), to ensure the boundary condition $\vec{\nabla} P = 0$ on the bond from
$\vec{r}$ to $\vec{r} + \vec{\delta}$. Eq. (5) is solved with the pressure equal to 1 on the interface between
the fluids, and to 0 on the external boundaries. After convergence, a new open
perimeter site is added to the displacing fluid region (and assigned $P = 1$), using
a stochastic probability which is proportional to $|\vec{\nabla} P|$ on the bond connecting that
site to the growing aggregate [8,16]. We believe that this procedure should imitate
the growth, since in our geometry growth occurs only at relatively few tips, and
it should not matter if these grow simultaneously or at randomly alternating time
steps.

Fig. 7 (see colour plate) shows a sequence of growth steps in a typical computer
simulation, chosen at the same aggregate masses as in Fig. 4. Clearly, the results of
our simulations are qualitatively very similar to the experimental results shown in
Fig. 4. To make this more quantitative, Fig. 8 (see colour plate) shows the two sets
of pictures superimposed on each other, when the mass was 213. For these three
times (aggregate masses of 30, 86 and 213), the numbers of overlapping sites are
60 %, 86 % and 77 % of the masses.

To appreciate these amounts of overlap, we repeated the numerical simulation
3 different times, with different random seeds for the random number generator
responsible for the random stochastic choices in the growth procedure. Typical
overlaps between two different simulations at the same times as above are 67 %,
75 % and 79 %, comparable to the overlaps between the simulation and the real
experiments! Similar overlaps were also found between each of the three simulations

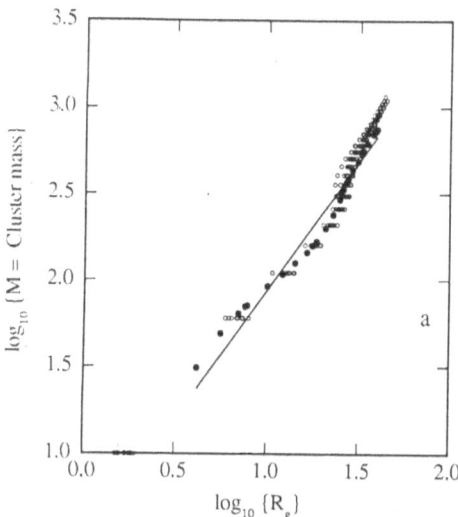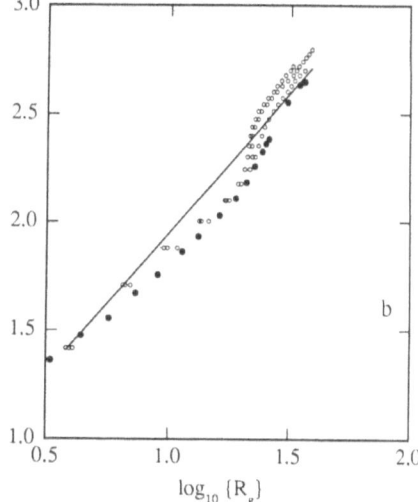

Figure 9: The mass of the fingers as function of the radius of gyration. a – slow flow, 1 ml/h. b – fast flow, 100 ml/h. The slopes of the straight lines are $D = 1.5$ and $D = 1.3$, respectively. Open circles show results of simulations, filled circles show experimental data.

and the experiment shown in Fig. 4 — the average overlaps were 66 %, 80 % and 78 % with errors of order $\pm$ 6 %.

These high values of overlap show that the geometry of the growing fingers is determined to a large extent by that of the underlying backbone. In fact, a finger which connects two sites on the backbone *must* pass through all the *singly connected sites* on the link between these two points [21]. It is now widely accepted that the number of these also grows with the distance L between the two end points, as [23] $L^{1/\nu}$. In fact, the actual overlap between different experiments is much larger than this, because the finger also usually picks the shortest path through "blobs" on the route.

The fact that percolation clusters have a finite order of ramification [24] implies that there is always a small finite number of sites which must be cut in order to isolate an arbitrarily large region of the cluster, independent of the length scale [7]. Apparently, this determines the paths of the fingers, and—through that—their detailed geometry. The fact that the geometry of the particular cluster dominates the growth is also reflected in the dependence of the fingers' mass on the radius of gyration. Fig. 9b shows points from the three simulations, as well as from the experiment. Although on the average all are consistent with the fractal dimension of 1.3 (full line), as was found in [17], all the data show systematic oscillations about the average, reflecting the particular geometry of our specific cluster. Similar lacunarity effects were discussed in Ref. [14]. Fig. 9a shows the same phenomenon (with 10 simulations) for a slow flow experiment.

6 Simulations of Slow Flows

As demonstrated by Fig. 6a, the growth of the fingers in the slow experiments does not depend on the pressure gradients in the viscous fluid. Instead, it is determined by *local* interfacial forces, rather than viscous ones. If one assumes that all the channels are identical, then one would expect equal growth probabilities at all the perimeter sites which allow flow, i.e. excluding those which surround isolated lakes of the invaded fluid. Since these allowed sites would have finite pressure gradients in the solution of Laplace's equation, $|\nabla P| \neq 0$, we can identify them by associating them with a growth probability

$$p_i \sim |\nabla P|^\eta \,, \quad \text{with } \eta \to 0, \tag{6}$$

where the parameter η was introduced in Ref. [8].

For homogeneous systems, the limit $\eta \to 0$ is usually associated [25] with the Eden growth model [26], in which *all* perimeter sites have the same growth probabilities. Indeed, when we simulated the model of Eq. (6) on a uniform lattice, we found compact aggregates, whose mass scales as the Euclidean area, $M \sim L^2$.

This equivalence, between the $\eta \to 0$ model and the Eden model, breaks down on the percolation cluster. Clearly, the Eden model could fill the whole cluster ($D \sim 1.9$), while the $\eta \to 0$ model is restricted to the backbone, and hence $D \leq 1.61$. In practice, the incompressibility of the displaced fluid creates many impenetrable regions on the backbone, and D is significantly lower.

In practice, our simulation ran as follows: We start with a seed of the invading fluid at the center. We then choose at random a perimeter bond of that seed, and add the site on the other side of the bond to the growing aggregate (invading fluid). We then identify all the 'finite' clusters of the invaded fluid, and mark their sites so that the bonds connecting them to the aggregate will never be chosen again. We continue in this manner until a boundary is reached.

Figure 10 (see colour plate) shows three stages of a simulation of this model on our cluster. Indeed, the aggregate is far from filling the backbone. In fact, averaging over several simulations on 35 different clusters of size 251×251 we estimate its fractal dimension to be ~ 1.5. Fig. 9b shows that, apart from the lacunarity effects, both our simulations and our experiments are consistent with this estimate.

As before, we find a large overlap between different runs of the simulation on the same cluster. At aggregate masses of 145, 379 and 543, typical overlaps are 78%, 76% and 80%. It is also rewarding to compare our simulations, Fig. 10, with the slow rate experiments, Fig. 5. The corresponding overlaps are 80 %, 80 % and 74 %, all $\pm$ 3 %.

Although the masses of the 'slow' fingers are somewhat higher than those of the 'fast' ones, we note that the geometry of Fig. 5 and Fig. 10 is also dominated by the underlying cluster geometry, particularly being constrained by the singly connected sites and the finite order of ramification.

7 Discussion

Our main results may be summarized as follows:

(a) The geometry of the percolation cluster, and in particular the finite order of ramification and the singly connected paths, determines the geometry of the viscous fingers on the cluster. This led to large overlaps between experiments and simulations.

(b) Fast flow experiments are dominated by viscous effects, and obey a quasi-one-dimensional Laplace equation. This is characterized by a non-linear $M(t)$ curve, Fig. 6b. The experiments are reproduced by the simulations with $\eta = 1$.

(c) Slow flow experiments are dominated by local capillary effects, and are identified by a linear $M(t)$ curve, Fig. 6a. They are reproduced by simulation with $\eta = 0$. However, unlike in homogeneous networks, these differ from the Eden model, which ignores the incompressibility of trapped fluid regions.

(d) Results on individual clusters deviate systematically from the average fractal mass relation $M \sim L^D$, reflecting the lacunarity of the specific clusters.

In the present study we looked only at the extreme case of percolation clusters at the percolation threshold. As demonstrated in Ref.[17], subtle crossover phenomena are expected as the concentration p increases above p_c. These remain for future studies.

An alternative crossover situation arises if the 'blocked' bonds are allowed to have a non-zero small permeability, $k_1 \ll 1$. As soon as k_1 becomes non-zero, new channels open between the spanning cluster and neighboring finite clusters, and the behavior again crosses over to that of the homogeneous system, as found by Chen and Wilkinson [13]. However, our results may still apply when k_1 is very small. Again, this crossover is also left for the future.

Apart from the percolation case, our results should apply to all other flows on fractal networks which have a finite order of ramification, i.e. have crucial 'hot' spots through which the flow must pass.

In conclusion, percolation clusters at p_c turn out to allow a satisfactory description of viscous fingering on ramified fractals, consistent with both experiments and simulations.

Acknowledgements

We wish to thank Geir Holm for expert Photographic work.

Work at Tel Aviv was supported by grants from the U.S.–Israel Binational Science Foundation, the Israel Academy of Sciences and Humanities and the Israel AEC.

Work in Oslo has been supported by VISTA, a research cooperation between the Norwegian Academy of Science and Letters and Den Norske Stats Oljeselskap a/s (STATOIL).

Bibliography

[1] P.G. Saffman and G.I. Taylor, Proc. Roy. Soc. London, Ser A **245**, 312 (1958).

[2] L. Paterson, J. Fluid Mech. **113**, 513 (1981).

[3] R. Lenormand and C. Zarcone, Phys. Chem. Hydro. **6**, 497 (1985).

[4] J. Nittmann, G. Daccord, and H. E. Stanley, Nature (London) **314**, 141 (1985).

[5] K. J. Måløy, J. Feder, and T. Jøssang, Phys. Rev. Lett. **55**, 2688 (1985).

[6] G. Daccord, J. Nittmann, and H. E. Stanley, Phys. Rev. Lett. **56**, 336 (1986), and refcerences therein.

[7] B. B. Mandelbrot, *The Fractal Geometry of Nature* (Freeman, San Francisco, 1982).

[8] L. Niemeyer, L. Pietronero, and H. J. Wiessmann, Phys. Rev. Lett. **52** 1033 (1984).

[9] M. Matsushita, M. Sano, Y. Hayakawa, H. Honjo, and Y. Sawada, Phys. Rev. Lett. **53**, 286 (1984).

[10] T. A. Witten and L. M. Sander, Phys. Rev. Lett. **47**, 1400 (1981); D. A. Weitz and M. Oliviera, Phys. Rev. Lett. **52**, 1433 (1983).

[11] L. Paterson, Phys Rev. Lett. **52**, 1621 (1984).

[12] See also L. P. Kadanoff, J. Stat. Phys. **39**, 267 (1985).

[13] J. Chen and D. Wilkinson, Phys. Rev. Lett. **55**, 1892 (1985).

[14] A. Kapitulnik, A. Aharony, G. Deutscher, and D. Stauffer, J. Phys. A **16**, L269 (1983).

[15] For example, A. Aharony, in *Multicritical Phenomena*, edited by R. Pynn and A. Skjeltorp (Plenum, New York, 1984), p. 309; A. Aharony, in *Directions in Condensed Matter Physics*, edited by G. Grinstein and G. Mazenko (World Scientific, Singapore, 1986), p. 1; D. Stauffer, *Introduction to Percolation Theory*(Taylor and Francis, London and Philadelphia, 1985).

[16] See also L. Nittmann and H. E. Stanley, Nature, (London) **321**, 663 (1986); J. D. Sherwood and J. Nittmann, Journal de Phys. (Paris) **47**, 15 (1986); A. J. DeGregoria, Phys. Fluids **28**, 2933 (1985).

[17] M. Murat and A. Aharony, Phys. Rev. Lett. **57**, 1875 (1986).

[18] B. Nienhuis, J. Phys. A **15**, 199 (1982).

[19] H. J. Herrmann, D. C. Hong, and H. E. Stanley, J. Phys. A **17**, L261 (1985). Recently D. Laidlaw, G. MacKay, and N. Jan [J. Stat. Phys. (to be published)] found $D_B = 1.61 \pm 0.02$.

[20] J. Bonnet and R. Lenormand, Rev. Inst. Fr. Pet. **42**, 477 (1977).

[21] H. E. Stanley, J. Phys. A**10**, L211 (1977).

[22] P. Meakin, H. E. Stanley, A. Coniglio and T. A. Witten, Phys. Rev. A **32**, 2364 (1985).

[23] A. Coniglio, Phys. Rev. Lett. **46**, 250 (1981); J. Phys. A **15**, 3829 (1982).

[24] Y. Gefen, A. Aharony, B. B. Mandelbrot ans S. Kirkpatrik, Phys. Rev. Lett. **47**, 1771 (1981); S. Kirkpatrick, AIP Conference Proceedings **58**, 79 (1974).

[25] P. Meakin, In *On Growth and Form*, edited by H. E. Stanley and N. Ostrowsky, (Martinus Nijhoff, Boston, 1986), p. 111.

[26] M. Eden, Proc. 4th Berkeley Symp. on Math. Stat. and Prob. **4**, 223 (1961).

SELF-AVOIDING WALKS BETWEEN TERMINALS

ON PERCOLATION CLUSTERS

Liv Furuberg[1], Amnon Aharony[1,2],
Jens Feder[1] and Torstein Jøssang[1]

[1] Institute of Physics, University of Oslo,
P. O. Box 1048 Blindern, 0316 Oslo 3, Norway
[2] School of Physics and Astronomy,
Tel Aviv University, Tel Aviv 69978, Israel

INTRODUCTION

Many natural structures exhibit fractal geometry, and much recent interest has been devoted to studying the physical properties of such structures [1]. In the fractal regime, many of these properties depend on the length scale L via power laws, e.g. $X(L) \sim L^x$. Often, one needs an infinite set of exponents, $\{x\}$, in order to fully characterize the structure [1]. We find that the various self-avoiding paths connecting two terminals on a percolating cluster scale with the Euclidean distance between the terminals. The different paths scale with different exponents forming a continuous spectrum.

Consider a dilute system, at the percolation threshold. A finite sample will have a finite probability to have a spanning percolating cluster, which connects two of its opposite edges. Flow between these edges can take place only via the backbone of this cluster, which has a fractal geometry [1]. In what follows, we choose two terminal points, on the opposite edges and on the spanning cluster, and study the dependence of various properties on the Euclidean distance L between them. We start by considering each connected bond on the cluster to represent a non-linear resistor : $\mid v \mid = r \mid i \mid^\alpha$. Here i is the current through the bond and v the voltage across it. With a total current I, through the terminal points, the total voltage V behaves as $V = R_\alpha(L)I^\alpha$. The resistance $R_\alpha(L)$ is expected to scale with the length scale as $R_\alpha(L) \sim L^{\tilde\zeta(\alpha)}$. [1,2]. Since the scaling relation for the resistance $R_\alpha(L)$ reduces to many relevant physical quantities for special values of α, the function $\tilde\zeta(\alpha)$ serves for characterizing the fractal structure of the cluster [1,2]. In particular, when $\alpha \to 0$ the current always goes via a self-avoiding walk (SAW). The shortest SAW is chosen when $\alpha \to 0^+$, and the longest SAW is chosen when $\alpha \to 0^-$. Thus $\tilde\zeta(0^+)$ and $\tilde\zeta(0^-)$ describe the scaling of the length of the shortest and longest SAWs between the terminals, $\mathcal{L}_{min} \sim L^{\tilde\zeta_{min}}$, $\mathcal{L}_{max} \sim L^{\tilde\zeta_{max}}$, $\tilde\zeta_{min} = \tilde\zeta(0^+)$, $\tilde\zeta_{max} = \tilde\zeta(0^-)$.

For simple symmetric fractal models (e.g. the hierarchical model proposed by de Arcangelis et al. [3]), all the SAWs have the same length, and $\tilde\zeta_{min} = \tilde\zeta_{max}$. This is not consistent with recent measurements of the asymmetry of percolation clusters [4] and with actual evaluations [2,5] of $\tilde\zeta(\alpha)$. The latter exhibit a discontinuity in $\tilde\zeta(\alpha)$ at $\alpha = 0$, indicating a different scaling of $\mathcal{L}_{min}$ and $\mathcal{L}_{max}$. Such a different scaling implies a non-trivial multifractal scaling of different SAWs on percolating clusters. The distribution of such SAWs is the subject of the present study. In

209

particular, we find a continuum of SAW exponents, $\tilde{\varsigma}_{min} \leq \tilde{\varsigma} \leq \tilde{\varsigma}_{max}$. In addition to giving more reliable values for the limiting exponents, $\tilde{\varsigma}_{min}$ and $\tilde{\varsigma}_{max}$, this continuum yields a novel type of a multifractal distribution.

The shortest SAW, $\mathcal{L}_{min}$, has been studied extensively before [6,7] , having realizations e.g. in the spreading of diseases in dilute forests or in the behaviour of the elastic backbone. Fluctuations in the spreading may imply propagation via longer SAWs.

SAWs have been used extensively to model non-branching polymers [8]. In the usual case, one fixes the number of steps of a SAW, $\mathcal{L}$, and studies the end-to-end distance, which scales as $< L >\sim \mathcal{L}^{\nu_{SAW}}$. The exponent is given very accurately by Flory's approximation, $\nu_{SAW} = 3/(d + 2)$ in d dimensions. Similar values were found for the same procedure on dilute networks [9]. Our approach is different: we count only SAWs that start from one terminal and reach the other one. Since we avoid traps, our model is probably related to infinite growth SAWs [10].

A detailed study of all the SAWs places $\tilde{\varsigma}_{min}$ and $\tilde{\varsigma}_{max}$ as limits of a whole function, which should improve their quantitative estimates. It also allows a better study of the scaling of the mean path, and of other averages over the SAWs. The distribution of SAWs may turn out to be useful in describing non-branching polymers on dilute networks.

2. ALOGRITHM

We generated finite $l \times l$ samples of site diluted square lattices, at the percolation threshold $p_c = 0.593$ ($\pm 1\%$). For samples which have spanning clusters, we chose pairs of terminals on their edges, at a distance L apart, and found all the SAWs connecting them. Since SAWs are confined to the backbone of the spanning cluster, we first eliminated the dead ends.

To find all the SAWs, we let a walker start at one terminal. The walker tries to enter the neighbour to the right. If that site belongs to the backbone, and if it has not yet been visited by this SAW, it is entered. If not, the walker tries to move upwards, if unsuccessful it tries to walk left etc. If the walker moved to a new site, it now searches for all the SAWs from that site to the exit terminal. Often, the walker is trapped and returns to the previous site to try another route. Reaching the end terminal, the length of the SAW is stored. The walker then goes one step back to test if the next way out of the last site also leads to the exit. Eventually the walker returns to the source terminal, having covered all the SAWs.

In practice the above procedure was made much more efficient by using the links-blobs stucture [11] of the backbone: All the SAWs must pass through the singly connected sites (SCS) of the backbone. It thus suffices to identify separately the SAWs in each blob, between SCSs. We identified the SCSs as the sites where biased walks [12] on the left and right perimeter of the backbone meet.

By recognizing doubly connected sites, and parts of the backbone connected to the rest at two sites only, we were able to divide the problem into even smaller, tractable parts. Also, we avoid to enter traps.

With these refinements, one can handle samples up to $L \sim 20$. We investigated 25 000 samples with $L = 9$, 40 000 with $L = 11$, 1 000 with $L = 17$ and 100 samples with $L = 22$. The corresponding average number of SAWs per sample were 60, $7 \cdot 10^2$, $5 \cdot 10^5$ and $9 \cdot 10^9$.

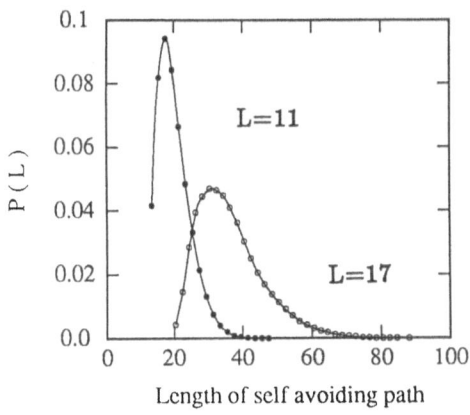

Figure 1: Probability distributions of SAWs on percolation backbones.

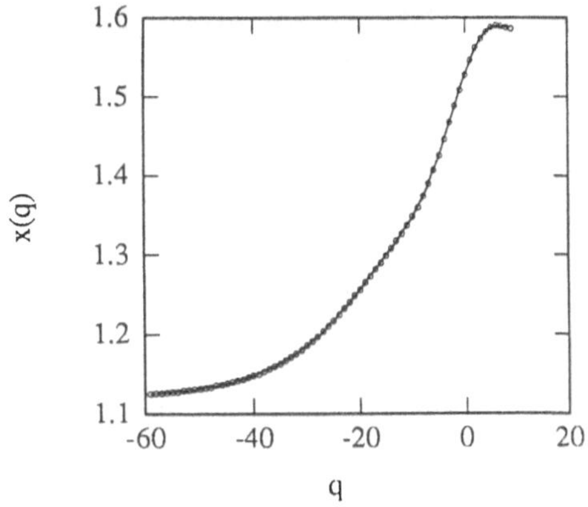

Figure 2: Fractal dimension $x(q)$ of the SAW which dominates Γ_q.

3. RESULTS

For each value of the end-to-end distance of the walks L, we found the probability distribution $P(\mathcal{L})$ of SAW lengths $\mathcal{L}$. Figure 1 shows two such distributions, for $L=11$ and $L=17$, normalized so that $\sum_{\mathcal{L}} P(\mathcal{L}) = 1$.

Clearly, $P(\mathcal{L})$ depends on L. To quantify this dependence, we calculated average moments of these distributions,

$$\Gamma_q(L) = \sum_{\mathcal{L}} P(\mathcal{L})\mathcal{L}^q \sim L^{\tilde{\tau}(q)} \tag{1}$$

Log–log plots of $\Gamma_q(L)$ versus L are straight, yielding $\Gamma_q(L) \sim L^{\tilde{\tau}(q)}$. For $q \to \infty$, the sum in (1) will be dominated by $\mathcal{L}_{max}$, so that $\tilde{\tau}(q)/q \to \tilde{\zeta}_{max}$. Similarly, $\tilde{\tau}(q)/q \to \tilde{\zeta}_{min}$ for $q \to -\infty$. For $q = 1$ we find the scaling exponent for the mean path: $< \mathcal{L} > \sim L^{\tilde{\tau}(1)}$, $\tilde{\tau}(1) = 1.52 \pm 0.03$ which is considerably larger than the value found by Kremer [9]: $1/\nu_{SAW} \simeq 1.34$, possibly due to the trap-avoiding effect.

The distribution is multifractal since $\tilde{\tau}(q)$ is not linear in q. The sum in (1) is dominated by its largest term at $\mathcal{L} = \mathcal{L}^*(q)$. We may use the method of steepest descent to find $\mathcal{L}_q^*$, which is defined by

$$\left. \frac{\partial \ln P(\mathcal{L})}{\partial \ln \mathcal{L}} \right|_{\mathcal{L}_q^*} = -q \tag{2}$$

If both $\mathcal{L}_q^*$ and $P(\mathcal{L}_q^*)$ behave as powers of L:

$$\mathcal{L}_q^* \sim L^{x(q)}, \quad P(\mathcal{L}_q^*) \sim L^{-y(q)} \tag{3}$$

, then $\tilde{\tau}(q) = qx(q) - y(q)$ and $d\tilde{\tau}/dq = x(q)$. The exponent $x(q)$ (Figure 2) thus represents the fractal dimension of the SAW which dominates Γ_q. It interpolates between $\tilde{\zeta}_{min}$ $(q \to -\infty)$, for which we estimate $\tilde{\zeta}_{min} = 1.12 \pm 0.04$ and $\tilde{\zeta}_{max}$ $(q \to \infty)$, for which we estimate $\tilde{\zeta}_{max} = 1.59 \pm 0.03$. The quoted errors are only statistical, ignoring systematic problems. These values are consistent with those in the litterature [2,6]. The fractal dimension of the longest path is approaching that of the underlying backbone: $D_{BB} = 1.61 \pm 0.02$ [13,14].

The exponent $y(q)$ represents the scaling of $P(\mathcal{L}_q^*)$. Our information on $\tilde{\tau}(q)$ and $x(q)$ can also be inverted into a curve of y vs x, which is analogous to the $f(\alpha)$ curves used for multifractality [15]. Note, however, that although the definitions of $y(x)$ resemble those of $f(\alpha)$, their signs are inverted. Unlike f, it is x that has the meaning of a fractal dimension in our case.

ACKNOWLEDGEMENTS

This research has been supported by VISTA, a research cooperation between the Norwegian Academy of Science and Letters and Den Norske Stats Oljeselskap a/s (STATOIL).

Work at Tel Aviv University was supported by grants from the U.S.-Israel Binational Science Foundation and the Israel Academy of Sciences and Humanities.

REFERENCES

[1] Aharony, A., these proceedings.

[2] Blumenfeld, R., Meir, Y., Harris, A. B. and Aharony, A., J. Phys. A19, L791 (1986)

[3] de Arcangelis, L., Coniglio, A. and Redner, S. , J. Phys. A18, L805 (1985)

[4] Grossman, T. and Aharony, A., unpublished

[5] Meir, Y., Blumenfeld, R., Harris, A. B. and Aharony, A., Phys. Rev. B (submitted)

[6] Havlin, S. and Nossal, R., J. Phys. A17, L427 (1984) and references

[7] Stanley, H. E., in 'On Growth and Form ', ed. H. E. Stanley and N. Ostrowsky (Martinus Nijhoff, Dordrecht, 1986), p. 21

[8] de Gennes, P. G., 'Scaling Concepts in polymer Physics' (Cornell University Press, Ithaca, N. Y. 1979)

[9] Kremer, K., Z. Phys. B45, 148 (1981)

[10] Kremer, K. and Lyklema, J. W., Phys. Rev. Lett. 54, 267 (1985)

[11] Stanley, H. E., J. Phys. A10, L211 (1977)

[12] Grossman, T. and Aharony, A., J. Phys. A19, L745 (1986)

[13] Herrmann, H. J., Hong, D. C., and Stanley H. E., J. Phys. A17, L261 (1985)

[14] Laidlaw, D., MacKay, G., and Jan, N., J. Stat. Phys. 46, 507 (1987)

[15] Stanley, H. E., these proceedings

DIFFUSION ON PERCOLATION CLUSTERS

Yigal Meir,[1] A. Brooks Harris [1,2] and Amnon Aharony[1]

[1]Raymond and Beverly Sackler Faculty of Exact Sciences
School of Physics and Astronomy
Tel Aviv University, Tel Aviv 69978, Israel

[2]Department of Theoretical Physics,
1 Keble Rd., Oxford, OX1   3NP, UK
Permanent address: Department of Physics,
University of Pennsylvania, Philadelphia, PA 19104, USA

1. INTRODUCTION

Diffusion on percolating clusters has attracted much attention since de Gennes' proposal[1] of the "ant in the labyrinth". It has recently been realized[2] that diffusion becomes anomalous for times shorter than a typical crossover time, τ, of order $\xi^{2+\theta}$, where $\xi \sim |p_c - p|^{-\nu}$ is the percolation correlation length and θ describes the scaling of the diffusion coefficient on the infinite percolating cluster above the threshold p_c, $D \sim \xi^{-\theta}$. On the infinite cluster, the mean square distance $\langle r^2 \rangle$ after t time steps behaves as $t^{2/(2+\theta)}$ for $1 \ll t \ll \tau$, and as Dt for $t \gg \tau$. τ is thus the time it takes to diffuse a typical distance ξ. Similar anomalous diffusion occurs on finite clusters, for distances short compared to the cluster size.

Here we report on a detailed investigation[3] of the anomalous diffusion time, τ. We calculate average moments of the time it takes a random walker to cover a finite cluster, and show that these times indeed scale, for various types of random walks, as $\tau \sim \xi^{2+\theta}$. In the process we relate these moments to resistive susceptibilities, on general networks.

II. BLIND AND MYOPIC ANTS

The probability $P_{ij}(t)$ that a random walker starting from site j, is at site i after t time steps obeys a master equation,

$$\frac{d}{dt}P_{ij}(t) = \Sigma_k \, [W_{ki}P_{kj}(t) - W_{ik}P_{ij}(t)] = -\Sigma_k \, M_{ik}P_{kj}(t), \tag{1}$$

where W_{ki} is the probability per unit time that the walker hop from site k to site i, and $P_{ij}(0) = \delta_{ij}$. Two main models have been used in the literature:[4] The "Blind" ant tries all the neighboring sites with equal weight, and hops only if the bond in question is occupied. Thus (up to an overall time scale), $W_{ki}^B = \frac{1}{2}\gamma_{ki}$, with $\gamma_{ki}=1$ if i and k are connected by an occupied bond, and $\gamma_{ki}=0$ otherwise. Thus, $M_{ik}^B = \frac{1}{2}(z_i\delta_{ik} - \gamma_{ik})$, where z_i is the number of occupied bonds intersecting site i. The "Myopic" ant chooses randomly from among *occupied* bonds a direction in which to hop. In contrast to the "Blind" ant, the "Myopic" ant always hops at each time step. Thus, $W_{ki}^M = \gamma_{ki}/z_i$, and $M_{ik}^M = \delta_{ik} - \gamma_{ik}/z_k = M_{ik}^B(2/z_k)$.

Since M_{ik}^B is symmetric, it can be diagonalized, with $\Sigma_j M_{ij}^B \phi_j^{(n)} = \lambda_n \phi_i^{(n)}$, and we find $P_{ij}(t) = \Sigma_n \phi_i^{(n)} \phi_j^{(n)} e^{-\lambda_n t}$. In particular, $P_{ij}(t) \rightarrow \phi_i^{(0)} \phi_j^{(0)} = 1/s(\Gamma)$ for $t \rightarrow \infty$, where $s(\Gamma)$ is the number of sites on the finite cluster Γ. To study the approach to equilibrium, we consider

$$F(\Gamma, t) = \Sigma_i [P_{ii}(t) - P_{ii}(\infty)] = \Sigma_n' e^{-\lambda_n t}, \tag{2}$$

where the primed sum does not contain $\lambda_o = 0$. The average k'th moment of the relaxation time is then defined, for $k \geq 1$, as

$$\chi_k^B(p) = \frac{1}{(k-1)!} \int dt \, t^{k-1} [F(\Gamma, t)]_{av} = [\Sigma_n' \lambda_n(\Gamma)^{-k}]_{av}, \tag{3}$$

where $[\ ]_{av}$ denotes averaging over finite clusters.

Although M^M is not symmetric, one can similarly find its eigenvalues μ_n and define $\chi_k^M(p) = [\Sigma_n' \mu_n(\Gamma)^{-k}]_{av}$.

III. RELATIONS TO RESISTANCES

Equation (3) may also be written as

$$\chi_k^B = [Tr \, \hat{G}^k]_{av}, \tag{4}$$

with $\hat{G}_{ij} = lim_{\omega \rightarrow 0}[G_{ij}(\omega) - \phi_i^{(0)} \phi_j^{(0)}/\omega]$, where $G(\omega) = (M^B + \omega I)^{-1}$ is the Green's function for solving Eq. (1) using the Laplace transform.

The matrix M_{ij}^B is exactly the one which appears in Kirchhoff's equations, when each bond of the cluster represents a unit resistor:

$$2 \Sigma_i M_{ki}^B V_i = I_k^{ext}, \tag{5}$$

where V_i is the voltage at site i and I_k^{ext} is the external current imposed at site k. For $I_k^{ext} = I(\delta_{ik} - \delta_{jk})$, the resistance between sites i and j is $R_{ij} = (V_i - V_j)/I$, which gives $2R_{ij} = \hat{G}_{ii} + \hat{G}_{jj} - 2\hat{G}_{ij}$. Solving this equation for $\hat{G}_{ij}$ we obtain $\hat{G}_{ij} = \bar{R}_i + \bar{R}_j - \bar{\bar{R}} - R_{ij}$, with $\bar{R}_i = \Sigma_j R_{ij}/s(\Gamma)$, $\bar{\bar{R}} = \Sigma_i \bar{R}_i/s(\Gamma)$. Using Eq. (4), we find

$$\chi_1^B = [\Sigma_{i,j\epsilon\Gamma} R_{ij}/s(\Gamma)]_{av} \equiv \chi_R, \tag{6}$$

where χ_R is the resistive susceptibility.[5] Similarly,

$$\chi_2^B = [\Sigma_{i,j\epsilon\Gamma} R_{ij}^2 - 2s(\Gamma) \Sigma_i \bar{R}_i^2 + s(\Gamma)^2 \bar{\bar{R}}^2]_{av}, \tag{7}$$

etc. Resistive susceptibilities are known to have a constant gap,[5]

$$[\Sigma_{i,j\epsilon\Gamma} R_{ij}^k/s(\Gamma)]_{av} \sim (p_c - p)^{-\gamma - k\zeta}, \tag{8}$$

where $\zeta = \mu - (d-2)\nu$ where μ describes the conductivity σ above p_c, $\sigma \sim (p-p_c)^\mu$. Similarly, moments of the cluster sizes scale with a constant gap exponent[6,7] Δ:

$$[s(\Gamma)^k]_{av} \sim (p_c - p)^{\beta - k\Delta}, \quad k \geq 1, \tag{9}$$

where $\Delta = D\nu = \beta + \gamma$. Here, D is the fractal dimensionality of the percolation clusters.

214

Combining these results, we find that asymptotically

$$\chi_k{}^B = A_k(p_c - p)^{\beta - k\Delta_\tau} , \; k \geq 1, \tag{10}$$

with the new gap exponent

$$\Delta_\tau = \Delta + \zeta = (2 + \theta)\nu, \tag{11}$$

in agreement with the original scaling argument[2] $\tau \sim \xi^{2+\theta}$.

Writing $M^M = M^B/Z$, with $Z_{ij} = \delta_{ij} \, z_i/2$, one can show[3] that

$$\chi_1{}^M = \tfrac{1}{4}[\, \Sigma_{i,j\epsilon\Gamma} \, z_i z_j \, R_{ij}/b(\Gamma) \,]_{av}, \tag{12}$$

where $b(\Gamma)$ is the number of bonds on Γ. For large clusters, $b(\Gamma)/s(\Gamma)$ approaches a (non universal) constant value, C_∞. Since on the average, $\langle z_i \rangle = 2b(\Gamma)/s(\Gamma)$, we expect that asymptotically $\chi_k{}^M \sim \chi_k{}^B C_\infty{}^k$. However, the two types of ants will have different corrections to this asymptotic behavior.

IV. EXPONENTS AND AMPLITUDES

The relations presented in the previous section allow an explicit evaluation of the moments χ_k. In one dimension, an exact solution yields

$$\chi_k{}^B(p) = D_k(1-p)^{-2k}[1 - k(1-p)+...], \tag{13}$$

$$\chi_k{}^M(p) = D_k(1-p)^{-2k}[1 - (k+2k/(2k+1))(1-p)+...],$$

with $D_1=2$, $D_2=16/3$, $D_3=128/3$, etc. Thus, $\Delta_\tau=2$, in agreement with $\beta=0$, $\gamma=1$, $\zeta=1$, $\theta=0$. Note that correction amplitudes are different for the two ants.

On a Cayley tree, with coordination number $(\sigma+1)$, we find

$$\chi_k(p) = C_k \, \frac{\sigma + 1}{\sigma} \left(\frac{\sigma-1}{\sigma} \right)^k (1 - \sigma p)^{1-3k}, \; k \geq 1, \; \sigma > 1, \tag{14}$$

with $C_1=1$, $C_2=4/5$, $C_3=47/10$, etc. Thus, $\Delta_\tau=3$, in agreement with the mean field values $\beta=1$, $\gamma=1$, $\zeta=2$, $\theta=4$, which should hold for $d \geq 6$.

In dimensions $1<d<6$, we used low concentration series, based on Eq. (4). We calculated $Tr\hat{G}^k$ for all clusters containing up to 11 bonds, and averaged over clusters.[3] Clearly, $\chi_1/\chi \sim (p_c-p)^{-\zeta}$, where $\chi \sim (p_c-p)^{-\gamma}$ is the usual percolation susceptibility, and $\chi_{k+1}/\chi_k \sim (p_c-p)^{-\Delta_\tau}$. We divided these series term by term,[9] obtaining new series which diverge at $p=1$ with the exponents $\zeta+1$ and $\Delta_\tau+1$, respectively. Our resulting estimates for the blind ant are summarized in Table I. Within the error bars, they all agree with values from the literature[7,8] and with Eq. (11). Results for the myopic ant are less accurate, but consistent with those in the Table. The new series, obtained above, also yield good estimates for universal amplitude ratios. Table I also contains our estimates for

$$S_{kl/ij} = \frac{A_k A_l}{A_i A_j} \frac{\Gamma(\gamma_i) \, \Gamma(\gamma_j)}{\Gamma(\gamma_k) \, \Gamma(\gamma_l)}, \tag{15}$$

where $\gamma_k = k\Delta_\tau - \gamma$, and $\Gamma(\gamma)$ is the gamma function. Again, the results for both ants agree with each other.

IV. GENERALIZATIONS AND CONCLUSIONS

Equation (12) hold for arbitrary z_i's as long as they are positive, uncorrelated, and are bounded away from 0 and ∞. Thus, we have shown that a large variety of "ants" and related models have the same asymptotic behavior. In particular, we expect spin waves in general Heisenberg ferromagnets, with different values of the spin S_i , to have the same asymptotic behavior as that of constant spin ferromagnets.

In conclusion, we obtained detailed quantitative relations between the asymptotic behavior of moments of the diffusion times and that of resistive susceptibilities, which confirm the scaling theory of Gefen *et al*[2] and which allow accurate numerical evaluations of the related exponents and amplitude ratios. It now remains to study in detail the difference between the various problem discussed above, which should show up in the correction terms.

Table I. *Series estimates of gap exponents and of amplitude ratios*

d	1 (exact)	2	3	4	5	>6 (exact)
ζ	1	1.31±.08	1.10±.06	1.04±.02	1.03±.02	1
Δ_τ	2	3.6 ±.3	3.30±.05	3.10±.03	3.04±.03	3
$S_{13/22}$	9/10	0.65±.15	0.75±.06	0.81±.02	0.84±.02	47/56
$S_{24/33}$	6/7	0.63±.06	0.67±.05	0.69±.04	0.71±.03	0.72±.03[a]

[a] Estimate for $d=20$.

ACKNOWLEDGEMENTS

This work was supported in parts by grants from the Israel Academy of Sciences and Humanities and the U.S.-Israel Binational Science Foundation.

REFERENCES

1. P. G. de Gennes, *La Recherche* 7, 919 (1976).
2. Y. Gefen, A. Aharony and S. Alexander, *Phys. Rev. Lett.* **50**, 77 (1983).
3. A. B. Harris, Y. Meir and A. Aharony, *Phys. Rev. B* (submitted).
4. C. D. Mitescu and J. Russeng, *Ann. Israel. Phys. Soc.* **5**, 81 (1983).
5. R. Fisch and A. B. Harris, *Phys. Rev. B* **18**, 416 (1983).
6. e.g. D. Stauffer, *Introduction to Percolation Theory* (Taylor and Francis, London 1985).
7. J. Adler, A. Aharony, Y. Meir and A. B. Harris, *J. Phys. A* **19**, 3631 (1986).
8. Y. Meir, R. Blumenfeld, A. Aharony, and A. B. Harris, *Phys. Rev. B* **34**, 3424 (1986).
9. Y. Meir, *J. Phys. A Lett.* (in press).

FLUCTUATION AND DISSIPATION ON FRACTALS: A PROBABILISTIC APPROACH

R. Hilfer[*] and A. Blumen[†]

[*]Department of Physics
University of California, Los Angeles
Los Angeles, CA 90024, USA

[†]Physikalisches Institut
Universität Bayreuth
D-8580 Bayreuth, Germany (W.)

The analogies between the diffusion problem and the resistor network problem as witnessed by the Einstein relation have been very important for analytical and numerical investigations of linear problems in disordered geometries (e.g. percolating clusters)[1]. This raises the question whether the resistor problem can be identified in a purely probabilistic context. An affirmative answer has recently been given and it was shown that the Einstein relation follows from a simple probabilistic argument[2,3] Here we present the results of a more general treatment.

We begin by considering conditional first passage probabilities in a Markov chain. From a relation for the corresponding generating functions we obtain once more the probabilistic analogue of the Einstein relation. We develop its interpretation and conclude by connecting it to the relation between conductivity and diffusion exponents[1].

Consider a homogeneous Markov chain with denumerable state space. Such a chain can be visualized as a walker (or particle) moving randomly between a countable number of states (sites). Homogeneous here means that the transitions of the walker from site i to site j are governed by single step transition probabilities which do not change with time. We will be interested in the first passage probability $F_{ij}^{(n)}$ that the walker will reach site j for the first time after n steps, given that he is at site i at time 0. In addition we introduce the conditional first passage probabilities $G_{ij}^{(n)}$ for starting at i at time 0 and reaching j for the first time at step n, conditioned on not having visited the elements of a given set S during the walk. The so called taboo set[4] S is restricted to be a finite subset of the state space. We define the generating functions

$$F_{ij}(z) = \sum_{n=1}^{\infty} F_{ij}^{(n)} z^n$$

and analogously $G_{ij}(z)$ for the conditional probabilities $G_{ij}^{(n)}$.

Let us consider the simplified case in which S={a} consists only of a single point. Then for i≠j the relation

$$G_{ij}(z) = \frac{F_{ij}(z) - F_{ia}(z)F_{aj}(z)}{1 - F_{ja}(z)F_{aj}(z)} \qquad (1)$$

can be derived from a straightforward probabilistic argument[5]. Note that $G_{ij}(1)$ is the conditional probability that the walker reaches j in one or more steps after starting from i at time 0. Thus Eq. (1) gives an explicit formula for this probability in the limit $z \to 1$. To take the limit we write $F_{ij}(z) = 1 - (1-z)f_{ij}(z)$ with $f_{ij}(z) = \langle T_{ij} \rangle + \sigma(1-z)$ where we have assumed that the mean first passage times $\langle T_{ij} \rangle$ between i and j are finite. One obtains

$$G_{ij}(z) = \frac{f_{ia}(z) + f_{aj}(z) - f_{ij}(z) - (1-z)f_{ia}(z)f_{aj}(z)}{f_{ja}(z) + f_{aj}(z) - (1-z)f_{ja}(z)f_{aj}(z)} \qquad (2)$$

and thence

$$q := G_{ij}(1) = \frac{\langle T_{ia} \rangle + \langle T_{aj} \rangle - \langle T_{ij} \rangle}{\langle T_{ja} \rangle + \langle T_{aj} \rangle} \qquad (3)$$

For $i=a=0$ this implies

$$\langle T_{00} \rangle = q \left[\langle T_{0j} \rangle + \langle T_{j0} \rangle \right] \qquad (4)$$

We argue that Eq. (4) is indeed a generalized analogue of the Einstein relation in a purely probabilistic context.

To identify the diffusion constant we write the relation $\langle r^2(t) \rangle \propto t$ as $\langle t(r) \rangle \propto r^2$ which can be justified using the invariance of Brownian motion under the transformation $t \to b^2 t$ and $r \to br$.[6] Here $\langle t(r) \rangle$ is the mean first exit time for the walker to leave a sphere of radius r around its starting point. For an inhomogeneous structure we then define a generalized r-dependent diffusion coefficient as $D(r) = r^2/\langle t(r) \rangle$. To identify the conductivity we introduce an external potential by assuming that the walker is absorbed with probability ρ at some boundary point B and subsequently replaced at 0. If N walkers start from 0 then Nq of them will reach B without having returned to 0. On the average $n = Nq\rho$ walkers will flow from 0 to B. If we identify q as the conductance and n/N as the probability current this is a statement of Ohms law. For a system of linear size L and cross section A we define the conductivity as $\sigma = qL/A$. Returning to the pure random walk picture we assume that the points 0 and B are a distance L apart and that $\langle T_{0B} \rangle \approx \langle T_{0B} \rangle$. We then get from Eq. (4) $\langle T_{00} \rangle = 2\sigma V \langle T_{0B} \rangle / L^2 \propto 2\sigma V/D$ where V denotes the corresponding volume. Hence we arrive at the Einstein relation $\sigma \propto D$. Taking ratios of the quantities in Eq. (4) for two systems whose linear sizes L, L' are scaled by a factor b, i.e $L' = bL$, and assuming that the limit $L \to \infty$ exists we obtain[2,3] from Eq. (4) the well known relation between diffusion and conductivity exponents.[1] In conclusion we remark that our results involve only quantities that are readily measured in simulations of diffusion in disordered media regardless of whether the systems behave fractally or not.

ACKNOWLEDGEMENT We gratefully acknowledge financial support from the Fonds der Chemischen Industrie and the Deutsche Forschungsgemeinschaft.

REFERENCES
1. Y.Gefen, A.Aharony, S.Alexander, Phys.Rev.Lett. 50 (1983), 77 ; C.K.Harris and R.B.Stinchcombe, Phys.Rev.Lett. 50 (1983), 1399; for a review see e.g. the contributions of H.E. Stanley or A.Aharony in: Scaling Phenomena in Disordered Systems, R.Pynn and A.Skjeltorp, Plenum Press, New York 1985
2. R.Hilfer, Renormierungsansätze in der Theorie ungeordneter Systeme, Verlag Harri Deutsch, Frankfurt a.M. 1986
3. R. Hilfer and A. Blumen, preprint
4. K.L. Chung, Markov Chains, Springer, Berlin 1967
5. R. Hilfer and A. Blumen, to be published
6. E.B.Dynkin and A.A.Juschkewitsch,Markoffsche Prozesse,Springer,Berlin 1969

TIME-DEPENDENT EFFECTS NEAR THE PERCOLATION THRESHOLD IN WATER-IN-OIL

MICROEMULSIONS

H.-F. Eicke[¶], S. Geiger[¶], R. Hilfiker[¶], F.A. Sauer[+] and H. Thomas[§]

[¶] Institut für Physikalische Chemie, Univ. Basel, CH-4056 Basel
[+] Max-Planck-Institut für Biophysik, D-6000 Frankfurt/Main
[§] Institut für Physik, Universität Basel, CH-4056 Basel

The three-component system water/surfactant(AOT)/oil(i-C$_8$) forms in a certain range of compositions and temperatures (see Fig. 1) thermo-dynamically stable microemulsions consisting of nanometer-sized water droplets (ND) covered by a monomolecular layer of surfactant and dispersed in oil /1/. The nanodroplet radius is controlled by the ratio w_0 = [H$_2$O]/[AOT] and has a typical value of 10 nm. The nanodroplets are spherically symmetric and rather monodisperse, as has been established both experimentally and theoretically (see e.g. /2/). The spontaneous formation and thermodynamic stability of these emulsions is due to the "ultra-low" interfacial tension of the order of $k_B T/4\pi R^2$ (R = radius of spherical water droplet) of the oil/water interface in the presence of surfactant(s). This is a consequence - to a first approximation - of a compensation of Coulombic, lateral dispersion and curvature energy contributions.

The interactions between the nanodroplets consist of a hard-core re-pulsion and a weak and short-ranged attraction at distances where the sur-factant tails overlap. The dispersions show an entropy-driven phase separation with a lower consolute point (see Fig. 1). In the single-phase region,

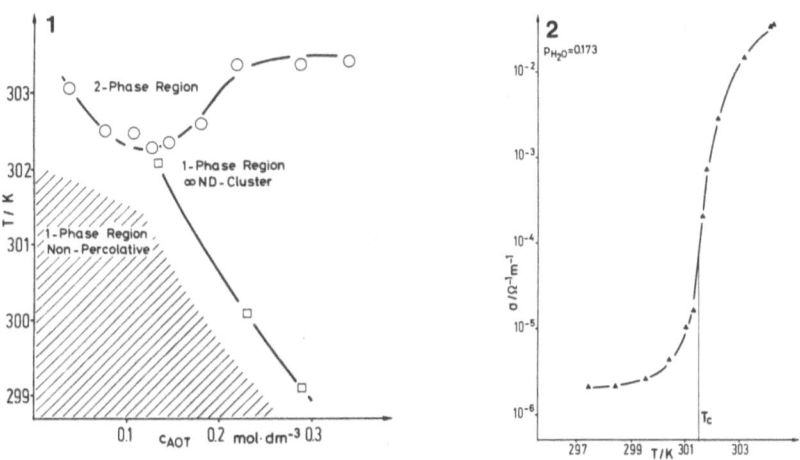

Fig. 1. Section of phase diagram of H$_2$O/AOT/i-C$_8$.
Fig. 2. T-dependence of ohmic conductivity.

one expects percolation phenomena to occur along a line in the concentrat-
ion-temperature diagram having a finite (negative) slope due to interaction
effects (Fig. 1). This figure and all following data refer to the same com-
Position w_0 = 60, C_{AOT} = 0.16 mol dm^{-3}.

The existence of a percolation threshold T_c is clearly demonstrated by
an increase of the ohmic conductivity σ of several orders of magnitude in a
narrow temperature interval (Fig. 2). We identify the percolation threshold
with the inflection point of the log σ against T curve. A plot of $\log|T-T_c|$
versus log σ is compatible with a scaling behaviour with critical exponents
s = -0.7 and μ = 2.06 /3/. It has been pointed out to us, however, that the
behaviour of this system may have to be described by a different exponent s'
below threshold /4/.

Charge transport in these systems has its origin in the dissociation
of the ionic surfactant (AOT = sodium di-2-ethylhexylsulfosuccinate) which
leads to the formation of a semi-diffuse double layer consisting of the
negatively charged surfactant heads and the free Na^+ ions in the interior
of the nanodroplets. In the non-percolative region (shaded area in Fig. 1),
exchange of surfactant ions and/or Na^+ ions between individual nanodroplets
yields an equilibrium distribution of charged nanospheres, which give rise
to the observed low conductivity of the order of $10^{-6} \Omega^{-1} m^{-1}$ (Stokes transport). As the percolation threshold is approached, the nanospheres form
larger and larger clusters, $n(ND) \rightleftarrows (ND)_n$, and the relatively free intra-
cluster charge transport leads to a sharp conductivity increase. Above the
percolation threshold, the conductivity is due to the (non-Stokes) charge
transport on the percolating cluster, and is only an order of magnitude
smaller than typical values for aqueous electrolyte solutions.

Valuable information on the charge transport mechanism and structural
changes near the percolation threshold is obtained from the time-resolved
current response to an applied voltage step /5/. It is found that the
current density $i(t) = i_0 + \Delta i(t)$ consists of two parts, an "instantaneous"
contribution i_0 responding to the applied voltage within the pulse rise time
of ∿ 20 ns, and a "slow" contribution $\Delta i(t)$ approaching a stationary value
after about 100-400 μs, depending on temperature and composition of the
system. The instantaneous part i_0 contains for $T < T_c$, all of the ohmic response σ_0 E, and shows a cubic nonlinearity of positive sign,

$$i_0(t) = \sigma_0 (T) E + b (T) E^3, \quad b (T) > 0. \qquad [1]$$

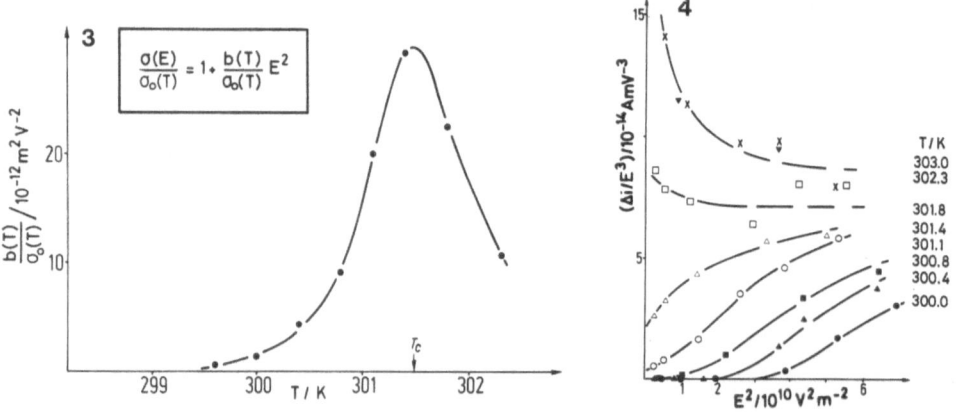

Fig. 3. T-dependence of cubic nonlinearity.
Fig. 4. Nonlinear field dependence of slow current response Δi.

220

We emphasize that the mechanism causing the conductivity increase as $T \uparrow T_c$ is very fast with a characteristic time scale well below 10^{-8} sec. The ratio $b(T)/\sigma_o(T)$ displays a pronounced maximum at a temperature which coincides with the percolation threshold T_c. (Fig. 3)

Well below T_c, two types of processes contribute to the non-linear charge transport: dissociation of a polarised cluster into charged fragments,

$$(ND)_n \rightleftarrows (ND)_{n-r}^+ + (ND)_r^-,\qquad\qquad [2]$$

and field-enhanced charge transfer between colliding clusters

$$(ND)_m + (ND)_n \rightleftarrows (ND)_m^+ + (ND)_n^-.\qquad\qquad [3]$$

For the former process, one obtains

$$b(T)/\sigma_o(T) = \frac{1}{4} p^2/(k_B T)^2\qquad\qquad [4]$$

where p is the dipole moment of the cluster $(ND)_{n-r}^+ (ND)_r^-$. If p is estimated for dimeric clusters, one obtains satisfactory agreement with Fig. 3 for temperatures $T < T_c - 2$ K. For increasing T, the model may be utilised to estimate the average cluster size as a function of temperature. With increasing cluster size, the finite intra-cluster conductivity begins to contribute, causing the maximum at T_c and the subsequent decrease of b/σ_o to zero, characteristic of the field-independent current transport on the percolating cluster.

The saturation value Δi of the slow response shows a strongly nonlinear field dependence (Fig. 4). For $T < T_c$ and $E \to 0$, the curves $\Delta i/E^3$ versus E^2 approach values close to zero, indicating that the contribution is zero and the cubic nonlinearity is very small. For $T > T_c$ there exists a small ohmic part producing the divergence at $E \to 0$. With increasing field strength, Δi becomes $\sim E^3$ independent of temperature. For $T < T_c$, the behaviour is well described by an ansatz of the form

$$\Delta i(E,\infty) = \frac{1}{2} c E^3 \left(1 + \tanh \frac{E^2 - E_o^2(T)}{c^*}\right)\qquad\qquad [5]$$

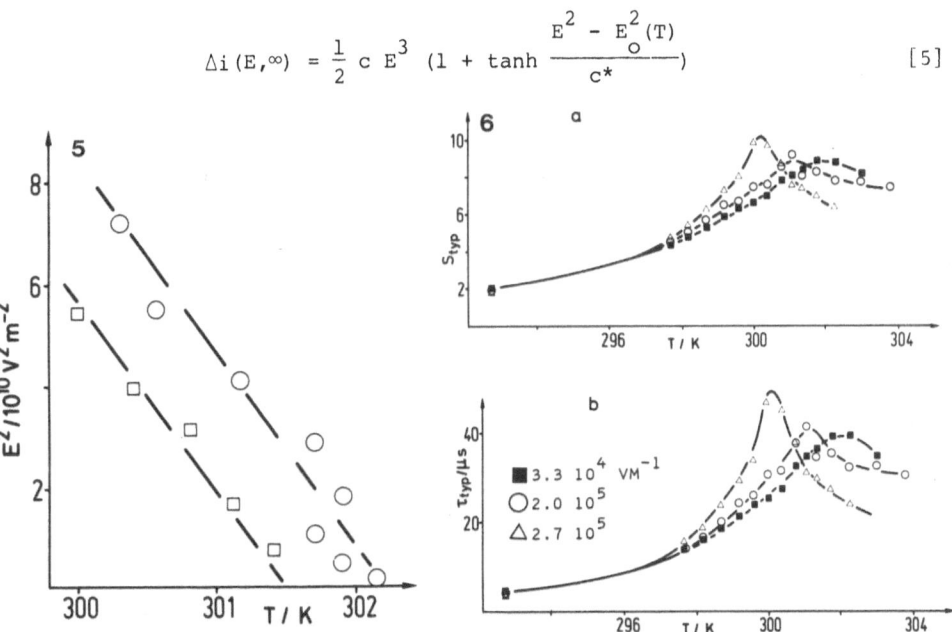

Fig. 5. T-dependence of parameter E_o (eq. (5)) /5/ and E-dependence of temperature T_o of τ_{max} /6/.

Fig. 6. T-dependence of typical cluster size (a) and relaxation time (b).

Fig. 5 shows the temperature dependence of the parameter E_0. The temperature at which E_0 extrapolates to zero agrees reasonably well with the percolation threshold.

The slow rise time and the strong nonlinearity of Δi suggest that this contribution is due to field induced changes of the structure and connectivity of the clusters.

Reorientations and structural changes of the clusters induced by an electric field also give rise to a huge electric birefringence with Kerr constants of the order $3 \cdot 10^{-11}$ V^{-2} m $/5/$. Time-resolved measurements show rise times in the same range of ~ 100 μsec as for the slow current response $\Delta i(t)$. We have analysed the results in terms of a model containing a typical cluster size $s_{typ}(T)$ and a relaxation time $\tau_{typ}(T)$ as model parameters, and find pronounced maxima at a characteristic temperature T_0 which decreases with increasing field strength (Fig. 6). The sharpening of the maxima with increasing field strength is probably due to increased life times of the nanodroplet clusters. The dependence of T_0 on E correlates nicely with the results of the slow current response (Fig. 5).

Information about the fractal structure of clusters is obtained from the dielectric response. Measurements in the frequency range between 100 kHz and 1 GHz show a pronounced increase both of the static dielectric constant ε_s and the dielectric relaxation time τ as the percolation threshold is approached (Fig. 7) $/7/$. This behaviour is incompatible with the existence of compact clusters consisting of closed-packed nanospheres, because the volume fraction covered by such clusters would be essentially constant, giving rise to a temperature-independent dielectric response. The results are a clear indication of the fractal nature of the clusters. As the temperature approaches T_c, the clusters become larger and at the same time more and more ramified, entrapping increasing amounts of oil of the dispersion medium which is bypassed electrically. This gives rise to the observed dielectric enhancement of ε_s. At the same time, the effective conductivity and dielectric constant of the cluster decrease, and the dielectric response of the dispersion becomes dependent on the capacitance of the dispersion medium between the clusters, which leads to the observed increase of the relaxation time.

A thorough theoretical description of these phenomena is a highly nontrivial problem, since the effective dielectric properties of the microemulsion depend on the spatial correlations between the nanodroplets, i.e. on the N-particle distribution function.

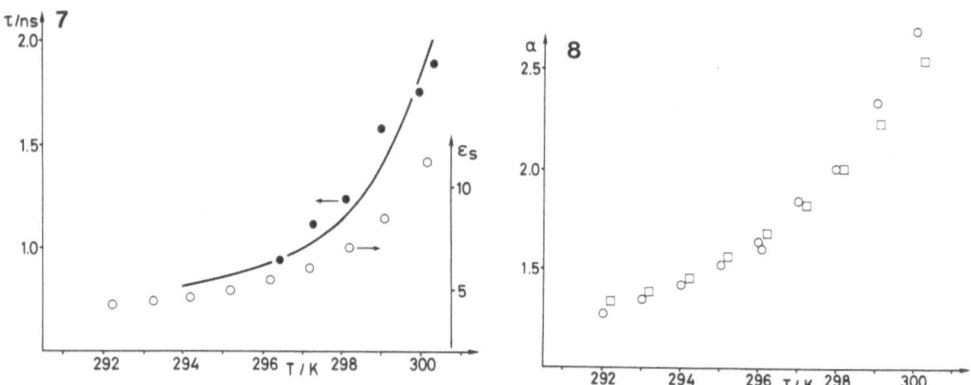

Fig. 7. T-dependence of static dielectric constant ε_s and dielectric relaxation time τ $/7/$.

Fig. 8. T-dependence of swelling factor α $/7/$. (⃝ from ε_s, □ from η).

We have constructed a simplified model in which the ramification of the clusters is described by a "ramification-factor" $\alpha = V_{cl}^{eff}/V_{ND}^{closed-packed}$ which determines effective intra-clusters constants $\sigma_{cl}(\alpha)$ and $\varepsilon_{cl}(\alpha)$. Application of effective-medium theory to a dispersion of such effective cluster particles in the remaining dispersion medium and comparison with the observed ε_s-values yields the ramification factor α as a function of temperature (Fig. 8). The results agree very satisfactorily with independent determinations from dynamic viscosity, Kerr-effect and light-scattering measurements /8/.

The relaxation times calculated using these α-values are also in good agreement with experiment (full line in Fig. 7). In conclusion, the reported experimental results on time-resolved current response and Kerr effect and on frequency-dependent dielectric response have yielded detailed insight into the structure of nanodroplet clusters and its change under an applied electric field.

References

1 Eicke, H.F. 1987 in "Interfacial Phenomena in Apolar Media", H.F. Eicke and G.D. Parfitt, ed. M. Dekker, New York
2 De Gennes, P.G. and Taupin, C. 1982, J. Phys. Chem. 86: 2294
3 Mitescu, C.F. and Musolf, M.J. 1983, J. Physique Lett. 44: L-679
4 Aharony and G. Grest, private communications.
5 Eicke, H.F., Hilfiker, R. and Thomas, H. 1986, Chem. Phys. Lett. 125: 295
6 Eicke, H.F., Hilfiker, R. and Thomas, H. 1985, Chem. Phys. Lett. 120: 272
7 Eicke, H.F., Geiger, S., Sauer, F.A. and Thomas, H. 1986, Ber. Bunsen-
ges. Phys. Chem. 90: 872
8 Hilfiker, R. and Eicke, H.F. 1987, J.C.S. Faraday Trans. I, in press

Acknowledgement

The authors are grateful to the Swiss National Science Foundation for financial support. They also would like to acknowledge useful discussions with A. Aharony and G. Grest.

TRACER DISPERSION: A NEW CHARACTERISTIC LENGTH SCALE

MEASUREMENT IN HETEROGENEOUS POROUS MEDIA

Groupe Poreux PC[+] (presented by J.P.Hulin and E.Charlaix at the 1987 Geilo meeting)
Laboratoire HMP - ESPCI
10 rue Vauquelin, 75231 Paris Cedex 05 France

INTRODUCTION

At the last 1985 Geilo meeting, dispersion of tracers in porous media has been suggested[1] by our group as a potentially very important tool to characterize the structure of heterogeneous porous media. Since then, many new theoretical and experimental results have been obtained, particularly on dispersion in heterogeneous systems[2] and on the influence of dead volumes[3], boundary layers[4] and recirculation zones[5,6]. We shall discuss here some experimental measurements obtained in our group in this perspective on two dimensional and three dimensional heterogeneous model media.

Without entering into a full discussion of the problem let us recall that at medium and high flow velocities the longitudinal dispersion coefficient[7] $K_{//}$ is controlled mostly by the geometrical dispersion[8]: this mechanism is associated with random variations of the fluid velocity from one pore to another. It is indeed experimentally observed in well connected porous media and its contribution to $K_{//}$ is proportional to the mean velocity U with:

$$K_{//} \cong U \, l_d \qquad (1)$$

the "dispersion length" l_d characterizes the path length after which the velocity of a tracer particle becomes decorrelated from its original value. l_d is a key output from the dispersion measurement and characterizes the spatial scale of the porous medium heterogeneities: dispersion is therefore a highly non-local process strongly influenced by long range correlations in the structure of the medium.

Another mechanism which is only effective in porous media at very low velocities is the Taylor-Aris[9] dispersion which is associated with velocity gradients inside individual pores and channels. The corresponding contribution to $K_{//}$ varies as U^2 with the mean velocity. Dispersion components proportional to U^2 or to U^α with $\alpha > 1$ may also appear due to the presence of stagnant zones or to flow recirculation effects.

In the present work, we analyse experimentally the relation between the degree of heterogeneity of two kinds of model porous systems and their dispersion characteristics. Measurements have first been performed on 2D systems of channels of different connectivities. In a second experiment, the correlation length of the flow field is varied by increasing the degree of compaction of 3D sintered glass beads samples.

TRACER DISPERSION IN TWO-DIMENSIONAL MICROMODELS.

2D analog models of porous media[10] allow easy visualizations as well as a perfect control of the flow geometry and, therefore, meaningful comparisons with numerical simulations. We have used two different models realized in Toulouse by C.Zarcone:
- a periodic fully connected square lattice (model I) with 160x160 1mm long channels of

---

[+] C.Baudet, E.Charlaix, E.Guyon, J.P.Hulin, L.Oger, P.Rigord, S.Roux

random width (0.2-0.8mm) representing a weakly disordered porous medium (Fig. 1a).
- a partly connected percolation hexagonal lattice (model II) of 1mm long, 0.5mm wide channels (Fig. 1b) representing a disordered system or a partly saturated medium (global size =183*167mm). The correlation length is $\cong$ 10 channel lengths (percolation parameter = 0.75).

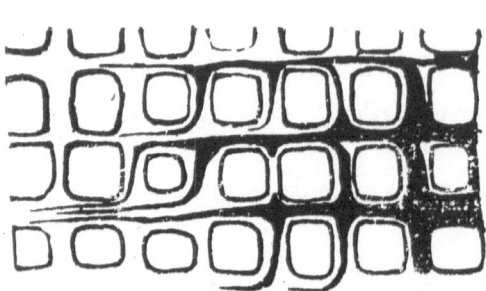

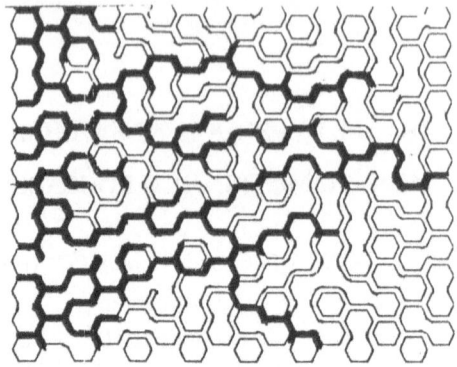

Figure 1a Close-up view of dye tracer particles paths inside a well connected 2D model with random channel widths (model I). Several dye fingers may coexist inside a single channel.

Figure 1b Close-up view of dye tracer particles paths inside a partly connected 2D hexagonal model with uniform channel widths (model II). Long dye fingers appear due to the flow field geometric disorder.

A permanent uniform flow is established between two opposite sides of the lattice and as tracer concentration variation front is established on the inlet side by rinsing the injection circuit. We use an ionic salt as a tracer and the mean concentration variation on the outlet side is obtained from the mean between two parallel wires laid along its whole length.

In model I, the dispersion curves are nearly Gaussian; for model II, a dispersion "tail" is visible at long times and a Gaussian fit is no more possible. We analysed these curves with a capacitance model[11] which allows to determine an asymptotic dispersion coefficient K_{as}; K_{as} is the value of the mean dispersion coefficient, for a sample of the same material but with an infinite size. In the rest of this paper, K_{as} is always used as the longitudinal dispersion coefficient $K_{//}$ for non Gaussian dispersion cases. Fig.2 shows the variation of the ratio $K_{//}/U$ with the mean transit velocity U in both models.

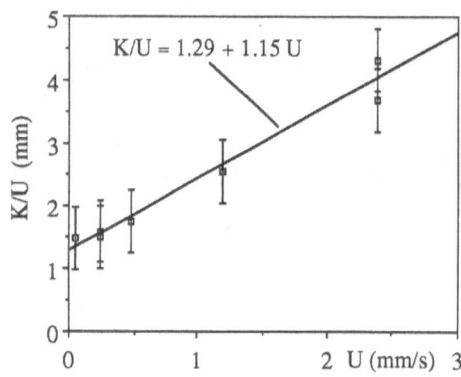

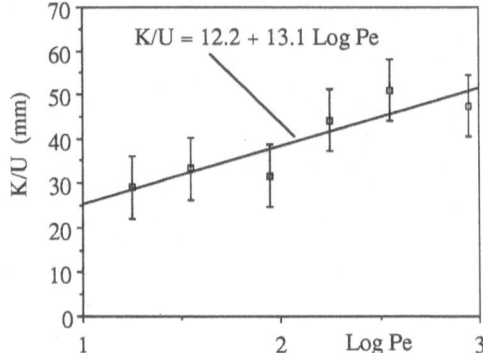

Figure 2a Variation with the mean velocity U of the ratio K/U (K= longitudinal dispersion coefficient) for the well connected 2D micromodel (model I). A significant Taylor-type component proportional to U is obtained.

Figure 2b Variation with the logarithm of the Péclet number Pe of the ratio K/U (K= asymptotic dispersion coefficient) for the partly connected hexagonal lattice (model II). K/U is of the order of the percolation correlation length.

In model I, one has approximately $K_{//}/U = U\tau + l$ where $l = 1.23$mm and $\tau = 1.15$s. τ is very close to the Taylor characteristic molecular diffusion time[9] τ across the mean half-width of the channels. This variation is rather unexpected but can be better understood by using a dyed tracer. At high flow rates, long dye fingers are observed (Fig.1a) and remain well defined along several mm; they do not mix at intersections and several fingers may coexist inside a same capillary because molecular diffusion is too slow to mix them. Therefore, for an individual finger, the process is analogous to Taylor dispersion, explaining the U^2 variation of K_{as}. The long length over which fingers subsist may be due to the use of a 2D model where the transverse dispersion of the flow lines is smaller than in a 3D system. At low flow rates, molecular diffusion uniformises the tracer concentration over the width of the capillaries; the process gets close to a mechanical dispersion with a perfect mixing at the intersections and $K/U \rightarrow 1.23$mm which is of the order of the channel length.

In model II, one has larger K_{as}/U values which can be approximated over the Péclet number Pe range investigated by:

$$K_{as}/U = 1 + C \ \log Pe \qquad (2)$$

where $1 = 12.2$ mm, $C = 13.1$ mm and $Pe = Ua/Dm$ (a = channel half-width, Dm = molecular diffusion time). Therefore, the dominant mechanism seems to be mechanical dispersion and the Log Pe term may be associated with low velocity zones. When a dyed tracer is used, the "fingers" present inside individual capillaries are much shorter and less significant than those associated with the different geometrical paths (Fig.1b).

The key result of this experiment is that both 1 and C are very close to the correlation length of the percolation structure: hence, we verify in this model system that $l_d = K_{as}/U$ measures a characteristic length scale of the heterogeneities, much larger than the size of an individual channel. Let us now verify this concept in a more realistic 3D model medium.

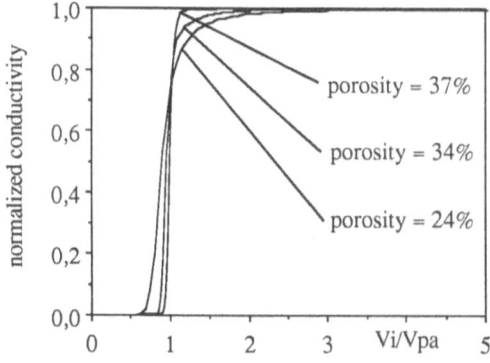 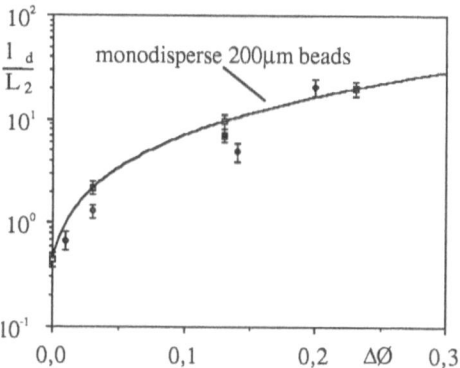

Figure 3 Variation of the normalized conductivity as a function of the ratio of the injected fluid volume and the accessible pore volume for a nonconsolidated sample (∅=37%) and two sintered samples (∅=34% and ∅=24%).

Figure 4 Variation of the ratio of the dispersion length l_d and the characteristic microscopic length L_2 with the porosity variation Δ∅ due to sintering for monodisperse (open symbols) and bidisperse (black symbols) samples.

TRACER DISPERSION IN SINTERED GLASS BEADS

We use sintered glass beads samples[12] refered to hereafter as Ridgefield sandstone. They are obtained in a large range of degrees of compaction (and therefore porosities) and of bead size distributions: this allows to study separately the influence of these various parameters. In addition, this material has been very well characterized by a large variety of electrical, acoustical and flow measurements. the first series of samples is prepared from monodisperse glass beads with a uniform 200μm diameter distribution: the porosity ∅ ranges from 37% (nonconsolidated beads) down to 14% (highly consolidated but well connected).

Qualitatively (Fig.3), when the degree of compaction increases, the slope of the variation of the tracer concentration at the sample outlet gets much smoother corresponding to a large increase of the dispersion coefficient. In addition, the dispersion curves which are Gaussian for the nonconsolidated materials display a "tail" at long times in the compacted materials and are not solutions of Eq.1 any more. Quantitatively, the ratio $K_{//}/U$ is nearly constant so that the geometrical dispersion mechanism appears to be dominant; therefore, one can define again a characteristic dispersion length scale $l_d = K_{//}/U$. l_d is of the order of the bead size in the non consolidated system and increases by a factor of 60 as ∅ goes from 37% to 14% (open square points in Fig.4).

To get closer to real systems, we have then tested samples prepared from bididisperse mixtures using beads with 300μm and 90μm diameters. First, for weakly compacted homogeneous mixtures, the coefficient of dispersion is low and experimental curves are nearly Gaussian: there is therefore no intrinsic dispersion anomaly due the bidispersity

227

contrary to previous observations[13,14] (probably associated with imperfect fillings). In order to analyse the influence of the compaction, one needs to compare l_d to a microscopic length scale characterizing the size distribution of beads. Recent electrical and acoustical measurements[15] on the same sintered samples have shown that the product $L_1 = C\,F\sqrt{k}$ is equal to either the nonconsolidated bead diameter (monodisperse samples) or to a weighted average L_2 of the diameters in the bidisperse mixtures (C is a constant of the order of 1, k the permeability and F the formation factor (ratio of the resistivity of the material filled with a conducting fluid and of that for the bulk fluid).This result is valid down to 10% porosities.

Using these results, we characterize dispersion by the ratio $l_d/L_1 \cong l_d/L_2$ (this eliminates the influence of the size distribution before consolidation); the degree of consolidation is represented by the difference $\Delta\varnothing$ between porosities before and after sintering. l_d/L_2 has very similar variations with $\Delta\varnothing$ for monodisperse and bidisperse systems (Fig.4). At low compactions in homogeneous systems l_d is of the order of the microscopic scale L_2; it gets 60 times higher at low porosities and varies little with the original mixture polydispersity.

CONCLUSION

We have shown experimentally on 2D and 3D model systems that tracer dispersion is a nonlocal measurement extremely sensitive to large scale heterogeneities of the material and depends little on the detailed microscopic structure. In a 2D model with a percolation lattice structure, we have been able to relate the characteristic dispersion length $l_d = K_{//}/U$ to the percolation correlation length. In sintered glass samples we have observed a large increase of l_d with the degree of compaction; this variation is independent on the polydispersity of the material and contrasts with more local measurements such as permeability or conductivity on which no such large variation is obtained. These results demonstrate the clear significance of dispersion as a characteristic heterogeneity length scale measurement

ACKNOWLEDGEMENTS The experiments on sintered glass beads have been performed with T.J. Plona at the Schlumberger-Doll Research Center which we wish to thank for their partial support of this research.

REFERENCES

1- Baudet, C., Charlaix, E., Clément, E., Guyon, E., Hulin, J.P. & Leroy, C. 1985 Scaling concepts in porous media. Proceedings of the NATO conference on "scaling phenomena in disordered systems" Geilo (Norway) April 10-21 1985, R.Pynn ed. Plenum (London)

2- Koplik, J., Redner, S. & Wilkinson, D.W. 1987 Transport and dispersion in random networks with percolation disorder to be published.

3- de Gennes, P.G. 1983 Hydrodynamic dispersion in unsaturated porous media, J. Fluid Mech. 136, 189-200

4- Koch, D.L. and Brady, J.F. 1985 Dispersion in fixed beds. J. Fluid Mech. 154, 399-427.

5- Guyon, E., Pomeau, Y., Hulin, J.P. and Baudet, C. 1987, Dispersion in the presence of recirculation zones to be published in Physica D.

6-Baudet, C., Chertcoff, R. & Hulin, J.P. 1987 Effets de désordre de structure sur la dispersion d'un traceur dans un milieu poreux modèle to be published in C.R. Acad. Sci.

7- Bear, J. 1972 Dynamics of fluids in porous media: Elsevier Publishing Co. (New-York). Chapter 10.

8- Saffman, P.G. 1960 Dispersion due to molecular diffusion and macroscopic mixing in flow through a network of capillaries. J. Fluid Mech. 7, 194-208.

9- Taylor, G.I. 1953 Dispersion of soluble matter in solvent flowing slowly through a tube. Proc. Roy. Soc. A 219, 186-203.

10- Lenormand, R., Zarcone, C. & Sarr, A. 1983 Mechanisms of the displacement of one fluid by another in a network of capillary ducts. J. Fluid Mech. 135, 337-353.

11- Coats, K.H. and Smith B.D. 1964 Dead end pore volume and dispersion in porous media, Soc.Pet.Eng.J. Trans.AIME 231, 73-84.

12- Charlaix, E., Hulin, J.P. & Plona,T.J. 1987 Experimental study of tracer dispersion in sintered glass materials of variable compaction to be published in Physics of fluids.

13- Guennelon, R., Zeiliguer,A. and de Cockborne, A.M. 1983 Effets texturaux sur la porosite et la dispersion hydrodynamique in Proceedings of the Colloquium on the "Variabilite spatiale des processus de transfert dans les sols", Avignon, 24-25 Juin 1982, Ed. INRA Publisher, les Colloques de l'INRA N°15, 133-138.

14- Lemaitre, J., Cintre, M., Troadec, J.P. and Bideau, D. Dispersion d'un traceur dans un mélange binaire de sphères C.R. Acad. Sci.(1986).

15- Guyon, E., Oger, L. and Plona, T.J. 1987 Transport properties in sintered porous media composed of two particle sizes submitted to J. Phys.D.

RELAXATION IN THE RANDOM ENERGY MODEL

G.J.M. Koper

Instituut-Lorentz
P.O. Box 9506
2300 RA Leiden
The Netherlands

Introduction

Derrida's [1] random energy model (REM) is defined as a system of a large number, $\mathcal{N}$, of energy levels that are independent random variables drawn from a Gaussian probability distribution. The model exhibits a critical temperature below which the free energy becomes a constant and the entropy vanishes. This indicates a completely frozen phase and is reminiscent of a spin glass phase in magnetic systems with random competing interactions [2]. The probability P_i^{eq} to find the equilibrated system in the level i with energy E_i is given by

$$P_i^{eq} = \frac{1}{Z} e^{-\beta E_i},\tag{1}$$

where the partition function is defined as

$$Z = \sum_{j=1}^{\mathcal{N}} e^{-\beta E_j}\tag{2}$$

and β is the inverse temperature (Boltzmann's constant $k_B = 1$). From these occupation probabilities it is clear that the lowest energy levels dominate the low temperature phase. The lowest N, N arbitrary, energy levels E_i will be written as

$$E_i = E_o + \epsilon_i \quad (i = 1, 2, \ldots, N),\tag{3}$$

where E_o is an extensive energy common to all levels and ϵ_i is a small non-extensive energy. By linearizing the Gaussian probability distribution of the full REM [1] one can show that in the frozen phase the ϵ_i are independent random variables distributed according to

$$\mathcal{P}(\epsilon) = \rho e^{\rho(\epsilon - \epsilon_c)} \theta(\epsilon_c - \epsilon)\tag{4}$$

(see also Ref. [3]). Here θ is the Heaviside step function and ϵ_c is a cut-off energy. The probability distribution defined by Eq. (4) can be used to calculate the physical properties of the REM for temperatures β^{-1} less than the transition temperature ρ^{-1} in the limit

$$N \to \infty, \quad \epsilon_c \to \infty, \quad Ne^{-\rho\epsilon_c} = v \text{ fixed,} \tag{5}$$

i.e. the limit in which the density of the levels ϵ_i at any given value of the energy becomes a constant.

Kinetic REM

For the dynamics of this model De Dominicis et al. [4] have postulated a master equation that describes the time evolution of the probability $P_i(t)$ of finding the system in level i at time t. It reads

$$\frac{dP_i(t)}{dt} = \sum_{j=1}^{N} W_{ij} P_j(t) - \sum_{j=1}^{N} W_{ji} P_i(t) \quad (i = 1, 2, \ldots, N). \tag{6}$$

The transition rates W_{ij} for going from level j to level i obey the detailed balance condition so that in the limit $t \to \infty$ equilibrium will be attained.

The solution of the master equation (5) is defined by means of a Green function $G_{ij}(t)$ which is the conditional probability of finding the system in level i at time t when it was initially, at $t = 0$, in level j. So, $G_{ij}(\infty) = P_i^{eq}$. We will consider the equilibrium autocorrelation function defined by

$$C(t) = \sum_{j=1}^{N} \{G_{jj}(t) - G_{jj}(\infty)\} P_j^{eq}. \tag{7}$$

When we associate with each level i an independent, randomly chosen, physical quantity M_i (e.g. a magnetization) then $C(t)$ is proportional to the equilibrium autocorrelation function of that quantity.

De Dominicis et al. [4] made a simple choice for the transition rates

$$W_{ij} = w_o e^{-\beta\epsilon_i}, \tag{8}$$

with w_o a constant, and with that choice one can find stretched exponential relaxation

$$C(t) \simeq \exp\left(-\left(\frac{t}{\tau_o(\beta)}\right)^{\frac{\rho}{\beta}}\right) \quad \text{as } t \to \infty, \tag{9}$$

where $\tau_o(\beta)$ is a temperature dependent relaxation time. We considered more general transition rates

$$W_{ij} = w_o \exp\left\{-\beta(1-q)\epsilon_i + \beta q\epsilon_j\right\}, \tag{10}$$

where q may range between 0 and 1, and calculated the asymptotic behaviour of the equilibrium autocorrelation function [5].

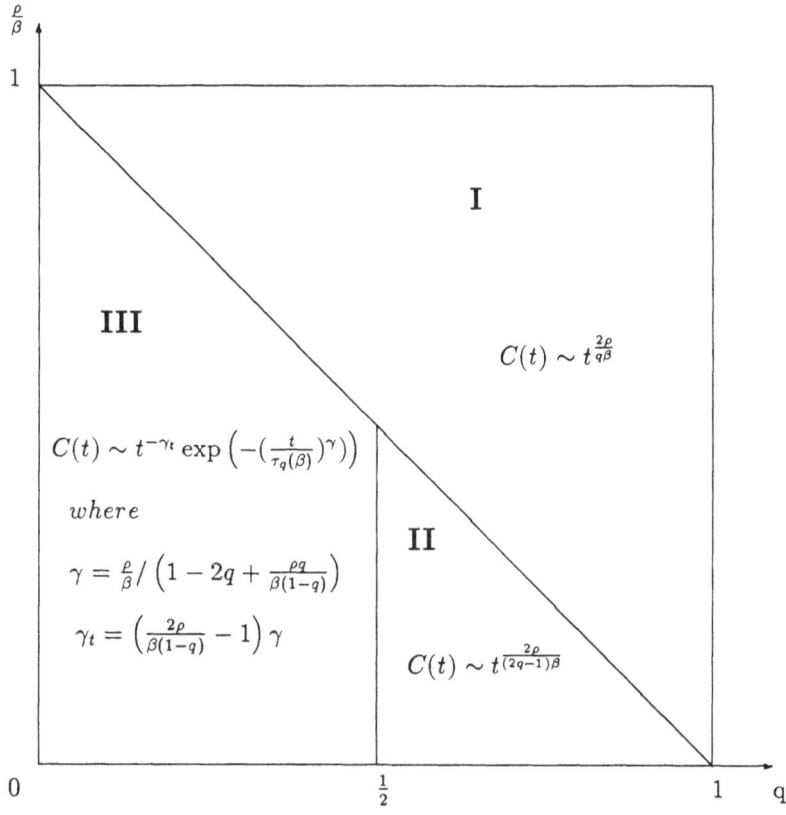

Figure 1: *Exact results for the asymptotic behaviour of the equilibrium autocorrelation function.*

There appear to be three regimes as is depicted in Fig. (1). In region **I**, where $\frac{\rho}{\beta} > 1 - q$ we find power law relaxation after scaling the transition rates with N in order to obtain finite results

$$w'_o = w_o N^{-1+\frac{\beta}{\rho}(1-q)} \tag{11}$$

If we do not scale the transition rates with N the relaxation is trivial. In region **II** we also find power law behaviour but with a different power. The limit $q \to 1$ for both regimes corresponds to the case that we considered elsewhere [6]. Finally, in region **III**, we find stretched exponential behaviour.

Final remarks

From our results we conclude that the choice of transition rates largely determines the kind of relaxation for this mean field model. This result is of particular interest since attempts have been made [7] to fit experimental relaxation data to the stretched exponential based on Eq. (9). Our calculation shows that more reflection is needed before one confronts experiments with a simple theory as the kinetic REM.

Part of this research was supported by the *"Stichting Fundamenteel Onderzoek der Materie" (FOM)*, which is financially supported by the *"Nederlandse Organisatie voor Zuiver Wetenschappelijk Onderzoek" (ZWO)*.

References

[1] B. Derrida, Phys. Rev. Lett. **45**,79 (1980); Phys. Rev. **B24**,2613 (1981).

[2] K. Binder and A.P. Young, Rev. Mod. Phys. **58**,801 (1986) and references therein.

[3] M. Mézard, G. Parisi, N. Sourlas, G. Toulouse and M. Virasoro, Phys. Rev. Lett. **52**,1146 (1984); J. Physique **45**,843 (1984).

[4] C. De Dominicis, H. Orland, and F. Lainée, J. Phys. (Paris) Lett. **46**,L463 (1985).

[5] G.J.M. Koper and H.J. Hilhorst, to be published.

[6] G.J.M. Koper and H.J. Hilhorst, to appear in Europhys. Lett.

[7] R. Hoogerbeets, Wei-Li Luo and R. Orbach, Phys. Rev. Lett. **55**,111 (1985); Phys. Rev. **B33**,6531 (1986); Phys. Rev. **B34**,1719 (1986).

SCALING THEORIES FOR ANOMALOUS DYNAMICS ON FRACTALS: FRACTONS

Amnon Aharony

School of Physics and Astronomy
Raymond and Beverly Sackler Faculty of Exact Sciences
Tel Aviv University, Tel Aviv 69978, Israel

Ora Entin-Wohlman

Laboratoire de Physique des Solides
Universite Paris-Sud
91405 Orsay, France

R. Orbach

Department of Physics, University of California
Los Angeles, California 90024, U.S.A.

1. INTRODUCTION

Many systems in nature exhibit different geometrical structures at different length scales. Although they are homogeneous (on the average) for large length scales, they exhibit *self-similarity* on short length scales. In the latter case such systems may be modeled by *fractal* structures.[1] On a fractal structure, all the physical properties behave as *powers* of the relevant length scale. This behavior crosses over to a *homogeneous* one, similar to that of usual bulk matter, at a *crossover length*, called ξ. Assuming that ξ is the only relevant length scale in the problem, the dependence on any other length L should arise via the ratio L/ξ. A typical intrinsic quantity χ will have the "fractal" behavior $\chi \sim L^X$ for $L \ll \xi$, and will crossover to a size independent behavior, $\chi \sim \xi^X$, for $L \gg \xi$. The dependence of χ on L and ξ usually has the *scaling* form

$$\chi(L,\xi) = L^X f_\chi(L/\xi). \tag{1.1}$$

Details of such crossovers, for various physical properties, have been the subject of much recent research.

Although there exist many physical systems with such crossover behavior, much of the detailed analysis has been devoted to dilute systems, near their percolation threshold.[2] As the concentration, p, approaches the threshold, p_c, the percolation connectedness correlation length ξ diverges as $\xi \sim |p-p_c|^{-\nu}$, and one has a fractal behavior of individual connected clusters on scales $L < \xi$. Particular attention has been devoted to the infinite incipient cluster, whose density grows for $p > p_c$ as $P_\infty \sim (p-p_c)^\beta$. Although most of our results are applicable to the general case, we shall often use the language of percolation theory.

In this paper we concentrate on *dynamical excitations*, (e.g., due to vibrational or spin-wave modes.) Such an excitation is characterized by a frequency, ω, and by a typical length, λ, and these are related via a dispersion relation, $\omega=\omega(\lambda)$. In the simplest situation, one expects a crossover between the "homogeneous" dispersion relation, for $\lambda \gg \xi$, to a "fractal" one, for $\lambda \ll \xi$. This crossover then occurs at the frequency $\omega_c = \omega(\xi)$. Much of this paper will discuss this crossover. We start with a short review, of the crossover in the mass density, conductivity and diffusion (Sec. 2). Using the analogy between the master equation for random walks and other linear problems, we then introduce the crossover from phonons to fractons (Sec. 3), and describe its effects on the density of states (Sec. 4). The last two sections then describe very recent results on the lifetime of such excitations and on the structure factor for scattering from them.

2. CROSSOVER SCALING

The clearest example of crossover concerns the *mass density* of sites on the percolating infinite incipient cluster.[3] In d (Euclidean) dimensions, this is expected to behave as

$$\rho(L) = L^{D-d} f_\rho(L/\xi), \tag{2.1}$$

where D is the *fractal dimensionality*.[1-3] The scaling function $\rho(x)$ has limiting properties such that $\rho(L) \sim L^{D-d}$ for $L \ll \xi$, independent of ξ, and $\rho = P_\infty \sim \xi^{D-d} \sim (p-p_c)^\beta$ for $L \gg \xi$.

The basic idea here is that for $L > \xi$ one can break the system into $(L/\xi)^d$ blocks, of linear size ξ. Within each of these, the behavior is self-similar, with $\rho \sim L^{D-d}$. The mass of each block is thus of order ξ^D, and that of the whole sample adds up to $\xi^D(L/\xi)^d \sim L^d P_\infty$.

A similar idea is applied to the *conductivity*, when each bond represents a unit resistor. For $L < \xi$, one believes that there exists practically only one conducting route between terminals at distance L. The resistance of such a route scales as $R(L) \sim L^{\tilde{\zeta}_R}$. Adding $(L/\xi)^d$ blocks of size ξ thus yields a conductance $\Sigma \sim (L/\xi)^{d-2}/R(\xi)$, or a conductivity $\sigma \sim \xi^{-\tilde{\mu}} \sim (p-p_c)^\mu$, with

$$\tilde{\mu} = \mu/\nu = d-2 + \tilde{\zeta}_R. \tag{2.2}$$

More generally, $\sigma(L) \sim L^{-\tilde{\mu}} f_\sigma(L/\xi)$.

The Einstein relation now enables us to consider the *diffusion coefficient* for random walks on the infinite incipient cluster.[4] This must behave as

$$D(L) = \sigma(L)/P_\infty(L) = L^{-\theta} f_D(L/\xi), \tag{2.3}$$

with $\theta=(\mu-\beta)\nu$. This implies that, in the fractal regime, the mean square distance travelled by the random walker after t steps scales as[4] $\langle r^2 \rangle \sim t^{2/d_w}$, or $t \sim r^{d_w}$, with the fractal dimensionality of the random walker

$$d_w = 2 + \theta = D + \tilde{\zeta}_R. \tag{2.4}$$

3. PHONONS AND FRACTONS

The motion of the random walker is governed by a master equation,

$$\frac{d}{dt} P_i(t) = \sum_j (W_{ji} P_j - W_{ij} P_i). \tag{3.1}$$

A similar linear equation governs the dynamics of many other properties of dilute systems. In particular, this is true for isotropic vibrational modes or for spin waves in antiferromagnets. In these cases, the time derivative is replaced by a second derivative. Therefore, the frequency of such an excitational mode behaves in the fractal regime as[5-7] $\omega \sim \lambda^{-(1+\theta/2)}$, where λ is a characteristic length scale of the mode. Since the fractal regime is very inhomogeneous, one expects such a mode to be localized, and one identifies λ as its localization length. These modes were called[5,6] "fractons". Note that the spatial disorder in the fractal regime is large, and the nature of the fracton states is quite different from those occurring in Anderson's weak localization.[8]

For $\lambda \gg \xi$, the dispersion relation should cross over to the linear phonon-like behavior, $\omega = c\lambda^{-1}$, where λ is an appropriate wave-length and c is the velocity of sound in the homogeneous regime. Combining the two limits, one is led to a scaling dispersion relation,[5,6]

$$\omega(\lambda^{-1}) = \lambda^{-(1+\theta/2)} f_\omega(\lambda/\xi), \tag{3.2}$$

with the phonon and fracton limits

$$\omega_{ph} \sim c\lambda^{-1} \sim \xi^{-\theta/2} \lambda^{-1}, \quad \lambda \gg \xi \tag{3.3}$$

$$\omega_{fr} \sim \lambda^{-(1+\theta/2)}, \qquad \lambda \ll \xi. \tag{3.4}$$

These two expressions coincide at a crossover frequency, $\omega_c \sim \xi^{-(1+\theta/2)}$.

Note that the whole discussion can be repeated for excitations governed by a first time derivative, as in Eq. (3.1). A useful example concerns spin waves in ferromagnets. Here, Eq. (3.2) is replaced by

$$\omega(\lambda^{-1}) = \lambda^{-(2+\theta)} \tilde{f}_\omega(\lambda/\xi), \tag{3.5}$$

and one can rephrase many of the following statements for this case.

4. DENSITY OF STATES

If we consider only the excitational modes on the infinite incipient cluster, then the phonon density of states per atom behaves as

$$N_{ph}(\omega) \sim \xi^{d-D} c^{-d} \omega^{d-1} . \tag{4.1}$$

Eq. (4.1) can also be written in the form $N_{ph}(\omega) \sim \omega^{\bar{d}-1} (\omega/\omega_c)^{d-\bar{d}}$, where

$$\bar{d} = 2D/(2 + \theta) \tag{4.2}$$

is the fracton dimensionality.[5,6] If ω_c is the only frequency scale in the problem, then the density of states should generally have the scaling form,[6,9,10]

$$N(\omega) = \omega^{\bar{d}-1} f_N(\omega/\omega_c) . \tag{4.3}$$

For the fracton regime, $\omega \gg \omega_c$, we expect $N(\omega)$ to become independent of ω_c, hence[5]

$$N_{fr}(\omega) = A\omega^{\bar{d}-1} \ . \tag{4.4}$$

Eq.(4.4) could also be derived directly, using the Laplace transform relation between the density of states and the probability that a random walker returns to the origin, which scales as[5] $1/r^D \sim 1/t^{\bar{d}/2}$.

At finite ξ, we expect a crossover from $N_{ph}(\omega)$, $\omega \ll \omega_c$, to $N_{fr}(\omega)$, $\omega \gg \omega_c$. Since we normalized $N(\omega)$ per atom, we expect that $\int_0^{\omega_D} N(\omega)d\omega = $ const., where ω_D is the Debye frequency. At p_c, this yields the amplitude in Eq. (4.4), $A = \bar{d}/\omega_D^{\bar{d}}$. For $p > p_c$, we expect $N_{fr}(\omega)$ to have the same value, given by Eq.(4.4), as long as $\lambda \ll \xi$, or $\omega \gg \omega_c$. However, for $\omega \ll \omega_c$, $N(\omega)$ becomes $N_{ph}(\omega) \sim \omega^{d-1}$, which is much lower than the contination of $N_{fr}(\omega)$ down to those frequencies (since $\bar{d} < d$). The only way to keep the total integral of $N(\omega)$ fixed is to have $N(\omega) > N_{fr}(\omega)$ near the crossover frequency ω_c. We therefore predict[9,10] the qualitative behavior shown in Fig.1, with an apparent step ΔN in $N(\omega)$ near ω_c. From our scaling form (4.3), this "step" is of order $\omega_c^{\bar{d}-1} \sim \xi^{-(D-1-\theta/2)}$, i.e., slowly decreasing as $p \to p_c$.

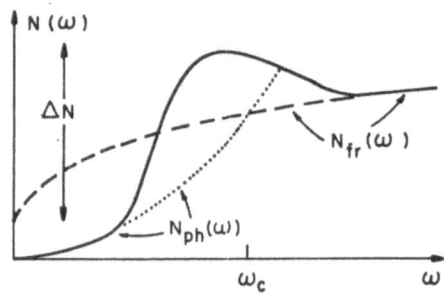

Fig.1: A sketch of the density of (isotropic) vibrational states near percolation.

Note that so far we discussed only the modes on the infinite incipient cluster. If we were to average over all clusters, then the fractal dimensionality of the random walk would change into $\bar{d}_w = (2+\theta)/(2-\beta/\nu)$. Similarly, the factor ξ^{d-D} would not appear in Eq. (4.1) and the fracton density of states, Eq. (4.3), would be replaced by[9]

$$\bar{N}(\omega) = \omega^{p-1} \bar{f}_N(\omega/\omega_c), \tag{4.5}$$

with

$$p = 2d/(2+\theta). \tag{4.6}$$

Although densities of state which exhibit the shape of Fig.1 should always accompany a crossover from phonons to fractons, they may also arise due to other causes (e.g., dimensional crossover from d=3 to d=2 or d=1. This generated some recent "warnings on fractal fashions".[11] These "warnings" have nothing to do with the systems discussed here. We emphasize again, that if the crossover frequency ω_c changes with the concentration as $\omega_c \sim \xi^{-(1+\theta/2)} \sim (p-p_c)^{\nu(1+\theta/2)}$ then there should be no doubt that the crossover is caused by fractons, and not by other mechanisms.[9,10]

Finally, two comments about $\bar{d}$. Using existing values of μ,β and ν, it was observed that $\bar{d}$ is very close to 4/3 for percolation clusters in $d\geq2$. Although this conjecture breaks down[12] very weakly in $d=6-\epsilon$, it remains controversial in $d=2$.[13] For other systems, e.g., lattice animals, the conjecture is known to break down slightly.[14] Published attempts to derive $\bar{d}$ from the geometry of "growth" sites remain inconclusive.[15,16]

The value $\bar{d}\simeq4/3$ arises only for scalar elasticity. When bending forces dominate the vibrational modes,[17] one should use a tensor version of Eq.(4.1), and one ends up with higher values of μ, i.e., lower values of $\bar{d}$. However, in many situations one still expects the scalar elasticity to dominate.[18]

5. LIFETIMES

The "pure" phonon states are described, in the fully homogeneous regime, as plane waves with wavelength λ and frequency $\omega=c/\lambda$. In the absence of any disorder, these have an infinite lifetime. In the presence of some weak disorder, a pure plane wave will decay with time, via scattering into other waves. For weak elastic scattering (e.g., from lattice inhomogeneities), the corresponding lifetime was given by Rayleigh,[19]

$$1/\tau(\omega) = \omega^{d+1} \ \omega_0^{-d}. \tag{5.1}$$

A direct way to derive Eq. (5.1) uses the Golden Rule,

$$1/\tau(\omega) = N(\omega)\left| V \right|^2, \tag{5.2}$$

where V is a matrix element for transition from the initial plane wave to another wave with the same frequency, and the bar denotes averaging over the random variables. The structural disorder involves matrix elements of $K_{ij}(\phi_i-\phi_j)^2\sim\bar{K}(\vec{\nabla}\phi)^2$, where ϕ is the atomic displacement. Since normal modes have an amplitude of order $\omega^{-1/2}$, since $|\vec{\nabla}\phi|\sim\phi/\lambda\sim\phi\omega/c$, and since in a region of size ξ one has $\bar{K}^2\sim\xi^D c^4$, we recover[20] Eq. (5.1), with $\omega_0=\omega_c$.

Using our standard scaling approach, we now rewrite Eq. (5.1) as [20]

$$1/\tau(\omega) = \omega \ f_\tau(\omega/\omega_c). \tag{5.3}$$

If, indeed, ω_c is the only frequency scale in the problem, then an extrapolation of Eq. (5.3) to the fracton regime is possible only if in that regime $f_\tau(x)$ becomes a constant, i.e.,

$$1/\tau(\omega) \sim \omega. \tag{5.4}$$

This is equivalent to the Ioffe-Regel criterion for localization,[21] in which λ is of the same order as the mean free path $\ell=c\tau$. Usually, one expects the weak scattering Rayleigh result (5.1) to break down at a Ioffe-Regel threshold, ω_{IR}, at which $\omega_{IR}\tau(\omega_{IR})\sim1$. Scaling implies that Eq. (5.4) must hold for all $\omega>\omega_c=\omega_{IR}$. Combining Eqs. (5.1) and (5.4) we thus expect the behavior shown in Fig. 2.

It is not trivial to identify what one means by the lifetime of the fracton states, which were already obtained as strongly localized states. One way to address this difficulty is to attempt to replace the strong scattering in real Euclidean space by a relatively weak scattering on an equivalent fractal geometry. Coarse grain the fractal space into units of size λ_{fr}. Since the geometry here is fractal, the strain $\vec{\nabla}\phi$ scales as $\phi/R(\lambda_{fr})$, where $R(x)$ scales as the resistance between points at a distance x apart, $R(x)\sim x^{\tilde{\zeta}_R}$. This follows from the analogies between the vibrational problem and that of the resistor network. Assuming that the pure fracton mode decays because of weak structural fluctuations within the scale λ_{fr}, and assuming that we can use the Golden Rule, we now find that[20]

$$1/\tau(\omega) \sim \omega^{\bar{d}-1} \ (\omega^{3-2\bar{d}})^2 \sim \omega^{5-3\bar{d}} \ . \tag{5.5}$$

This result agrees with the Ioffe-Regel one, Eq. (5.4), only if $\bar{d}$ has the value 4/3! Thus, if $\bar{d} = 4/3$ we recover Fig. 2.

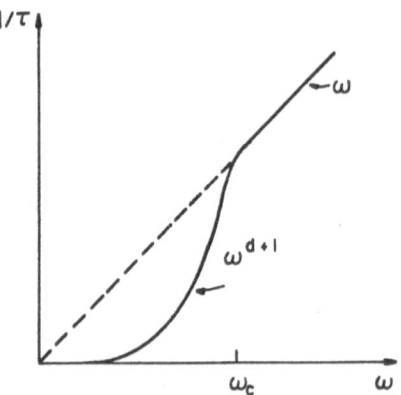

Fig.2: Schematic plot of $1/\tau$, crossing over from Rayleigh scattering ($1/\tau \sim \omega^{d+1}$) to the Ioffe-Regel behavior ($1/\tau \sim \omega$).

If $\bar{d} \neq 4/3$, then Eq. (5.5) must be written as

$$1/\tau(\omega) = \omega(\omega/\omega_{IR})^{4-3\bar{d}} \ , \tag{5.6}$$

with a new, non-critical frequency ω_{IR}. For the more common case, $\bar{d} < 4/3$, we then expect a crossover from Eq. (5.1) to Eq. (5.6) at ω_c, given by $(\omega_c/\omega_0)^d = (\omega_c/\omega_{IR})^{4-3\bar{d}}$. If ω_{IR} is smaller than the Debye frequency, then for $\omega > \omega_{IR}$ Eq. (5.6) would imply that $\omega\tau(\omega) < 1$, which is unacceptable. A possible resolution of this difficulty is shown in Fig. 3, in which we speculate that the Ioffe-Regel behavior (5.4) is recovered for the high frequency fractons. Thus, the excited levels somehow rearrange themselves to resume the value $\bar{d} = 4/3$!

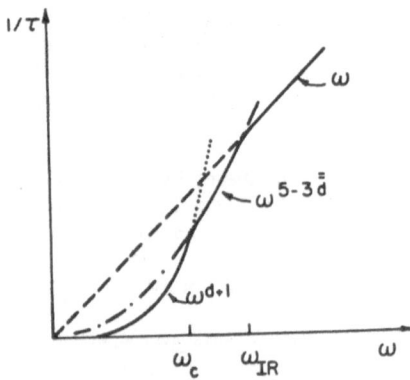

Fig.3: Schematic plot of $1/\tau$, when $\bar{d} < 4/3$.

Brillouin scattering experiments on silica aerogels[22] seem to yield the phonon dispersion law (3.3) and the Rayleigh lifetime (5.1), with $\omega_c = \omega_0$, as expected if our scaling (5.3) holds. Unfortunately, these data do not extend beyond the "phonon" regime. On the other hand, the Ioffe-Regel prediction (5.4) seems to be consistent with recent neutron scattering data from the dilute antiferromagnet[23] $Mn_{0.5}Zn_{0.5}F_2$.

6. STRUCTURE FACTOR FOR SCATTERING

The scattering experiments mentioned above contain much more information on the scattering structure factor $S(q,\omega)$. In the present section we briefly summarize our recent calculations[24] of $S(q,\omega)$. An effective medium calculation of $S(q,\omega)$ was recently done in Ref. 25.

The scattering structure factor $S(q,\omega)$ is given by an imaginary part of the retarded Green's function, as a function of the scattered particle's energy $\hbar\omega$, for various values of its wave vector q,

$$S(q,\omega) = -2(1-e^{-\hbar\omega/kT})^{-1} \, \text{Im} \, G^R(q,\omega), \qquad (6.1)$$

with

$$G^R(q,\omega) = \sum_\alpha \frac{1}{2\rho\omega_\alpha} I_\alpha(q) \left[\frac{1}{\omega-\omega_\alpha+i\eta} - \frac{1}{\omega+\omega_\alpha+i\eta} \right], \qquad (6.2)$$

where ω_α is the frequency of the α'th vibrational mode and

$$I_\alpha(q) = \int d^d r \, d^d r' \, e^{i\vec{q}\cdot(\vec{r}-\vec{r}')} \, \phi_\alpha(\vec{r})\phi_\alpha^*(\vec{r}') \,, \qquad (6.3)$$

where $\phi_\alpha(\vec{r})$ is the mode's wavefunction. In Eq. (6.2), η is an infinitesimal positive number, representing the imaginary part of the frequency. We attempt to evaluate the structure factor by replacing it by the excitation inverse lifetime, $1/\tau(\omega)$, discussed in the previous section. It then remains to evaluate $I_\alpha(q)$ (Eq. (6.3)).

For phonons, $I_\alpha(\vec{q}) \sim \delta(\vec{q}-\vec{k}_\alpha)$, where $\vec{k}_\alpha$ is the phonon wave vector, as long as $q\xi \ll 1$. Thus the phonon part $(\omega_\alpha < \omega_c)$ of the sum in Eq. (6.2) contributes

$$S_{ph}(q,\omega) \sim \frac{1}{cq} \left[\frac{\eta_{ph}(cq)}{(\omega-cq)^2+\eta_{ph}^2(cq)} - \frac{\eta_{ph}(cq)}{(\omega+cq)^2+\eta_{ph}^2(cq)} \right], \quad q\xi \ll 1 \,, \qquad (6.4)$$

where $\eta_{ph}(cq)$ is given by Eq. (5.1), and thus $\eta_{ph}(cq) \ll cq$. For $q\xi \ll 1$ we therefore expect a very sharp structure factor, peaked at $\omega = cq$ (Fig. 4a), associated with the phonon excitations. At $q\xi \gg 1$, $I_\alpha(q)$ for $\omega_\alpha < \omega_c$, namely of the phonon regime, is very small, decaying as q^{-D}, and its contribution to S is neglected.

For fractons, $\phi_\alpha(\vec{r})$ is localized over a range $|\vec{r}| < \lambda_{fr}$. Thus, we can estimate $I_\alpha(q)$ by λ_α^D if $|\vec{q}| < 1/\lambda_\alpha$, and by a quick decay to zero for larger $|\vec{q}|$. Using $\lambda_\alpha \sim \omega_\alpha^{-2/(2+\theta)} \sim \omega_\alpha^{-\bar{d}/D}$, we have $N(\omega_\alpha)I_\alpha(q) \sim \omega_\alpha^{-1}$ for $|\vec{q}| < 1/\lambda_\alpha$, and we approximate $I_\alpha(q)=0$ for $|\vec{q}| > 1/\lambda_\alpha$. Therefore, the sum over the fracton part of Eq. (6.2) becomes

$$S_{fr}(q,\omega) \sim \int_{\omega_x}^{\omega_D} \frac{d\omega_\alpha}{\omega_\alpha} \frac{4\omega\eta(\omega_\alpha)}{(\omega_\alpha^2-\omega^2)^2+2\eta(\omega_\alpha)^2(\omega^2+\omega_\alpha^2)+\eta(\omega_\alpha)^4} , \qquad (6.5)$$

with

$$\omega_x = \max\{\omega_c, q^{1+\theta/2}\} , \qquad (6.6)$$

and with $\eta(\omega_\alpha) = 1/\tau(\omega_\alpha)$ approximated by Eq. (5.4) or Eq. (5.6).

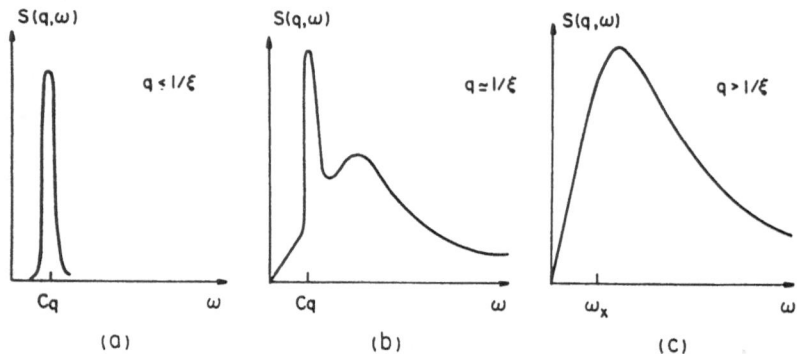

Fig. 4: Structure factor for scattering from (a) phonons; (b) phonons and fractons; (c) fractons.

Note that Eq. (6.5) has non-zero values for all frequencies. In particular, note that $S_{fr}(q,\omega) \sim \omega$ for $\omega \ll \omega_x$, independent of any details on $\eta(\omega_\alpha)$. Also, for $\omega > \omega_D$ we have $S \sim 1/\omega^3$. Fig. 4c shows a schematic curve of $S(q,\omega)$, dominated by the fracton contribution for $q \gg 1/\xi$. We typically find that S is peaked around 30%-50% above ω_x.

When q is near $1/\xi$, we expect the phonon and the fracton parts in $S(q,\omega)$ to add up. This yields curves like Fig. 4b. It is very gratifying that very similar curves were observed experimentally,[23] and obtained numerically, within the effective medium approximation.[25] Fits to Eq. (6.5) may yield much better values for $\eta(\omega_\alpha)$, hopefully distinguishing between Eqs. (5.4) and (5.6).

7. CONCLUSION

Scaling theory, and the assumptions of a single length scale (ξ) related to a single frequency scale (ω_c), prove to have very powerful predictive powers. Detailed experimental checks of these predictions could confirm the assumptions of the fracton model, or show how it could be extended.

ACKNOWLEDGEMENTS

We enjoyed discussions on experiments with R. J. Birgeneau and E. Courtens. This research was supported in part by grants from the National Science Foundation, through grant 84-12898, the U.S.-Israel Binational Science Foundation and the Israel Academy of Sciences and Humanities.

REFERENCES

1. B. B. Mandelbrot, The Fractal Geometry of Nature (Freeman, San Francisco, 1982).
2. For a recent review on percolation theory, see A. Aharony, in Direction in Condensed Matter Physics, edited by G. Grinstein and G. Mazenko (World Scientific, Singapore, 1986), p.1.
3. e.g. A. Kapitulnik, A. Aharony, G. Deutscher and D. Stauffer, J. Phys. A16, L269 (1983).
4. Y. Gefen, A. Aharony and S. Alexander, Phys. Rev. Lett. 50, 77 (1983).
5. S. Alexander and R. Orbach, J. Phys. (Paris) Lett. 43, L-625 (1982).
6. S. Alexander, Ann. Israel Phys. Soc. 5, 149 (1983).
7. R. Rammal and G. Toulouse, J. Phys. (Paris) Lett. 44, L-13 (1983).
8. S. Alexander, Physica 140A, 397 (1986).
9. A. Aharony, S. Alexander, O. Entin-Wohlman and R. Orbach, Phys. Rev. B31, 2565 (1985).
10. S. Alexander et al., Phys. Rev. B28, 4615 (1983).
11. J. A. Krumhansl, Phys. Rev. Lett. 56, 2696 (1986); J. Maddox, Nature 322, 303 (1986).
12. A.B. Harris, S. Kim and T. C. Lubensky, Phys. Rev. Lett. 53, 743; 54, 1088 (E) (1984).
13. D. C. Hong et al., Phys. Rev. B30, 4083 (1984); J. G. Zabolitzky, ibid, 4077; H. J. Herrmann et al., ibid, 4080; C. J. Lobb and D. J. Franck, ibid, 4090; R. Rammal et al., ibid, 4087; J. W. Essam and F. M. Bhatti, J. Phys. A18, 3577 (1985).
14. S. Havlin et al., Phys. Rev. Lett. 53, 178 (1984).
15. F. Leyvraz and H. E. Stanley, Phys. Rev. Lett. 51, 2048 (1983).
16. A. Aharony and D. Stauffer, Phys. Rev. Lett. 52, 2368 (1984); for a recent review, see A. Aharony, in Scaling Phenomena in Disordered Systems, ed. by R. Pynn and A. T. Skjeltorp (Plennum, N. Y., 1985), p. 289.
17. I. Webman and Y. Yantor, in Kinetics of Aggregation and Gelation, ed. by F. Family and D. P. Landau (North Holland, Amsterdam, 1984), p. 133.
18. S. Alexander, J. Phys. (Paris) 45, 1939 (1984).
19. Lord Rayleigh, Theory of Sound (McMillan, London, 1986), Vol. II.
20. A. Aharony, S. Alexander, O. Entin-Wohlman and R. Orbach, Phys. Rev. Lett. 58, 132 (1987).
21. A. F. Ioffe and A. R. Regel, Progress in Semiconductors 4, 237 (1960); N. F. Mott, Phil. Mag. 19, 835 (1969).
22. E. Courtens et al., Phys. Rev. Lett. 58, 128 (1987).
23. Y. J. Uemura and R. J. Birgeneau, Phys. Rev. Lett. 57, 1947 (1986).
24. A. Aharony, S. Alexander, O. Entin-Wohlman and R. Orbach, in preparation.
25. R. Orbach and K. W. Yu, J. Appl. Phys. XX, XXXX (1987).

FRACTON INTERPRETATION OF THERMAL CONDUCTIVITY OF AMORPHOUS MATERIALS

O. Entin-Wohlman

Lab. de Physique des Solides, 91405 Orsay, France*

R. Orbach

Dept. of Physics
University of California, Los Angeles, CA 90024, USA

ABSTRACT

It is shown that the fracton model for the vibrational spectrum of amorphous systems predicts a linear temperature dependence of the thermal conductivity at high temperatures. The mechanism proposed is phonon-assisted hopping of localized fractons and the linear dependence arises from the occupation number of the phonons participating in the process.

1. INTRODUCTION

The thermal properties of amorphous materials exhibit a nearly universal behaviour, very different from that observed in crystals[1]. One has at low temperatures[2] (T < 5K) a specific heat, C, that grows linearly with the temperature T and a thermal conductivity $\kappa \sim T^2$. At higher temperatures (typically up to 10K), the thermal conductivity flattens off into a "plateau" regime, and there is an excess specific heat over the Debye T^3-contribution (usually observed as a bump in C/T^3). At higher temperatures (T > 10K) the thermal conductivity rises slowly with increasing temperature[3].

The lowest temperature regime has been studied very carefully. It is explained[4] in terms of two-level-systems (TLS). There is no generally accepted explanation for the origin of the plateau in κ and the bump in the specific heat, as well as for the higher temperature behaviour of the thermal conductivity. There is also relatively little detailed experimental data available for the latter temperature regime, primarily because of the need for radiative corrections to the determination of κ.

There have been recent suggestions linking the temperature dependence of the thermal conductivity with resonant scattering of phonons from a large peak in the density of non-propagating states at a suitable energy. Karpov and Parshin[5] identify these additional excitations as quasilocalized anharmonic oscillators, whose density of states exhibits Van Hove singularity. In their picture, the plateau in κ is due to resonant scattering of thermal phonons off the anharmonic oscillators. At higher temperatures they find a linear increase, κ ∿ T, arising from the contribution to κ of prethermal phonons (of energy much less than T) which are scattered resonantly by the TLS. The linear temperature dependence is attributed to a reduction in the relative level population of the TLS, proportional to T^{-1}. Thus, this model ascribes the linear temperature dependence of κ above the plateau to a linear decrease in the scattering rate of the heat-carrying states. This appears to be inconsistent with the ultrasonic attenuation in glasses[6], which is observed to increase monotonically with increasing temperature. More recently, Randeria and Sethna[7] have identified the resonant modes of the orientational glass KBr:KCN as the small angular oscillations, "librations" of the individual cyanides. They find that the "traditional" scattering mechanisms (i.e., scattering of phonons off the TLS, Rayleigh scattering) are insufficient to explain the thermal conductivity data in the plateau range, and they propose additional scattering of the phonons by the librational modes. They are able to fit the plateau regime, but the specific heat of the additional modes has a serious quantitative discrepancy with the data.

These explanations are related to the specific properties and parameters of the materials involved. Another approach[8], based upon summation of maximally crossed diagrams involves Rayleigh elastic scattering of the phonons and derives a frequency at which the phonon diffusion coefficient vanishes (i.e., Anderson mobility edge of the phonons[9]) to explain the plateau in κ. However, it does not offer an explanation for the slow increase of κ at temperatures above the plateau[10].

It has been argued previously[11,12] that the thermal properties reflect a crossover behaviour from extended phonon excitations at sufficiently low frequencies to localized fracton excitations at higher frequencies. This idea was first suggested to explain the specific heat and thermal conductivity data of epoxy resins[13]. A surprising feature of the data is the finding that analysis of the κ-data in terms of the kinetic formula (i.e. assuming κ(T) is due to the mode of frequency comparable to T) yields a precipitious drop in the phonon mean free path at about the plateau temperature. Such behaviour was also noted for glasses[2,14], where it was found that the effective mean free path at the plateau temperature was comparable to the thermal phonon wavelength. This suggests localized states according to the Ioffe-Regel criterion[15]. Recent experimental evidence[16] shows that this situation occurs for wave lengths whose magnitude is determined by some structural correlation length, which is larger (by about an order of magnitude) than the microscopic, unit cell size.

The physical picture that emerges from the general features of the thermal properties of amorphous materials is a follows. At about a frequency comparable to the plateau temperature the vibrational modes are localized. The finding that there is an excess specific heat in this range suggests an increase in the vibrational density of states at about that frequency[13]. This, in turn, implies that the localization of the modes is not in the Anderson sense. At the Anderson localization edge the vibrational density of states is still given by its scattering-free form[9]. Moreover, the mean free path and the wave length retain there their separate identities, the mean free path being longer than the phonon wave length[8]. It follows that the localized vibrational modes at frequencies at about and above the plateau temperature are substantially above the Anderson frequency edge, and their nature of localization is different from the Anderson localization[17]. Their special property is that they have peculiar eigenstates described by a single length scale. These excitations were termed fractons[18]. Our purpose here is to show that the fracton model predicts a linear temperature dependence of the thermal conductivity above the plateau temperature, and that this prediction is not sensitive to the detailed assumptions of the model. We shall argue that the linear increase is dominated by phonon-mediated hopping of localized fractons. The predicted linear temperature dependence seems to have been confirmed by experiments[19] over a very wide temperature range[20]

2. FRACTON-HOPPING CONTRIBUTION TO THE THERMAL CONDUCTIVITY

The fracton model is based upon crossover behaviour from phonon excitations at low frequencies ($\omega < \omega_c$) to localized fracton excitations at higher frequencies ($\omega > \omega_c$), where ω_c is the crossover frequency. An amorphous material is homogeneous macroscopically, and is therefore homogeneous on relatively large length scales. The short-range properties reflect the disorder and the large local inhomogeneities. The transition between the two regimes defines a crossover length, ξ. The crossover frequency is related to ξ by the relationship

$$\omega_c \sim \xi^{-D/\bar{\bar{d}}},$$

where D is the fractal dimensionality and $\bar{\bar{d}}$ is the fracton dimensionality[18]. The long wave length, or low frequency vibrational modes, are phonons. The short length scale, or higher frequency harmonic modes, are fractons. They are strongly affected by the disorder, and are therefore qualitatively different from the phonons[17]. In particular, they exhibit a different frequency dependence of the vibrational density of states[18],

$$N_{fr}(\omega) \sim \omega^{\bar{\bar{d}}-1}, \tag{2.1}$$

and different dispersion[21]

$$\omega \sim \lambda_\omega^{-D/\bar{\bar{d}}}, \tag{2.2}$$

relating the localization length of the fracton wave function to the frequency.

Consider now the contribution of the localized fracton modes to the thermal conductivity. Quite generally, the thermal conductivity of the harmonic modes is

$$\kappa(T) = \int d\omega N(\omega)\, C(\omega/k_B T) D(\omega,T), \tag{2.3}$$

where $N(\omega)$ is the vibrational density of states, $C(x) = x^2 e^x/(e^x-1)^2$ is the single oscillator specific heat and $D(\omega,T)$ is the diffusion coefficient for the vibrational mode of frequency ω at temperature T. We want to investigate the thermal conductivity at temperatures above the plateau temperature i.e., $k_B T > \omega_c$. The picture we suggest is that the anharmonic vibrational interaction allows for hopping transitions between different fracton states localized at spatially distinct sites. The diffusion coefficient has the form

$$D(\omega,T) \sim R^2(\omega)/\tau(\omega,T), \tag{2.4}$$

where R is the mean hopping distance and τ is the hopping time.

We first evaluate the hopping time. To this end, we calculate the fracton lifetime caused by the anharmonic process phonon-fracton ↔ fracton. We shall refer to this process as phonon-assisted fracton hopping.

The conventional anharmonic interaction is

$$\mathcal{H} = \frac{\gamma}{V} \int d\vec{r} (\vec{\nabla} \ \vec{u})^3 , \tag{2.5}$$

where γ is the anharmonic interaction constant, V is the volume and $\vec{u}$ the displacement operator. As usual, we expand the displacement in terms of the vibrational normal modes

$$u(\vec{r}) = \sum_{\alpha} \sqrt{\frac{T}{2\rho\omega_{\alpha}}} (\varphi_{\alpha}(\vec{r})b_{\alpha} + \varphi_{\alpha}^{*}(\vec{r})b_{\alpha}^{+}) , \tag{2.6}$$

dropping polarization directions. Here ρ is the average mass density, φ_{α} is the vibrational wave function for the αth mode, ω_{α} its frequency and b_{α} and b_{α}^{+} are the annihilation and cre- ation operators of the αth mode, respectively. Inserting (2.6) into (2.5), the lifetime is extracted from the transition probability per unit time for the state α' to combine with α'' to yield the state α.

The phonon-assisted fracton hopping is generated by the process in which a fracton at the state α omits or absorbs a phonon at state α' and goes into another fracton state α''. Thus, ω_{α} and $\omega_{\alpha''}$ are above ω_c, while $\omega_{\alpha'} \leq \omega_c$. We estimate the fracton hopping rate by assuming that $\omega_{\alpha} \sim \omega_{\alpha''}$ for the frequencies of the fractons involved, while $\omega_{\alpha'} \sim \omega_c$ for the phonon participating in the process. It is then straightforward, though cumbersome, to show that[20] the hopping rate of a fracton at state α is

$$\frac{1}{\tau_{\alpha}} \sim \frac{\gamma^2}{2mc^2} N_{ph}(\omega_c) \left(\frac{\omega_{\alpha}}{\omega_0}\right)^{4q-2} k_B T \exp\left[-\left[\frac{|\vec{R}_{\alpha} - \vec{R}_{\alpha''}|}{\lambda_{\alpha}}\right]^{d_{\phi}}\right] . \tag{2.7}$$

The terms appearing in this expression are as follows: The first factor describes the anharmonic interaction strength in dimensionless form; here c is the velocity of sound. The second term is the phonon density of states at the crossover frequency[21],

$$N_{ph}(\omega) \sim \omega^{d-1} , \tag{2.8}$$

where d is the Euclidean dimensionality of the system. The third factor results from the derivatives of the fracton wave functions [Eqs. (2.5 and (2.6)]; assuming the super-localized form for the fracton wave function[20],

$$\varphi_{\alpha}(r) \sim \left(\frac{1}{r}\right)^{\frac{d-D}{2}} \left(\frac{1}{\lambda_{\alpha}}\right)^{\frac{D}{2}} \exp\left[-\frac{1}{2}\left(\frac{r}{\lambda_{\alpha}}\right)^{d_{\phi}}\right] , \tag{2.9}$$

where d_{ϕ} is the super-localization index and λ_{α} is the frequency-dependent localization length [Eq. (2.2)], the local strain $\partial\varphi_{\alpha}/\partial r$ will in general behave as ω_{α}^{q}, $q = dd_{\phi}/D;$. Combined with $\omega_{\alpha}^{-1/2}$ which appears in the normal mode expansion [Eq. (2.6)], the square of the matrix element gives rise to $(\omega_{\alpha}/\omega_0)^{4q-2}$, where ω_0 is the upper cutoff upon the vibrational spectrum. The next factor in (2.7), $k_B T$, is just the phonon occupation number, which is linear in the temperature because we are considering the temperature range where $k_B T > \omega_c$. Finally, the exponential factor describes the overlap integral between two (localized) fracton wave functions, of about the same frequency ω_{α}, located around the sites

$\vec{R}_\alpha$ and $\vec{R}_{\alpha''}$, respectively [see Eq. (2.9)]. The distance $|\vec{R}_\alpha - \vec{R}_{\alpha''}|$, which appears in the exponent is the hopping distance. We now use a Mott-type estimate to find this distance.

Given that there is a fracton state, of frequencey w_α, there is a probability unity to find another fracton state of frequency $w_\alpha \pm w_c$, at a distance $R(w_\alpha)$ given by

$$N_{fr}(w_\alpha) \, w_c [R(w_\alpha)]^D \sim 1, \qquad (2.10)$$

where $N_{fr}(w)$ is the fracton density of states [Eq. (2.1)]. Eq. (2.10) thus gives us the most probable hopping distance. Note that it is valid when the hopping distance is shorter than ξ, as for such length scales the relevant dimensionality is D. When the hopping distance is longer than ξ, we replace[20] D by the Euclidian dimensionality d in Eq. (2.10). This, however, will not affect the result we shall obtain for the thermal conductivity.

The hopping distance is a decreasing function of the density of states, and therefore of the frequency w_α. However, from Eqs. (2.1), (2.2) and (2.10),

$$[\frac{R(w_\alpha)}{\lambda_\alpha}]^D \sim \frac{w_\alpha}{w_c} > 1, \qquad (2.11)$$

and thus the hopping distance of the fracton is always larger than its localization length.

It follows from these considerations[20] that the most probable hopping rate is

$$\frac{1}{\tau_\alpha} \sim \frac{\gamma^2}{2mc^2} N_{ph}(w_c)(\frac{w_\alpha}{w_c})^{4q-2} k_B T \exp[-[\frac{w_\alpha}{w_c}]^{d_\phi/D}]. \qquad (2.12)$$

and it decreases exponentially as the fracton frequency w_α is increased. This is the most crucial factor in the evaluation of the thermal conductivity. We see that the diffusion coefficient [Eq. (2.4)] will also be proportional to the exponential factor appearing in (2.12), and to the temperature T. Namely,

$$D_{fr}(w_\alpha, T) \sim (\lambda_\alpha w_\alpha^{1/D})^2 \, w_\alpha^{4q-2} k_B T \exp[-[\frac{w_\alpha}{w_c}]^{d_\phi/D}], \qquad (2.13)$$

where the first factor comes from the square of the hopping distance. The fracton hopping contribution to the thermal conductivity [Eq. (2.3)] is thus

$$\kappa(T)|_{hop} \sim k_B T \int_{w_c}^{k_B T} dw_\alpha N_{fr}(w_\alpha)(\lambda_\alpha w_\alpha^{1/D})^2 \, w_\alpha^{4q-2} \exp[-[\frac{w_\alpha}{w_c}]^{d_\phi/D}]. \qquad (2.14)$$

Because of the exponential factor, the integral is dominated by the lower bound of the integration, i.e., by the lowest frequency fractons near the crossover frequency w_c. As a result, the integral in (2.14) is approximately temperature-independent, and the contribution of the phonon-assisted frac-ton hopping to κ is linear in the temperature.

3. CONCLUSIONS

The fracton model suggests the following picture for the temperature dependence of the thermal conductivity. At low temperatures, κ is due to propagating phonons. This contribution saturates at about the plateau temperature. At higher temperatures, our model predicts an <u>additonal</u> contribution to κ. It is dominated by hopping of fractons which is mediated by low frequency phonons, and it is linear in the temperature. The linear temperature dependence results from the occupation number of the phonons, and thus it does <u>not</u> depend upon the dimensionality of the system.

Our findings are confirmed by recent measurements of Oliveira and Rosenberg, carried out on epoxy resins[19,22]. They find a linear tempera-ture dependence up to ~ 100K. In particular, extrapolation of the linear dependence back to T = 0 yields intersection of the ordinate at the value of κ at the plateau. A more recent experimental paper[23] reports thermal conductivity measurements of various glasses up to room temperature. The fracton interpretation is rejected in that paper. Instead, the authors explain their data as heat transport through a random walk of the thermal energy between atoms. The analysis yields a T^2-dependence for temperatures well below the Debye temperature[10], and obtains a rollover from saturation at the Debye energy. Our approach predicts a linear temperature dependence for κ, and a glance at the data exhibited in Ref. 23 will show that such a dependence is observed up to nearly 100K in nearly every sample examined. The saturation above this temperature may well be due to the anharmonic interactions decreasing the fracton lifetime, and thence dimishing the fracton contribution to the thermal conductivity.

ACKNOWLEDGEMENTS

This research has been supported by the U.S. National Science Foundation under grant DMR 84-12898, and by the Fund for Basic Research administered by the Israel Academy of Sciences and Humanities.

REFERENCES

* Permanent address: School of Physics and Astronomy, Tel Aviv University, Tel Aviv 69978, Israel.
1. For a review, see Amorphous Solids: Low Temperature Properties, ed. by W.A. Phillips (Springer, Berlin 1981).
2. R.C. Zeller and R.O. Pohl, Phys. Rev. <u>B4</u>, 2029 (1971).
3. J.J. De Yoreo, R.O. Pohl and G. Burns, Phys. Rev. <u>B32</u>, 5780 (1985).
4. P.W. Anderson, B.I. Halperin and C.M. Varma, Phil. Mag. <u>25</u>, 1 (1972); W.A. Phillips, J. Low Temp. Phys. <u>7</u>, 351 (1972).
5. V.G. Karpov and D.A. Parshin, Sov. Phys. JETP <u>61</u>, 1308 (1985) [Zh. Eksp. Teor. Fiz. <u>88</u>, 2212 (1985).
6. S. Hunklinger and W. Arnold, Physical Acoustics, ed. by W.P. Mason and R.N. Thurston, <u>12</u>, 155 (1976); J.T. Krauss and C.R. Kurkjian, J. Am. Ceram. Soc. <u>51</u>, 226 (1968); C.K. Jones, P.G. Klemens and J.A. Rayne, Phys. Lett. <u>8</u>, 31 (1964).
7. M. Randeria and J.P. Sethna, preprint (1987).
8. E. Akkermans and R. Maynard, Phys. Rev. <u>B32</u>, 7850 (1985).
9. S. John, H. Sompolinsky and M.J, Stephen, Phys. Rev. <u>B27</u>, 5592 (1983).

10. S. Alexander, O. Entin-Wohlman and R. Orbach, in "Phonon Scattering in Condensed Matter ", eds. A.C. Anderson and J.P. Wolfe (Springer, Berlin, 1986) p. 15.

11. S. Alexander, C. Laermans, R. Orbach and H.M. Rosenberg, Phys. Rev. $\underline{B28}$, 4615 (1983).

12. R. Orbach and H.M. Rosenberg, LT-17 Proceedings, ed. by U. Eckern, A. Schmid, W. Weber and H. Wühl (Elsevier Science Publishers B.V., Amsterdam, 1984), p. 375.

13. S. Kelham and H.M. Rosenberg, J. Phys. $\underline{C14}$, 1737 (1981); A.J. Dianoux, J.N. Page and H.M. Rosenberg, Phys. Rev. Lett. $\underline{58}$, 886 (1987).

14. M.R. Zaitlin and A.C. Anderson, Phys. Rev. $\underline{B12}$, 4475 (1975).

15. A.F. Ioffe and A.R. Regel, Prog. Semicond. $\underline{4}$, 237 (1960); N.F. Mott, Phil. Mag. $\underline{19}$, 835 (1969).

16. J.E. Graebner, B. Golding and L.C. Allen, Phys. Rev. $\underline{B34}$, 5696 (1986).

17. S. Alexander, Physica $\underline{140A}$, 397 (1986).

18. S. Alexander and R. Orbach, J. de Physique Lett. $\underline{43}$, L625 (1982).

19. A.K. Raychaudhuri, Ph.D. Thesis, Cornell University (1980), unpublished; J.E. de Oliveira and H.M. Rosenberg, private communication (1986).

20. S. Alexander, O. Entin-Wohlman and R. Orbach, Phys. Rev. $\underline{B34}$, 2726 (1986).

21. S. Alexander, Ann. Isr. Phys. Soc. $\underline{5}$, 149 (1983).

22. See Fig. 2 of Ref. 20.

23. D.G. Cahill and R.O. Pohl, Phys. Rev. $\underline{B35}$, 4067 (1987).

FRACTON EXCITATION IN SILICA SMOKE-PARTICLE AGGREGATES

D. Richter*, T. Freltoft** and J.K. Kjems**

* Institut Laue-Langevin
  F-38042 Grenoble Cedex, France

**Risø National Laboratory
  DK-4000 Roskilde, Denmark

INTRODUCTION

The dynamic properties of random networks are not very well understood [1]. Alexander and Orbach [2] first pointed out that the thermal excitation spectra are strongly influenced by the fractal structure of such systems and they introduced a new dynamical exponent, d_s, to describe the vibrational density of states, $Z(\omega) \propto \omega^{d_s-1}$ for the fracton modes. In normal, homogeneous systems, d_s corresponds to the Euclidian dimension. It was also shown [3] that an anomalous enhancement of the density of states, the so-called fracton edge, could be expected at the crossover between the homogeneous, long wavelength phonon regime and the fracton regime at shorter wavelengths. Neutron scattering experiments on epoxy resins [4,5] and vitrious silica [6] have been interpreted in these terms. In the analysis it is presumed that the fractal nature originates from the chemical bonding network rather than the mass distribution which is quite uniform on length scales that exceed the atomic distances. A recent Brillouin scattering experiment [7] has shown the long wavelength phonon excitations near the expected cross-over to the fracton regime in aerogel samples of different densities. In a neutron scattering experiment [8] on a dilute antiferromagnet the similar magnetic excitations were followed through the cross-over region.

Here we present a spectroscopic determination of the vibrational density of states for ramified clusters with known fractal geometry on the length scale corresponding to the wavelength of the excitations that was probed. The clusters are formed as smoke-particle aggregates of small silica particles. The fractal nature of the geometrical structure has been established by small angle neutron scattering [9] and it can be characterized by a fractal dimension, $d_f = 2.6\pm0.1$ up to length scales of order 30 nm.

EXPERIMENT

The experiment was performed on SiO_2 aggregates commercially available under the trade name Cab-O-Sil and Alfasil [10]. They are produced by the process of flame hydrolysis in which $SiCl_4$ is burned to give a snowlike product in which the basic smallparticle units are amorphous SiO_2 spheres roughly 4 nm in diameter. The incoherent neutron scattering from this system can be dramatically enhanced by studying a hydroxylated sample. This technique was first applied by Richter and Passel [11,12] may be used to obtain the density of states. They exploited the large incoherent cross section of hydrogen and enhanced the incoherent cross section of the sample by attaching H-atoms to the SiO_2 surfaces.

The hydrogen covering of the SiO_2 surface is accomplished by soaking the aggregates in H_2O. During hydroxylation, a surface layer of adsorbed water also forms, but it was removed again by baking the sample at 110°C. The OH groups are, however, much more tightly bound and will first be removed after extensive baking at 700-800°C as was done for a reference sample for 24 hrs (to a residual vapor pressure below 10^{-3} mbar) [12].

The measurements were performed with cold neutrons on a time-of-flight spectrometer IN5 at the Institut Laue-Langevin high-flux reactor in Grenoble. A chopper system supplied a pulsed monochromatic beam of $\lambda = 0.5$ nm (3 meV) incident neutrons, and detectors were placed at the fixed scattering angles $2\theta = 54°$, $67.5°$, $88°$, $107.3°$, and $124°$ covering the q range from 11 nm^{-1} to 22 nm^{-1} at zero energy transfer.

RESULTS

In the quasiharmonic approximation the incoherent neutron scattering experiment measures an amplitude weighted density of states

$$\left(\frac{d^2\sigma}{d\Omega dE}\right)^{inel}_{inc} = \frac{\sigma_{inc}}{4\pi}\frac{k'}{k}\frac{N}{4M} q^2 \exp\left[-\frac{1}{3} q^2 \langle u^2\rangle\right] \frac{Z(\omega)}{\omega} \left\{\coth\left(\frac{\hbar\omega}{2k_BT}\right) \pm 1\right\} \quad (1)$$

The consistency of the measured spectra $I(q,\omega)$ with $\frac{\partial^2\sigma}{\partial\omega\partial\Omega}$ of Eq. (1) can be examined through their q and T variation. In order to test the q dependence of the measured intensities at a given temperature, $\ln[I(q,\omega)/q^2]$ is calculated for each scattering angle. This function is then plotted vs. q^2 for each energy. Eq. (1) predicts a linear relationship in such a plot. This is fulfilled, by the data as can be seen from the inset in Fig. 1.

Extrapolating to q = 0, we obtain:

$$\exp\left[\lim_{q\to o} \ln\{I(q,\omega)/q^2\}\right] \propto \frac{k'}{k} \frac{Z(\omega)}{\omega} \left\{\coth\left(\frac{\hbar\omega}{2k_BT}\right) \pm 1\right\}$$

$$\propto \frac{k'}{k} \frac{Z(\omega)}{\omega 2} k_BT \qquad k_BT \gg \hbar\omega$$

$$(2)$$

From Fig. 1 it is clear that $Z(\omega)$ strongly deviates from the normal Debye behavior, $Z(\omega) \sim \omega^2$. In the energy range from 0.2 meV to 2 meV $Z(\omega)/\omega^2$ follows a power-law with the exponent -0.9 ± 0.05 for T = 265 K, and for T = 136 K we find an exponent of -1.2 ± 0.1. There is no trend towards saturation in the low energy limit indicating that the cross-over to the long wavelength phonon regime is not within the reach of the present experiment.

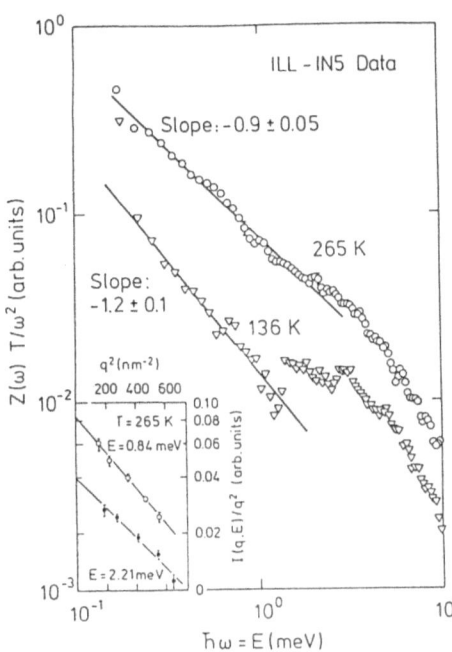

Fig. 1. Double-log plots of $\overline{Z(\omega)T/\omega^2}$ extracted from extrapolating the measured scattering curves to q = o for T = 265 K and T = 136 K. In the insert is shown two of typical extrapolations leading to the two points with error bars on the 265 K data. The slopes correspond to $Z(\omega) \propto \omega^p$; $p = 1.20\pm0.1$ at 136 K.

DISCUSSION

After having established that the inelastic intensity observed in our scattering experiment originates from vibrational excitations, we now remark on the relevance for the fracton problem: We commence with the data taken at 136 K where the peak observed at 1.5 meV corresponds to the lowest small particle mode of a SiO_2-sphere of 25 nm radius. We estimate that $\hbar\omega_{min} = \hbar c a_{10}/R = 1.65$ meV, where c = 4200 m/s is the average sound velocity in fused silica, a_{10} is the first zero of the derivative of the first order spherical Besselfunction and R is the particle radius. Excitations of frequency less than ω_{min} must correspond to the network modes and in this frequency range we measure the density of states of the network excitations. From the slope of $Z(\omega)/\omega^2$ vs. ω of -1.2 we derive a spectral dimension $d_s = 1.8\pm0.1$ (see Eq. (1)). At T = 265 the slope decreases or the spectral dimension increases to $d_s = 2.1\pm0.05$. It is surprising to find a significant temperature dependence of the exponents and simple treatments of anharmonic effects cannot explain this. Furthermore, the high temperature value of $d_s = 2.1$ is larger than the thereoretical limit for localization $\emptyset$ and, taken at face value, this would indicate a profound temperature effect on the nature of the excitations.

To our knowledge these measurements constitute the first direct observations of fracton density of states that exhibits strong deviations from the normal Debye deviations spectrum. The experimental results of $d_s = 1.8$ and $d_s = 2.1$ are both considerably larger than the value $d_s = 4/3$ postulated as universal for percolation clusters by Alexander and Orbach [2]. It also exceeds the value of 1.56 which can be calculated for a Sierpinsky gasket. Our results indicate that there is considerable scope for further investigations both experimental and theoretical to establish the nature of the vibrational modes in random networks and to understand their dynamic stability, the role of anharmonicity and their thermodynamic properties in general.

REFERENCES

1. Scaling Phenomena in Disordered Systems edited by R. Pynn and A. Skjeltorp (Plenum Press, New York, 1985).
2. S. Alexander and R. Orbach, J. Phys. (Paris) Lett. $\underline{43}$, L-625 (1983).
3. A. Aharony, S. Alexander, O. Entin-Wohlman, and R. Orbach, Phys. Rev. $\underline{B31}$, 2565 (1985).
4. H.M. Rosenberg, Phys. Rev. Lett. $\underline{54}$, 704 (1985).
5. R. Orbach in ref. 1. pp. 335-359.
6. U. Buchenau, N. Nücker, and A.J. Dianoux, Phys. Rev. Lett. $\underline{53}$, 2316 (1984) and A.J. Dianoux, J.N. Page and H.M. Rosenberg, Phys. Rev. Lett. $\underline{58}$, 886 (1987).
7. Eric Courtens, Phys. Rev. Lett. $\underline{58}$, 128 (1987).
8. Y.J. Uemura and R.J. Birgeneau, Phys. Rev. Lett. $\underline{57}$, 1947 (1986).
9. T. Freltoft, J.K. Kjems, and S.K. Sinha, Phys. Rev. $\underline{B33}$, 269 (1986).
10. Cab-O-Sil (grade M5) is a trademark of Cabot Corporation; Alfasil was stock no. D 89376 (Alfa Products 1981 Catalogue).
11. D. Richter and L. Passell, Phys. Rev. Lett. $\underline{44}$, 1593 (1980).
12. D. Richter and L. Passell, Phys. Rev. $\underline{B26}$, 4078 (1982).

EVIDENCE FOR PHONON-FRACTON CROSSOVER IN SILICA AEROGELS

BY BRILLOUIN-SCATTERING MEASUREMENTS

E. Courtens[†], J. Pelous*, R. Vacher*, and T. Woignier*

[†]IBM Research Division, Zurich Research Laboratory
8803 Rüschlikon, Switzerland
* Laboratoire de Science des Matériaux Vitreux
Université des Sciences et Techniques du Languedoc
F-34060 Montpellier, France

INTRODUCTION

Systems crossing over from homogeneity to self-similarity at a correlation length ζ are expected to exhibit rather peculiar elastic properties.[1] Phonons of well-defined frequency ω and wavevector q can propagate in such media if their wavelength $\lambda = 2\pi/q$ is larger than ζ. In contrast, a crossover to fractons, the vibrational excitations characteristic of fractal structures,[2,3] should occur as $q\zeta$ increases towards 1.

In this paper, we demonstrate that silica aerogels can have a fractal structure with a fractal (Hausdorff) dimension $D = 2.40\pm0.03$ over an extended range of dimensions L, $\zeta > L > a$, where a is the size of the small homogeneous grains from which the fractal structure is built. The correlation lengths of our samples range from 30 to 300 Å, depending on their macroscopic density ρ. Brillouin scattering, which allows probing acoustic excitations whose wavelengths are as small as 2000 Å, is an appropriate tool for studying the crossover from phonons to fractons in these porous solids. We summarize results giving evidence for such a crossover,[4] and briefly discuss the significance of the fractal and spectral dimensions extracted from the scaling of acoustic dispersion branches.

SAMPLE PREPARATION AND SMALL-ANGLE NEUTRON SCATTERING

Silica gels were prepared from solutions of tetramethoxysilane (TMOS) in methanol to which 4 moles of H_2O per mole of TMOS were added. The hydrolysis, polycondensation, and gelification of the present samples took place under initially neutral conditions. After long aging at 55°C, the alcogels were hypercritically dried to obtain "aerogels".[5] The final density was adjusted by initial dilution of the reactants with methanol. The densities of these aerogels, derived from the ratio of sample weight to geometrical volume, range from 95 to 400 kg/m^3.

Small-angle neutron-scattering (SANS) experiments were carried out on the PACE spectrometer at the Laboratoire Léon Brillouin at Saclay

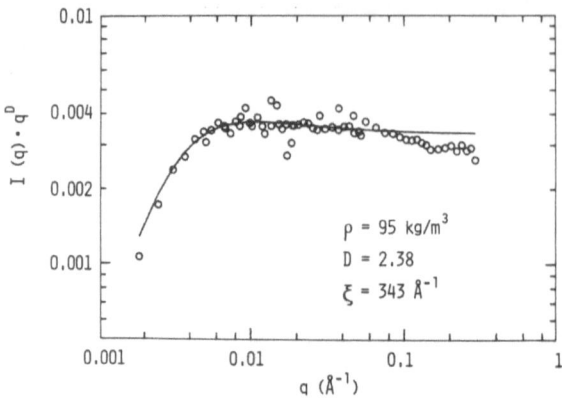

Fig. 1. SANS results on the lightest sample. The curve is a fit to the
theoretical $I(q)$ with the values indicated. It emphasizes that
$I(q) \times q^D$ = constant is only an approximation. The deviations
are within statistical and systematic errors.

(France). The experiments were performed with cold neutrons, at wave
vectors q ranging from ~0.002 to ~0.3 Å^{-1}, on samples of ~2 mm
thickness. Detailed results are published elsewhere.[6] The theory[7,8] for
the scattered intensity $I(q)$ predicts that $I(q) \propto q^{-D}$ in the regime
$1/a > q > 1/\xi$, and that $I(q) \sim$ constant for $q < 1/\xi$. Intensity data on our
lightest sample are presented in Fig. 1. This is a logarithmic plot of
$I(q) \times q^D$ vs. q. This uncommon way of presenting SANS results greatly
enhances all deviations. It emphasizes crossovers from the fractal to
other regimes, and increases the visibility of small experimental
irregularities. Fig. 1 demonstrates that the lightest sample has a
fractal structure over about two orders of magnitude in L, and down to
~3 Å. The fractal behavior is also observed on the other samples, over
a reduced range of L, as ξ decreases and a increases with increasing ρ.[6]
We find that the variation of ξ with ρ scales according to

$$\xi \propto \rho^{1/D-3} , \tag{1}$$

with the *same* value of D as that obtained from the fit of each individual curve $I(q)$. The fact that all samples have a fractal structure
with the same D, and that ξ obeys (1), establishes that all aerogels
are *mutually* self-similar.[6] This property is essential to allow scaling
of the acoustic properties measured on the whole set of samples.

BRILLOUIN SCATTERING RESULTS

The dispersion curves $\omega(q)$ of longitudinal phonons at room temperature were obtained from Brillouin-scattering measurements on six samples
of different ρ. Several scattering angles from backscattering to near
forward configurations, corresponding to wavevectors q from ~0.001 to
~0.0025 Å^{-1}, were investigated. Details on the experimental setup and
the procedure used for extracting data have been given elsewhere.[4] The
backscattering spectra shown in Fig. 2 illustrates the remarkable
changes occurring in the position and line width of Brillouin lines as
a function of the aerogel density. With decreasing ρ, the longitudinal
phonon phase velocity v softens, while the corresponding elastic
scattering width increases.

The analytical form of the acoustic branches in such media has not
yet been well established. A simple way to take into account the
curvature of the dispersion curves is to fit them with a sine law,

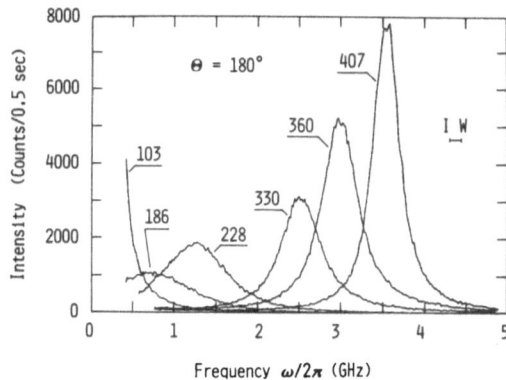

Fig. 2. Backscattering spectra for six densities, in kg/m³. IW is the
full instrumental width. The region near $\omega = 0$, affected by
strong elastic scattering, was removed for clarity. From Ref. 4.
© 1987 The American Physical Society.

$$\omega = (2v/\pi l)\, \sin(\pi q l/2) \; . \tag{2}$$

Here, l is a characteristic length, which must be proportional to ξ if
we assume, as for the simplest crossover model, that the behavior
depends on one length scale only. Therefore, l should scale with ρ
according to (1). The fits gave l, and a determination of v in the
phonon limit, $q l << 1.4$. The scaling of l is found in good agreement with
that of ξ.[4,6] v can be scaled with an exponent x defined by
$v/v_a = (\rho/\rho_a)^x$. This exponent is related to D and $\bar{\bar{d}}$, independently of
the elasticity model, by

$$x = (D - \bar{\bar{d}})/\bar{\bar{d}}(3 - D) \; . \tag{3}$$

It must be noted that only in the case where the velocity v_a and density ρ_a at the scale of the grains are sample independent, can one
extract x direct from a scaling of the velocities. Otherwise, an
effective exponent x' is obtained. If $x = x'$, (3) then gives $\bar{\bar{d}} = 1.25$.[4]

Using the SANS results, an alternative scaling approach can be
followed to analyze these data. It uses the wavevector at crossover
$q_{co} = 1/\xi$, and the crossover frequency $\omega_{co} = v \times q_{co}$. The former is
determined by the SANS results. Assuming a scaling of the dispersion
branches with ρ, these should all collapse to a single curve when
plotted as ω/ω_{co} vs. q/q_{co}. This is shown in Fig. 3. The parameter

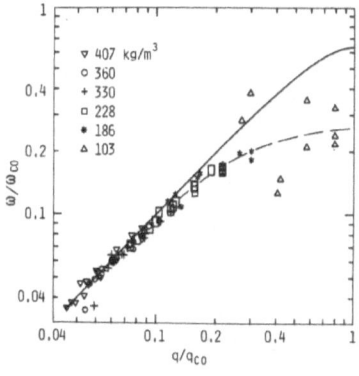

Fig. 3. Scaling of the dispersion curves with $x' = 1.4$, for six samples
as indicated by the various symbols. The solid line is the sine
law. The dashed one is a guide to the eye.

adjusted to obtain scaling is the exponent x' describing the experimental ρ dependence of the velocity. We deduce $x' = 1.4\pm0.2$, independently of any hypothesis on the form of the dispersion branches. Also shown in Fig. 3 is the sine law. We observe that, although the experimental results depart from a sine variation for $q/q_{co} > 0.1$, the scaling with a sine gave correct values for D and x'. From these results, with D = 2.40 and $x = x'$, one finds using (3), $\bar{\bar{d}} = 1.3\pm0.1$.

The linewidth Γ can similarly be scaled. Two interesting conclusions are obtained from a plot of Γ/ω_{co} vs. ω/ω_{co}, not shown here for lack of space:
(1) Γ is proportional to ω^4. This dependence, expected for Rayleigh scattering of plane waves by the irregular network of the gel, is observed practically up to ω_{co};
(2) Γ is equal to ω not far from crossover.
These two observations are in excellent agreement with expectations.[9]

DISCUSSION

A remarkable result of the SANS experiment is that D = 2.40 ± 0.03 for a *large series* of samples, and that those are *mutually* self-similar.[6] This dimension is close to that of the infinite cluster in percolation (D $\simeq$ 2.5). The mutual self-similarity is presumably related to the reaction mechanisms and to cluster-cluster aggregation in the high density regime. The effective value of $\bar{\bar{d}}$ is much closer to 4/3 than to 0.9. The former applies to scalar elasticity,[2,10] while the latter is valid for the more usual tensorial elasticity.[11] The occurrence of an effective exponent with a scalar value can be given various tentative explanations. One is that the gels really have scalar elasticity, owing to large internal stresses related to shrinkage during preparation. Another is that $x \neq x'$. However, the value of x needed to obtain $\bar{\bar{d}} = 0.9$ is twice the measured x'. It seems somewhat unlikely that such a large effect could be produced solely by the ρ dependence of v_a and ρ_a. Finally, it is possible that the scaling depends on more than one length scale. Several models of more complicated scalings do already exist. We hope to be able to return to this subject in later publications.

REFERENCES

1. A. Aharoni, S. Alexander, O. Entin-Wohlman, and R. Orbach, Phys. Rev. B31:2565 (1985).
2. S. Alexander and R. Orbach, J. Phys. Lett. 43:1625 (1982).
3. R. Rammal and G. Toulouse, J. Phys. Lett. 44:L13 (1983).
4. E. Courtens, J. Pelous, J. Phalippou, R. Vacher, and T. Woignier, Phys. Rev. Lett. 58:128 (1987).
5. M. Prassas, J. Phalippou, and J. Zarzycki, J. Mater. Sci. 19:1656 (1984).
6. R. Vacher, T. Woignier, J. Pelous, and E. Courtens, to be published.
7. S. K. Sinha, T. Freltoft, and J. Kjems, in: "Kinetics of Aggregation and Gelation," F. Family and D. P. Landau, eds., Elsevier, Amsterdam (1984).
8. J. Teixeira, in: "On Growth and Form," H. E. Stanley and N. Ostrowski, eds., Nijhoff, Dordrecht (1986).
9. A. Aharony, S. Alexander, O. Entin-Wohlman, and R. Orbach, Phys. Rev. Lett. 58:132 (1987).
10. S. Alexander, J. Phys. 45:1939 (1984).
11. I. Webman and G. S. Grest, Phys. Rev. B31:1689 (1985).

CRITICAL BEHAVIOR OF ELASTIC STIFFNESS MODULI IN A RANDOM NETWORK OF RIGID AND NONRIGID BONDS

Edgardo R. Duering and David J. Bergman

The Raymond and Beverly Sackler Faculty of Exact Sciences, School of Physics and Astronomy, Tel Aviv University, Tel Aviv 69978, Israel

For many years no research was done on elastic properties near the percolation threshold of a composite medium, even though at the same time the electrical properties of such media were being investigated very thoroughly. This was largely due to the mistaken view that the critical behavior would be the same for both sets of properties. Since this misconception was corrected[1-3], the field has attracted a more proportionate degree of attention. The important pioneering papers include Refs. 1-7.

Here we review some of the work done recently at Tel Aviv University in this field, which involves simulations of two-dimensional (2D) elastic networks that are random mixtures of normal elastic bonds and totally rigid bonds.[8-10] The simulations were performed on radnom honeycombe networks in the form of long L x N strips (N >> L) using a transfer matrix method, as described in Ref. 8. We simulated networks at P_c (where $\xi = \infty$) as a function of the strip width L, and we also simulated networks with $p < P_c$ (where $\xi < \infty$) in order to investigate the dependence of elastic properties on ξ/L. The main results were as follows: as the percolation threshold of the rigid bonds P_c is approached from below, all of the stiffness moduli diverge with the same characteristic exponent

$$C_{11} \sim (P_c - P)^{-S}$$

$$\mu \sim (P_c - P)^{-S}$$

$$S \stackrel{\sim}{=} 1.30$$

This is very good evidence that the value of this exponent is the same as that which characterizes the divergence of the conductivity in the analogous normal conductor-superconductor random network. Near P_c, the ratio μ/C_{11} was found to depend on the ratio ξ/L between the percolation correlation length ξ and the linear sample size L, but not on the microscopic force constants. Its value varies from about 0.67 when $\xi >> L$ to about 0.46 when $\xi << L$, and this translates to a Poisson ratio σ that varies from -0.33 to +0.08. These values are unusually low: no naturally occurring solid has ever been found that has a negative σ, and most of them have values within the rather narrow range 0.30 - 0.35.

More recently we have investigated the scaling behavior of the stiffness moduli in a network which is a mixture of bonds of two types - a stiff component (but with non-infinite force constants) and a soft component (but with non-zero force constants). In Figs. 1 and 2, we show results for the scaled modulus $C_{11}/L^{S/\nu}$ as a function of each of the scaling parameters

$$\tilde{m} \equiv \frac{m_1/m_2}{L^{-(T+S)/\nu}} \quad ; \quad \tilde{k} \equiv \frac{k_1/k_2}{L^{-(T'+S)/\nu}}$$

where k_i, m_i are the stretching and bending force constants of the soft bonds (i=1) and the stiff bonds (i=2). Although we had originally expected both of the ratios m_1/m_2 and k_1/k_2 to scale in the same way, we in fact found that while m_1/m_2 scales as $L^{-(T+S)/\nu}$, where $T \cong 3.96$, the other ratio k_1/k_2 scales as $L^{-(T'+S)/\nu}$ where $T' \cong 1.30$, i.e., the same as the conductivity critical exponent.

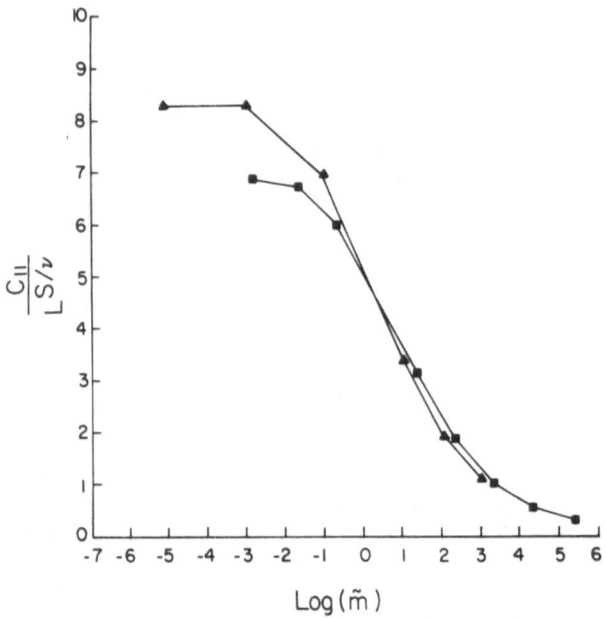

Fig. 1: Values of $C_{11}/L^{S/\nu}$ plotted vs. $\tilde{m}$ with T = 3.96 for L=10 (■) and L=40 (▲), when k_1/m_1=15.82 (C_{11}=2.82) and k_1/k_2=10^{-9}.

In Fig. 3 we show $C_{11}/L^{S/\nu}$ as a function of $\tilde{m}$ for the case when $k_1/k_2 = m_1/m_2$ and for different values of L, k_1, m_1. We note that the result is almost independent of the value of C_{11} in component No. 1. This means that the scale of C_{11} for the mixture is entirely determined by the shear modulus μ_1 of component No. 1. We may summarize these results by writing the following scaling form for C_{11} of the mixture

$$C_{11} \cong A \, \mu_1 \, L^{S/\nu} \, F\left[\tilde{m}, \, \tilde{k}, \, \frac{k_1}{m_1}\right] ,$$

where A is a numerical factor of order 1.

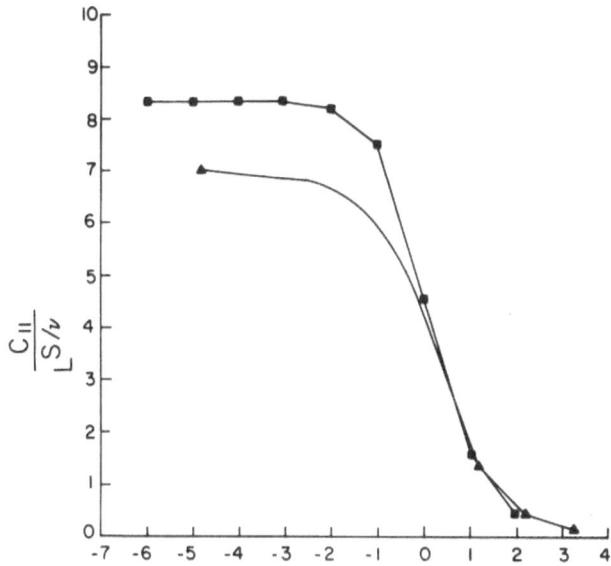

Fig. 2: Values of $C_{11}/L^{S/\nu}$ plotted vs. $\tilde{k}$ with $T = 1.3$ for $L = 10$ (▲) and $L = 40$ (■), for $k_1/m_1 = 15.82$ ($C_{11} = 2.82$) and $m_1/m_2 = 10^{-9}$.

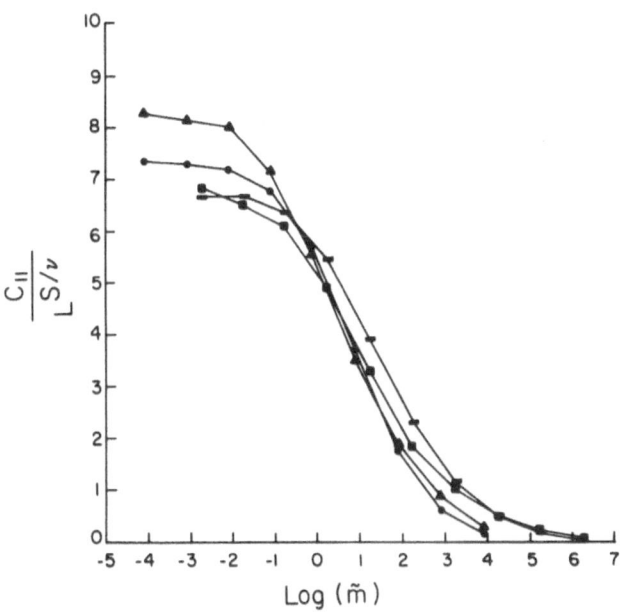

Fig. 3: Values of $C_{11}/L^{S/\nu}$ plotted vs. $\tilde{m}$ for different values of L, k_1 and m_1, at $p = p_c$. The points are the results of numerical simulations with $k_1/m_1 = 15.82$ ($C_{11}, = 2.82$, $\mu_1 = 1$) for $L = 10$ (▲) and $L = 40$(■), and $k_1/m_1 = 1.5$ ($C_{11} = 1.625$, $\mu_1 = 1$) for $L = 10$ (•) and $L = 40$ (▬).

This research was supported in part by a grant from the Israel Academy of Sciences and by a grant from the Center of Absorption in Science.

REFERENCES

1. D. J. Bergman and Y. Kantor, Phys. Rev. Lett. 53, 511 (1984); see also D. J. Bergman in Proceedings of the Workshop on Physics of Disordered Systems, Santa Barbara, California, July, 1983 (unpublished).
2. S. Feng and P. N. Sen, Phys. Rev. Lett. 52, 216 (1984).
3. Y. Kantor and I. Webman, Phys. Rev. Lett. 52, 1891 (1984).
4. D. J. Bergman, Phys. Rev. B31, 1696 (1985).
5. S. Feng, P. N. Sen, B. I. Halperin and C. J. Lobb, Phys. Rev. B30, 5386 (1984).
6. L. Benguigui, Phys. Rev. Lett. 53, 2028 (1984).
7. S. Feng and M. Sahimi, Phys. Rev. B31, 1671 (1985).
8. D. J. Bergman, Phys. Rev. B33, 2013 (1986).
9. D. J. Bergman and E. Duering, Phys. Rev. B34, 8199 (1986).
10. E. Duering and D. J. Bergman, unpublished.

1/f NOISE IN RANDOM SYSTEMS: RECENT PROGRESS

R. Rammal

AT&T Bell Laboratories
Murray Hill, NJ
and
CRTBT-CNRS
Grenoble

I. INTRODUCTION

Most electrical conductors exhibit low frequency electrical noise when subjected to a constant bias. This so called "excess" or "1/f noise" depends on the bias and, at sufficiently low frequencies, can be much larger than the usual Nyquist-Johnson equilibrium noise. The origin of this noise, and of its particular inverse frequency dependence (1/f), has been a mystery for a number of years.[1] Many models have been proposed, but no universal mechanism has been found. The evidence seems to suggest that *non diffusive* defect motion is at the origin of the phenomenon in many metal films (see below section 5), but in certain cases, other mechanisms must be invoked.[2] For example, in niobium films, the resistance is modulated by hydrogen impurity diffusion while in superconducting films near the transition, temperature diffusion is involved.[3] In the latter two cases the spectrum of fluctuations does not have the canonical 1/f dependence, but is nevertheless a form of low-frequency noise, which is commonly referred to as "diffusion noise".

Low-frequency noise of metal-insulator mixtures has recently been the subject of experimental investigations. The focus has been on the divergence of the magnitude of the noise near the percolation threshold.[4] The main motivation was recent theoretical predictions, for the resistance noise near threshold,[5] based on the assumption that the noise is uncorrelated between the different pieces of the material making up the percolating network. The first part of this lecture will be devoted to this problem: resistance (1/f) noise, diffusion noise and Nyquist noise in percolating structures. In the second part, we consider the *equilibrium* magnetic 1/f noise in spin glasses, recently studied by different groups.[6] Low-frequency noise in charge-density-wave systems has also been the subject of recent studies, but for this topic, we direct the reader to the relevant literature.[7]

II. DIFFUSION NOISE and NYQUIST NOISE

Thermal (Nyquist-Johnson) noise is governed by the resistance of the system. Similarly "diffusion noise" is controlled by the diffusion properties (anomalous or not) of the considered medium. Therefore, there is nothing new to be learned there, at least from a fundamental point of view. For practical purpose we shall discuss first Diffusion and Nyquist noise in percolating systems and, more generally, on pure fractal networks. The reason is the following one. When the diffusion length is less than the percolation correlation length, one is in the fractal regime, where it is known that the self-similar geometry of the system leads to anomalous diffusion.[8] This, in turn, modifies the spectrum of electrical fluctuations[9,10] leading in some limits to an unlikely mechanism for 1/f noise.

In what follows two classes of problems are considered: diffusion and Nyquist noises. In both cases, the noise is assumed to be measured across planar electrodes of size L^{d-1} separated by a distance L on a portion of an infinite piece of material (percolating clusters or pure fractal networks). Low-frequency diffusion noise arises when, for example, the resistance r of a small piece of material is modulated by a diffusion variable n, i.e. $\delta r = \delta n(dr/dn)$. The resistance fluctuations in that case are directly related to the diffusion properties of the system. The same problem arises in the context of the Nyquist noise of a resistor network whose nodes are connected to ground through a capacitor. Indeed, in such a case, the voltages obey a diffusion equation.[9] The most significant situation is probably the first one, where the resistance fluctuations are caused by a diffusion process and measured through voltage fluctuations in the presence of a constant applied current (i.e. in the configuration where 1/f noise is usually observed).

A. Diffusion Noise

Clearly the measured quantity (power noise spectrum) is given by

$$S(\omega) \sim \int_0^\infty dt \cos \omega t <\Delta N(o)\Delta N(t)> / N^2 \tag{2.1}$$

where N denotes the number of sites enclosed within the electrodes, while $\Delta N(t)$ is the fluctuation of the diffusing quantity averaged over the sites enclosed within the electrodes at time t. When the frequency is low enough that the diffusion length is much larger than the size of the measuring region, $\Delta N(t)/\Delta N(o)$ is simply equal, in the random walk analogy, to the conditional probability of return to the origin, i.e. to one of the sites within the measurement region.

For the sake of clarity we shall discuss two different cases: the pure fractal case $L \ll \xi$ and the percolation case $L \ll \xi$. Here ξ refers to the correlation length $\xi \sim (p-p_c)^{-\nu_p}$.

Fractal Case In this case, two typical frequencies enter: $\omega_p \sim D_o \, \xi^{-1/\nu_{RW}}$ and $\omega_L \sim D_o L^{-1/\nu_{RW}}$ where ν_{RW} is the so called random walk exponent.[8] The corresponding behavior of $S(\omega)$ is as follows.

In the low frequencies range: $\omega_p \ll \omega \ll \omega_L$, one has two different regimes

$$S(\omega) \sim L^{-\bar{d}+1/\nu_{RW}} \quad \text{if} \quad \tilde{d} > 2$$

$$\sim \omega^{\tilde{d}/2-1} \qquad \text{if} \quad \tilde{d} < 2 \tag{2.2}$$

where we have used the notations: $\tilde{d}$ = spectral dimension, $\bar{d}$ = fractal dimension, the random walk exponent $\nu_{RW} = \tilde{d}/2\bar{d}$. This result reduces to the well known Euclidean case where $\tilde{d} = \bar{d} = d$. In the latter there is a power law divergence at $d = 1$ and logarithmic corrections at $d = 2$. For percolation clusters ($\tilde{d} \approx 4/3 < 2$) there would be a divergence in all dimensions.

In the opposite limit $\omega \gg \omega_L$, we have also a simple result:

$$S(\omega) \sim \frac{1}{L^{\bar{d}+1}} \omega^{-1-\nu_{RW}} \tag{2.3}$$

(For point electrodes the result of Ref. 9 applies for $\tilde{d} < 2$).

Both (2.2) and (2.3) are actually simple consequences of (2.1). Consider first (2.2) and use $<[\Delta N(o)]^2> \sim L^{\bar{d}}$ and $N \sim L^{\bar{d}}$. The conditional probability of return to the origin is known:[8] $L^{\bar{d}}(Dt)^{-\tilde{d}/2}$ and then when $\tilde{d} < 2$, (2.1) leads to (2.2). When $\tilde{d} > 2$ the walk is not recurrent, but the number of visited sites within a region L scales as the time it takes to travel the distance, i.e. $L^{1/\nu_{RW}}/D$. Hence $\int_0^\infty \Delta N(t)dt \approx \Delta N(o)L^{1/\nu_{RW}}/D$. Substituting in (2.1) one obtains (2.2).

In the high-frequency regime, the time is so short that only the random walkers near the boundary can exit the system. Hence, with Σ the surface of the system and $D_o^{1/2}t^{\nu_{RW}}$ the distance traveled by random walkers in a time t, one obtains:

$$1 - \Delta N(t)/\Delta N(o) \approx (D_o^{1/2}t^{\nu_{RW}})\Sigma/N$$

Substituting in (2.1), one obtains

$$\int_0^\infty dt \cos \omega t \frac{<\Delta N(t)\Delta N(o)>}{N^2} \approx \frac{<[\Delta N(o)]^2>}{N^2} \tag{2.4}$$

With $\Sigma \sim L^{\bar{d}-1}$, $N \sim L^{\bar{d}}$ and $<[\Delta N(o)]^2> \sim L^{\bar{d}}$ one deduces (2.3) immediately.

Percolation Case Here one has to integrate on length scales over which the system is Euclidean or fractal. Three regimes for the diffusion noise in the percolation case can be considered in turn.

At very low frequencies $\omega \ll D(p)/L^2$, where $D(p) \sim D_o(p-p_c)^{t-\beta}$ one recovers the familiar frequency dependence of Euclidean lattices:

$$S(\omega) \sim \begin{cases} \dfrac{\xi^{\beta/\nu}L^{2-d}}{D(p)} \sim (p-p_c)^{-t}L^{2-d} & \text{if } d > 2 \\[3mm] \dfrac{\xi^{\beta/\nu}}{(D(p))^{d/2}}\, \omega^{d/2-1} \sim (p-p_c)^{-\beta+\frac{d}{2}(\beta-t)} & \text{if } d < 2 \end{cases} \qquad (2.5)$$

In the intermediate regime: $D(p)L^{-2} \ll \omega \ll D(p)\xi^{-2}$ one obtains

$$S(\omega) \sim \frac{\xi^{\beta/\nu}D^{1/2}(p)}{L^{d+1}\omega^{3/2}} \sim (p-p_c)^{t/2-3\beta/2}/L^{d+1}\omega^{3/2} \qquad (2.6)$$

Finally, at $\omega \gg \omega_p$, $S(\omega)$ assumes the same frequency dependence as in the pure fractal case

$$S(\omega) \sim \xi^{\beta/\nu}\frac{D_o^{1/2}}{L^{d+1}}\,\omega^{-1-\nu_{RW}} \sim \frac{(p-p_c)^{-\beta}}{L^{d+1}}\,\omega^{-1-\nu_{RW}} \qquad (2.7)$$

The result (2.5) follows similarly as before, with the following substitutions in the above reasoning $(\bar{d} = d - \beta/\nu)$

$$N \approx \xi^{\bar{d}}(L/\xi)^d \sim \xi^{-\beta/\nu}L^d$$

$$<[\Delta N(o)]^2> \sim \xi^{-\beta/\nu}L^d$$

and

$$\Delta N(t) \approx \Delta N(o)L^d[D(p)t]^{-d/2} \text{ when } d < 2 \text{ and } \int_o^\infty \Delta N(t)dt \approx \Delta N(o)L^2/D(p)$$

when $d > 2$.

Similarly, in the high-frequency Euclidean regime of percolation clusters, $\nu_{RW} = 1/2$, $D \sim D(p)$ and $\Sigma \sim \xi^{d-1}(L/\xi)^{d-1}$ while $N \sim \xi^d$ $(L/\xi)^d$ and $<[\Delta N(o)]^2> \sim \xi^{-\beta/\nu}L^d$, so that one obtains (2.6). In the very-high-frequency regime $\omega \gg \omega_p$, where the fractal structure reveals itself, Σ, N and $<[\Delta N(o)]^2>$ take the same value we just discussed, but $\nu_{RW} \neq 1/2$ and D are independent of p, yielding the results (2.7). The exponent $1+\nu_{RW}$ of $S(\omega)$ in that limit is always between 1 and 3/2, so it belongs to the category of "1/f noise", which often includes cases like $\omega^{-\alpha}$ with $\alpha > 1$.

B. Nyquist—Johnson Noise

This part of the noise can be calculated using the fluctuation dissipation theorem for instance.

Fractal Case For $L \ll \xi$, the same cutoffs ω_p and ω_L appear. At low frequencies $\omega_p \ll \omega \ll \omega_L$, one recovers

$$S(\omega) \sim L^{-\beta_L} \tag{2.8}$$

i.e. the same size dependence as for the resistance. Here $-\beta_L = 2 - d + t/\nu$ is the resistance exponent.[8]

In the high-frequency regime, only a layer one diffusion length deep contributes to $S(\omega)$:

$$S(\omega) \sim \frac{L^{1-\bar{d}}}{D_0^{1/2}} \, \omega^{-1+\nu_{RW}} \tag{2.9}$$

(The result for point contacts is different and has been given in Ref. 9).

Percolation Case Here $L \gg \xi$ and three cases must be distinguished.

At very low frequencies $\omega \ll D(p)/L^2$, $D \sim (p - p_c)^{t-\beta}$ and $S(\omega)$ is proportional to the dc resistance

$$S(\omega) \sim \frac{\xi^{\beta/\nu}}{D(p)} L^{2-d} \sim (p - p_c)^{-t} L^{2-d} \tag{2.10}$$

(An extra factor $(p - p_c)^{2\beta}$ arises for point electrodes, Ref. 9).

In the intermediate regime, $D(p)L^{-2} \ll \omega \ll D(p)\xi^{-2}$, one recovers the Euclidean high frequency behavior

$$S(\omega) \sim \frac{\xi^{\beta/\nu}}{D(p)^{1/2}} L^{1-d} \, \omega^{-1/2} \sim (p - p_c)^{-\beta/2 - t/2} L^{1-d} \omega^{-1/2} \tag{2.11}$$

(with an extra factor $(p - p_c)^{2\beta}$ for point electrodes)

Finally at $\omega \gg \omega_p$ one obtains

$$S(\omega) \sim \frac{L^{1-d} \xi^{\beta/\nu}}{D_0^{1/2}} \, \omega^{-1+\nu_{RW}} \sim L^{1-d}(p - p_c)^{-\beta} \omega^{-1+\nu_{RW}} \tag{2.12}$$

III. 1/f NOISE IN A METAL-INSULATOR MIXTURE

While the physical origin of 1/f noise is still uncertain,[1] this noise is independent of the resistance of the sample and is in opposition with thermal or diffusion noise. Furthermore, two experimentally established results are very well known.

i) 1/f noise is caused by resistance fluctuations, i.e. it is observed as voltage fluctuation $\delta V(t)$ which are proportional to the constant applied current I.

ii) voltage fluctuations between two segments of the same sample are uncorrelated over distances larger than a few microns.

Recent progress in the study of 1/f in mixtures was motivated by the following question: How does the magnitude of 1/f noise diverge near the percolation threshold, and is this divergence governed by a new exponent or by an exponent related to previously defined ones?

To be consistent with properties i) and ii), the following model has been used. Assume that each resistance fluctuates in time around an average r, independently from the other resistances. In other words, if α and β are two of the resistances of the network, $r_\alpha = r + \delta r_\alpha$, $<\delta r_\alpha> = 0$ and $<\delta r_\alpha \delta r_\beta> = \rho^2(\omega)\delta_{\alpha,\beta}$, where the brackets stand for ensemble average. Here each of the elementary fluctuating resistances is assumed to have an identical spectrum $\rho(\omega)$, but the exact frequency dependence of this spectrum is not important for the following. To compute the magnitude of the resistance noise, it is useful to proceed as follows. First recall that the total resistance may be calculated from

$$RI^2 = \sum_\alpha r_\alpha i_\alpha^2 = r\sum_\alpha i_\alpha^2 , \tag{3.1}$$

where I is the total input current and i_α is the current in branch α. To compute the overall resistance fluctuations of the network, we use well known Theorems in network sensitivity analysis. Cohn's theorem, which is a direct consequence of Tellegen's theorem, shows that to linear order in the fluctuations

$$I^2 \delta R = \sum_\alpha \delta r_\alpha i_\alpha^2 \tag{3.2}$$

Normalizing the input current to unity, one obtains

$$< \delta R(t)\delta R(o) > = \rho^2(t) \sum_\alpha i_\alpha^4 \tag{3.3}$$

More generally, suppose that one is interested in higher order cumulants of the resistance fluctuations. By analogy with (3.3), these would be obtained from

$$< \delta R^n >_c = < \delta r>_c \sum_\alpha i_\alpha^{2n} \tag{3.4}$$

where the subscript c indicates cumulant average.

A. The Infinite Set of Exponents

By means of (3.3), lower and upper bounds for $S_R \equiv \, <\delta R \delta R> /R^2$ may be found[5]

$$1/N_b \leq S_R/s \leq r/R \quad ;s \equiv \rho^2(\omega)/r^2 \,, \tag{3.6}$$

where N_b is the total number of conducting branches in the network. The lower bound is obviously reached on regular Euclidean networks: $S_R \simeq sL^{-d}$, where L denotes the length scale of the system, of Euclidean dimension d. However, for a self-similar network, one can show that $S_R/s \simeq L^{-b}$, where b is a new exponent. In general, the exponent b is neither related to the fractal dimension $\bar{d}$ nor to the resistance exponent defined by $R(L) \sim L^{-\beta_L}$. Using (3.6), one deduced that $-\beta_L \leq b \leq \bar{d}_B$, where $\bar{d}_B$ denotes the fractal dimension of the backbone.

The exponent b has been calculated for different families of fractal structures. In particular, for percolation clusters, one can show that $b = 1$ at $d = 1$ and $b = 2$ at $d \geq 6$. For finite $2 \leq d \leq 6$, b has also been calculated using different methods (transfer-matrix method, real space renormalization group, effective medium theory, etc).

The existence of b leads to a singular behavior of the noise close to percolation threshold p_c. In fact, close to p_c, one can show that for a finite sample (L^d), $S_R(p,L)$ obeys the following scaling behavior:

$$S_R(p,L) = L^{-b} f(\xi_p/L) \tag{3.7}$$

where $\xi_p \sim p - p_c^{-\nu_p}$ is the percolation length, and $f(x)$ denotes a scaling function describing the fractal-to-Euclidean crossover: $f \quad (x \gg 1) \sim 1$ and $f(x \gg 1) \sim x^{d-b}$. This implies $S_R(p,L) \sim L^{-d}(p-p_c)^{-\nu_p(d-b)}$, which leads to the following behavior of the normalized noise

$$s(p)/s \equiv S_R(p,L)/S_R(p=1,L) \simeq (p-p_c)^{-\kappa} \tag{3.8}$$

with $\kappa = \nu_p(d-b)$.

Numerical values of b will be found in Ref. 11. For instance, at $d = 2$: $b = 1.16 \pm 0.02$ which, with $\nu_p = 4/3$, implies that $\kappa = 1.12 \pm 0.02$. In general κ appears as an increasing function of d since it starts from $\kappa = 1$ at $d = 1$ and reaches the value $\kappa = 2$ at $d \geq 6$.

The study of the resistance fluctuations has lead the authors of Ref. 5 to introduce a hierarchy of exponents x_n. That hierarchy is defined with the help of the geometrical factors which relate, within the independent resistor model, higher-order cumulants of the overall resistance fluctuations to the cumulants of the individual resistance fluctuations. According to (3.4), one defines x_n in the fractal regime by

$$G_{2n} = \sum_{\alpha} i_{\alpha}^{2n} \sim L^{-x_n} \qquad (3.9)$$

For percolation problems, one has : $x_o = -\overline{d}_B$, $x_1 = \beta_L$, $x_2 = b + 2\beta_L$, ...etc. The first members of that hierarchy reproduce well known dimensions (fractal, spectral, ...etc.) introduced previously in the physics of fractals or percolation problems. Furthermore, the physical meaning of $\{x_n\}$ is quite clear. For 2D percolation problems, one has the following estimation for x_n : $-x_o = 1.65 \pm 0.02$, $-x_1 = 0.986 \pm 0.01$, $-x_2 = 0.818 \pm 0.01$, $-x_3 = 0.771 \pm 0.01$, $-x_4 = 0.757 \pm 0.01$ and $-x_5 = 0.75 \pm 0.01$. At large n, one can show that $x_\infty = 1/\nu_p$ is valid at all d ($\nu_p = 4/3$ at $d = 2$). The argument can be understood from the scaling of singly connected bonds.

In addition to direct implications for experiments on real materials (see below) the introduction of the infinite set of measurable exponents $\{x_n\}$ was important from a theoretical point of view. In the random resistor problem, this implies that at least one infinite family of exponents is necessary to describe the underlying fractal structure. Another fashion to understand the meaning of x_n is provided by (3.9). For each integer n, x_n describes the behavior of the moment i^{2n} of the current distribution in the network. Considered as a function of n, x_n is a convex increasing function of n. Furthermore one can show that[12] knowing i^{2n} is sufficient to be able to deduce i^{2q} for an arbitrary q. Changing variables from i^2 to $y \equiv -\ln(i^2)$ one then notes that, when (3.9) holds, this means that the Laplace transform of $p(y)$ (with $p(i^2)di^2 = p(y)dy$) behaves as

$$\overline{i^{2q}} = \int_0^\infty dy e^{-qy} p(y) = D_q L^{-x_q + x_o} \qquad (3.10)$$

where D_q is independent of L. One can then find $p(y)$ by inverting the Laplace transform of (3.10). In the limit $\alpha \equiv -\ln(i^2)/\ln L$ finite and $\ln(L) \to \infty$, one finds in the stationary phase approximation ($x_q - x_o$ is convex and an increasing function of real q).

$$p(\alpha) = C(\alpha, \ln L) L^{f(\alpha) - \overline{d}_B} \qquad (3.11)$$

where C is only weakly dependent on lnL, $\overline{d}_B = -x_o$ is the fractal dimension of the backbone and where $f(\alpha)$ is the Legendre transform of x_q : $f(\alpha) = q\alpha - x_q$, $\dfrac{\partial x_q}{\partial q} = \alpha$.

Notice that $f(\alpha)$ does not contain more information than x_q. Usually, (3.11) is postulated and then (3.10) is deduced. In the recent literature where infinite sets of exponents have arisen in many fields related to fractals, this kind of presentation is called: Multi fractal, fractal measures, etc. For this we direct the reader to Ref. 13 (where $\tau(q)$ refers to x_q) and references therein.

B. Continuum Percolation and Experiments

Careful studies of $1/f$ noise in metal-insulator mixtures have been reported in the recent literature.[4] Close to p_c, both the resistance R and the power spectrum S_R have been shown to diverge in perfect agreement with Ref. 5: $R \sim (\Delta p)^{-t}$ and $S_R/R^2 \sim (\Delta p)^{-\kappa}$. The direct plot of S_R vs R leads to $S_R \sim R^Q$ where $Q \equiv 2 + \kappa/t$ is the noise-versus-resistance exponent. Large values of Q have been obtained ranging from $Q = 3.9 \pm 0.2$ in 3D (carbon-wax) and $Q = 4$ (clumped evaporated gold films subjected to ion milling), to $5.4 \leq Q \leq 8.1$ (in Cr, Al, In films where the metal was removed by sandblasting). Similar enhancement of t has also been reported.

The measured exponents are actually outside the bounds found for Q in the lattice percolation theory. With use of the known bounds for the exponents: $b = d - \kappa/\nu = d - t(Q-2)/\nu$, $-\beta_L \leq b \leq -2\beta_L - 1/\nu$, it is easy to obtain: $2.82 \leq Q \leq 3.05$ in 2D ($\nu = 4/3$, $-\beta_L = 0.973$) and $2.84 \leq Q \leq 2.85$ in 3D ($\nu = 0.88$, $-\beta_L = 1.16$). The effective medium theory gives the value $Q_m = 3$ for Q.

The enhancement of the transport exponents such as t and κ can actually be understood within the framework of the continuum percolation models. The simplest model is provided by the following probability distribution $p(g)$ of bond conductances in the equivalent lattice model: $p(g) = (1-p)\,\delta(g) + ph(g)$. Here $h(g)$ is a continuous normalized function. The "Swiss-cheese" class of models is actually a possible realization[14] of $p(g)$, with an anomalous distribution $h(g) \sim g^{-\alpha}$ ($\alpha < 1$) near $g = 0$. For this class of models, the conductivity exponent is given by $t(\alpha) = (d-2)\nu + 1/(1-\alpha)$ for $0 \leq \alpha < 1$ and $t(\alpha) = (d-2)\nu + 1$ for $\alpha \leq 0$.

The calculation of κ has been carried out for the continuum percolation models,[15] by assuming ("Swiss-cheese" models): $g_\ell \sim \delta_\ell^u$, $s_\ell \sim \delta_\ell^{-v}$ where $u = 1/(1-\alpha)$ and v is the exponent relating the relative noise s_ℓ of bond ℓ to the neck width δ_ℓ. For instance, $u = d - \dfrac{3}{2}$ and $v = d - \dfrac{1}{2}$ for spherical holes in a conducting medium. Depending on the values of u and v, the following results have been found ($\kappa' = $ continuum percolation, $\kappa = d\nu + v$)

a) $u < 1$ and $v + 2u < 1 : \kappa' = \kappa$,

b) $u < 1$, $v + 2u > 1$ and $v - 2u < 1 : \kappa' = \kappa + (v + 2u - 1)/u$

c) $u > 1$, $v + 2u > 1$ and $v - 2u < 1 : \kappa' = \kappa + (v+1)$

d) $u > 1$, $v + 2u > 1$ and $v - 2u > t : \kappa' = \kappa + ?(\text{unknown})$

These expressions are actually different from the effective medium results $Q_m = 2 + v/u$ at $u > 1$ and $Q_m = 3 + (v-1)/u$ at $u < 1$, which are expected to be correct far from p_c.

At $d = 2$, where $u = 1/2$, $v = 3/2$, one obtains (case b): $t = 1$, $\kappa' = 3.16$ and $Q = 5.16$. Similarly at $d = 3$, where $u = 3/2$ and $v = 5/2$, one gets: $t = 2.38$, $\kappa' = 5.14$ and $Q = 4.16$. Among the available estimates of Q, only the 3D data seem to fit with these predictions ($t = 2.3 \pm 0.4$ and $Q = 3.7 \pm 0.2$). Note further that in all cases, an enhancement of Q is obtained in the continuum models. A definitive comparison in 2D would require the simultaneous measurement of both exponents, t and κ.

C. Non Linear Networks

The extension of the previous results to more complicated situations have been the object of recent works. In the context of nonlinear random resistor networks, the hierarchy $\{x_n\}$ has been shown for instance to generate a continuum of hierarchies, indexed by a real number α. The "exponent" α describes here circuit elements with the non-linear I-V characteristic $v = ri^\alpha$. In that case (3.2) and (3.3) are generalized as follows[11] $(\alpha > o)$

$$\delta R = \sum_\ell \delta r_\ell (i_\ell/I)^{\alpha+1}$$

and

$$S_R = \frac{<\delta R \delta R>}{R^2} = \frac{\rho^2}{r^2} \sum_\ell i_\ell^{2(\alpha+1)}/(\Sigma_\ell i_\ell^{\alpha+1})^2$$

Interesting results relative to this problem will be found in Ref. 16. Some limiting cases, relative to particular values of α are discussed in Ref. 17.

IV. 1/f NOISE in SPIN GLASSES

The observation of magnetic noise in ferromagnets (below T_c) is traditionally associated[18] to the rearrangement of magnetic domains induced by a varying magnetic field (Barkhausen noise). Near T_c, critical slowing down is also known[19] to produce a magnetic noise $(\sim\omega^{-\alpha})$ due to critical magnetic fluctuations. The existence in spin glasses of a 1/f like spectrum of thermodynamical magnetic fluctuations extending down to ultralow frequencies has been predicted in numerous numerical works,[20] as a consequence of a very broad spectrum of relaxation times.

Magnetic noise[6] studies correspond to a very recent way of exploring the dynamic properties of spin glasses without any excitation field. Typical materials which have been studied are mainly insulating spin glasses: Cs Ni Fe F_6, Cd $In_{0.3}$ $Cr_{1.7} S_4$ and Eu Sr S, from 4.2K to 30K and in the frequency range: $10^{-2}Hz < \omega < 10^3Hz$. The questions which can be addressed by the noise measurements can be summarized as follows.

i)   nature of 1/f noise in spin glasses: equilibrium (stationary) part of the spin glass dynamics versus the nonstationary part (ageing effect)

ii) validity of the Fluctuation Dissipation Theorem (FDT) in the spin glass phase (at $\omega = 0$); the later is strongly suspected to be non-ergodic.

iv) Deviations from pure $1/f$ behavior.

iii) implications for the time dependence of response functions.

Before discussing these different questions, let us remark that a pure $1/f$ power spectrum cannot be accepted as a description of the noise frequency dependence over the whole frequency range. First, its fluctuation power integral diverges. Second, and equivalently, a $1/f$ power spectrum corresponds via the FDT, to a frequency independent $\chi''(\omega)$ and a logarithmic relaxation function: both dynamical laws are incompatible with the Kramers-Kronig relations. Thus, one is led to introduce cut-off frequencies at low and high frequencies. Accurate measurements of the stationary noise spectra have shown indeed small deviations from pure $1/f$ behavior.

A. Equilibrium Part of the Spin Glass Noise

It is important to notice first that, contrary to $1/f$ noise in conductors, the observed noise in spin glasses is an equilibrium noise (see below). From the careful studies of this spectrum, two contributions seem to appear. The first is a stationary part, observed at short time scales: $t < t_w$, where t_w refers to the so called waiting time. The second part, which appears at frequencies lower than t_w^{-1} is called the non stationary (or ageing) part. In fact as t_w increases, the low frequency part of the noise spectrum decreases whereas the high ω part remains insensitive to t_w. This is the reason why the spectra for high frequencies are called the stationary part.

The two contributions to the noise spectrum have in general different frequency dependence. The high ω part follows $1/\omega$ and describes the fluctuations at equilibrium (stable or metastable state). The non stationary part corresponds to a more rapid decay in ω and describes probably the relaxation towards the nearest metastable state. This separation between two contributions to $S(\omega)$ at a given $T < T_c$ reflects the same behavior observed in the thermoremanent magnetization relaxation, where ageing effects are known. In this respect, the main interest of the noise experiment is probably a direct confirmation of these old observations.

B. Fluctuation Dissipation Theorem

The fact that the spectra $M^2(\omega)$ for high ω are stationary allows for a direct test of the fluctuation dissipation Theorem in the spin glass phase. From FDT, the power spectrum of magnetic fluctuations can be expressed as a function of $\chi''(\omega)$ (out of phase susceptibility) by

$$S(\omega) \simeq \overline{M^2(\omega)} = \frac{2k_B T}{\pi V} \frac{\chi''(\omega)}{\omega} \tag{4.1}$$

where V is the volume of the sample. A close agreement between $\chi''(\omega)$ and $\frac{\omega}{T} \overline{M^2(\omega)}$ has been observed in a wide range of frequencies and temperatures. All the results obtained lead to the same conclusion that the FDT is obeyed in spin-glasses. This is at first sight rather surprising since the spin glass phase is described by numerous theoretical works as non ergodic.[21] It can be argued that, in fact, stationarity means, not that we have an equilibrium state, but rather that in the same characteristic experimental times, equivalent parts of the whole phase space are explored during different measurements.

C. Implications of FDT

The FDT relates the internal spontaneous fluctuations of a system at thermal equilibrium to its response to the action of an external field. It has only been established for ergodic and linear systems (linear response theory). In its fundamental form, it relates the response function $X(t)$ to the autocorrelation function which due ergodicity is identical to the time averaged auto correlation function $C(t)$

$$X(t) = \frac{-1}{k_B T} \frac{d}{dt} C(t) \qquad (4.2)$$

Consequently, the relaxation function (i.e. the response to a field step) can be written

$$\phi(t) = (C(t) - C(\infty))/k_B T \qquad (4.3)$$

In an ergodic magnetic system $C(\infty) = 0$ and then $\phi(t) \equiv \sigma(t)$ is the ratio of thermoremanent magnetization $M_{TRM}(t)$ to the applied field H. In that case (4.1) is a simple consequence of (4.2) in addition to the linearity of the system and the stationarity of the random function (Wiener-Kintchine Theorem). In case of full ergodicity, one deduces in particular

$$S(\omega) \equiv \int dt \cos \omega t . C(t) = k_B T \int dt \cos \omega t \sigma(t) \qquad (4.4)$$

$$X''(\omega) = \frac{\pi}{2} \omega \int dt \cos \omega t \, \sigma(t) = \frac{\pi \omega}{2 k_B T} \int dt \cos \omega t C(t) \qquad (4.5)$$

A direct consequence is that a relaxation function $\sigma(t) \sim t^{-\alpha}$, $t > t_0$ would correspond to $X''(\omega) \sim \omega^\alpha$ and a power spectrum $S(\omega) \sim \omega^{\alpha-1}$ at $\omega t_0 \ll 1$. Similarly, a relaxation function $\sigma(t) \sim 1 - at^{-\alpha}$ below t_0 would correspond to $X''(\omega) \sim \omega^{-\alpha}$ and a power law spectrum $S(\omega) \sim \omega^{-(1+\alpha)}$ for $\omega t_0 \ll 1$. Non vanishing values of α would indicate a deviation from a pure $1/f$ spectrum and this is actually the case in real systems.

D. Deviations From 1/f Power Spectrum

Accurate measurements of $S(\omega)$ of several materials have shown small deviations from pure $1/f$ behavior. However, in all cases (4.1) has been observed to be obeyed and the following picture for $X''(\omega)$ emerged. At every temperature

T, there is a frequency $\omega_o(T)$ at which $\chi''(\omega)$ goes through a widely spread maximum, with

$$\chi''(\omega) \sim (\frac{\omega}{\omega_o})^{\alpha(T)} \chi''(\omega_o) , \quad \omega \ll \omega_o(T) \tag{4.6}$$

$$\sim (\frac{\omega}{\omega_o})^{-\beta(T)} \chi''(\omega_o) , \quad \omega \gg \omega_o(T)$$

where $\alpha(T)$ and $\beta(T)$ are two T-dependent exponents $(\ll 1)$. Both $\alpha(T)$, $\beta(T)$, and $\omega_o(T)$ increase with T below T_c (close to T_c, $\alpha \lesssim 0.2$) indicating a deviation from a pure $1/f$ spectrum. Furthermore, (4.6) yields for the frequency dependence of the noise an expression which is an integrable function of the frequency

$$\int_o^\infty \overline{M^2(\omega)}/\omega \sim \int_o^\infty \chi''(\omega)d\omega/\omega \sim \chi''(\omega_o) (\frac{1}{\alpha}+\frac{1}{\beta}) \tag{4.7}$$

Above T_c, the existence of a critical slowing down, with a maximum relaxation time $\tau(T)$ diverging at T_c, leads to a constant noise spectrum at $\omega\tau(T) \ll 1$. Equivalently, $\chi''(\omega) \sim \omega$. Of course, a direct check of this prediction calls for a very low frequency measurement of $S(\omega)$ close to T_c.

V. ORIGIN of 1/f NOISE

It is clear that an $1/f$ spectrum $S(\omega) \sim 1/\omega$ can only appear between two cutoffs: $\omega_m \leq \omega \leq \omega_M$. Furthermore, the occurrence of such a spectrum in various situations excludes any universal origin for the noise. Formally, $S(\omega) \sim 1/\omega$ is equivalent to the existence of a very broad spectrum of relaxation times: $p(\tau) \sim 1/\tau$, i.e. with long tail at large τ. Such an assumption is more or less visible in a large class of models[22] which leads to an "explanation" of $S(\omega) \sim 1/\omega$. However, it is important to realize that in all cases, the "exponent" α in the so called $1/f$ spectrum $S(\omega) \sim \omega^{-\alpha}$ cannot be viewed as a very well defined number.[23] Very often, α can depend on: i) temperature T, which is a natural consequence of thermal activation process or/and ii) frequency ω, which can be attributed to $\log \omega$ corrections (in some cases!). At least for dirty conductors, α appears to depend strongly on the frequency interval used to define[24] it: $\alpha(\omega) \equiv -d \ln S(\omega)/d \ln \omega$, with $0.8 \leq \alpha(\omega) \leq 1.4$. Furthermore $\alpha(\omega)$ has a non-monotonic variation as function of ω and T. This aspect of the $1/f$ spectrum seems to exclude a very simple picture to explain its origin. However, as recalled in the previous sections, two experimentally well established results are known: a) $1/f$ noise is caused by resistance fluctuations and b) voltage fluctuations between two segments of the same sample are uncorrelated over distances larger than a few microns.

Recent progress in localization theory[25] seems to provide a natural explanation for 1/f noise and this is in perfect agreement with the experimental facts. The basic idea is the sensitivity of the conductance of a disordered metal to the motion of a single atom. For instance, it was shown[25] that the residual conductivity of samples of small dimensions (1 micron at 1K) is highly sensitive to small changes in the impurity potential. The conductivity of a film of finite thickness changes, for example, when an impurity is moved a finite distance, regardless of the film dimensions. The high sensitivity of the conductance to a change in the position of the impurities provides an explanation for the property a). If the impurity jumps a distance r_o over a time scale τ, the conductance will change by the amount $e^2/\hbar$ in a time $t \sim \tau(\ell^2/L^2)$. Here $\ell =$ mean free path, $L =$ dimension of the sample in the direction of the current. The low diffusion coefficient of the impurities $D \sim r_o^2\ell^2/L^2\tau$ can lead to large time scale for the fluctuations of the conductance. For instance, with $r_o^2 \sim 10^{-15}$ cm^2, $\ell \sim 10^{-6}$ cm, $L \sim 10^{-4}$cm and $D \sim 10^{-19}$ cm^2/s one obtains $\tau \sim 1$ sec.

Property b) is supported by recent investigations[26] of the voltage fluctuations between two segments of the same sample. For a more detailed discussion we direct the reader to Ref. 26. In this respect, a very promising subject for further studies would be the magnetoresistance of a spin glass. In that case,[25] the magnetic field has an effect on the scattering potential (for the electrons) which is caused by the exchange interaction of localized spins with the conduction electron spins. Accordingly, the conductance of a single sample fluctuates as a function of the magnetic field H with a characteristic field scale H_s, which is related to the spin-spin correlation function. A comparative study of the magnetic noise and the electrical noise may be of some help to understand the physics of spin glasses.

REFERENCES

1. See P. Dutta, and P. M. Horn, Rev. Mod. Phys. **53**: 492 (1981).
   F. N. Hooge, T. G. M. Kleinpenning, and L. K. J. Vandamme, Rep. Prog. Phys. **44**: 479 (1981).

2. J. Pelz and John Clarke, Phys. Rev. Lett. **55**: 738 (1985)
   R. H. Koch, J. R. Lioyd, and J. Cronin, Phys. Rev. Lett. **55**: 2487 (1985)
   D. M. Fleetwood and N. Giordano, Phys. Rev. B **31**: 1157 (1985)
   J. H. Scofield, J. V. Mantese and W. W. Webb, Phys. Rev. B**32**: 736 (1985)

3. J. H. Scofield and W. W. Webb, Phys. Rev. Lett **54**: 353 (1985).
   M. B. Ketchen and John Clarke, Phys. Rev. B **17**: 114 (1978)

4. G. A. Garfunkel and M. B. Weissman, Phys. Rev. Lett. **55**: 296 (1985);
   R. H. Koch, R. B. Laibowitz, E. I. Alessandrini, and J. M. Viggiano, Phys. Rev. B **32**: 6932 (1985)
   J. V. Mantese and W. W. Webb, Phys. Rev. Lett. **55**: 2212 (1985)
   D. A. Rudman, J. J. Calabrese and J. C. Garland, Phys. Rev. B **33**: 1456 (1986)
   C. C. Chen and Y. C. Chou, Phys. Rev. Lett. **54**: 2529 (1985)

5. R. Rammal, J. Phys. (Paris) Lett. **46**: 129 (1985)

   R. Rammal, C. Tannous and A. M. S. Tremblay, Phys. Rev. A **31**: 2662 (1985)
   R. Rammal, C. Tannous, P. Breton and A. M. S. Tremblay, Phys. Rev. Lett. **54**: 1718 (1985)

6. M. Ocio, H. Bouchiat and P. Monod, J. Phys. (Paris) Lett. **46**: 647 (1985)
   H. Bouchiat, Thesis, Orsay (1986)
   P. Refregier, M. Ocio and H. Bouchiat, Europh. Letts. xxx (1987)
   M. Alba, H. Bouchiat, J. Hamman, M. Ocio, and Ph. Refregier, preprint (1987)

7. For a recent collection of papers on CDW transport, see "Charge Density Waves in Solids," Lecture Notes in Physics, vol. 217, ed. G. Hutiray and J. Solyom (Springer, Berlin, 1985)

8. R. Rammal and G. Toulouse, J. Phys. (Paris) Lett. **44**: 13 (1983)
   S. Alexander and R. Orbach, J. Phys. (Paris) Lett. **43**: 623 (1982)

9. R. Rammal, J. Phys. (Paris) Lett. **45**: 1007 (1984)

   R. Rammal, in Sixth International Symposium on "Fractals in Physics" ICTP, Trieste, Italy, edited by L. Pietronero and E. Tosatti (North-Holland, Amsterdam, 1985)

10. A. Hansen and M. Nelkin, Phys. Rev. B **33**: 649 (1986)
    K. W. Yu, Phys. Lett. **115 A**: 161 (1986)
    B. Fourcade and A. M. S. Tremblay, Phys. Rev. B **34**: 7802 (1986)

11. R. Rammal, in "Physics of finely divided Matter" (Ed. N. Boccara, M. Daoud, Springer-Verlag 1985) p. 118

12. B. Fourcade, P. Breton, A. M. S. Tremblay, preprint 1987
    C. Castellani, L. Peliti, J. Phys. A **19**: L 429 (1986)

13. L. DeArcangelis, S. Redner and A. Coniglio, Phys. Rev. B **34**: 46455 (1986)

14. B. I. Halpern, S. Feng, P. N. Sen, Phys. Rev. Lett. **54**: 2391 (1985)

15. R. Rammal, Phys. Rev. Lett. **55**: 1428 (1985)
    A. M. S. Tremblay, S. Feng and P. Breton, Phys. Rev. B **33**: 2077 (1986)

16. R. Rammal and A. M. S. Tremblay, Phys. Rev. Lett. **58**: 415 (1987)

17. R. Blumenfeld and A. Aharony, J. Phys. A **18**: L 443 (1985)

18. See for a review H. Bittel, Physica, B **83**: 6 (1976)

19. J. C. Angles d' Auriac, R. Maynard and R. Rammal, J. Stat. Phys. **28**: 309 (1982); J. C. Angles d' Auriac, Thesis Grenoble (1980)

20. R. Rammal, Thesis, Grenoble (1981)
    N. Sourlas, Europhys. Lett. **1**: 189 (1986)

21. See for a review, K. Binder and A. P. Young, Review Mod. Phys. xxx (1986)

22. See for instance E. Marinari, P. Paladin, G. Parisi, and A. Vulpiani, J. de Physique (Paris) **45**: 657 (1984)

23. M. Nelkin and A. M. S. Tremblay, J. Stat. Phys. **25**: 253 (1981).

24. R. D. Black, P. J. Restle and M. B. Weissman, Phys. Rev. B **28**: 1935 (1983).

25. B. L. Altshuler and B. Z. Spivak, JETP Lett. **42**: 447 (1985)
    B. L. Altchuler, JETP Lett. **41**: 648 (1985)
    S. Feng, P. A. Lee and A. D. Stone, Phys. Rev. Lett. **56**: 1960 (1986)

26. A. Benoit, C. P. Umbach, R. B. Laibowitz and R. A. Webb, to be published
    R. Rammal and B. Doucot, to be published.

THE GLASS TRANSITION OF POLYMERS

S.F. Edwards

Cavendish Laboratory, Madingley Road, Cambridge CB3 0HE, UK

T.A. Vilgis

Max Planck Institute of Polymer Physics, Postfach 3148
D-6500 Mainz, W-Germany

1. INTRODUCTION

The glass transition is not limited to special types of materials.
Every class of material can be tansformed in an armorphous solid (without
crystallinity) if the experimental parameters are adjusted to the dynamics
of the system. Consider therefore two extreme examples. Consider the system consisting of spherical molecules such as rare gases. The hopping time
of the spheres is very short and the dynamics extremely fast. Nevertheless
it has been shown that such fluids undergo a glass transition from the
super cooled melt if one cools the system with a quenching rate of

$$q \sim 10^{12} \ Ks^{-1}.$$

This is only possible at present in computer simulation. On the other hand
consider polystyrene which consists of very large molecules. The dynamics
of such a system is much more complicated in comparison to spheres because
there are many degrees of freedom. It is most significant that the overall, i.e., the centre of mass diffusion of a single molecule is small.
Polystyrene undergoes always a glass transition (this is true only for
atactic polystyrene). For other chemical structures one has to "quench"
with

$$q \sim 1 \ Ks^{-1}.$$

For such observations one can think of the glass transition to be a universal phenomena were details of the chemical structure are ruled by the
minimum cooling rate q for which a glass phase is observed.

In this paper we want to stress features characteristic of the glass
transition in general. The most significant points we want to discuss are:

1. The last divergence of the transport or inverse transport properties,
    such as viscosity, inverse diffusion constant, and relaxation times.

2. The extreme broad relaxation phenomena of the stress, modulus etc.
3. The quantitative definition of the term cooperativity.
4. Influence of external parameters on T_g.

It has been recognised that the relaxation time follows an unusual law in physics described by the Vogel-Fulcher Law (VF):

$$\eta, \; D^{-1}, \; \tau \sim \exp \left\{ \frac{A}{T-T_0} \right\} \qquad\qquad (1.1)$$

This law has many names. In polymers it is often called the Williams Landel Ferry (WLF) law. Much useful information is contained in Ferry's book[1]. The law is not valid over the whole temperature range (see the following figure).

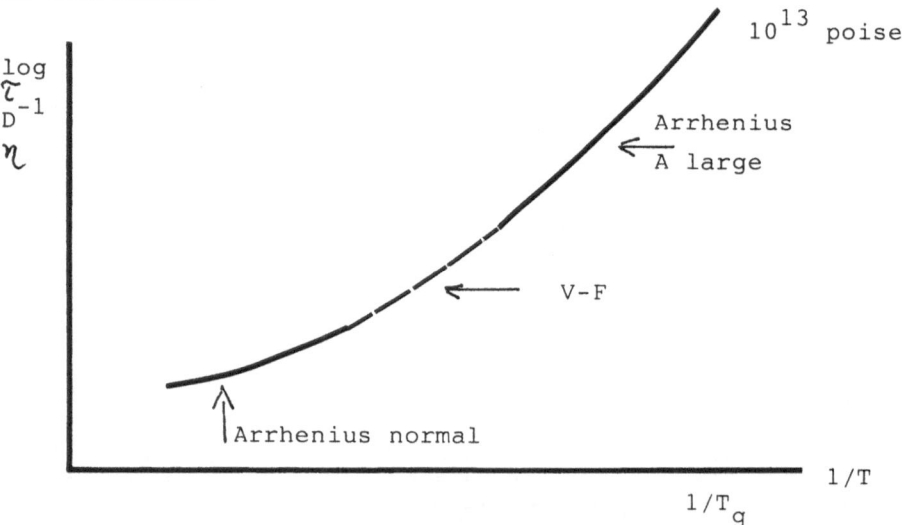

Figure 1

The divergence given by (1.1) comes at T_0, where T_0 is a temperature below the freezing temprature T_g and the empirical rule is

$$T_0 \simeq T_g - (20 \sim 30)^0 .$$

The physical meaning of T_0 is still unclear and we want to try to clarify this point. The VF law is much stronger than the critical slowing down in phase transition phenomena where, by scaling arguments, the relaxation time is given by

$$\tau \sim |T - T_0|^{-\nu z} \qquad\qquad (1.2)$$

where ν is the correlation length exponent i.e. $\xi \sim (T - T_0)^{-\nu}$ and z the dynamical exponent

$$\tau \sim \xi^z .$$

There have been attempts to fit data for freezing transitions (glass transition, spin glass transiton) by eq. (1.2) and it turned out that νz is very high

$vz \sim 10, \ldots 20$

which seems very unphysical. This again indicates a physical significance to the VF law.

Another peculiar point lies in the relaxation properties. An empirical law was found long ago by Kohlrausch and in the early seventies recovered by Williams and Watts in their studies of the broadening of relaxation processes[2]. For example, measurement of the dielectric constant shows a much larger half width in the imaginary part compared to the Debye process.

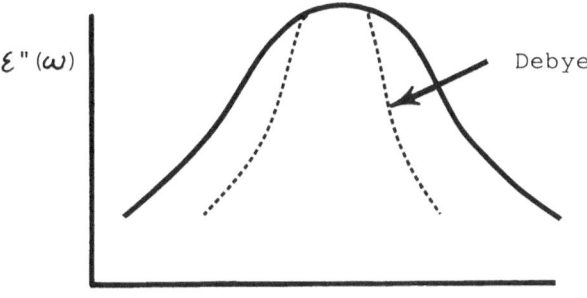

Figure 2

Empirically this is described by the KWW law

$$\varphi(t) \sim e^{-(t/\tau)^{\beta}} \tag{1.3}$$

where φ can be any quantity which relaxes, i.e.

$$\varphi(t) = \frac{\varepsilon(t) - \varepsilon(\infty)}{\varepsilon(0) - \varepsilon(\infty)} \tag{1.4}$$

There are many ways to find laws like (1.3). Most of them are collected by Ngai in his review.

But there has been no convincing explanation for a unique value for β. We doubt that there is more to the physics of (1.3) than a wide relaxation spectrum due to different physical processes and a more relevant question is how are both laws KWW and VF linked together in general, and what is the relationship to cooperativity.

It is believed that the glass transition exhibits a large amount of cooperative motion as the system is close to T_g. This might be indicated as well by the VF law which is the crossover to the Arrhenius behaviour right at T_g but with an extremely high activation energy (see Fig. 1). This activation energy is so high that it could hardly be attributed to only one molecule and the phenomenological interpretation is that there are cooperative regions of some linear size diverging at some temperature

$$\xi \sim (T - T_g)^{power}. \tag{1.5}$$

Another quite general question is the state of time-temperature superposition principle. This says if one measures a physical quantity D(t) at some temperature T and if the measurement is repeated at some temperature

T_1, the quantity $D(t)$ can be resolved by a "shift factor" a_T i.e. $D(T,t)$ is not a function of two variables but only of a combination of both i.e.

$$D(T,t) = D(\frac{t}{a_T})$$ (1.6)

where a_T is often given by

$$a_T = \frac{c_1 + T_g}{T - T_g + c_2}$$ (1.7)

near T_g and

$$a_T = \frac{\Delta E}{T} - \frac{\Delta E}{T^\star}$$ (1.8)

far from T_g. Hence (1.7) is of the Vogel-Fulcher form.

The remarks above are quite general and apply for all sorts of glasses, but there are still challenges, especially in polymer physics and materials science concerning the glass temperature T_g itself. For polymers the glass transition temperature depends on the molecular weight and an empirical rule is given by Flory and Fox:

$$T_g(L) = T_g(\infty) - \frac{\text{constant}}{L} ,$$ (1.9)

where L is the length of the molecules i.e. the molecular weight. The agreement of (1.9) with experiments is not particularly good, but it gives an estimate of $T_g(L)$. Hence (1.9) tells us that there is not a very significant dependence on the molecular weight unless L is small. This point should be investigated using the knowledge of polymer dynamics in melts which has recently emerged.

A last empirical point we want to discuss is the mixing rule in plasticization. Mixing two glasses forming polymers together, the new T_g is often given by

$$\frac{1}{T_g(\text{mix})} = \frac{\phi_1}{T_{g_1}} + \frac{\phi_2}{T_{g_2}}$$ (1.10)

to zeroth order. The ϕ's are the volume fractions of the polymers.

In this paper we are going to give our views on these problems which have developed by considering simple models wherein essentials are omitted, but essential features, we believe, are included. The next section studies the result of the rod glass and shows what can be done by this simple "toy model" before looking at polymers in the third section.

2. THE ROD GLASS AS A SIMPLE MODEL
2.1 VF and KWW

Very recently we reported on a model exhibiting most of the points in the introduction[3]. The results are all in agreement with the real world. The idea is very simple. Take a dense solution of hard inflexible rods as a model system.

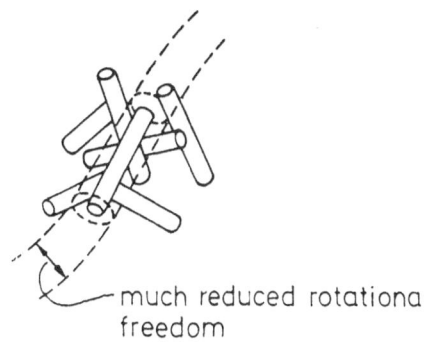

much reduced rotational
freedom

The reason why we took rods was the advantage of simple geometry, no internal degrees of freedom and very slow dynamics, so that we have no problems with high quenching rates, so that one does not have to worry about thermodynamics. If the solution of the rods is dense the concentration

$$c \stackrel{\sim}{\scriptstyle <} d^2 L$$

and severe constraints are acting on the rods. For example, they cannot move rotationally and they can only make progress along their length. Such a solution has been called entangled. Suppose now we have sugh a solution of highly entangled rods. A rod can slide between the entangling rods until it meets rods which block it.

$$D^{-1} = D_0^{-1} (1 + \alpha)$$
from this type of obstacle

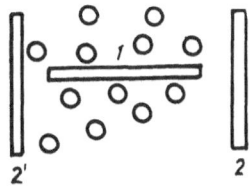

Figure 3

Rod(1) can only diffuse if rod (2) or (2') moves out of the way. Rod (2) is entangled also and can only move along its length and there will be another one (with probability α^2)

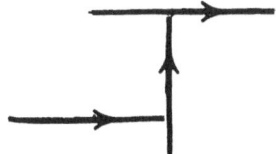

$$D^{-1} = D_0^{-1} (1 + \alpha + \alpha^2)$$

Figure 4

and so on. Finally, drawing the rods at right angles and in a plane purely to give a simple diagram, one will find diagrams like

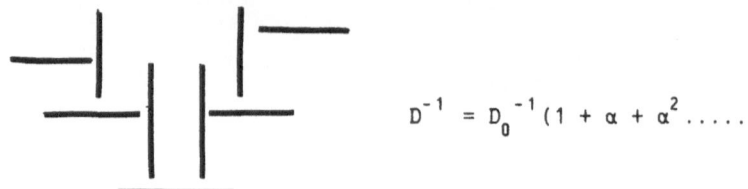

$$D^{-1} = D_0^{-1} (1 + \alpha + \alpha^2 \ldots \ldots$$

Figure 5

$$D^{-1} = D_0^{-1} \sum_0^\infty \alpha^n \tag{2.1}$$

giving a concentration dependent diffusion constant

$$D = D_0 (1 - \alpha) \tag{2.2}$$

where

$$\alpha \sim \varepsilon (cdL^2)^{3/2} \tag{2.3}$$

where d is the diameter of the rod, L its length and c the concentration. More detail is given in ref. 4. Eq. (2.1) gives now a diffusion constant equal to zero if

$$\alpha = 1 \text{ i.e. } \varepsilon cdL^2 = 1.$$

Clearly the assumption was that all rods are in a random configuration saying that for all times the unit vectors n_α satisfy

$$\langle n_\alpha \cdot n_\beta \rangle = \delta_{\alpha\beta} \tag{2.4}$$

where α and β are rod labels.

This assumption seems to be crude as it is well known that the most favourable state in the dense rod ensemble is a liquid crystalline configuration, but we assumed here that the relaxation to a nematic state is very slow and the random state is a metastable one. This is borne out by detailed calculation, since it depends on the front factor ε. The result (2.2) can be modified to include cooperativity. Again the details are given in Ref. 4. The diagram in eq. (2.1) is connected with independent motion, say if one rod is able to move then another one can make progress, i.e.

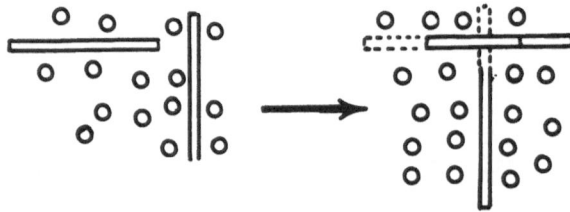

Figure 6

On the other hand if this is not possible and all rods are locked as given
by diagrams of the type (2.1), motion is still possible, but only in a total
cooperative manner, for example

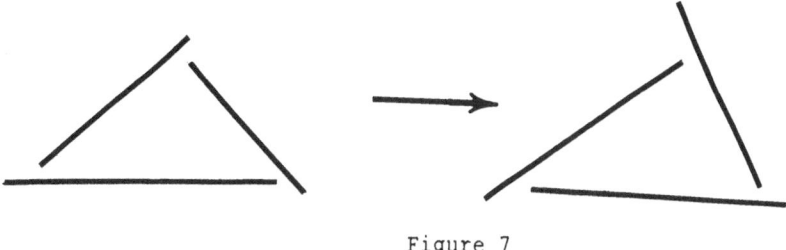

Figure 7

When the loop is made out of n rods it gives rise to an additional contribution:

$$(\alpha - \alpha_1 + \frac{\alpha_1}{\sqrt{n}})^n \qquad\qquad (2.4)$$

Figure 8

and the VF law appears

$$D \sim D_0 \exp{(-\frac{\alpha_1^2/4}{1 - \alpha_2})} \qquad\qquad (2.5)$$

where

$$\alpha_2 = \alpha - \alpha_1. \qquad\qquad (2.6)$$

The parameter α_1 contains generalized constants and a minimum loop size.
Physically the complete solution has been done in Ref. 4 by mapping the
problem onto the selfavoiding walk problem.

It can be shown that the number of rods moving cooperatively in the loops are given by

$$\bar{n} \sim (1 - \alpha_2)^{-2} \qquad\qquad (2.7)$$

and the size of the loop is therefore

$$\xi^2 \sim \bar{n}L^2 \sim \frac{L^2}{(1-\alpha_2)^2} \quad . \qquad\qquad (2.8)$$

Thus ξ diverges if $\alpha_2 \to 1$ as the phenomenological interpretation requires.

So far we have used α to denote the expression cdL^2, a combination of constants well known in liquid crystal theory since Onsager's work. However, its role in our theory is much more general, for example "d" can be T dependent through the fact that Van der Waals forces appear in the form $\exp(-E/nT)$ whereas hardcore forces do not contain T. Thus, at the level of a <u>model</u> one can regard $\alpha = 1$ as $T = T_g$ or $P = P_g$ etc.

Until one studies a detailed physical case there is no purpose in false verisimilitude. Thus the above model says now that the VF law is a direct consequence of cooperativity.

Turning now to the relaxation behaviour, it can be shown that for the rod model the stress relaxation follows, to first order, the law is

$$S(t) \sim e^{-(t/\tau)^{1/2}} \qquad\qquad (2.9)$$

This law has been found by using the diagrams of the type in Fig. 6. Cooperativity as indicated by (2.6) does <u>not</u> change (2.9) drastically, giving rise to logarithms

$$S(t) \sim e^{-(t/\tau)^{1/2}}[1 + const.\log(t/\tau)] \qquad\qquad (2.10)$$

This leads us to the conclusion that cooperativity is responsible for the VF behaviour but <u>not</u> for the relaxation phenomena.

2.2 A Mixture of Rods of Different Length

In order to simulate the result for mixtures of two glass forming liquids eq. (1.10) we consider a mixture of rods with two different lengths but not of different chemical species so that no thermodynamic interaction is present. For the following study we stay in the simple mean field theory without cooperativity to get an estimate of the freezing temperature. As we have noted above, the fact that zero diffusion at some concentration corresponds to zero diffusion at some temperature

$$D = D_0(1 - c_0/c) \qquad\qquad$$
$$\qquad\qquad\qquad\qquad (2.11)$$
$$D = D_0(1 - T/T_0)$$

implies a correspondence between T_0 and c_0 (or P_0 etc.) and is simply assumed here.

Now let us assume that we are near c_0 and the volume fraction of the shorter rods is ϕ_1 and that of the longer ones is $\phi_2 = (1 - \phi_1)$.

A further assumption is that the lengths of the rods are not too different in order to avoid further complications. (For example if the short rods are too short no tube constraint is present and one has simply Brownian motion of two types of particles).

Consider now a test rod which is one of the species (and remains the same for all further considerations). Now we argue on the same level as in the mean field theory and use graphs like Figs. 3, 4, 5. Again if no rod acts as a barrier we find

$$D = D_0. \tag{2.12}$$

If one barrier is on it could be a short one or a long one. The time for the short one is shorter of course while the long rod acts for a longer time as barrier,

$$\tau \sim L^2/D$$

and one can use the same type of self consistency argument as in the Edward & Evans theory and use the graphical expansion of the Edwards & Vilgis paper.

The corresponding diagram for one barrier would then be

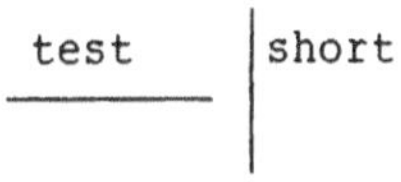

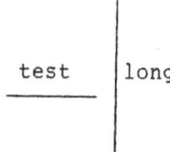

test   short

Figure 9

with probability $\phi_1 \alpha_1$ and $\alpha_1 = cdL_1^2$ $\qquad$ (2.13)

test   long

Figure 10

with probability $\phi_2 \alpha_2$ and $\alpha_2 = cdL_2^2$ $\qquad$ (2.14)

(c is the overall concentration $c = N_1 + N_2/N$). For one barrier this gives

$$D^{-1} = D_0^{-1} (1 + \phi_1 \alpha_1 + \phi_2 \alpha_2) \tag{2.15}$$

Now two barriers are on. Again this gives various possibilities

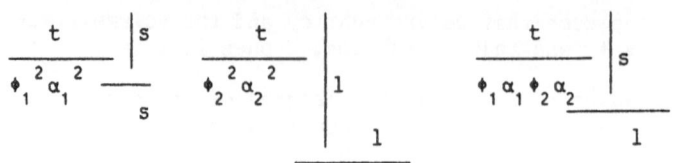

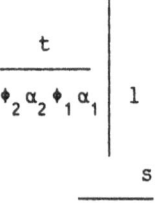

Figure 11

And the contribution from two barriers is given by

$$\phi_1^2 \alpha_1^2 + \phi_2^2 \alpha_2^2 + 2\phi_1 \phi_2 \alpha_1 \alpha_2 . \tag{2.16}$$

The way is now clear and for three barriers one has the possibilities

and so on

Figure 12

giving

$$\phi_1^3 \alpha_1^3 + \ldots = (\phi_1 \alpha_1 + \phi_2 \alpha_2)^3 \tag{2.17}$$

and for n rods one has the contribution

$$(\phi_1 \alpha_1 + \phi_2 \alpha_2)^n \tag{2.18}$$

Giving the series

$$D^{-1} = D_0^{-1} \sum_0^{\infty} (\phi_1 \alpha_1 + \phi_2 \alpha_2)^n \tag{2.19}$$

$$= D_0^{-1} (1 - \alpha_1 \phi_1 - \alpha_2 \phi_2) \tag{2.20}$$

Transforming it back by

$$\alpha_i = c/c_{01}$$

we get

$$\frac{1}{T_0} \sim \frac{\phi_1}{c_{01}} + \frac{\phi_2}{c_{02}} \tag{2.21}$$

which corresponds to the empirical law (1.10).

288

Clearly this argument holds only if <u>no</u> thermodynamic foces between the rods are present (i.e. the Huggins parameter is zero) otherwise (2.21) should be modified.

Again for the mixture of rods we expect VF behaviour which is now trivial since

$$\alpha_{eff} = \alpha_1 \phi_1 + \alpha_2 \phi_2 \qquad\qquad (2.22)$$

replaces the α in the VF calculation for single rods and no other change in the analysis occurs.

In the next section we shall move on to flexible polymers.

3. COOPERATIVE MOTION IN POLYMERS

In random flight polymers we expect the same behaviour. Although a rigorous attack on the problem is difficult we give a few arguments which have to be confirmed later. First we have to argue that the same type of cooperativity happens in flexible chains. Here it is not the whole chain which moves cooperatively like in the case of rods but only portions of the chains, which shuffle around cooperatively. Roughly there are several ways to make progress.

One can argue first that there is some motion of defects along the chain in the way used by de Gennes when he derived the first results on reptations[5]. Here we allow the defects to jump from one chain to another whenever the chains are close together.

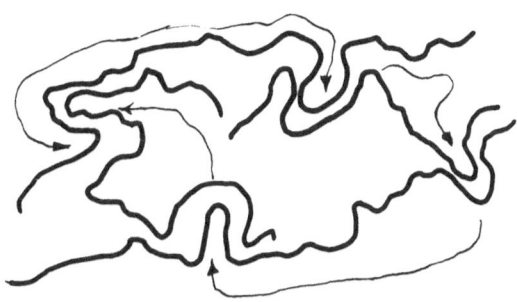

Figure 13

The defect can now with some probability go along a loop made out of several chains, four in the above figure, or with some other probability stay within one chain and contribute to the classical reptation motion. Already at this stage we have to divide the model into two cases:

I. $M < M_c$ Low Molecular Weight

In this case the dynamics of all single chains depends upon the molecular weight of the chains, say by their length L. For this regime we expect the freezing temperature T_g to be a function of the chain length

$$T_g = T_g(L).$$

T_g will then be a function of the density of polymer present and of the length. Experiments show that this is indeed the case. The length dependence is given by the empirical relationship (1.9). The concentration dependence is also found by experiment, where the T_g is lowered if the concentration of polymer is decreased, i.e. by adding low molecular weight additives.

II. $M > M_c$ High Molecular Weight

From the dynamical point of view this problem is well understood (see for example the book of Doi and Edwards[5]). The basic point here is that by introducing a new length scale into the problem the dynamical behaviour has been changed completely. The new length scale, a, is the distance between entanglements (the radius of the enclosing tube which also is the step length of the "primitive path" i.e. the topological skeleton of the polymer). In this regime the freezing of large scale motion is the freezing of motion on the scale of the primitive path step length, and the long term reptative motion of the whole chain is not a vital constitutive of the glass temperature.

Hence T_g is a function of a and the density of material, and also additional parameters like chainstiffness, κ, etc.

$$T_g \ (a(c,l), \ \kappa.....).$$

But a is a function of the density itself, so that the effect of a diluent acts on both a and c, and we expect a different concentration dependence of T_g above M_c. $T_g(L)$ is roughly given by

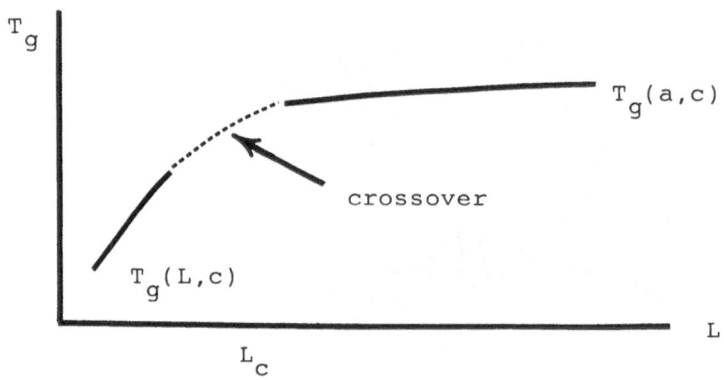

Figure 14

Relaxation

When the glasses are formed large scale motion is frozen and local rearrangements might be possible. How can this be modelled? Suppose we have a chain $M < M_c$ which undergoes a glass transition. The viscosity, for example is given by

$$\eta(\omega) \ = \ \Sigma \ \frac{1}{p^2} \ \frac{\tau}{\beta^2 + i\omega\tau} \ . \tag{3.1}$$

The relaxationspectrum is given by

$$H(p) = p^{-2} \tag{3.2}$$

which looks like

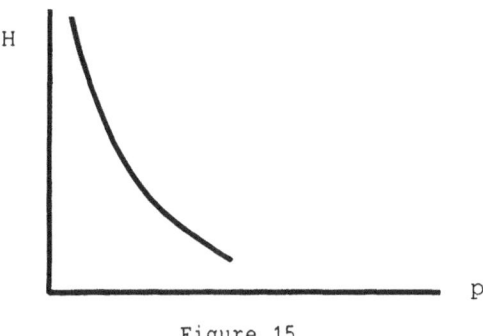

H

p

Figure 15

and it is dominated by the smallest p which is $p = 1$ in the Rouse case. Small p's are responsible for the large scale motion and the portion of chain moving in mode p is roughly L/p.

For polymers near T_g the very long modes are weakly present, and the longest mode in abundance will be related to the primitive path step length. Thus we might expect

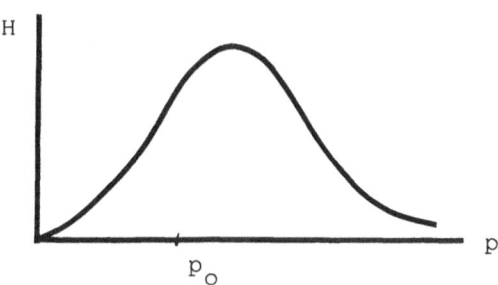

H

p_0

p

Figure 16

with some maximum at p_0 and modes with $p \ll p_0$ are not important. To suggest a model one can try

$$H(p) = \frac{p^2}{(p_0^2 + p^2)^2} \tag{3.3}$$

which comes back to the Rouse case for $p_0 \sim 0$ (or alternatively one can say that if p is not confined to the integers but used as a continuous variable, p_0 is just $2\pi/L$). The viscosity is then

$$\eta(\omega) = \int_0^\infty \frac{p^2 \, dp}{(p^2 + p_0^2)^2} \frac{\tau}{(i\omega\tau + p^2)} \tag{3.4}$$

giving

$$\eta(\omega) = \frac{\tau\pi}{p_0} \frac{1}{(p_0 + \sqrt{i\omega\tau})^2} \tag{3.5}$$

and the translational diffusion constant becomes zero. This is now in the form $\eta = \eta(t/\tau a_T)$ as the time temperature super position suggests.

The structure (3.5) appears in the literature in the high frequency studies of short chain melts by Barlow, Erginsav and Lamb (the BEL equation)[6]. We are here suggesting that it may be usefully extended more generally.

4. THE VOGEL-FULCHER LAW

Another question would be if there is a general argument for the behaviour of the VF law in temperature. Normally in the high temperature range the diffusion constant follows an Arrhenius law

$$D \sim D_0 e^{-\Delta E/uT} \qquad (4.1)$$

where ΔE is the activation energy. Again as in the previous section 2 let us study the inverse diffusion constant. This follows a law like

$$D^{-1} \sim e^{\Delta E/uT} \qquad (4.2)$$

for the motion of a single particle. If we consider now the motion of particles the activation energy is roughly $n\Delta E$ and we have

$$D^{-1} \sim e^{n\Delta E/uT}. \qquad (4.3)$$

All particles can be considered to have neighbours so that we have a co-ordination number z giving additional terms:

$$D_n^{-1} \sim (z)^{-n} e^{n\Delta E/\kappa T} \qquad (4.4)$$

or

$$D_n^{-1} \sim e^{-(\log z - \frac{\Delta E}{\kappa T}) n} \sim e^{-u(1 - \frac{T_0}{T})} \qquad (4.4a)$$

with

$T_0 = \dfrac{\Delta E}{\kappa \log z}$. Integrating over all n produces

$$D \stackrel{\sim}{=} (1 - \frac{T_0}{T}) \qquad (4.5)$$

The problem now is, can we find an argument to give

$$D_n^{-1} \sim e^{-(1 - \frac{T_0}{T})n + b\sqrt{n}} \qquad (4.6)$$

Clearly if we follow the previous rod model, cooperation will appear in chains which must close, and the $\sqrt{n}$ is characteristic of the "random walk" of the cooperation.

As the polymer melt is cooled (or otherwise has its density increased) it passes through the glass transition. Glass formation is easy for polymerized materials because the basic diffusion process is reptation which involves the slow Brownian motion of the chain down the tube i.e. along the primitive path (with new tube being created by an emerging end, or the old tube being destroyed by a retracting end). Reptation is a tiny part of the many different degrees of freedom of the polymer, but it is the motion which allows creep. At the glass transition creep ceases, reptation ceases, but the other degrees of freedom keep going albeit more slowly as the temperature diminishes or/and the density increases. An easy way to visualize the entanglements of the polymer is to represent them by slip links attaching it to the matrix of other polymers. We can now imagine that the slip link points tighten to the extent that slipping is no longer possible. This could be due to an increased constriction, or it could be due to slow local crystallinity developing in the chain, making it possible for the chain to slip through the slip link. Or it could be due to the creation of quite new constrictions along the length of the chain. For whatever reason our picture now is of a polymer "pinched" at a series of points along its length. Some of these pinches will be weak, and as the material is stressed will open, permitting an easier stress strain relationship than that at zero stress and strain. This is familiar with all frictional systems, the initial modulus is higher than that at finite strain. But other pinches are very strong and will hold. This now means that the primitive path segments will extend under stress, but will do so irreversibly as the polymer is being drawn over potential barriers by the stress.

A picture to illustrate this would be

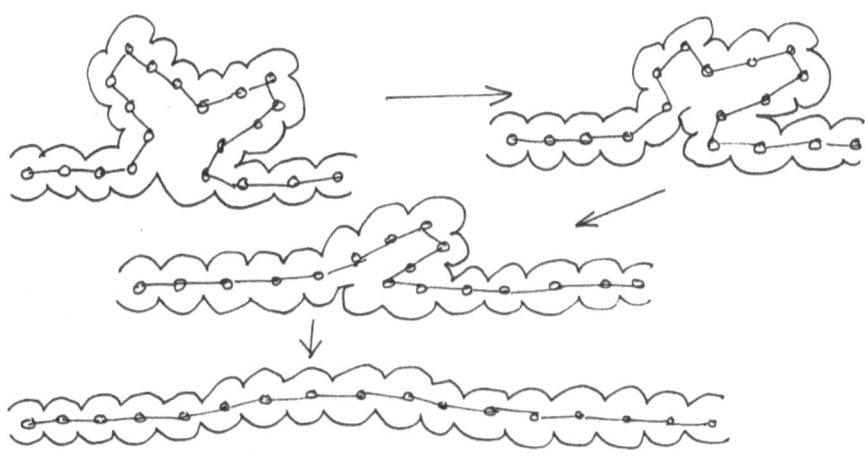

Figure 17

where the slack in the primitive path in the first picture is pulled out to reach the last picture. A rough contour of the potential energy is sketched around the polymer and this field is climbed to move from one picture to the next.

If the primitive path of the polymer has N_{pp} step lengths a, and the polymer itself has N of step length l, then the end to end distance

$$R^2 = Nl^2 = N_{pp}a^2 .$$

Hence a segment is fully extended when the strain is

$$\lambda = \sqrt{\frac{N}{N_{pp}}} = \frac{a}{l}$$

where now a can be shorter than in the melt because of the effects (extra pinches, chain thickened by crystalline region etc.) of the colder denser environment.

When one passes this strain the material will craze. This crazing itself can be studied by the above concepts since it can be regarded as the consequence of the checking of cracks by the full extension of chains which have not up to that point been fully extended'. Thereafter the material has no further capability of strain and suffers brittle fracture.

6. THE RECOVERY PROCESS

Suppose we study a glass which is strained but not fractured, and heat the glass back through the glass transition to the melt. We have argued that the "slip links", or the "tube" are consequences of the topology of the entanglements. The slack of the chain which is taken up on straining, and the development of pinches and crystallinity, do not violate this topology. It follows that on returning to the melt the topological specification of the material is unaltered except in as much as some withdrawal may have taken place at the ends of the chains:

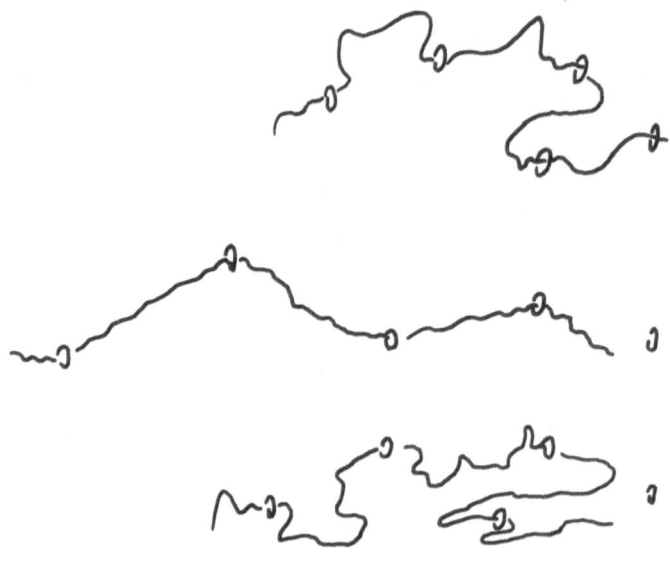

Figure 18

In Fig. 18 from the initial (a) to the fully extended (b) the chain has slipped out of the last link so that on heating its return to (c) will not be to the identical shape (a) but one which retains a memory of the shape (b).

These remarks include the effects of patches of crystallinity for crystallinity retains the topological specification of the chain except near the end points. We therefore find a simple explanation of the well know recovery process[9].

The difficulty with the above picture is in making it quantitative. For melts Doi and Edwards argued that an effective friction coefficient, and the step length a gave a total specification of the visco-elasticity of the melt or to be more precise it is possible to produce a theory of visco-elasticity with only these two paramters, and one naturally explores that theory before embarking on adding new parameters to overcome failings of the simplest theory.

When we come to study the glassy state, the stress strain relationship is already non-linear in its initial phase, and the only easy prediction is that if there are no extra pinches or crystalline accretions, the full extension of the glass up to fracture will be the same as that of the rubbers. This must, however, be an overestimate and more parameters appear describing all the various concepts described in this paper. It is clear that the glass will never be as simple as the melt.

ACKNOWLEDGEMENT

The authors would like to thank Dr. Athene Donald for helpful discussions.

REFERENCES

1. J.D. Ferry, "Viscoelastic Properties of Polymers", Wiley and Sons, New York (1980).
2. R. Kohlrausch, Pogg. Ann. (3) 12, 1974; G. Williams and D.C. Watts, Trans. Farad. Soc. 66, 80 (1970).
3. S.F. Edwards amd T.A. Vilgis in "The Physics of Disordered Materials", D. Adler and H. Fritsche (eds), Plenum Press, New York (1985) and Physica Scripta, T13, 7-16 (1986).
4. S.F. Edwards and K.E. Evans, JSC Farad. Trans. II, 78, 113 (1982).
5. J.P. de Gennes, J.C.P. 55 572 (1971). See also "Scaling Concepts in Polymer Physics", Cornell University Press (1979) and M. Doi and S.F. Edwards in "Polymer Dynamics", Oxford University Press (1986).
6. A.J. Barlow, A. Erginsav and J. Lamb, Proc. Roy. Soc. A298, 481 (1967) and A309, 473 (1969). See also J. Lamb in "Molecular Motion in Liquids", J. Lascombe (ed), Reidel, Dordrecht 1974 and J. Rheol. 22, 317 (1978) and also ref. 1 above, page ch. 15.
7. K.E. Evans and A.M. Donald, Polymer 26, 101 (1985) where other references are given.
8. M. Doi and S.F. Edwards, JCS Farad. Trans. II 74, 1789;1802 (1978)
9. R.W. Truss, P.L. Clarke, R.A. Duckett and I.M. Ward, J. Pol. Sci. (Pol. Physics) 22, 191 (1984).

TIME DEPENDENT NONLINEAR RESPONSE OF A DIPOLAR GLASS

K. B. Lyons and P. A. Fleury

AT&T Bell Laboratories
Murray Hill, NJ 07974

The competition between the development of long range order and the freezing in of disorder at low temperatures may be particularly delicate in dipolar systems. The ordering of randomly placed electric dipoles in an incipient ferroelectric host can be affected by controlling concentration, temperature, electric field, pressure or uniaxial stress. The dynamical behavior of such systems may be experimentally probed on time scales ranging from picoseconds to hours, so they offer the prospect for detailed study of this important competition.

In this paper we describe experiments on both the linear and nonlinear dielectric response of a dipolar glass. By combining results of inelastic light scattering with dielectric relaxation data, we obtain a quantitative measure of the dipolar dynamics spanning more than 15 decades in frequency. We find clear evidence from the linear response for a cooperative dynamic regime, suggesting a transition temperature to a glassy dipolar state. Our experiments on the nonlinear dielectric response, initiated in an attempt to clarify the divergent behavior near T_g expected for a spin glass, revealed a number of surprises. Most importantly the polarization response is found to be a nonanalytic function of electric field, revealing the inadequacy of the usual nonlinear susceptibility description. Moreover, we find that the response of a sample cooled in zero field depends on the waiting time between cooling and field application in a manner reminiscent of, but differing in detail from, that which has been observed for spin glasses.[1]

The system of interest here, $KTa_{0.991}O_3$, is a simple cubic perovskite lattice ($KTaO_3$) containing a dilute concentration ($x=0.009$) of isoelectronic symmetry breaking (SB) defects (Nb^{5+}). The host lattice is centro-symmetric at all temperature. It is, however, potentially unstable to a ferroelectric distortion and exhibits a soft zone-center infrared active TO phonon. Experimentally, at low temperature, the Nb lies off the Ta site along one of the eight equivalent [111] directions. It is a symmetry breaking, relaxing defect, able to hop among the eight available sites.

As T decreases, the Nb hopping slows down and the host lattice correlation length increases, thus causing the Nb related dipolar clusters (the polarization clouds surrounding one or more Nb moving cooperatively) to grow. As the clusters grow, their relaxation slows precipitously. For $x=0.009$, the mean Nb-Nb distance is about 5 unit cell diameters. Our earlier studies using defect induced first order Raman spectroscopy[2] show that the range of symmetry breaking which extends from each defected cell reaches this value at temperatures of order a few K. Hence, the competition is very delicately balanced in this system.

Two samples electroded along [100] and [111] were cut from neighboring portions of a single crystal boule of $KTa_{0.991}Nb_{0.009}O_3$. The [100] sample was the same one used in the previous dielectric study.[3] The light scattering data[4] displayed in Figure 1 are obtained in zero field. A complete description of that experiment has appeared elsewhere, so we shall only summarize it here.

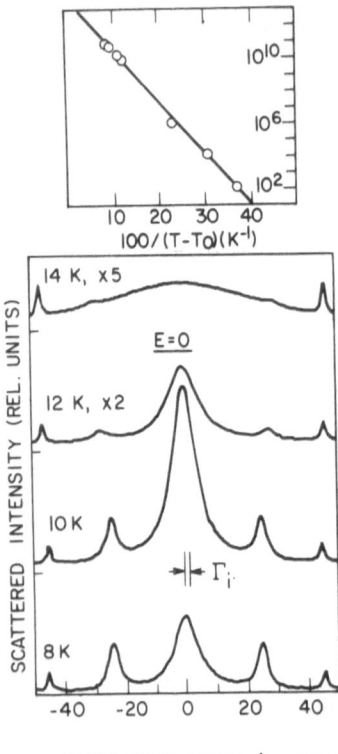

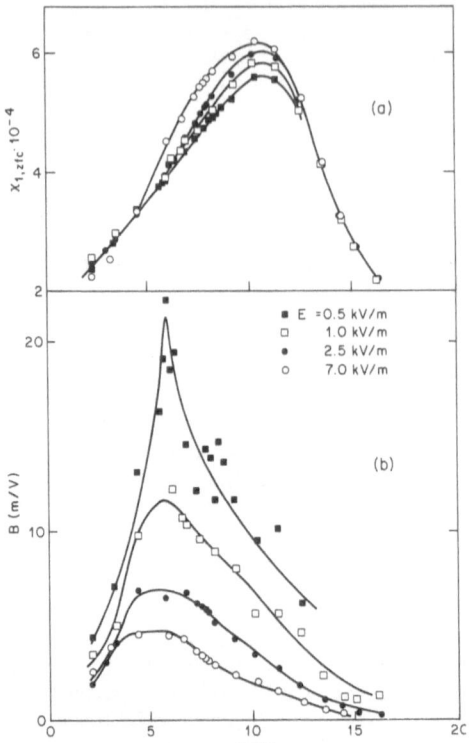

Fig. 1: The relaxation times shown in the Arhennius plot are extracted from dielectric measurements[3] (1 MHz and below) and from Brillouin spectra shown in the bottom half ($>$ 1 GHz). All results shown are obtained in zero DC field.

Fig. 2: Plots of the linear susceptibility $\chi_{1,zfc}$ and the Rayleigh parameter B, as defined in the text, for various applied field amplitudes. Lines are guides to the eye. All data taken immediately upon cooling.

In the Raman spectrum, an underdamped soft-mode peak appears below 14 K, which has been taken as evidence of a ferroelectric transition[5]. However, this mode appears far above the temperature (6 K) where the χ_1 dielectric data[3] indicate a maximum. Evidently, between 6 and 14 K, the Nb ions are hopping sufficiently rapidly that they exhibit an average on-site position on the time-scale of the dielectric data ($>10^{-6}$ sec), but exhibit clear symmetry breaking on the time-scale of the soft mode vibration ($\sim 10^{-12}$ sec). A strong electric field dependence of the soft mode intensity indicates that only a fraction of the sample bulk is in fact distorted for E=0.

The Brillouin data[4] reveal a strongly temperature dependent quasielastic feature, which obviously couples to certain acoustic modes. It is well described as a sum of two Lorentzians, one of which has a width the same as the instrument. The latter component has been removed in the spectra displayed in Fig. 1. Since the elastically scattered light is removed by the I_2 absorption cell, this component must be dynamic but narrow compared to 1 GHz. The width of the broader central peak component (Fig. 1) changes very rapidly above about 10 K.

Clearly, there is an average characteristic relaxation time which increases dramatically as temperature is lowered from 14 to 6 K (Fig. 1). The light scattering and earlier linear dielectric data are all compatible with glassy relaxation of the Vogel-Fulcher form[6]

$$\tau_c^{-1} = \nu_0 \exp[-E_a/(T - T_0)] \ , \tag{1}$$

where τ_c is the average relaxation time, E_a an activation energy, ν_0 an attempt frequency, and T_0 the temperature of a transition to a glassy state. Fig. 1 shows good agreement with Eq. (1) over nine decades in frequency with the physically reasonable parameters $\nu_0 = 300 \text{cm}^{-1}$, $E_a/k = 70$ K, and $T_0 = 3.0$ K.

Thus the linear response of this system already indicates glassy behavior with an extrapolated relaxation time divergence indicated at a small but nonzero temperature, $T_0 = 3$ K. Although the form Eq. (1) breaks down before T_0 is reached, it has been suggested that this extrapolated transition should bear a strong analogy to a spin glass transition. Verification of the spin glass analogy requires that the non-linear susceptibilities, $\chi_n(\omega)$, exhibit singularities[7,8] at a well-defined dipolar glass temperature. Careful measurement of the nonlinear polarization response is a key to elucidating the competition between long range order and dipolar glass behavior. We have measured the polarization directly using the circuit sketched in Fig. 2. The input voltage trace was used to determine the zero phase point, and then the digitized signal voltage V_{sig} was Fourier analyzed for its harmonic content, $P = \sum[F_n \sin(n\omega t) + G_n \cos(n\omega t)]$. The coefficients $\{F_n, G_n\}$ may be related to the material parameters $\{\chi_n\}$ through $P/\epsilon_0 = \sum \chi_n E^n$ (where $\epsilon_0 = 8.85 \cdot 10^{-12}$ for mks units). At sufficiently low voltage, we expect simply $F_n \propto E_0^n \chi_n$. In more generality, the term E^n in (2) will contribute to all harmonics equal to and less than n. We have performed the analysis outlined above for harmonics up to 11, at a nominal frequency of 10 Hz, with various field amplitudes. The absolute values obtained for the nonlinear susceptibilities are never found to be independent of the input voltage amplitude, even for voltage amplitudes of 4 V/cm. The inescapable conclusion is that P(E) is not an analytic function.

Similarly we find that values of the linear susceptibility obtained from standard analysis may be quantitatively incorrect. Also, we measured the quadrature voltage at the fundamental, G_1, over the frequency range from 10 Hz to 10 kHz. Over a wide range of temperature, extending far above that where χ_1 peaks (6-8 K), we find a value which is nearly frequency independent. This behavior persists, albeit weakly, as high as 45 K. No simple relaxation mechanism can account for such a weak frequency dependence.

All of these observations suggest the presence of switchable dipolar states with a distribution of coercive fields which extends all the way to zero. These may be viewed as activated two-level systems. This model predicts that the *initial* polarization response to a field imposed on the system after zero field cooling will contain a component *quadratic* in the field, and that $P(E)$ is not an analytic function as E is cycled. We have extended these arguments to show that the odd harmonics, with an ac applied field, will exhibit quadrature components which all increase as E_0^2, rather than E_0^n as usually expected. The harmonics we measure all behave as E_0^α, with $1.6 < E_0 < 2.3$.

We expect the initial response to a field E, during the first half cycle after zero field cooling, to obey the Rayleigh relation:[9]

$$P/\epsilon_0 = \chi_{1,zfc} E + sgn(E) B E^2,$$

(2)

where the term $\chi_{1,zfc}$ accounts for the linear response of the lattice and of those switchable units for which the barrier W is sufficiently small, $W \lesssim kT$. Fits of Eq. (2) to our data yield the parameters $\chi_{1,zfc}$ and B as shown in Fig. 2

Note the approach to a cusp-like behavior near $T \sim 5$ K as $E \to 0$. Note also the displacement between the peaks in B and $\chi_{1,zfc}$ as functions of temperature. This displacement makes it clear that they arise from different phenomena (B due to cluster reversal). $\chi_{1,zfc}$ is relatively insensitive to E_0, and its values lie considerably below those obtained by the usual ac technique.

In order to verify the activated hopping mechanism we measured the relaxation of the polarization after application of a dc step voltage on a sample cooled in zero field. The resulting polarization was recorded from 10^{-3} sec out to 200 sec. Over this range we find an excellent parametrization to be $P/\epsilon_0 = P_0 + P_1 \ln(t)$.

The simple model used here fails to reproduce one important aspect of our data. Namely, the decrease in B for $T < 6$ K suggests a *nonuniform* density of states as a function of E_c, while the polarization relaxation results suggest a *uniform* distribution as a function of W. We thus conclude that E_c is not simply proportional to W.

The measurements of $\chi_{1,zfc}$ and B reported above, as well as the logarithmic growth of P following a voltage step, are all obtained with essentially zero waiting time after establishment of the temperature. We found, however, that the value of B depended on the waiting time, especially near the cusp at low fields in Fig. 2. Since this was reminiscent of the "waiting time effect" which has been observed in spin glasses, we attempted to perform the same type of experiment in the KTN system. In Fig. 3 we show two traces obtained with the sample immersed in liquid helium, with waiting times of 1.0 and 10.0 min between immersion and the application of the voltage step. Since

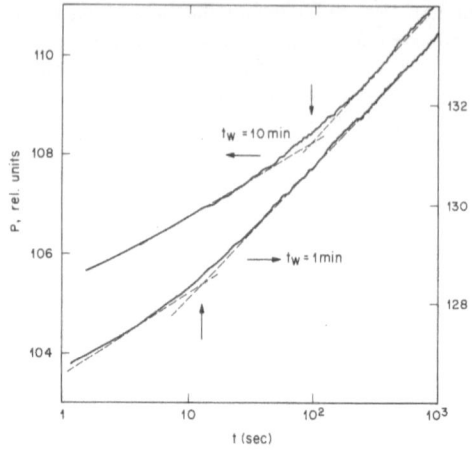

Fig. 3: Semilog plots of polarization response to a voltage step of 14 V/cm
following zero-field cooling. Values of t_w are indicated, the time
elapsed between cooling and field application. Vertical arrows indicate approximate shoulder positions.

the shoulder marks the transition from the relaxation characteristic of the equilibrated states to
that of the nonequilibrated ones, it is expected that at low fields it should occur at a time comparable to the waiting time t_w, as observed. We have also found, in other measurements, that use of
higher fields affects the position and clarity of the shoulder. Indeed, a field of 70 V/cm is sufficient
to remove it entirely.

In summary we have shown that the linear response of Nb:$KTaO_3$ relaxes as T is lowered in a
manner consistent with a Vogel-Fulcher behavior, extrapolating to a fully frozen glassy state below
$T_0 = 3$ K. The nonlinear dielectric response, however, is not simply that expected of a spin glass.
Rather, even at the lowest electric fields applied the polarization is nonanalytic in the field, even
well above the temperature at which $\chi\prime$ peaks. Furthermore, fields as low as a few V/cm are sufficient to distort apparently linear dielectric measurements.[3] A model based on activated two-level
systems describes our data with a barrier distribution which extends to zero energy. The polarization response after application of a step dc voltage, although largely logarithmic in time, exhibits
waiting time effects akin to those observed in spin glasses. It is thus apparent that some analogy
exists between our results and those in spin glasses. However, the deviations from spin glass
behavior in our observations for KTN make it clear that the dipolar glass is more complicated in
both its dynamic and nonlinear behavior than any model thus far formulated.

The authors gratefully acknowledge discussions with D. Huse, H. Bouchiat, L. Levy, R. Rammal, and P. Littlewood.

The crystals used for this study were kindly provided by D. Rytz, of Hughes Research. The
authors also gratefully acknowledge discussions with D. Huse, H. Bouchiat, L. Levy, R. Rammal, and
P. Littlewood.

References

1. P. Svedlindh, P. Granberg, P. Nordblad, L. Lundgren, and H. S. Chen, Phys. Rev. B **35**, 268
   (1987).
2. H. Uwe, K. B. Lyons, H. L. Carter, and P. A. Fleury, Phys. Rev. B **33**, 6436 (1986).
3. G. A. Samara, Jpn. J. Appl. Phys. Suppl. **24**, Pt. 2, 80 (1985).
4. K. B. Lyons, P. A. Fleury, and D. Rytz, Phys. Rev. Lett. **57**, 2207 (1986).
5. U. T. Hochli and H. E. Weibel, and L. A. Boatner, Phys. Rev. Lett. **39**, 1158 (1977).
6. M. Cyrot, Phys. Lett. **83A**, 275 (1981).
7. H. Bouchiat, J. Physique **47**, 71 (1986).
8. L. Levy and A. Ogielski, Phys. Rev. Lett. **57**, 3288 (1986).
9. Lord Rayleigh, Phil. Mag. **23**, 225 (1887).

SLOW RELAXATION OF ORGANIC LOW TEMPERATURE GLASSES

Werner Köhler and Josef Friedrich

Physikalisches Institut
Universität Bayreuth
Postfach 101251, D-8580 Bayreuth

ABSTRACT

The technique of persistent spectral hole burning provides a sensitive tool for examining slow relaxation processes in amorphous solids. Photochemical holes relax on a logarithmic timescale. Temperature cycle experiments lead to irreversible hole filling and broadening. This behaviour can be understood within the frame of two level systems. The dispersion of relaxation rates and the distribution of barrier heights can be determined.

INTRODUCTION

Glasses are not in a global minimum of their free energy. They are disordered on microscopic scales. Transitions between local minima of their free energy are possible and allow for structural relaxation processes. The timescales of such processes cover many orders of magnitude due to a distribution of the microscopic parameters.

Photochemical holeburning provides means to examine both very fast dynamic processes and very slow ones on the timescale of seconds to weeks. The latter will be addressed in this article.

PHOTOCHEMICAL HOLEBURNING

Dye molecules doped in organic glasses show inhomogeneously broadened optical absorption bands, reflecting the statistical fluctuations of the amorphous host matrix. Narrow band laser excitation at a frequency ω_L within the inhomogeneous absorption band allows only the small fraction of molecules to absorb which have their transition energy close to the laser frequency. After the absorption of a photon a dye molecule may undergo with a certain probability a photochemical reaction (photochemical holeburning) or the solvent cage may rearrange (photophysical holeburning). In either case the molecule will change its absorption frequency. The consequence is a narrow dip within the inhomogeneous absorption spectrum - a persistent spectral hole. The hole spectrum is defined as the change of the absorption spectrum caused by the holeburning process. For a detailed derivation see ref. 1.

EXPERIMENTAL

The samples used were quinizarin, tetracene and dimethyl-s-tetrazine (DMST) in alcohol glass. The concentrations were typically 10^{-4} mole/l. Quinizarin and DMST undergo photochemical holeburning while the photoreaction of tetracene is of photophysical nature. Despite the differences of the photoreactions all systems show a similar characteristic relaxation behaviour on logarithmic timescales. In this article we will only look at tetracene in greater detail. For more information about the other systems see ref. 3-7.

To monitor long time relaxation at constant temperature a hole was first burnt and then observed over a period of almost one week. Hole areas and widths were determined by fitting the data to a Lorentzian. This experiment was done both in protonated and perdeuterated alcohol.

In a second kind of experiment a hole was burnt at low temperature T_b. Then the temperature was raised to a higher value and cycled back to T_b where the hole was measured again. This was done again and again while increasing the temperature in a stepwise fashion[3].

RESULTS

1. At constant temperature the hole area decreases linearly on a logarithmic timescale, independent whether the alcohol is deuterated or not (Fig. 1).
2. The holewidth increases linearly on a logarthmic timescale in the protonated sample whereas it remains nearly constant in the deuterated sample (Fig. 1).
3. In the temperature cycle experiments the hole area (always measured at the burning temperature T_b) decreases with increasing cycling temperature (Fig. 2).
4. In both experiments the shapes of the holes remain Lorentzian.

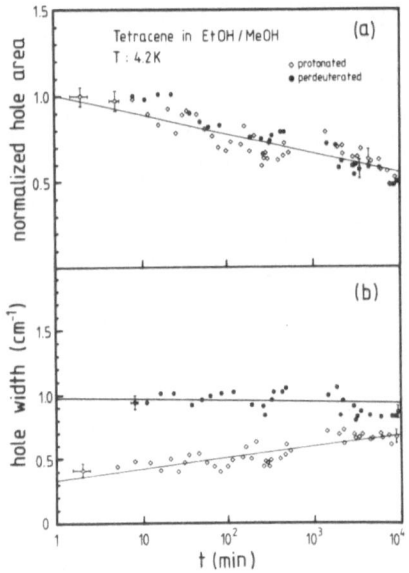

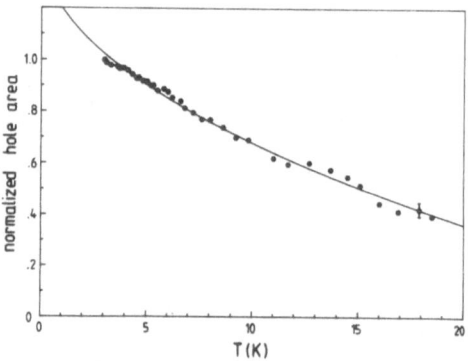

Fig. 1. (a) Hole area and (b) hole width of tetracene in protonated and perdeuterated alcohol as a function of time

Fig. 2. Hole area of tetracene as a function of cycling temperature

DISCUSSION

The tunneling model

In the following paragraph the results stated above will be explained within the frame of the tunneling model. The concept of two level systems (TLS) was first introduced by Anderson, Halperin and Varma[8] and by Phillips[9] to meet the anomalous low temperature properties of the specific heat, the thermal conductivity and the ultrasonic attenuation of glasses.

Glasses are not in thermal equilibrium and, hence, only in a local minimum of their free energy. A transition between two adjacent minima can be viewed as a transition between the two minima of a double minimum potential. Neglecting higher levels the double mimimum potential can be described as a TLS. Let us now assume that the photoreaction may also be described in terms of TLS, denoted as TLSp in contrast to the intrinsic TLS of the glass. Furthermore, we assume that the characteristic parameters of the photoreactive TLSp are distributed in a similar manner as that of the intrinsic TLS, reflecting the randomness of the matrix. Transition are possible either thermally activated − as it is the case in the temperature cycle experiment − or by tunneling through the barrier even at low temperatures.

For the tunneling relaxation rates R Jäckle derived the following distribution[10].

$$p(R) = \frac{\overline{P}}{2} \frac{1}{R\sqrt{1-R/R_{max}}}$$

Long time relaxation of the hole area

The hole are diminishes because molecules relax from the product to the educt state. It is proportional to the number of molecules that are still in the product state. To calculate the area $A(t)$ one has to integrate the above distribution $p(R)$ from the slowest relaxation rate R_{min} to a rate R which corresponds to the current experimental time t. R_{min} is a cut-off value which is introduced both from physical and mathematical reasons since it corresponds to the maximum barriers, which have to be finite, and allows normalization of $p(R)$.

Normalization to the first measured hole area immediately after the burning at time t_1 (corresponding rate R_1) yields

$$A(t)/A_1 \simeq 1 - [\ln R_1/R_{min}]^{-1} \ln R_1 t$$

This is exactly the observed logarithmic time dependence. The slope factor is determined by the minimum relaxation rate. We found a dispersion of the relaxation rates R_1/R_{min} on the order of 10^{10} with R_1 being on the order of minutes and determined by the experiment.

Spectral diffusion

To explain the observed line broadening we use the concept of spectral diffusion which was used by Klauder and Anderson[11] for spin-1/2-systems. TLS are formally equivalent to spin-1/2-systems. A dye molecule ('A-spin') is coupled to an ensemble of TLS ('B-spins') via electric or strain fields. The energy of the A-spin − the absorption frequency of the dye molecule − is determined by the configuration of the surrounding B-spins. When the B-spins flip the energy of the A-spin is shifted away from its former value. In turn this leads to a broadening of the spectral hole. For dipolar interaction it can be shown that the diffusion kernel is a Lorentzian with a width γ_D proportional to the number of B-spins having flipped on odd number of times. Since both the burnt hole and the diffusion kernel are Lorentzian the convolution is again a Lorentzian with a width $\gamma = \gamma_o + \gamma_D$. This is in agreement with result 4). Thus, the spectral diffusion can easily be measured by monitoring the broadening of a photo-chemical hole. To get the width γ_D one has to calculate the number of TLS which have relaxed since the hole was burnt. With similar arguments as for the hole area one gets a logarithmic time dependence for the hole width:

$$\gamma = \gamma_0 + B \left[\ln \frac{R_{max}}{R_{min}} \right]^{-1} \ln R_{max} \, t$$

R_{max} is the maximum rate and B contains unknown system parameters.

From the fact that there is an isotope effect on the line broadening while there is none on the hole filling we conclude that there are two different sets of TLS involved. From the pronounced isotope effect on the spectral diffusion it follows that the responsible TLS are connected with proton tunneling in the alcohol matrix. The photoreactive TLSp are assumed to originate from reorientation of the whole tetracene molecule[3].

Barrier distribution

While the experiments discussed above are sensitive to the distribution of relaxation rates, the temperature cycle experiments can be used to sample the distribution of barrier heights. During a temperature cycle - the sample is kept at temperature T for a time τ - only those TLS can relax which have a barrier height less than $V_0 = kT \ln \nu_0 \tau$. ν_0 is an attempt frequency on the order of the zero point vibration of the double well. Within the tunneling model the constant distribution of the tunneling parameter λ allows to calculate the distribution $p_V(V_0)$ of the barriers:

$$p_V(V_0) \sim V_0^{-1/2}$$

To calculate the relative hole area A(T) as a function of the cycling temperature T one has to integrate $p_V(V_0)$ over this TLS which have not yet relaxed.

$$A(T) \sim 1 - \sqrt{kT \ln \nu_0 \tau / V_{max}}$$

V_{max} is the maximum barrier height. From the fit in fig. 2. V_{max} turns out to be on the order of 1000 cm^{-1}. It should be recalled that the hole filling samples the photoreactive TLSp. The TLS of the matrix are also sensitive to temperature changes which results in a thermally induced line broadening. For more details see ref. 3.

ACKNOWLEDGEMENT

The work was supported by the Deutsche Forschungsgemeinschaft, Sonderforschungsbereich 213.

REFERENCES

1. J. Friedrich and D. Haarer, Ang.Chemie 96: 96 (1984, Int.Ed.Engl. 23:113 (1984)
2. W. Köhler, W. Breinl, and J. Friedrich, J.Phys.Chem. 89:2473 (1985)
3. W. Köhler, J. Meiler, and J. Friedrich, Phys.Rev. B35 (1987)
4. W. Breinl, J. Friedrich, and D. Haarer, Chem.Phys.Lett. 106:487 (1984)
5. W. Breinl, J. Friedrich, and D. Haarer, J.Chem.Phys. 80:3496 (1984)
6. J. Meiler, J. Friedrich, Chem.Phys.Lett. (1987), accepted
7. J. Friedrich and D. Haarer, in: "Optical Spectroscopy of Glasses", I. Zschokke ed., Reidel, Berlin 1986, p.149
8. P.W. Anderson, B.I. Halperin, and C.M. Varma, Philos.Mag. 25:1 (1971)
9. W.A. Phillips, J.Low Temp.Phys. 7:351 (1972)
10. J. Jäckle, Z.Phys. 257:212 (1972)
11. R.J. Klauder and P.W. Anderson, Phys.Rev. 125:912 (1962)

SPIN DYNAMICS IN DILUTE SYSTEMS

R. Rammal

AT&T Bell Laboratories
Murray Hill, NJ
CRTBT-CNRS
Grenoble

I. INTRODUCTION

At the percolation threshold, both the infinite cluster and its backbone are self-similar (fractals). This self-similarity also holds at concentrations p near p_c for length scales which are smaller than the percolation length ξ_p. For $\ell < \xi_p$, the number of bonds (or sites) on the infinite cluster scales as ℓ^{d_p}, where $d_p = d - \beta_p/\nu_p$ denotes the fractal dimensionality and β_p, ν_p are the percolation transition critical exponents. The influence of the scale invariance on the physics of fractals is now very well known. For instance, anomalous behavior of physical properties described by linear problems (classical diffusion, spectrum of the discrete [Laplacian operator, localization, etc.) has been shown to occur on fractals.[1,2]

In addition to linear problems, the static critical behavior of spin models on dilute lattices is now well established.[3] For instance, Ising spins model on such a dilute lattice is defined by the Hamiltonian:

$$H = - \sum_{(ij)} J_{ij} \, \sigma_i \sigma_j \qquad (1.1)$$

where $J_{ij} = J$ (if both the nearest-neighbor sites i and j are occupied) or $J_{ij} = O$ (otherwise). Here $\sigma_i = \pm 1$ are Ising spins. Other models: Potts, Heinsenberg, XY, .. etc. are defined similarly. The model (1.1) is relevant for dilute magnets and is known to have no long-range order at $p < p_c$, and orders at $T_c(p)$ for $p > p_c$, with $T_c(p) \to o$ as $p \to p_c$. Near the multicritical point $p = p_c$, $T = o$. the critical line is given by: $\exp(-2J/T_c(p)) \sim (p - p_c)^\phi$, where the crossover exponent ϕ has been shown to be equal to unity for all dimensions. Accordingly, at $p = p_c$ ($\xi_p = \infty$), the spin-spin correlation length scale as $\xi_T \sim \exp(2K\nu_p) \sim \xi_1^{\nu_p}$, i.e. $\nu_T = \nu_p$. Here $K = J/T$ and ξ_1 is the correlation length for the one-dimensional (1D) Ising model. The other thermal exponents associated with the zero-temperature $(T_c(p_c) = 0)$ ferromagnetic transition on the infinite cluster are given in terms of the percolation exponents $(\beta_T = 0$, $\nu_T = \nu_p$, $\eta_T = 2 - d_p$, etc.).

The purpose of this paper is to review the most recent results relative to the spin dynamics described by (1.1). These results should describe recent neutron-scattering experiments performed on a 2D material.[4] The spin dynamics associated with continuous degrees of freedom can actually be described as a linear problem. Therefore, there is nothing new to be learned there, at least from a fundamental point of view. The main known results relative to this problem will be summarized in Section II. Sections III and IV will be devoted to discrete degrees of freedom (Ising, Potts, etc.) models. Here a natural formulation of a new dynamic scaling hypothesis arises and will be discussed in relation with the so-called glassy dynamics.

II. DILUTE HEINSENBERG MAGNETS

The dynamic scaling for dilute Heisenberg magnets can be understood through a simple generalization of the principle of dynamic scaling to multicritical behavior. The basic idea is the existence of a single correlation length and a single relaxation time to which all other lengths and times scale. Accordingly, we write the Fourier transform of the order-parameter correlation function $C(\vec{k}, \omega)$ as[5]

$$C(\vec{k}, \omega) = \frac{2\pi}{\omega(\vec{k}, \xi)} C(\vec{k}) F(k\xi, \omega/\omega(\vec{k}, \xi)) \tag{2.1}$$

where ξ is the correlation length, and $\omega(\vec{k}, \xi)$ is the typical relaxation (or spin wave) frequency associated with the order parameter. $\omega(\vec{k}, \xi)$ has the following scaling form

$$\omega(\vec{k}, \xi) = k^z f(k\xi) \tag{2.2}$$

where z is the dynamical exponent. Near the multicritical point $p = p_c$ and $T = 0$, the correlation length can be written as

$$\xi = [\Delta p]^{-\nu_p} X(\epsilon/\Delta p^\phi) \tag{2.3}$$

where $\Delta p = p - p_c$, and $\epsilon = k_B T/J$ is the appropriate temperature variable. ν_p is the percolation-correlation length exponent and ϕ denotes the percolation-thermal crossover exponent. The asymptotic forms for the function $X(y)$ are: $X(y \ll 1) = X_o$ and $X(y \gg 1) = X_\infty y^{-\nu_p/\phi}$. As a result, we have

$$\xi = \chi_{o} (\Delta p)^{-\nu_p} \quad \text{at} \quad \epsilon \rightarrow o, \ \Delta p \neq o \tag{2.4}$$

$$= \chi_{\infty} \ \epsilon^{-\nu_T} \quad \text{at} \quad \Delta p = o, \ \epsilon \rightarrow o$$

(here $\nu_T = \nu_p/\phi$).

A. Heisenberg Ferromagnets

Consider the case of a ferromagnet below the transition temperature $T_c(p)$. The low-temperature spin wave mode has the frequency proportional to k^2, and can be generally written as

$$\omega = k^2 f^- (k\xi) \tag{2.5}$$

with $f^- (y \gg 1) = f^-_\infty$ and $f^- (y \ll 1) = f^-_o \ y^{2-z}$. Noting that at $T = o$, $\omega = D(p)k^2$ with $D(p) \sim (\Delta p)^{t-\beta_p}$, one can easily see that z is given by

$$z = 2 + (t - \beta_p)/\nu_p = 1/\nu_{RW} = (2 + \theta). \tag{2.6}$$

Here ν_{RW} is the so called random-walk exponent on percolating clusters. These exponents can be related to diffusive regime above the transition temperature. Writing $\omega(k) = D'k^2$, one finds for $\Delta p \simeq o$ and $\epsilon \ll 1$

$$D' \sim \epsilon^{\theta \nu_p/\phi_H} \tag{2.7}$$

where ϕ_H is the percolation-thermal crossover index for the Heisenberg system. On the other hand, for $\Delta p \lesssim o$ and $\epsilon \simeq o$, one has

$$D' \simeq |\Delta_p|^{(z-2)\nu_p} \tag{2.8}$$

which corresponds to diffusion in the large finite cluster. The numerical values of z as given by (2.6) are respectively: $z = 2.72$ at $d = 2$ ($\phi_H = 1.5$, $\nu_p = 4/3$) and $z = 3.64$ at $d = 3$ ($\phi_H = 1.2$, $\nu_p = 0.86$).

Note that the same argument leads, for a planar ferromagnet, to the following expression for z:

$$z = 1 + (t - \beta_p)/2\nu_p \tag{2.9}$$

and $z = 1.82$ at $d = 3$. It is important to notice that the argument leading to (2.6) and (2.9) depends upon identifying the zero-temperature correlation length with ξ_p. Furthermore at $d = 2$, isotropic magnets do not have long-range order for $T > o$. The result (2.6) is then valid for weakly anisotropic systems, in which a part of the critical region shows the isotropic behavior before crossing over to the anisotropic behavior.

B. Dilute Antiferromagnet (AF)

In a dilute AF, the hydrodynamic spin-wave mode governing transverse order-parameter correlations is given by

$$\omega = cq \sim (p - p_c)^{(t+\tau)/2} \sim \xi^{-\left(1 + \frac{t+\tau}{2\nu_p}\right)} \, (q\xi) \tag{2.10}$$

where τ is an exponent governing the divergence of the transverse susceptibility $\chi_\perp$ of an antiferromagnet: $\chi_\perp \sim (p - p_c)^{-\tau}$ ($\tau \simeq 0.5$ at $d = 3$). From (2.10), we identify

$$z = \left[1 + \frac{(t+\tau)}{2\nu_p} \right] \tag{2.11}$$

and write generally

$$\omega = k^z \, f^-(k\xi)$$

For large k, we get a critical mode $\omega \simeq k^z$. Considering the hydrodynamic regime above $T_c(p)$, we know that sublattice magnetization relaxes at a k-independent rate, which should depend on ξ as

$$\omega \sim \xi^{-z}$$

Using the expression (2.4) of ξ, one obtains

$$\omega \sim |\Delta p|^{z\nu_p} \quad , \quad \epsilon \simeq o, \ \Delta p < o$$

and

$$\omega \sim \epsilon^{\nu_p z/\phi_H} \quad , \quad \Delta p = o, \ \epsilon \ll 1$$

The numerical estimate of z in (2.11) is $z = 2.22$ at $d = 3$. The above arguments cannot be applied to Ising spins dynamics in dilute magnets the reason being the importance of domain-wall excitations in that case (non linear problem). For continuous degrees of freedom, the expressions of the exponent z involve just the diffusion exponent on the percolation clusters (linear problems) and this contrasts with both Ising and Potts dynamics.

III. ISING SPINS DYNAMICS

The critical Ising spin dynamics near the percolation threshold p_c have been worked out using the stochastic kinetic Ising model on percolation clusters.[6,7] Within the framework of Glauber dynamics,[8] the flip rate of a spin at site i is given by $\Gamma_i = \tau_0^{-1} [1 - \sigma_i \text{ th } (K \sum_j \sigma_j)]$, where the sum is taken over neighbors j of i. Critical Glauber dynamics at $p = 1$ have been studied previously.[9] The fractal structure of the percolation clusters leads to unusual critical dynamics in sharp contrast with the Euclidean regime (i.e. at $p = 1$). This can be seen in a low-temperature approach with use of the concept of thermal activation and energy-barrier description.

Consider a finite cluster at threshold ($p = p_c$), containing $s \sim \ell^{dp}$ spins. The relaxation problem is, in principle, solvable, involving a diagonalization of a $2^s \times 2^s$ matrix, leading to a spectrum of relaxation times. The size dependence of the relaxation time (actually this is the longest of the spectrum) of a given cluster is the relevant information. On the other hand, the equation of evolution for $\sigma_i(t)$ involves multispin correlation functions and a rigorous analysis of the problem is very difficult (even at high temperature). Therefore, only a low-temperature approach can be used to analyse this problem.

A. Energy—Barriers—Scaling

At low temperature ($\xi_T \gg \ell$), the thermal-relaxation rate is dominated by the (free) energy barrier for single-spin flip dynamics. Near $T = o$, entropy can be neglected and the energy barrier for the (dominant) reversal process from the state up ($\sigma_i = +1$, all i) to the state down ($\sigma_i = -1$, all i) can be defined as the s spins follows. We reverse the s-spins in a given sequence and find the energy barrier (maximum energy) for that sequence. There are s! sequences (paths) between the states up and down and the cluster energy barrier V_s is defined as the smallest energy barrier among the s! paths. The optical path associated with such a "minimax" sequence is very well defined but not unique in general. With this definition, V_s can be calculated for a cluster of arbitrary shape. For instance, for Euclidean blocks, we have $V_s \sim 2J.s^{1-1/d}$.

For percolation clusters, the average barrier has been shown numerically to scale as ℓns. More precisely[6]

$$V_s/2J = A\ell ns + B \qquad (3.1)$$

with $A = 1.058 \pm 0.05$, $B = 0.024$ in 2D and $A = 0.92 \pm 0.03$, $B = 0.065$ in 3D. Furthermore, the domain-wall energy barrier $W_s \geq V_s$ has also been shown to scale with size as (3.1), only the numerical factors are different. It is important to notice that (3.1) refers to the energy barrier for a typical cluster of size s. A more precise analysis would take into account the fact that the individual barrier V and then the relaxation rate depend on the shape of the cluster. Therefore, for a given size s, there is a probability distribution $p_s(V)$ for the energy barriers and then V_s, as given by (3.1), must be viewed as the average (or typical) energy barrier.

The scaling form of barriers (3.1) is actually a direct consequence of the scale invariance of the percolation clusters. To see this, let us consider a renormalization-group procedure on the infinite percolation cluster ($p = p_c$)

which forms a block spin from λ^{d_p} spins with a length scale change by a factor λ. The generic renormalization group recursion relation can be written as $K^{(n+1)} = K^{(n)} \, F(K^{(n)})$, where $K^{(o)} = K$. In the strong-coupling limit, $F(K)$ approaches a constant and, to first order in $1/K^{(n)}$, $F(K^{(n)}) = \lambda^x - 1/K^{(n)}$, where x is a real number. Near the critical region, one can assume [6] that the average energy barrier $V^{(n)}$ has the following recursion equation: $V^{(n+1)} = V^{(n)} + K^{(n)} \, T$. The iteration equation for $K^{(n)}$ and $V^{(n)}$ can easily be solved leading to coupling and barrier at length scale $\ell = \lambda^n$. It turns out that only $x = o$ can describe correctly the thermal critical behavior of percolation clusters, $T_c = o$ and $\ln \xi_T \sim J/T$. In this case $K_\ell = K_o - \ln\ell/\ln\lambda$ and $V_\ell \sim J(\ln\ell/\ln\lambda)$ in agreement with (3.1).

Note that $x > o$ corresponds to the case where $T_c \neq o$ exists and the energy barrier at length scale $\ell < \xi_T$ ($\xi_T \sim |T - T_c|^{-\nu}$) assumes again a logarithmic size dependence near the critical point T_c. Therefore, aside from the case $x < o$ where V_ℓ becomes independent of ℓ at large ℓ, the logarithmic law $V_\ell \sim \ln\ell$ appears as the most general one.

B. Singular Dynamic Scaling

The phenomenological scaling analysis above provides a convenient framework to extract the temperature dependence of equilibrium (relaxation) time scale $\tau(T) \sim \exp(V(T)/T)$. Here $V(T)$ denotes the largest thermal barrier, associated with length scale $\ell \sim \xi_T$. For instance at $x < o$, $\tau(T)$ is given by a simple (Arrhenius) law with $V(T \sim o) \sim J$. In contrast, near $T_c \, (x > o)$, $\tau(T) \sim \xi_T^z$ reproducing the known form of the dynamic-scaling hypothesis. Here z denotes the dynamic critical exponent, relating correlation length and relaxation time in the critical region.

In the case of the percolation clusters ($x = o$), one obtains a temperature-dependent thermal barrier $V(T) \sim J^2/T$ at low temperature. This leads to $\ln\tau(T) \sim T^{-2}$ which can also be cast as $\tau(T) \sim \xi_T^z$ but with a temperature dependent dynamical exponent $z = z(T) \sim 1/T$. Such a result can be viewed as a violation of the standard dynamic scaling where z is assumed to have a constant value.[9] The prediction of a singular dynamic scaling on percolation clusters has been checked numerically for both Ising and Potts models. For instance for a given cluster of size s, $\tau(s, T)$ can be extracted from a nonlinear relaxation function

$$\tau(s, T) = \int_o^\infty dt \, \Phi(t) \tag{3.2}$$

where $\Phi(t)$ denotes the following non-linear relaxation function

$$\Phi(t) = (M(t) - M(\infty))/(M(o) - M(\infty))$$

The initial state for the magnetization $M(t)$ is either an ordered state $M(o) = 1$ or a disordered state $M(o) = 1$. The relaxation time $\tau(s, T)$ as defined by (3.2) exhibits a crossover between a low temperature regime

$$\ell n\,\tau(s, T) \sim V/T \tag{3.3}$$

and a singular critical regime

$$\ell n\,\tau(s, T) \sim (A\ell n\,\xi_T^{d_p} + B)/T \sim a/T^2 + b/T \tag{3.4}$$

where $a = 4J^2\,d_p\,\nu_p$ and $b \simeq J$. The crossover temperature T^* is defined by $s \sim \xi_{T^*}^{d_p}$, i.e. $T^*(s)/2J \simeq \nu_p\,d_p/\ell n\,s$. Close to $T^*(s)$, corrections due to entropy barriers become important but they have been neglected here.

Further numerical works have confirmed (3.4), which is usually written as

$$\ell n\tau = A'(\ell n\,\xi_T)^2 + B'(\ell n\,\xi_T) \tag{3.5}$$

For Ising systems at $d = 2$, the numerical estimates are:[10] $A' = 0.51 \pm 0.05$ and $B' = 3.25 \pm 4.1$. Similarly, for $q = 3$ Potts model:[11] $A' = 0.28 \pm 0.15$ and $B' = 3.35 \pm 0.88$. For $d = 3$ Ising systems, the prediction of (3.4) has been checked on large samples,[12] (90^3) on the line $\xi_T = \xi_p$ below p_c.

IV. BARRIERS DISTRIBUTION AND RELAXATIONAL DYNAMICS

Because of dilation invariance, fractals and percolation clusters lead themselves particularly conveniently to scaling approaches. Thanks to some simple composition rules, the size dependence of the energy barriers can be calculated directly in some cases.[6] Using the composition rules and the scale invariance of percolation clusters, the probability distributions $p_s(V)$ of barriers has been calculated for large $s \sim \ell^{dp}$. Global distribution of barriers is simply deduced from the cluster sizes distribution n_s. At threshold, $n_s \sim s^{-\tau}$ for large s and the following result[6] has been derived

$$\pi(V) \equiv \int_o^\infty s\,n_s\,\delta(V - V_s)\,ds = \lambda \exp(-\lambda V) \tag{4.1}$$

where $\lambda = (d - d_p)/d_pA$ and $\tau = 1 + d/d_p$. This result holds also at $p \neq p_c$ up to $V \leq V^*$, where V^* corresponds to the cutoff $s^* \sim \xi_p^{dp}$ in the sizes-distribution n_s. A straightforward alegbra leads to the following power law distribution for the relaxation times $\tau \simeq \exp(V/T)$

$$\pi(\tau) = \lambda T/(\tau/\tau_o)^{1 + \lambda T} \tag{4.2}$$

which exhibits a long-time tail particularly at very low temperatures. Note that (4.2) can be used to generate "1/f" noise in dilute magnets: $\omega^{-1 + \lambda T}$. Using the results above, different relaxational dynamics can be studied. At low temperature the spin dynamics are dominated by thermally activated processes, associated to jump rate scale $\Gamma_s = \Gamma_o \exp(-V_s/T)$ where Γ_o^{-1} denotes an elementary time scale. Γ_s depends on s at very low T: $T \ll T^*(s) = \dfrac{2J}{\ell n s}$ $d_p \nu_p$ (= the crossover temperature). At $T \gg T^*(s)$, the relaxation time $\tau(T)$ is independent of s and assumes the following form

$$\ell n \tau(T) = \frac{d_p}{\nu_p} (\ell n \, \xi_T)^2 \tag{4.3}$$

instead of the usual form of the dynamic scaling $\tau(T) \sim \xi_T^z$. In the following we shall denote $\varsigma(T) \equiv \dfrac{2J}{T} \, d_p$, the "dynamic exponent" describing the size dependent of the relaxation time

$$\tau_s = \tau_o \, s^{\varsigma(t)/d_p} \tag{4.4}$$

(Note that this form for τ_s contrasts sharply with the corresponding expression: $\tau_s = \tau_o \cdot s \, \xi_T$ in the case of 1D chains).

Let us consider now the expressions of the time-dependent correlation function in equilibrium, and the time-dependent magnetization starting from a uniformly magnetized initial state:

$$C(t) = \sum_s s^2 \, n_s(p) \cdot \int dV \, p_s(V) e^{-t\Gamma} \equiv g_c(t) C(o) \tag{4.5}$$

$$M(t) = \sum_s s \, n_s(p) \cdot \int dV \, p_s(V) e^{-t\Gamma} \equiv g_M(t) M(o) \tag{4.6}$$

Here $n_s(p)$ refers to the probability distribution of clusters size and $\Gamma = \Gamma_o \, e^{-V/T}$ is the relaxation rate for a cluster of size s. The characteristic decay times τ_c and τ_M are defined by: $\tau_i = \int_o^\infty g_i(t) dt$, $i = M, c$.

Below threshold ($p < p_c$), the expressions of τ_c and τ_M are given by[6]

$$\tau_c \sim \tau_M \sim \tau_o \cdot \xi_p^{\varsigma(T)} \tag{4.7}$$

As expected both τ_c and τ_M diverge at $p \to p_c^-$. More interesting is the scaling of τ_i as implied by (4.7):

$$\ell n \tau_i / \tau_o = \frac{d_p}{\nu_p} \, \ell n \, \xi_T \cdot \ell n \, \xi_p \tag{4.8}$$

to be contrasted with the 1D result:[13]

$$\ell n \, \tau_i = \ell n \, \tau_p + \ell n \, \xi_T \tag{4.9}$$

In both cases, this expression for τ_i is only valid at $\xi_T \gg \xi_p$ and crossovers to: $2\ell n\, \xi_T$ at $d = 1$ and $(\ell n\, \xi_T)^2$ at $d \geq 2$ for $\xi_T \ll \xi_p$. The expressions of $g_c(t)$ and $g_M(t)$ are easy to calculate.[6] For instance, at $\xi_T \ll \xi_p$ one obtains:

$$g_c(t) \sim \exp[-(t/\tau_o)^{1-n}] \quad , \quad t \gg \tau_c \qquad (4.10)$$

where the exponent $1 - n = 1/(1 + \varsigma(T)/d_p)$ of the stretched exponential decay function is actually a function of temperature. More precisely, $n = n(T) \simeq 1$ at $T \ll 2J$ and $n \simeq o$ at $T \gg 2J$. At low T, $1 - n(T) \sim T$ and this follows from thermally activated processes. Indeed, for such processes $T\ell nt$ is the natural scaling variable at low T. It is important to notice that (4.10) seems to be "observed" in a large number of glassy materials. In this respect, it is very important to realize that (4.10) is only valid at $t \gg t_c$, where "rare" large clusters dominate the dynamics. The derivation of (4.10) does not involve any ad hoc hypothesis on the dynamics mechanism and this contrasts with all previous attempts. Both results (3.1), (4.2), (4.4) and (4.10) are actually direct consequences of scale invariance.

V. NON UNIVERSALITY OF THE DYNAMICS

i) The power law expression (4.4) for the relaxation rate Γ_s, which is implied by the dilation-symmetry, appears as the basic element in the relaxation behavior. The first consequence of this result is the formulation of a new dynamic scaling hypothesis $\ell n\, \tau(T) = \psi(\ell n\, \xi_T)$. Standard hypothesis appears as a special case: $\psi(u) = zu$. For dilute ferromagnets, $\psi(u)$ is a quadratic function of its argument. It this respect, it is interesting to notice that relaxation times $\tau(T)$ such as $\ell n\, \tau(T) \sim T^{-2}$ have been observed in a large number of systems:[14] ionic conductors, short range spin glass models $(d = 2, T_c = o)$, ... etc. Following this line of ideas, it would be very interesting to compare this law with the similar behavior observed on kinectic coefficients and relaxation times near the glass temperature of glasses. Actually, the plot of $\ell n\, \tau$ vs $1/T$ is very suggestive: straight line below T_g and a convex curve (parabola ?) at $T \gg T_g$. Is T_g a simple crossover temperature between two relaxation regimes?

ii) The singular dynamic scaling, i.e. $z(T) \sim 1/T$ for Ising spin dynamics found here is actually not accidental. Aside artificial models which appeared recently in the literature, two simple models exhibit actually such a singular behavior, i.e. z is a function of the model parameters. The first violation of dynamic scaling has been pointed out in the framework of the mean field theory of the Ising spin-glass transition. In this case the exponent z has been found[15] to be non universal, except at $T = T_c$ and $H = o$; z is a nontrivial function of H close to the critical line $T_c(H)$. Moreover, throughout the spin glass phase $(T < T_c(H))$, z depends on both T and H. A more simple model where z is found to be non universal is the ordered periodic Ising chain made of unit cells with positive interactions $J_1, J_2, ...$

J_n. In that case, one can show[16] that $z = 1 + \dfrac{\text{Max}\{J_i\}}{\text{min}\{J_i\}}$. The usual result is recovered for $J_1 = J_2 = \cdots = J_n$, where $z = 2$.

iii) From the examples discussed above, one can conclude that for systems having $T_c = o$ (with or without quenched disorder), a "trivial" violation of dynamic scaling can take place. Furthermore, disordered systems with $T_c \neq o$ can also exhibit such a non standard behavior. In addition to spin glasses, this is also the case of random field models[17] and other systems. However, this is not the rule. Some systems (e.g 3D Ising spin glasses) can still exhibit the conventional dynamic scaling, with a very well defined exponent z at transition. A classification of these different situations would be the next step towards the understanding of the dynamics of glassy materials.

REFERENCES

1. R. Rammal, J. Stat. Phys. **36:** 547 (1984).

2. S. Alexander, Ann. Isr. Phys. Soc. **5:** 149 (1983).

3. For a review on dilute magnetism, see R. B. Stinchcombe in Phase Transitions and Critical Phenomena, edited by C. Domb and J. L. Lebowitz (Academic Press, New York, 1983) Vol. 7.

4. G. Aeppli, H. Guggenheim, and Y. J. Uemura, Phys. Rev. Lett. **52:** 942 (1980); G. Aeppli et al., to be published (1987).

5. E. F. Shender, Sov. Phys. JETP **43:** 1124 (1976); D. Kumar, Phys. Rev. B30: 2961 (1984).

6. R. Rammal and A. Benoit, Phys. Rev. Lett. **55**, 649; R. Rammal, J. Phys. (Paris) **46:** 1837 (1985); R. Rammal and A. Benoit, J. Phys. (Paris) Lett. **46:** 667 (1985).

7. C. L. Henley, Phys. Rev. Let. **54:** 2030 (1985).

8. R. J. Glauber, J. Math Phys. (N.Y) **4:** 294 (1963).

9. For a recent review, see M. Suzuki, in Dynamical Critical Phenomena and Related Topics, edited by C. P. Enz, Lecture Notes in Physics, Vol. 104 (Springer, New York, 1979) p. 75. See also G. F. Mazenko, ibid p. 92.

10. S. Jain, J. Phys. A **19:** L 57 and L 667 (1986).

11. S. Jain, R. B. Stinchcombe and E. J. S. Lage, preprint (1987).

12. D. Chowdhury and D. Stauffer, J. Phys. A **19:** L 19 (1986).

13. D. Dhar in Stochastic Processes: Formalism and Applications, Lecture Notes in Physics (Springer, Berlin) 1983, Vol. 184, p. 130.

14. U. Larsen, Phys. Lett. **105A:** 307 (1984).

15. H. Sompolinsky, unpublished.

16. J. C. Angles d'Auriac and R. Rammal, preprint 1987.

17. J. Villain, J. Phys. (Paris) **46:** 1843 (1985).

SPIN GLASS THEORY

Giorgio Parisi

Dipartimento di Fisica, II Universita' di Roma
"Tor Vergata" and INFN, sezione di Roma, Italy

A spin glass is a disordered magnet: in a typical example 50% of the bonds (randomly chosen) between two spins are ferromagnetic while the others are antiferromagnetic. From the experimental point of view[1] spin glasses are characterized by the presence of a low temperature phase in which the relaxation time toward equilibrium is very large (as in real glasses). The magnetic susceptibility in the presence of a time dependent magnetic field depends on the frequency also in the region of a few Hertz.

From the theoretical point of view these properties may be connected to the practical impossibility of finding numerically the ground state of a spin glass (if the number of spins is not very small). Indeed, according to the conventional wisdom, the number of operations any algorithm needs to find the ground state increases exponentially with the volume (the problem is NP complete in more than 2 dimensions). There is an exponentially large number of spin configurations which are local ground states, in the sense that their energy increases if only one spin is flipped. Although it is very easy to find efficient algorithms for finding local minima, it is very difficult for the algorithm to find the global minimum (i.e. the true ground state).

Numerical simulations[2] suggest that different minima (of the energy at zero temperature or of the free energy at finite temperature) are macroscopically different from each other. This situation is rather unprecedented in statistical mechanics: macroscopically different configurations with the same total free energy are normally present at the point where a first order transition happens (which is an isolated point). For spin glasses this phenomenon is present for a large range of temperatures and magnetic fields, where there is no first order phase transition from the thermodynamic point of view. A strong effort has been needed to find the appropriate theoretical framework to study these systems.

The origin of this peculiar behaviour is the presence of frustration: it is not possible for all pairs of spins connected by a ferromagnetic (antiferromagnetic) bond have the same (the opposite) sign. Different terms of the Hamiltonian push in different directions, and the number of possible compromises is very high. For example in the Ising case the Hamiltonian is

$$H = -1/2 \sum_{i,k} J_{i,k} \sigma_i \sigma_k - h \sum_i \sigma_i \qquad (1)$$

where h is the magnetic field and the J's are the coupling between the spins. It is clear that it is not always possible to find configurations of the σ's such that all terms are positive.

There are many systems in which different terms of the Hamiltonian (or different constraints) are in competition with each other. This happens very frequently when the system is complex. Typical examples are real glasses, other NP complete problems like the traveling salesman or the matching problem, protein folding, biological organization, prebiotic evolution, neural networks and so on. At the present moment many of the concepts that have been developed in the study of spin glasses start to be useful also for these systems and it is quite possible that in the long run the applications of these ideas beyond solid state physics will be the most interesting ones.

In recent years much progress has been made and after much effort a self-consistent mean field approximation has been obtained[3] (sometimes this construction goes under the name of broken replica theory). This theory should be exact for weak long range forces or when the dimensions of the space become very large (fluctuations are neglected). The ideas involved in this construction are rather different from those of the mean field approximation for other models and take care of the peculiar properties of spin glasses. In this note we will not show how these ideas have been developed. We will also skip most of the technical details and concentrate our attention on a few important physical results.

We suppose a system whose Hamiltonian $H_J [\sigma]$ depends on the configuration $[\sigma]$ of the system and on some control variables J, which are distributed according to a probability distribution P[J]. For each choice of the J's we can compute the partition function

$$Z_J = \sum_{\{\sigma\}} \exp \{-\beta H_J[\sigma]\} \qquad (2)$$

and the free energy density is

$$F_J = -1/(\beta N) \ln Z_J = -1/(\beta N) \ln \{\sum_{\{\sigma\}} \exp \{-\beta H_J[\sigma]\}\}, \qquad (3)$$

where N is the total number of variables σ's. Standard statistical mechanics deals with the problem of computing F at given J. Here we suppose that the J's are not known, but that they are random variables and that their probability distribution is known. The J's are called quenched variables. Physically the necessity of computing the free energy F at fixed J's results from the fact that the changes of the J's happen on a time scale which is infinitely larger than the time scale characterizing the changes in the σ's.

We are interested in computing the average value of the free energy density:

$$F = \sum_J P[J] \; F_J = \overline{F_J} \; . \tag{4}$$

The replica method was originally proposed as a trick to simplify the computation of eq. (4). Later on it was found that the replica method (when the replica symmetry is broken) is very powerful in coding in a simple and compact way quite complex properties of the system.

In a nutshell the basic idea of the replica method is very simple. We start with some preliminary definitions:

$$Z_n = \sum_J P[J] \; \{Z_J\}^n = \overline{(Z_J)^n}, \tag{5}$$

$$F_n = -1/(\beta n N) \; \ln Z_n.$$

Using relations $\sum_J P[J] = 1$ and $A^n \approx 1 + n \ln(A)$ for $n \approx 0$, it is evident that

$$\lim_{n \to 0} \; F_n \equiv F_0 = \overline{F}. \tag{6}$$

For integer n we can write:

$$(Z_J)^n = \prod_{a=1}^{n} \; \sum_{\{\sigma^a\}} \exp \{-\sum_{a=1,n} H[\sigma^a]\}, \tag{7}$$

where we have introduced n replicas of the same system. Indeed the partition function of n replicas of the same system is the partition function of the original system to the power n. The variables $\sigma^a{}_i$ carry two indices: the upper one denotes the replica and goes from 1 to n and the lower one denotes the site and goes from 1 to N. At the end we must perform the two limits $n \to 0$ and $N \to \infty$ (the two limits are supposed to commute).

The replica trick consists of the following steps: we use equation (7) for integer n to define the function F_n for integer n; we extend this function to an analytic function of n and finally we compute $F_0 = \overline{F}$. If the partition function is expanded in power of β, F_n is a polynomial in n and the method is quite safe. Problems will arise in the low temperature region.

We can apply this method to the case of infinite range spin glasses (the Sherrington-Kirkpatrick model[4]) where the J's are independent Gaussian random variables with zero mean and variance 1/N:

$$\overline{J_{i,k}} = 0, \quad \overline{J_{i,k}^2} = 1/N, \quad J_{i,k} = J_{k,i}. \tag{8}$$

The thermodynamic limit is obtained when N goes to infinity. The factor $1/N$ in eq. (8) has been chosen in such a way that at fixed β the total energy is proportional to N and therefore the energy density is N-independent.

For fixed J's we expect that in the high temperature phase the local magnetizations $m_i \equiv <\sigma_i>$ are different from zero at non-zero magnetic field and they become zero when the magnetic field goes to zero. On the other hand, we naively expect that in the low temperature region there should be some freezing of the spins in the position which is mostly favoured energetically, hence the m_i should be different from zero also at h=0. It is evident that m_i depends on the J's and it will be sometimes positive and sometimes negative (a detailed computation shows that $m_i = 0$ at h=0) so it is convenient to characterize the system in terms of the quantity

$$q_{EA} \equiv 1/N \sum_{i=1}^{N} m_i^2 \;, \tag{9}$$

i.e., the Edward Anderson order parameter[5]. It is possible to show that using a good definition of the m's q_{EA} does not depend on the J's in the limit $N \to \infty$. We can thus conclude that q_{EA} is also equal to $\overline{m_i^2}$. Summarizing, q_{EA} should be equal to zero at h=0 for a temperature greater than the critical temperature (T_c), while it should be different from zero at low temperature.

If we apply eq. (5) to the infinite range case, after some simple computations we find that

$$Z_n = \int d[Q] \exp \{-N A[Q]\} \tag{10}$$

$$A[Q] = -n\beta^2/4 + 1/4 \sum_{a=1,n} \sum_{b=1,n} Q_{a,b}^2 - \ln Z[Q]$$

$$Z[Q] = \sum_{\{S\}} \exp \{-\beta H[Q,S]\}$$

$$H[Q,S] = \sum_{a=1,n} \sum_{b=1,n} Q_{a,b}^2 S_a S_b - h \sum_{a=1,n} S_a$$

$$F_n = -1/\beta \min A[Q] \quad \Rightarrow \quad \partial A/\partial Q_{a,b} = 0,$$

where the matrix Q is an n x n symmetric matrix, with zeroes on the diagonal, and the integral is done on all non-diagonal elements of the matrix $\dot{Q}$ from $-\infty$ to ∞. The sum over $\{S\}$ goes over the 2^n configurations of the variables S_a, a=1,n ($S_a=\pm 1$).

The function A[Q] is left invariant when we exchange some of the lines (or the rows) of the matrix Q and therefore the group of permutations of n elements (P_n or S_n) is a symmetry of the problem (all replicas are equivalent!). This group is often called the replica group.

For positive n the minimum of A can be found and the matrix Q has the following form

$$Q_{a,b} = q \ \forall \ a,b \ a \neq b \quad (Q_{a,a} = 0 \ \forall \ a). \tag{11}$$

This form of the matrix Q is the only one which is left invariant by the action of the replica group and it is therefore the natural solution, usually named the replica symmetric solution.

If we analytically continue the solution of the equation dA/dq=0, up to n=0, we find the following equation for q

$$q = \int dz/(2\pi)^{1/2} \exp(-z^2/2) \ th^2(\beta q^{1/2}z + \beta h), \tag{12}$$

where the integral over z goes from $-\infty$ to $+\infty$.

At zero magnetic field eq.(11) has only the solution q=0 for $1/\beta \equiv T > T_c=1$; for $T < T_c$ there is another solution (the physical one) which is different from zero. In conclusion there is a phase transition at T=1, h=0 while there is no transition at h≠0.

Everything seems perfect. Unfortunately, detailed computations show that the entropy (which in a discrete system is non-negative by definition) becomes negative at low temperature and at zero temperature we get $S(0) = -1/(2\pi) \approx -.17$. In this case, the replica method leads to a disaster!

It is evident that the saddle point method can be applied correctly only if the eigenvalues of the Hessian matrix $(\partial^2 A/\partial Q \partial Q)$ are non-negative, as it can be checked by computing the correction to the saddle point result[6]. At low temperatures we must look for a new solution of the equation

$$\partial A/\partial Q = 0, \tag{13}$$

whose Hessian has no negative eigenvalues.

If the matrix Q does not have the form shown in eq.(11), it is not left invariant by the action of the replica group and we have to specify in detail the value of the different elements of the matrix.

We will always write formulae which are correct for integer n and make the analytic continuation at n=0 only at the end. In other words we will consider some matrices $Q^{[n]}{}_{a,b}$ which depend on n in a simple way. We will compute all the quantities we need for integer n and only at the end we will perform the continuation of the results at n=0.

The 0 x 0 matrix $Q^{(0)}{}_{a,b}$ is defined in terms of the matrices $Q^{(n)}{}_{a,b}$ for all values of n and therefore the space of all 0 x 0 matrices is an infinite dimensional space!

At the present moment only one reasonable form of the matrix is known. There is a widespread agreement that this choice is the correct one because the results are very satisfactory. It is quite possible that for systems which are different from spin glasses (e.g. for real glasses), different choices of the matrix $q_{a,b}$ are the correct ones.

A natural suggestion for the matrix Q consists in dividing the n replicas into n/m groups of m replicas. (Of course n must be a multiple of m, i.e. n/m must be an integer.) We set $Q_{a,b}=q_0$, if a and b belong to the same group, and $Q_{a,b}=q_1$, if a and b belong to different groups (we do not consider the $Q_{a,a}$'s which are identically equal to 1). In other words

$$Q_{a,b} = q_0 \quad\quad \text{if} \quad I(a/m) = I(b/m) \quad\quad\quad\quad (14)$$

$$Q_{a,b} = q_1 \quad\quad \text{if} \quad I(a/m) \neq I(b/m),$$

where $I(x)$ is an integer valued function whose value is the smallest integer which is greater than or equal to x.

Each line of the matrix has m-1 off-diagonal elements which are equal to q_0 and n-m which are equal to q_1 (the total number of off diagonal elements is n-1, i.e. -1 in the limit n→0). According to eq. (14) we have in the limit n→0:

$$P(q) = m \, \delta(q-q_0) + (1-m) \, \delta(q-q_1) \quad\quad\quad\quad (15)$$

where the function P(q) is defined as

$$P(q) = 1/[n(n-1)/2]\sum_{a,b} \delta(q_{a,b}-q) . \quad\quad\quad\quad (16)$$

We will see later that the function P(q) must be non-negative and this is possible only if:

$$0 \leq m \leq 1 \quad\quad\quad\quad (17)$$

It is obvious that (if we exclude the two uninteresting cases m=0 and m=1) m cannot be an integer and satisfy the inequality (17). On the other hand the limit n→0 is obtained by doing an analytic continuation in m and nothing seems to forbid that in such a process m may take non-integer values.

After some computations we find the explicit form of the function $A(q_0,q_1,m)$. It is easy to check that when m is set to 0 or to 1 we recover (as we should) the free energy of the replica symmetric approach with $q= q_0$ or q_1 respectively.

A careful analysis shows that the free energy $A(q_0,q_1,m)$ must be maximized with respect to all the variables, i.e. q_0, q_1 and m. If a maximum is found in the region 0<m<1, the results of this approach must be better than those obtained assuming unbroken replica symmetry. The true free energy takes values greater than those obtained from the unbroken replica appproach.

The function $A(q_0,q_1,m)$ can be easily computed numerically with high precision. Maximizing A with respect to the q_0,q_1, and m one finds that, in the whole region where the replica symmetry should be broken (i.e. at sufficiently low temperatures), m is neither zero or one ($m\to0$ both for $T\to0$ and $T\to1$ at zero magnetic field).

The properties of the solution are very satisfactory:
a) the zero temperature internal energy at zero magnetic field is $-.7652$ in good agreement with the numerical data ($-.765\pm.01$).
b) the zero temperature entropy at zero magnetic field has collapsed from $S(0)\approx-.16$ to $S(0)\approx-.01$.

We are clearly on the right track although we have not found the final answer.

The final form of the matrix Q is thus the following: we introduce a set of integer numbers m_i ($i=0,\dots,k+1$) such that $m_0=0$ and $m_{k+1}=n$ and m_i/m_{i+1} is an integer (for $i=1,\dots,k+1$). We can divide the n replicas in n/m_k groups of m_k replicas; each group of m_k replicas is divided in m_k/m_{k-1} groups of m_{k-1} replicas and so on....

The off-diagonal elements of the matrix Q are thus given by:

$$Q_{a,b}= q_i \text{ if } I(a/m_i) \neq I(b/m_i) \text{ and} \qquad (18)$$

$$I(a/m_{i1}) = I(b/m_{i+1}), \quad i= 0,\dots,k,$$

where the q_i's are a set of k+1 real parameters. For k=1 we recover the previous example and for k=0 we recover the unbroken replica symmetry theory.

An easy computation shows that

$$P(q)= \sum_{i=0,k} (m_i-m_{i+1}) \,\delta(q-q_i). \qquad (19)$$

Eq. (19) for $P(q)$ is positive only if the m's satisfy the conditions

$$0 \leq m_{i+1} \leq m_i \leq 1. \qquad (20)$$

Detailed arguments suggest that we should look for the maximum of the free energy as a function of the q_i's and the m_i's (from now on we will assume that conditions (20) are satisfied).

In order to keep track of the parameters q and m it is convenient to introduce the function $q(x)$ defined as

$$q(x)= q_i \text{ if } m_{i+1}<x<m_i. \qquad (21)$$

There is a one-to-one correspondence between the piecewise constant function with k discontinuities and the parameters q and m (if eq. (21) holds).

If the function $q(x)$ is a monotonic function (as we shall see it must be for physical reasons) the relation between the functions $P(q)$ and $q(x)$ is the following:

$$dx/dq = P(q) \tag{22}$$

where $q(x(q)) = q$.

In the limit $k \to \infty$ the function $q(x)$ becomes arbitrary (any reasonable function can be approximated by a piecewise constant function) and

$$\lim_{n \to 0} \{n^{-1} \Sigma_{a,b=1,n} Q^k_{a,b}\} = -\int_0^1 dx\, q^k(x) \tag{23}$$

In this formulation the free energy becomes a functional of $q(x)$, ($A[q]$) which must be maximized with respect to $q(x)$ and in this way we find the correct solution of the infinite range model.

The replica formalism is far from physical intuition and therefore it is convenient to recast the previous result using a more familiar language. In principle all the previous results can be obtained without using replicas. To this end we analyze the concept of a pure equilibrium state. If we consider a simple three dimensional ferromagnetic Ising model (with zero magnetic field) at a temperature greater than the critical one, there is only one pure equilibrium state and the spins are disordered. Below the critical temperature there are two translational invariant pure equilibrium states, one with positive magnetization, the other with negative magnetization.

Generally speaking a pure equilibrium state is any state which is at local equilibrium and whose connected correlation functions go to zero at large distance; the expectation value in the state α will be denoted by $<>_\alpha$. It is clear that the linear combination of two pure equilibrium states will be a mixed equilibrium state. According to the conventional wisdom, a system, which is not at equilibrium at initial time, evolves toward a pure equilibrium state and the time needed to go from one equilibrium state to another equilibrium state is exponentially large.

From the technical point of view a pure equilibrium state is characterized by the clustering property: all the connected correlations functions go to zero at large distance. Equivalently intensive quantities do not fluctuate. In a nutshell a pure equilibrium state corresponds to our intuitive idea of a normal equilibrium state. All equilibrium states have the same free energy density and states with higher free energy density are metastable.

We have already seen that spin glasses may arrange the directions of their spins in many different ways and all these arrangements give similar values of the free energy; in other words there exist many different ground states of spin glasses which are nearly degenerate in free energy. It is natural to assume that an Ising spin glass has many more equilibrium states than a ferromagnetic Ising model[3]. The configurations which have lower free energy correspond to pure equilibrium states while those configurations which have a higher free energy density in the infinite volume limit correspond to metastable states.

We face the new problem of characterizing the ensemble of equilibrium states. The *a priori* probability w_α for the state α to appear in the ensemble is:

$$w_\alpha \propto \exp(-\beta \, F_\alpha), \tag{24}$$

where F_α is the free energy of the state α.

The second question is to study how all these states differ from one another, more precisely to define a distance between these states. A simple possibility for the distance between the state α and the state β is the following:

$$d^2{}_{\alpha\beta} = 1/N \sum_i (m_i{}^\alpha - m_i{}^\beta)^2, \tag{25}$$

where N is the total number of spins, the sum over i is from 1 to N and $m_i{}^\alpha$ is the average magnetization of the spin i in the state α, i.e. $m_i{}^\alpha = \langle \sigma_i \rangle_\alpha$. In the same way we can define the overlap between two states α and β as;

$$q_{\alpha\beta} = 1/N \sum_i m_i{}^\alpha \, m_i{}^\beta. \tag{26}$$

If all the states have the same overlap with themselves (as should happen in spin glasses), i.e. $q_{\alpha\alpha} = q_{\beta\beta} \equiv q_{EA}$, the distance is very simply related to the overlap:

$$d^2{}_{\alpha\beta} = 2 \, (q_{EA} - q_{\alpha\beta}). \tag{27}$$

It is possible to define the probability distribution of overlap $P(p)$ [8]:

$$P(q) = \overline{P_J (q)} = \sum_{\alpha\beta} w_\alpha \, w_\beta \, \delta(q_{\alpha\beta} - q), \tag{28}$$

where the bar denotes the average over different samples with different values of the couplings J.

In other words $P(q)$ is the probability of finding two states with overlap p, weighting each state with its probability of appearing in the ensemble. Only states which have a non-zero probability ($w_\alpha \neq 0$) contribute to eq. (28).

The physical interpretation of replica symmetry breaking is based on the highly non- trivial result [8] that the two functions $P(q)$, defined in eqs. (15) and (28), coincide.

In the ferromagnetic Ising model at non-zero magnetic field $P(q)$ is a delta function because there is only one equilibrium state. Exactly at zero magnetic field the function $P(q)$ contains two delta funtions, one at $q=m^2$, the other at $q=-m^2$, where m is the spontaneous magnetization.

In spin glasses in the mean field approximation the funtion $P(q)$ at high temperatures is a delta function whereas at low temperatures at non-zero magnetic field it has two delta functions with a smooth region in between. In this case, where the function $P(q)$ is not a simple delta function, we say that the replica symmetry is broken. The transition is very smooth from the thermodynamic point of view and only the second derivative of the specific heat is discontinuous.

In the usual approach all possible probability distributions with more than one overlap can be computed from the function $P(q)$. The free energy can be computed as a functional of $P(q)$ and the function $P(q)$ can be found by looking for the maximum of free energy. The first interesting result is that $P_J(q)$ fluctuates from sample to sample[9].

$$P_J(q_1) \ P_J(q_2) = 1/2 \ P(q_1) \ P(q_2) + 1/2 \ \delta(q_1-q_2) \ P(q_1) \quad (29)$$

This result indicates that only a few states give an important contribution to the function $P_J(q)$ and their relative weight and overlap fluctuate from sample to sample.

The most interesting and unexpected result concerns the probability distribution of the overlap of three or more states. One finds that the distance satisfies the ultrametric inequality[9]:

$$d_{\alpha\beta} \ < \ \max \ (d_{\alpha\gamma}, \ d_{\beta\gamma}) \ \forall \gamma. \quad (30)$$

As is well known to mathematicians, this inequality implies that the space of states, can be divided into clusters of states; each cluster of states may be further divided into subclusters and so on.....; a cluster can be characterized as the set of states whose distance from a given (but arbitrary) state of the cluster is smaller than a given distance.

A more intuitive definition of an ultrametric distance is that two spheres are identical if they have a point in common. As a consequence, if we consider a random walk in the space of the states of the system and the length of each step is less than or equal to δ, after M steps the distance between the initial point and the final point will also be less than or equal to δ. The whole cluster can be reached in one step and there is nothing else to explore at distances not greater than δ.

This result is extremely important because it may be the basis for the existence of many different scales of relaxation times. Indeed let us consider a system which evolves in time and relaxes toward equilibrium with many different scales $\tau_1 << \tau_2 << \tau_3 << \ .. << \tau_n$. If the system is in a given configuration at time 0, we call R_k the region of configuration space that the system may explore in time τ_k.

All the points in configuration space which belong to R_k but not to R_{k-1} are said to be at distance k.

It is evident that this definition of distance satisfies the ultrametric inequality eq. (30). Let us suppose the contrary, i.e. there are two configurations (α and β) at distance k and there is another configuration (γ) at distance k-1 from both α and β. Now the times needed to go from α to γ and from γ to β are both equal to τ_{k-1} and the estimated time to go from α to β via γ is about $2\tau_{k-1}$ in contradiction with the inequality $\tau_{k-1} \ll \tau_k$.

This result is not surprising: many different times of relaxation are naturally present if the free energy landscape in configuration space contains many valleys which are separated by high mountains. The time to go from one valley to an other valley is the exponential of the height of the lower saddle between the two valleys.

The division of configuration space in valleys is clearly ultrametric: more precisely if the distance between two configuration is defined as the height of the highest saddle one must cross doing the most convenient trip between the two configurations, this definition of distance satisfies the ultrametric inequality (30).The real surprise is that the space of equilibrium states of a spin glass is ultrametric using the natural definition of the distance.

The dynamics of spin glasses at finite temperature should be the following: at relatively short times the system evolves exploring the configuration space corresponding to one pure equilibrium space, while at much larger times (e.g. for times proportional to $\exp\{(Nd)^{1/4}\}$) the system may arrive at other equilibrium states at distance d from the original one. A careful study of the dynamics is very important[10]. Very interesting result have been obtained, in particular it has been stressed that the behaviour of the system may be rather different if the system starts really from equilibrium or from a state which is slightly off equilibrium[11].

Another result which is very interesting is that the specific form of the equilibrium states is very sensitive to external parameters. A small change of the external magnetic field, of order $1/N^{1/2}$, is sufficient to completely upset the microscopic details of the equilibrium states. The equilibrium states at two different magnetic field have very small overlap[3].

All these results certainly hold when fluctuations are neglected. This approximation is good when the dimensions of the space are very large and the prediction of this approach compare well with the results of the numerical simulation in infinite dimensions, i.e. for the Sherrington and Kirkpatrick model[12].

If we decrease the dimension the situation is less clear. The most likely scenario, although infinitely many different possibilities are open, is that the picture does not change for dimensions greater than D_C. At the lower critical dimension D_C, the critical temperature becomes zero and there is no transition to a spin glass phase. However if the coherence length ξ is large and if the relaxation time is proportional to exp $(\xi^{1/4})$, it is possible that the relaxation time becomes very large (seconds, days, years...) and for reasonably long observation times the system seems to be in the spin glass phase.

In three dimensions the situation is confused: many data can be fitted as $(T - T_C)^{-\upsilon}$ with a reasonable value of T_C. However, the same data can also be fitted as exp (A/T^3)[13]. It is clear that we cannot decide between the two different options by looking at the value of the χ^2 of the fit. It is difficult to find a crucial test.

A very neat way of solving the problem consists of computing the corrections induced by fluctuations in the framework of the present theory and comparing them to the experimental and numerical data. If the agreement is good, the value of the lower critical dimension should be the theoretical one. Unfortunately such computation is very difficult and progress has been made very slowly[14].

References

1. For a nice short experimental review of spin glasses see J. A. Mydosh, in "Disordered Systems and Localization", ed. by C. Castellani, C. Di Castro and L. Peliti, Springer-Verlag (1981).
2. N. Sourlas LPTENS preprint 85/9 (1985).
3. G. Parisi, Phys. Rev. Lett. 43, 1754 (1979), J. Phys. A13, 1101, 1887, L115 (1980). For a recent theoretical review see M. Mezard, G. Parisi, M. Virasoro "Spin Glass Theory and Beyond", World Scientific, Singapore (1987).
4. D. Sherrington and S. Kirkpatrick, Phys. Rev. Lett. 35, 1792 (1975); Phys. Rev. B17, 4384 (1978).
5. S.F. Edwards and P.W. Anderson, J. Phys. F5, 965 (1975); F6, 1927 (1976).
6. J.R.L. de Almeida and D.J. Thouless, J. Phys, A11, 983 (1978).
7. M. Mezard, G. Parisi, M. Virasoro, Eur. Phys. Lett. 1, 105 (1986)
8. G. Parisi Phys. Rev. Letters 50, 1946 (1983).
9. M. Mezard, G. Parisi, N. Sourlas, G. Toulouse and M. Virasoro, Phys. Rev. Lett. 52, 1156 (1984); J. Physique 45, 843 (1984)
10. H. Sompolinsky and A. Zippelius, Phys. Rev. Lett. 50, 1297 (1983) and references therein.
11. A. Houghton, S. Jain, A. P. Young, Phys.Rev. B25, 2630 (1983).
12. A. P. Young, Phys. Rev. Lett. 43, 1206 (1983); K. Namoto, Hoikkaido University Preprint (1986).

14. C. De Dominicis and I. Kondor J. Physique Lett. <u>46</u>, L-1037 (1985) and references therein.

GRAPH PARTITIONING AS A SPIN GLASS PROBLEM

David Sherrington

Physics Department
Imperial College
London SW7 2BZ
UK

ABSTRACT

It is shown how some complex problems in graph partitioning can be studied by an extension of techniques developed for the theory of spin glasses.

An idealization of many practical optimization exercises is the problem of the partitioning of the vertices of a randomly connected graph into groups, such that each group contains an equal number of vertices and the number of edges connecting different groups is a minimum[1]. This problem has much in common with that of finding the ground state of a spin glass[2]. Both involve quenched disorder and frustration[3]. Furthermore, the magnetic analogues of such equipartitioning problems have effectively infinite-ranged exchange, thereby providing fertile ground for the transplantation of techniques cultivated for modern spin-glass mean-field theory[4]. Below are described briefly some aspects of this cross-fertilization.

Let us label the N vertices of a graph by {i} and the connectivity by an N x N matrix $\underline{a}$ with elements $a_{ij} = a_{ji} = 1,0$ depending on whether edge (ij) is present or absent. For simplicity let us further restrict explicit discussion to bipartitioning; a parameter $\sigma_i = \pm 1$ thereby suffices to indicate whether vertex i is in the first group ($\sigma_i = +1$) or the second ($\sigma_i = -1$). The number of cross-edges connecting the two groups is

$$N_{ce} = \sum_{(ij)} a_{ij} \{(1 - \sigma_i\sigma_j)/2\} \tag{1}$$

and the problem is its minimization subject to the equipartition condition

$$\sum_i \sigma_i = 0 \tag{2}$$

Interpreting the σ_i as Ising spins, $\sigma_i = \pm 1$ now corresponding to spin up or spin down, this minimization condition is equivalent to one of minimizing the ground state energy of the random ferromagnet of Hamiltonian

$$H = - J \sum_{(ij)} a_{ij} \sigma_i \sigma_j \tag{3}$$

subject to a constraint of zero overall magnetization. We shall discuss the problem in this latter form. The two problems are related by

$$E/J = 2N_{ce} - N_{ed} \tag{4}$$

where E is the ground state energy of the ferromagnet problem and N_{ed} is the total number of edges. Our interest here is in situations in which the a_{ij} are randomly distributed and the minimization problems are NP-complete[1].

In the present paper the equipartition problem will only be considered statistically, in the sense that rather than studying specific instances we shall be concerned with the evaluation of the minimum energy averaged over an ensemble of systems with different specific connectivities but with the same rules for randomly choosing these connectivities. Furthermore, the approaches described tackle the problem of evaluating the ground state energy using techniques of statistical mechanics. Explicitly, an artificial temperature parameter T is introduced so that a free energy F can be defined by

$$F = - kT \ln Z \tag{5}$$

where Z is a partition function

$$Z = Tr' \exp(- H/kT) \, , \qquad (6)$$

Tr' denoting a sum over all $\{\sigma_i\}$ subject to constraint (2), and the ground state energy E is obtained via the limiting procedure

$$E = \lim_{T \to 0} F(T) \qquad (7)$$

The ensemble averaging corresponds to studying $\bar{E}$ (or equivalently $\bar{F}$, $\overline{\ln Z}$) where the bar indicates the ensemble average. Note that, as in the spin-glass problem[4], Z itself cannot be averaged directly. We shall further be interested in the thermodynamic limit, as the number of vertices N approaches infinity.

Within the general set of random graphs we consider the union of two interesting sub-sets. The first concerns the scaling of the average valence (vertex coordination number) with N. We refer to cases in which the average valence is proportional to N as extensive, those in which it is independent of N as intensive. We restrict explicit discussion to these two cases, although the procedures used for the case we call "extensive" apply more generally to valences increasing monotonically with N. The second sub-division concerns whether (i) each vertex is connected to exactly the same number of other vertices, a situation we refer to as fixed valence, or (ii) each pair of vertices has an equal and independent probability of being connected - we refer to this as average valence. No further restrictions are made on the random connectivity.

All the above sub-classifications can be treated by replica theory[4,5,6] and, furthermore, because there are no lattice restrictions and the connectivity is effectively infinite-ranged, steepest descents can be applied exactly.

For the cases of average valence, the free energy is given by[7]

$$\bar{F} = - kT \lim_{n \to 0} \frac{1}{n} \left\{ Tr'_n \exp \left[\sum_{m=1}^{\infty} \beta^m J^m C_m \sum_{\{\alpha_m\}} (\sum \sigma_i^{\alpha 1} \sigma_i^{\alpha 2} .. \sigma_i^{\alpha m})^2 \right] - 1 \right\} \qquad (8)$$

where the C_m are functions of the probability p that an edge is present and β is used as shorthand for $(kT)^{-1}$. The restricted trace is now over all replicas, each with a constraint

$$\sum_i \sigma_i^\alpha = 0 \quad . \tag{9}$$

Because of this constraint the term m = 1 can be eliminated from the summation on m. Furthermore, it is convenient to remove the constraint (9) in favour of an extra effective penalty Hamiltonian contribution[2]

$$H_{eff} = \lambda \sum_\alpha (\sum_i \sigma_i^\alpha)^2 \tag{10}$$

where λ is independent of N. In the usual fashion $\exp(...(\sum_i \sigma_i^{\alpha_1}...\sigma_i^{\alpha_m})^2)$ may be replaced by an auxiliary variable $Q^{\alpha_1 \cdots \alpha_m}$ with a local interaction, steepest descents yielding self-consistent (mean-field) equations for the Q and an equation for $\bar{F}$ in terms of the Q.

In the case of extensive average valence, p is independent of N and if βJ is chosen to scale as $N^{-1/2}$ then the m sum in (8) terminates after m = 2 and the problem is exactly equivalent[7] to that of the Sherrington-Kirkpatrick[6] spin glass. In terms of the total number of edges, $N_{ed} = pN/2$, there results[7]

$$\frac{N_{ce}}{N_{ed}} = \frac{1}{2} \left\{ 1 - c \left(\frac{1-p}{pN}\right)^{1/2} \right\} + O(N^{-1}) \tag{11}$$

where c = 1.5266 ± 0.0002.

For intensive average valence, p = z/N, the C_m scale as N^{-1} and all m ≥ 2 are relevant, yielding an infinite set of order parameters $Q^{\alpha_1 \cdots \alpha_m}$; m = 2,... ∞[8,9]. Within a replica-symmetric (RS) approximation

$$Q_m = \int dh\, P(h)\, (\tanh\beta h)^m \quad , \tag{12}$$

where Q_m denotes the m-th order RS order function, and the P(h) satisfy self-consistency equations[8,9].

$$P(h) = \sum_{k=0}^{\infty} \alpha^k \frac{e^{-\alpha}}{k!} \int dh_1 \ldots dh_k \, P(h_1) \ldots P(h_k)$$

$$\delta(h - \beta^{-1} \sum_{i=1}^{k} \tanh^{-1}(\tanh\beta J \tanh\beta h_i)) \qquad (13)$$

$$\int dh \, P(h) \tanh\beta h = 0 \qquad (14)$$

Furthermore, Q_m can be identified as the average of the mth moment of the local magnetization,

$$\overline{Q} = \langle\sigma_i\rangle^m. \qquad (15)$$

For fixed valence z, it turns out that for the extensive case the result (11) of Fu and Anderson continues to hold[10], but for intensive valence a new situation arises. Nevertheless, despite the extra complications of the latter case it too can be solved in terms of self-consistent fields within a replica-symmetric ansatz[9,11]. In this case the order parameter Q_m satisfies the equation

$$Q_m = \int db \, \pi(b)(\tanh\beta b)^m \qquad (16)$$

with $\pi(b)$ given self-consistently by

$$\pi(b) = \int (\prod_{i=1}^{z-1} db_i \pi(b_i)) \delta(b - \beta^{-1} \sum_{i=1}^{z-1} \tanh^{-1}(\tanh\beta J \tanh\beta b_i)). \qquad (17)$$

Q_m does not, however, now correspond to the average mth moment of the local magnetization. Rather, that is given by

$$\overline{\langle\sigma_i\rangle^m} = \int dh \, P(h)(\tanh\beta h)^m \qquad (18)$$

where

$$P(h) = \int (\prod_{i=1}^{z} db_i \pi(b_i)) \delta(b - \beta^{-1} \sum_{i=1}^{z} \tanh^{-1}(\tanh\beta J \tanh\beta b_i)) \qquad (19)$$

These results can be interpreted in terms of a Bethe lattice picture in which $\pi(b)$ gives the effective field distribution due to descendents and $P(h)$ that due to all neighbours[11]. In fact, such a Bethe lattice picture was devised[12] independently of the replica procedure, the graph partitioning problem being qualitatively mapped onto a Bethe lattice of ferromagnetically interacting spins, with random boundary conditions ensuring zero magnetization; formulation in terms of fields due to descendents is natural on Bethe lattices[13].

Finally, we note that as well as the above analytic treatments, the problem of the equi-bipartitioning of random graphs with intensive fixed valence has been studied numerically using Monte Carlo simulation[14], for valences between 3 and 20 and system sizes up to N = 4000. These studies have shown that the cost/ground state energy is self-averaging and given empirically by

$$
\frac{N_{ce}}{N_{ed}} = \frac{1}{2} \left\{ 1 - c \left(\frac{1 - \alpha/N}{\alpha + (c^2 - 2)(1 - \alpha/N)} \right)^{1/2} \right\} , \tag{20}
$$

within a few percent of the values obtained from the RS theory. The numerical simulations have also shown evidence for non-selfaveraging of magnetization overlap distributions

$$
P(q) = Z^{-2} \sum_{ss'} \exp(-\beta(E_s + E_{s'})) \delta(q = N^{-1} \sum_i \sigma_i^s \sigma_i^{s'}), \tag{21}
$$

where the s label microstates of energy E_s, and a suggestion of ultrametricity in overlap space for large enough z. Such features are indicators of hierarchical complexity of the free energy[4]. Analytic evidence for such properties requires a treatment beyond replica-symmetric theory, currently available via the SK mapping for the case of extensive valence but not as yet for the intensive limit.

CONCLUSION

We have demonstrated the relationship between some problems in the equipartitioning of random graphs and the mean field theory of spin glasses. In fact, the techniques and concepts of spin glass theory are finding many applications outside the confines of magnetism, one other prime example being application to neural networks[15]. It is particularly interesting to note that in several of these extensions, including those described above, infinite-ranged interactions are the norm and it may be that it is in such problems that the sophisticated machinery of modern spin glass mean field theory[4] will find its best application.

ACKNOWLEDGEMENT

The author's work on graph partitioning has been carried out in collaboration with Drs. J. Banavar, N. Sourlas, W. Wiethege and K.Y.M. Wong. He thanks them all for their help and stimulation.

REFERENCES

1.   N. Christofides, "Combinatorial Optimization", Wiley, New York
     (1979).
     C. H. Papadimitriou and K. Steglitz, "Combinatorial Optimization",
     Prentice-Hall, Englewood Cliffs (1982).
2.   S. Kirkpatrick, C. D. Gelatt and M. P. Vecchi, Science $\underline{220}$, 671
     (1983).
3.   G. Toulouse, Commun. in Phys. $\underline{2}$, 115 (1977).
4.   G. Parisi, this volume.
5.   S. F. Edwards and P. W. Anderson, J. Phys. $\underline{F5}$, 965 (1975).
6.   D. Sherrington and S. Kirkpatrick, Phys. Rev. Lett. $\underline{35}$, 1792 (1975).
7.   Y. Fu and P. W. Anderson, J. Phys. $A\underline{19}$, 1605 (1986).
8.   I. Kanter and H. Sompolinsky, Phys. Rev. Lett. $\underline{58}$, 164 (1987).
9.   M. Mézard and G. Parisi, Europhys. Lett. $\underline{3}$, 1067 (1987).
10. W. Wiethege and D. Sherrington, J. Phys. $A\underline{20}$, L9 (1987).
11. K. Y. M. Wong and D. Sherrington, to be published in J. Phys. A.
     (1987).
12. D. Sherrington and K. Y. M. Wong, to be published in J. Phys. A.
     (1987).
13. M. F. Thorpe, in "Excitations in Disordered Systems" ed.
     M. F. Thorpe, Plenum, New York (1981), p.85.
14. J. R. Banavar, D. Sherrington and N. Sourlas, J. Phys. $A\underline{20}$, L1
     (1987).
15. J. J. Hopfield, Proc. Nat. Acad. Sci. USA $\underline{79}$, 2554 (1982).
     D. J. Amit, H. Gutfreund and H. Sompolinsky, Ann. Phys. $\underline{173}$, 30
     (1987).

RANDOM LATTICE IMPURITIES

IN THE ZERO-DENSITY LIMIT

Jan Naudts

Physics Department
University of Antwerpen
Belgium

INTRODUCTION

Consider the geometrical object formed by impurity sites, randomly distributed on a square lattice. If the density p is non-vanishing then the fractal dimension d_f equals the dimension of the lattice. But in the limit of zero density one cannot exclude that fractal behaviour with a dimension strictly less than 2 occurs.

Preliminary numerical results on square lattices with impurity concentrations around 5 percent show that there is indeed scaling behaviour in an intermediate range of lengthscales. The fractal dimension derived from the numerical data is about 4/3.

BOX-COUNTING

Let X denote the set of impurity sites. A standard procedure to determine the fractal dimension d_f of X is the box-counting algorithm: consider a large square part of the lattice containing 1^2 sites; partition it into smaller squares of side s; count how many of the small squares contain at least one impurity site; finally plot the result in a log-log plot as a function of the size s of the squares. If the set X is self-similar[1] then the plot shows a straight line, the slope of which equals $- d_f$.

Let n(1,s) denote the number of non-empty squares of side s, and let n(1) denote the number of impurity sites in the large square of side 1. Then the expression for the fractal dimension is

$$d_f = - \lim_{s,1 \to \infty} (\ln n(1,s) - \ln n(1)) / \ln s \qquad (1)$$

In the present case the set X is a configuration of random sites with density p. One can calculate the average value of the quantities appearing in (1). The number of squares of side s is $1^2/s^2$. The probability of an empty square is $(1 - p)$ to the power s^2. Hence on the average one has

$$d_f = - \lim_{s,1 \to \infty} (\ln (1 - (1 - p)^{s^2}) - \ln s^2 p) / \ln s = 2$$

339

There is no range of s-values where the quantity $n(1,s)$ exhibits scaling behaviour with an exponent different from 0 or 2. In other words the analysis of the random impurity configuration with the box-counting algorithm does always indicate a fractal dimension equal to 0 or 2.

FRACTAL DIMENSION

Recently, an improved definition of the fractal dimension has been proposed[2]. Its main characteristics are:
- instead of partitioning the lattice into squares all of the same size, one allows a covering of the fractal lattice by squares of different sizes; an optimal covering has to be sought;
- the new definition coincides with the box-counting result whenever the self-similarity is present;
- in the same way as for the Hausdorff definition (which applies to continuous fractals) a measure m_d is introduced for each possible dimension d; the fractal dimension d_f is the unique value of d where the measure becomes non-trivial.

The practical way to calculate d_f goes as follows. One starts by covering all impurity sites within the large square of side 1 by squares of sides s_i, all larger than or equal to some minimal side s. Overlap of the squares is allowed. Then one seeks the optimal covering, which is the one minimising the expression $\sum_i (s_i/s)^d$. Now the measure m_d is given by

$$m_d(s,1) = (1 / n(1)) \min \sum_i (s_i/s)^d \qquad (2)$$

Scaling behaviour, when present, manifests itself as a straight line in a log-log plot of $m_d(s,1)$ as a function of s. From the slope of the line one deduces the fractal dimension $D(d)$. In principle it still depends on the choice of the dimension d. The fractal dimension is obtained from the relation $D(d_f) = d_f$.

NUMERICAL RESULTS

The first investigation concerns a square lattice of 30 x 30 lattice sites containing 51 randomly choosen impurity sites. The measure m_d is calculated for the value of $d = 2$. The result is shown in the figure. In the range of values $s = 5$ to 8 a powerlaw dependence is observed with exponent 1.31. For smaller values of s the covering problem decomposes into independent subproblems, indicating zero-dimensional behaviour. For larger values of s the total surface of the optimal covering tends to a constant, indicating 2-dimensional behaviour.

In a subsequent investigation a 55 x 55 Ising configuration at inverse temperature $\beta = 0.48$ was taken from the literature[3]. The 191 Ising spins which are misaligned with respect to the direction of the spontaneous magnetisation form the visible part of the fractal lattice. The measure $m_d(s,1)$ was calculated for $d = 2$ (see the figure). A powerlaw with exponent 1.33 is observed in the range $s = 8$ to 13 (the deviation for $s = 9$ is probably due to numerical difficulties).

In both cases a fractal dimension $D(2) \simeq 4/3$ is observed. By lowering the value of the dimension d below 2 a further increase of the dimension $D(d)$ is possible. Hence the value of d_f can be slightly higher than $D(2)$.

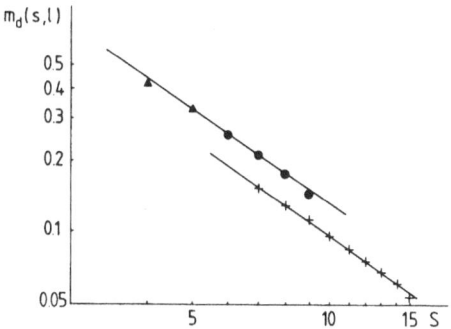

Fig. 1. The measure $m_d(s,1)$ as a function of s. Triangles and
filled circles are results for a random configuration
of 30 x 30 sites; crosses are results for an Ising spin
configuration of 55 x 55 sites (d = 2; the ● are obtained
by exact optimisation, the + and ▲ are heuristic results).

THE SET-COVERING PROBLEM

A straightforward numerical evaluation of the fractal dimension (using
the definition of ref. 2) is rather cumbersome as it requires a minimisation
over a large number of coverings. This kind of problem is well known in computer science[4]. In the present case it turns out that the minimisation is
quite feasible in one dimension, i.e. the numerical evaluation of the fractal dimension of subsets of Z can be done easely. On the other hand the
problem in dimensions 2 and higher is probably NP-complete, and the computing time needed for an exact minimisation increases exponentially with the
size of the problem.

The complete-search algorithm used for the results presented in the
figure is known as a combination of reduction and backtracking. For each of
the round points in the figure between ten and hunderds of hours of CPU time
on a VAX730 computer was used. For the other points heuristic search methods
were used (including optimisation by hand!). More intelligent algorithms are
needed in those cases to complete a complete search within a reasonable
amount of (computer) time.

Remark that the present results suggest the possibility that the out-
come of some of the hard optimisation problems can be predicted because of
a powerlaw dependence on some model parameter.

REFERENCES

1. B. B. Mandelbrot, "Fractals," Freeman, San Fransisco (1977); "The
   fracteal geometry of nature," Freeman, San Fransisco (1982).

2. J. Naudts, Dimension of Fractal lattices, preprint (1987).

3. K. Binder and H. Muller-Krumbhaar, Phys. Rev., B9:2328 (1974).

4. M. M. Syslo, N. Deo and J. S. Kowalik, "Discrete Optimization Algorithms," Prentice-Hall, Inc., Englewood Cliffs, N. J. (1983).

GLASSY RELAXATION IN CHARGE DENSITY WAVE SYSTEMS

R. Rammal and P. B. Littlewood

AT&T Bell Laboratories
Murray Hill, NJ
and
CRTBT-CNRS
Grenoble

I. INTRODUCTION

Slow thermal relaxation in random systems and glasses has been an important field of study for many years. While such properties have received much attention in glasses,[1] spin-glasses[2] and random field magnets,[3] the behavior of pinned charge-density-wave (CDW) systems has received rather less study.[4] Since the commonly used Fukuyama-Lee-Rice (FLR) model[5] for sliding CDW's is similar to that of a random-field XY model and therefore possesses many metastable states, one expects that processes involving thermal hopping over free energy barriers between such states would be the dominant mechanism at long times, or low temperatures.

There are a large number of experiments[4] which have demonstrated the existence of metastable states of the CDW and their importance for the understanding of transient, short-time relaxation phenomena. These include: history-dependence in the transient response to a square-wave electric field pulse, history-dependence (with both field and temperature) of the CDW wave-vector and correlation length; thermal and electrical history dependence of the ohmic conductivity σ_o, the "freezing" of a CDW polarization at low temperature and its subsequent slow relaxation;[6] an unusual power-low dependence of the dielectric function $\epsilon(\omega)$. In only a few experiments involving the time dependence of σ_o is there clear evidence for relaxation of the CDW between metastable states induced by thermal activation over barriers. The reason for this is twofold:

(i) Because most CDW systems are weakly-pinned the correlation length induced by disorder is very long. In three dimensions this means that the barriers between metastable states are likely much larger than thermal energies $\sim k_B T$, so that relaxation effects are small.

(ii) In semiconducting CDW compounds (i.e. all materials except NbSe$_3$). The viscosity inhibiting CDW motion becomes very large at low temperatures.[7] Thus even linear response (not involving hopping over barriers) occurs on a time scale of seconds or longer[6] at low temperatures and the hopping regime is difficult to reach.

In principle, the clearest experiments would involve the relaxation of the polarization P(t) following the application of an electric field. Such experiments have been performed, but typically on time scales within the linear response regime.[8] Clearer evidence for thermal relaxation comes from measured time-dependence in the ohmic conductivity[4] (following logt or t^{-x}) just after changes in temperature or field.

In this lecture, an analysis of the behavior of the FLR model at low T is presented. In general it is shown that there is no universal behavior, but that the relaxation may show some generic features under typical experimental conditions. Because the relaxational behavior depends in detail on the distribution of pinning strengths, as well as the initial conditions, we will limit the discussion to a particular experiment. The relaxation of the polarization P(t) following the turning-off of an electric field exceeding, or close to, the threshold field for sliding E$_T$. In general there can be as many as four different physical regimes explored with increasing time.

1. Initial relaxation to the nearest metastable state (Regime I).

2. First escape over low energy barriers (Regime II).

3. Relaxation down a hierarchy of barriers (Regime III).

4. Final approach to the "ground" state (Regime IV).

The only detailed treatment we shall give of regimes III and IV will be within mean-field theory.[9] While this may not be appropriate for real systems, it emphasizes clearly how important is the role played by the distribution of pinning strengths.

II. CLASSICAL MODEL

A. Formulation of the model

We employ here the classical elastically-deformable model of an incommensurate CDW interacting with quenched, random pinning centers.[5] The CDW is represented by a periodic component with wavevector $\vec{Q}$ and a collective charge-density component ρ_c, giving

$$\rho(\vec{r}) = \rho_c + \rho_o \cos{(\vec{Q} \cdot \vec{r} + \phi(\vec{r}))}. \tag{2.1}$$

Neglecting amplitude fluctuations, and assuming a short range interaction with impurities distribution at random sites $\vec{R}_i$ leads to a Hamiltonian for phase fluctuations alone

$$H = \int d^d \vec{r} \, [\frac{1}{2} K \, |\nabla \phi|^2 + \sum_i V_i \, \delta(\vec{r} - \vec{R}_i) \rho_o \cos(\vec{Q} \cdot \vec{r} + \phi(\vec{r}))]$$

$$(2.2)$$

The dynamics (at $T = o$) are specified by the overdamped equation of motion

$$\lambda \dot{\phi} = - \frac{\delta H}{\delta \phi} + \frac{\rho_c E}{|\vec{Q}|} \qquad (2.3)$$

where E is the applied field, and λ a damping coefficient representing viscous forces acting on the CDW. The moving CDW carries a current density given by $j(t) = \rho_c \, \dot{\phi}/|\vec{Q}|$. At this point, it is important to stress that the parameters entering here have a microscopic origin, and therefore have some temperature-dependence, which must not be confused with the effects of thermal fluctuations in the model itself. This is particularly important in the case of semiconducting CDW materials at low T, where λ shows a strong T-dependence, being roughly proportional to the low field resistivity $\rho(T) \sim e^{\Delta/T}$, all other parameters being much more weakly (if at all) temperature-dependent.[7]

B. Characteristic Length Scales and the Role of Temperature

It was appreciated some time ago that random impurity pinning, however weak, would always lead to loss of long range order in the FLR model in dimensions $d < 4$. Employing a scaling argument to roughly equate the two terms in equation (2.2) on a length scale of order the correlation length ξ_o one obtains[5]

$$\xi_o^{(4-d)/2} = \kappa \, \rho_o^{-1} <V^2>^{-1/2} c^{-1/2} \qquad (2.4)$$

in the weak pinning limit, when $\xi_o \, c^{1/d} \gg 1$ and c is the impurity concentration. This leads to an estimate of the pinning energy per unit volume of $E_p \sim \kappa \, \xi_o^{-2}$, and an estimate of the threshold field $E_T \sim \kappa \, |Q| \, \rho_c^{-1} \, \xi_o^{-2}$. In general E_T is very small and the correlation length is very long (weak-pinning description). Although the pinning energy per unit volume is small if ξ_o is large, the pinning energy of a correlated region is of order $\kappa \, \xi_o^{d-2}$. Because regions of the CDW of order the correlation length must move more-or-less coherently, this energy sets the scale of the characteristic height of barriers between metastable states. With a large value of ξ_o this characteristic barrier height will be much larger than $k_B T$ in dimensions $d > 2$ but much smaller for $d < 2$. This makes it clear that thermal effects should be much stronger in one and two dimensions than in three dimensions.[10] In the following, we shall be concerned with dimensions larger than two, where the effective temperature can be assumed small.

C. Metastable States and Linear Response

Although the average barrier heights are large, the pinning of the CDW is not uniform, so that there is a broad distribution of barriers between states. Concomitant with this broad distribution of barriers is a broad distribution of local curvatures of the quadratic free energy minimum about a given metastable state. For example, suppose we look at small fluctuations $\Psi(\vec{r}, t)$ about some metastable (static) solution $\phi_0(\vec{r})$ of equations (2.2) and (2.3). Then to linear order in Ψ, the equation of motion becomes

$$\lambda \dot{\Psi} = \kappa \nabla^2 \Psi + \tilde{V} (\vec{r}, \{\phi_0(\vec{r})\}) \Psi + \frac{\rho_c}{|Q|} h(t) \tag{2.5}$$

where the effective potential is

$$\tilde{V} = \sum_i V_i \rho_0 \cos (\vec{Q} \cdot \vec{r} + \phi_0(\vec{r})) \, \delta(\vec{r} - \vec{R}_i) \tag{2.6}$$

and we have added a small perturbing field h(t). The potential $\tilde{V}$ has a non-zero average $(\sim -E_p)$, but will also have large spatial fluctuations. $\tilde{V}$ clearly depends in detail on the metastable configuration $\phi_0(\vec{r})$. Because (2.5) is a linear problem, it can be analyzed in terms of eigenmodes $\Psi_\nu (\vec{r}, t) = \Psi_\nu(\vec{r}) \exp \left[-\dfrac{\nu t}{\lambda} \right]$, which can be determined by a diagonalization of equation (2.5). Thus we obtain for the dielectric response function

$$\epsilon(\omega) = \frac{\rho_c}{|Q|} < \frac{\partial \Psi(\omega)}{\partial h(\omega)} > = \frac{\rho_c^2}{|Q|^2} \int_0^\infty d\nu \, n(\nu) \frac{[\int d^d \vec{r} \, \Psi_\nu(\vec{r})]^2}{i \lambda \omega + \nu} \tag{2.7}$$

The eigenmodes $\Psi_\nu(\vec{r})$ are always localized on a length scale of order ξ_0 for low energy modes. The dielectric constant $\epsilon(o)$ can be very large; if we set $<\nu^{-1}> \sim E_p^{-1}$, we have $\epsilon(o) \sim \rho_c |Q|^{-1} E_T^{-1}$, which yields values of order $10^7 - 10^9$ for the small values of $E_T \simeq 100$ mV/cm typical in CDW materials.

The most important feature of equation (2.7) is the distribution $n(\nu)$; both in mean-field theory, and in one-dimensional numerical simulations, it is found that the generic behavior of $n(\nu)$ is a broad distribution extending to zero in energy. Such a gapless distribution is consistent with experimental measurements of $\epsilon(\omega)$, which have been fit to the empirical form

$$\epsilon(\omega) = \epsilon_0 / [1 + (i\omega\tau)^{1-\alpha}]^\beta \tag{2.8}$$

This is consistent with (2.7) if we choose a distribution

$$n(\nu) \sim \nu^{1-\alpha} \, g(\nu/\nu_0) \tag{2.9}$$

where $g(x \ll 1) = 1$, $g(x \gg 1) \sim x^{-(1-\alpha)(1+\beta)}$, and $\tau = \lambda/\nu_o$, with $\nu_o \sim E_p$. There is a number of comments which should be made about equations (2.7-2.9), viz.

(i) Neither the existence of a "gapless" distribution of $n(\nu)$ nor their precise value of α is a universal feature. There are a large number of metastable states $(\approx \exp(c(L/\xi_o)^d))$ in a system of linear dimensions L with $c \sim 1$, most of which have a gap in the excitation spectrum. However, gapless states are "generic", in the sense that: the application of a dc field will always drive the system into such a state at a field $E_o < E_T$[9] (the precise value of E_o depends on the initial state, and corresponds to the onset of hysteresis); and after turning off an electric field above threshold the CDW will initially relax to such a configuration.

(ii) The gapless states are marginally stable, in that all non-linear response functions (e.g. $<\dfrac{\partial^2 \Psi(o)}{\partial h(\omega) \partial h(-\omega)}>$) diverage as the driving frequency ω tends to zero. The dc response is given by

$$< \frac{\partial \Psi}{\partial h}\Big|_{DC} > = \lim_{\omega \to o} <\frac{\partial \Psi(\omega)}{\partial h(\omega)}> + \text{ singular terms} \qquad (2.10)$$

where the singular terms arise on account of the vanishing range of stability of the low modes of energy $\nu \to o$. Because these modes are localized, a simple picture can be given for this. We choose a local coordinate $\eta(\equiv \Psi_o(r)$ for a mode of energy $\nu)$, and expand the local energy to cubic order in η about the minimum,

$$U(\eta) = \frac{1}{2}\nu\eta^2 + \frac{1}{3}b\eta^3 + h\eta \qquad (2.11)$$

Because the local potential contains a periodic component, one expects that $|b| \sim E_p$, while ν may be small. From (2.11) we can see easily that modes for which $\nu^2 < 4hb$ will be unstable, giving $\eta(h) \sim 0(1)$, and thus a contribution to the d.c. susceptibility (Eq. 2.10) of

$$\lim_{h \to o}\frac{\partial}{\partial h}\int_o^{\sqrt{4hb}} d\nu\, n(\nu) \sim h^{-\alpha/2} \qquad (2.12)$$

If one requires the distribution $n(\nu)$ to be robust under small changes in d.c. field, it is necessary to choose $\alpha = o$. This is true within mean field theory, and the argument given above suggests that it may be true in all dimensions. Whether the inconsistency between the measured values of $\alpha\,(0.3 \sim 0.5$ from $\epsilon(\omega))$ and the value $\alpha = o$ predicted above is a result of a flaw in the argument, or is a result of the experimental data not reaching the asymptotically low-frequency regime is not known. An important assumption of equation (2.11) is that the low energy modes are strongly localized and do not interact.

The relation between "nearby" metastable states has been seen to be quite simple.[10] Because one obtains an identical metastable state by setting $\{\phi_i\} \rightarrow \{\phi_i + 2\pi n\}$, "nearby" metastable states are related to each other by local motions of 2π in regions of order the correlation length ξ_o.

As a corollary to the existence of a distribution of local modes ν, is that there is a corresponding distribution of local barriers Δ to nearby metastable states. From equation (2.11), we have $\Delta = \nu^3/6b^2$ (at h = o), leading to a distribution of barrier heights.

$$F(\Delta) \sim \Delta^{-(1+\alpha)/3} \tag{2.13}$$

for small barriers Δ. We see that even though the average barrier height may be much larger than $k_B T$, we are always guaranteed to have a large number of barriers with arbitrarily small Δ.

D. Inclusion of Thermal Effects

The effects of finite temperature can be included in equation (2.3) by adding a thermal noise term $\eta(t)$: $<\eta(t)> = o$, $<\eta(t)\eta(t')> = D\delta(t-t')$. Equation (2.3) can then be used to derive a Fokker-Planck equation for the probability distribution $P(\{\phi(\vec{r})\}, t)$. This is rather more general than we need, because at low T we may assume that the phases are always to be found in the vicinity of some metastable state $\{\Phi(\vec{r})\}_i$. We write

$$\frac{\partial P}{\partial t}(\{\Phi\}_i, t) = \sum_j [W(j|i)P(\{\Phi\}_j, t) - W(i|j)P(\{\Phi\}_i, t)] \tag{2.14}$$

where $W(i|j)$ is the transition probability for hopping from state i to state j. The transition will be restricted to small local motions, (of order 2π) of regions of size the correlation length, for which we have

$$W(i|j) = \gamma \exp[-\beta \Delta(i \rightarrow j)] \tag{2.15}$$

with γ a microscopic rate constant, $\beta = (k_B T)^{-1}$ and $\Delta(i \rightarrow j)$ is the barrier energy between states i and j. A detailed description of the metastable states is only possible within mean field theory (Regimes III and IV).

III. REGIME I

Let us consider the initial rapid relaxation process from a time-dependent sliding state at a finite d.c. field $E < E_T$ to the initial stationary (metastable) state reached after the field is turned to zero. This process does not involve thermal hopping over barriers but simply a viscous relaxation down the energy surface to the first metastable minimum. We discuss this process because it is much in evidence in the experiments and because it defines the initial configuration from which further thermal relaxation will take place. From equations (2.2) and (2.3) we have

$$\lambda \dot{\phi} = \kappa \nabla^2 \phi + \rho_0 \sum_i V_i \sin\left(\vec{Q} \cdot \vec{r} + \phi(\vec{r}, t)\right) \delta(\vec{r} - \vec{R}_i) + \frac{\rho_c E(t)}{|Q|} \qquad (3.1)$$

and we choose $E(t) = E\theta(-t)$, where $\theta(x)$ is the Heaviside step function.

Because a complete solution of equation (3.1) is not available, in order to go further we make a large assumption, that the initial state $\phi(\vec{r}, t = 0)$ is close to the final metastable state $\Phi(\vec{r})$ reached at long times, so that we may use linear response theory about the final state to describe relaxation. Numerical simulations[10] provide a justification for this assumption. The linear response analysis will yield, for the time dependence of the polarization

$$P(t) \sim \int d\nu\, n(\nu) f(\nu)\, e^{-\nu t/\lambda} \qquad (3.2)$$

where $f(\nu)$ is the initial $(t = 0)$ occupation of the modes of energy ν. A rough estimation[11] of $f(\nu)$ leads to the following result: $f(\nu) \simeq f(\nu/\nu_1)$ where $\nu_1 \sim \nu_0$ and $f(x \ll 1) = 1$, $f(x \gg 1) \sim 1/x$. Using equation (2.9) to evaluate (3.2), one obtains the relaxing current

$$\tau I(t) = \tau \frac{\partial P(t)}{\partial t} \sim -(t/\tau)^{-1+\beta(1-\alpha)} \quad , \quad t/\tau \ll 1, \qquad (3.3)$$

$$\sim -(t/\tau)^{-3+\alpha} \quad , \quad t/\tau \gg 1,$$

which is a crossover from a small to a large (negative) power law on a time scale τ. This result gives actually a reasonably good fit to available data for $P(t)$: the values of α and β are extracted from measurements of $\epsilon(\omega)$. We note that it is practically indistinguishable from a stretched exponential from

$$P(t) = P_0 \exp\left[-(t/\tau)^{1-n}\right] \qquad (3.4)$$

for $n = 1 - \beta(1 - \alpha)$ if the value of n is small, and for $t/\tau < 10$. Such an empirical formula is widely used in the general context of glassy relaxation, without any clear justification (for a discussion see Refs. 6 and 8).

IV. REGIME II

Assume now that following the early-time relaxation, the system is found in a marginal metastable state with a distribution of barriers $F(\Delta)$ given by equation (2.13). Once the small barrier has been overcome, the rate of hopping back will be negligible. From (2.14) we have

$$\frac{\partial}{\partial t} F(\Delta, t) = -\gamma F(\Delta, t) e^{-\beta \Delta} \tag{4.1}$$

so that

$$F(\Delta, t) = F(\Delta, o) \exp\left[-\gamma t\, e^{-\beta \Delta}\right] \tag{4.2}$$

Here we have reset the origin of time, assuming that T is so small that the two regimes I and II do not overlap. Because the system is polarized, we have

$$I(t) \sim \frac{d}{dt} \int d\Delta F(\Delta, t) \tag{4.3}$$

$$\sim -\gamma \int_o^\infty d\Delta \ \Delta^{-(1+\alpha)/3} e^{-\beta \Delta} \exp\left[-\gamma t\, e^{-\beta \Delta}\right]$$

In this expression, $\Delta_o = \beta^{-1} \ell n(\gamma t)$ acts as lower cutoff and then ($\Gamma(a, x) = $ incomplete gamma function)

$$I(t) \sim -\gamma \beta^{(\alpha-2)/3} \Gamma\left(\frac{2-\alpha}{3}, \ell n(\gamma t)\right) \tag{4.4}$$

For large $\ell n\, \gamma t \gg 1$, one obtains

$$I(t) \sim -\gamma \beta^{(\alpha-2)/3} (\gamma t)^{-1} [\ell n(\gamma t)]^{-(1+\alpha)/3} \tag{4.5}$$

The annealing of small barriers will lead to a lower cutoff at $\nu_o \sim \Delta_o^{1/3}$ in the integral of $\epsilon(\omega)$ (equation (2.7)) and then to a reduction of $\epsilon(\omega)$ with increasing time. At low frequencies $\nu_o > \omega > \omega_s$ where $\omega_s \sim \gamma e^{-\beta \Delta_o}$, $\Delta_o \sim E_p \xi_o^d$, the dielectric function should decrease with waiting time t_w as

$$\epsilon'(\omega, o) - \epsilon'(\omega, t_w) \sim [\beta^{-1} \ell n(\gamma t_w)]^{(1-\alpha)/3} \tag{4.6}$$

V. MEAN FIELD THEORY AT FINITE TEMPERATURES

As we mentioned above, because of the difficulty of defining metastable states in a finite range model a direct calculation is not straightforward. In mean field theory, there is a simple description of the metastable states.

A. Mean Field Theory at T = O

In the mean field model[9], the local phase ϕ_j at site j is pinned by a potential to a preferred value of the phase θ_j. The coupling between different phases ϕ_j is taken to be of infinite range,

$$H = \frac{1}{2} \sum_j (\bar{\phi} - \phi_j)^2 + \sum_j V_j f(\phi_j - \phi_j) \tag{5.1}$$

with the average phase $\bar{\phi} = M^{-1} \sum_j \phi_j$, $(M \rightarrow \infty)$ and an equation of motion

$$\lambda \dot{\phi}_j = \bar{\phi} - \phi_j - V_j f'(\phi_j - \theta_j) + E \tag{5.2}$$

where the prime denotes a derivative with respect to the argument. We choose θ_j to be uniformly distributed over $[-1/2, 1/2]$, and the potential $f(x)$ to be periodic and symmetric:

$$f(x) = f(x + 1) = f(-x) \tag{5.3}$$

The static solutions of equation (5.2) can be written in the form

$$\phi_j = \bar{\phi} + E + g(\bar{\phi} + E - \phi_j, V_j) \tag{5.4}$$

where the function $g(x) = g(x + 1)$ is periodic and given by the solution of

$$g(x, V) + V f'(x + g(x, V)) = o \tag{5.5}$$

at an electric field

$$E = - \int dV \rho(V) \int_{-1/2}^{1/2} dx\, g(x, V) \tag{5.6}$$

introducing a distribution $\rho(V)$ for V. If V is large enough, the solutions $g(x, V)$ of equation (5.5) will be multiply-valued, which allows for the existence of many metastable states satisfying (5.6). The threshold field is given by choosing always the lowest solution $g_{min}(x, V)$, viz.

$$E_T = - \int dV \rho(V) \int_{-1/2}^{1/2} dx\, g_{min}(x, V) \tag{5.7}$$

As $E \rightarrow E_T$ from below, only a single parameter family of stable solutions exists. Without loss of generality, $f(x)$ can be chosen as a repeated parabola

$$f(x) = \frac{1}{2} \min [(x - n)^2], \tag{5.8}$$

with n integer or zero, so that there are $2N + 1$ branches of $g(x, V)$, given by

$$g_n(x, V) = - (V/V + 1)(x - n) \; ; \; -N < n < N \tag{5.9}$$

where $|g_n(x, V)| < V/2$, so that N is the largest integer less than $(1 + V/2)$.

For some given (θ, V) the local minima of the phase are at $(x = \bar{\phi} + E - \theta)$

$$\phi = \bar{\phi} + (V/V+1)\,(n-x) \qquad (5.10)$$

and the corresponding energy is just

$$U_n = \frac{1}{2}(V/V+1)\,(n-x)^2\,. \qquad (5.11)$$

The barriers to hopping between wells are

$$\Delta\,(n \rightarrow n+1) = \frac{1}{4}V - \frac{1}{2}(V/V+1)\,(x-n), \qquad (5.12)$$

and

$$\Delta(n \rightarrow n-1) = \frac{1}{4}V + \frac{1}{2}(V/V+1)\,(x-n).$$

B. Mean Field Theory at Low T

We define $P_n(x, V, t)$ to be the probability that well n is occupied for a given x, V. Then,

$$\frac{d}{dt}P_n\,(x, V, t) = -\,(\nu_n + \mu_n)\,P_n + \nu_{n-1}P_{n-1} + \mu_{n+1}P_{n+1} + \dot{\bar{\phi}}\,\frac{\partial P_n}{\partial x} \qquad (5.14)$$

The last term follows from the definition of $x = \bar{\phi} + E - \theta$. The transition rates are given by:

$$\mu_n\,(x, V) = \gamma\exp\,[-\,\beta\Delta\,(n \rightarrow n+1)], \qquad (5.15)$$

$$\nu_n\,(x, V) = \gamma\exp\,[-\,\beta\Delta(n \rightarrow n-1)].$$

P_n satisfies the boundary conditions

$$P_{n+1}\,(\frac{1}{2}, V, t) = P_n\,(-\frac{1}{2}, V, t) \qquad (5.16)$$

and is normalized so that

$$\sum_{n=-N}^{N} P_n\,(x, V, t) = 1 \qquad (5.17)$$

The time dependence of P_n arises both from hopping over barriers and from a slow drift induced by the average current $\dot{\bar{\phi}}$. The system of equations (5.14) is closed by the self-consistency condition (in obvious analogy to equation (5.6))

$$E = - \int dV \, \rho(V) \int_{-1/2}^{1/2} dx \sum_n g_n(x, V) P_n(x, V, t) \qquad (5.18)$$

In particular, if E is time-independent, we have:

$$\frac{dE}{dt} = 0 = \int dV \, \frac{\rho(V)V}{V+1} \left\{ \dot{\bar{\phi}} + \int_{-1/2}^{1/2} dx \sum_n P_n(x, V, t) (\nu_n - \mu_n) \right\} \qquad (5.19)$$

Notice that these equations are invariant under a change of the average phase $\bar{\phi}$, and only the average current $\dot{\bar{\phi}}$ enters explicitly.

In general, the resolution of equation (5.14) leads to an inhomogeneous Volterra equation for the current $\dot{\bar{\phi}}$: the polarization decay depends explicitly on the initial conditions.

VI. REGIME III

We consider the behavior of $P_n(t)$ when the local phases are hopping down a distribution of barriers, but are not close to equilibrium state. This regime will only be important in mean field theory if there are many branches of g_n, i.e. $V \gg 1$. In contrast with regime II, which was controlled by the distribution of small barriers, regimes III and IV are controlled by the behavior of $\rho(V)$ at large V.

A low T $(\beta \gg 1)$ and for large V, we have a number of strong inequalities viz.

$$\mu_n \gg \mu_{n+1}, \, \nu_n \gg \nu_{n-1}, \qquad (6.1)$$

$$\nu_n(x, V) \gg \mu_n(x, V) \text{ for } n - x > 0, \qquad (6.2)$$

$$\mu_n(x, V) \gg \nu_n(x, V) \text{ for } n - x < 0$$

Equations (6.2) guarantee that the hopping of an individual phase will be undirectional. Such a hierarchical picture has been used[12] to describe general relaxation mechanisms in glasses; we have here a particular case of the "rules" to determine the number of states at each level, and their transition rate. As expected the depolarization current depends in details on the initial conditions. For the sake of simplicity, we consider the initial choice of $\bar{\phi}(t = 0) = 0$. The result for $\dot{\bar{\phi}}(t)$ will depend also on the distribution $\rho(V)$ for large V. Two generic cases will be considered here: a power law distribution and an exponential distribution.

Power Law distribution of $\rho(V)$ If we take
$$\rho(V) \sim V_o^p/V^{(p+1)}, \tag{6.3}$$

then

$$\frac{d\Phi}{dt} = -(p\beta t)^{-1}\{(4/\beta)^{-p}(\ell n\gamma t)^{-p} - (E_T - \bar\phi + (2/\beta)\ell n\,\gamma t)^{-p}\} \tag{6.4}$$

In particular for $t \gg t_o$ (where $\ell n\gamma t_o \sim (\beta V_o)^{-1}$), the current is given by

$$\frac{d\bar\phi}{dt} = (p\beta t)^{-1}(4/\beta)^{-p}(\ell n\,\gamma t)^{-p} \tag{6.5}$$

We note that (except for the case $p = o$), the mean-field coupling is important only at intermediate times $t \simeq t_o$; the relaxation is similar to that obtained in the absence of coupling between wells.

Exponential distribution of $\rho(V)$ With an exponential distribution of $\rho(V)$

$$\rho(V) = \frac{1}{V_o}\exp(-V/V_o), \tag{6.6}$$

one has $E_T \sim V_o/2$ and a similar complicated expression for $d\bar\phi/dt$. The surprising feature of these results is the weak role played by the coupling between the wells - we would have obtained the same results if the mean field coupling had been neglected in the relaxation process. If a bias field close to E_T were applied, such a result would not be expected.

VII. REGIME IV

The approximation of "undirectional" hopping will fail when the system approaches the lowest energy state, because there may be many such states with nearly equal energies, but with large barriers between them. Close to the ground state, we include only the three states $n = \pm 1, 0$. A detailed analysis[11] implies that the final approach to equilibrium depends on whether $\rho(V)e^{\beta V/\Delta}$ vanishes at large V. For the particular case of an exponential distribution, $\rho(V) = V_o^{-1}\exp(-V/V_o)$, the change in the long-time behavior marks a critical temperature $\beta_c = 4/V_o$. We can see this straightforwardly by calculating the d.c current flowing in a small but finite field E. For $t \to \infty$, we find

$$\overset{\cdot}{\bar\phi} = E/A \qquad, \quad \beta < \beta_c \tag{7.1}$$

and

$$\dot{\phi} = (E/A') \, t^{(\beta_c/\beta - 1)} \quad , \quad \beta > \beta_c \tag{7.2}$$

where A, A' are constants. Because we find a zero steady-state current in response to a small d.c field, this suggests that there may be a finite threshold field at low temperatures, vanishing at $\beta = \beta_c$. As is typically found for problems involving activation over barriers, an exponential distribution of barriers is the marginal case. For all distribution $\rho(V)$ which fall off more slowly than an exponential, we would find finite threshold field at all finite temperatures. Another fashion to understand the meaning of β_c is the following one. If there is no d.c bias applied, then for $\beta > \beta_c$, it does not make sense to discuss relaxation to an equilibrium ground state: this will never be reached in a finite time. With a distribution $\rho(V)$ which falls off faster than exponentially for large V (so that $\beta_c = 0$) or with an exponential distribution at a temperature above the critical temperature, an exponential decay will dominate at very long times.

The existence of a finite threshold field at a finite temperature is likely to be an artifact of the mean field approximation, as can be seen by rederiving this answer in a simple fashion. The local field E_i necessary to produce a finite current $d\phi_i/dt$ at site i is just

$$E_i = \frac{d\phi_i}{dt} \, \gamma^{-1} \, e^{\beta\Delta_i} \tag{7.3}$$

where Δ_i is the local barrier height close to the ground state. Because in the mean field theory the local barrier height (close to the ground state) depends only the local pinning potential ($\Delta = V/4$ here), one can average equation (7.3) to obtain

$$E = \gamma^{-1} \, \frac{d\bar{\phi}}{dt} \int dV \rho(V) e^{\beta V/4} \tag{7.4}$$

We recover immediately the result that there is a finite threshold field if $\rho(V)e^{\beta V/4}$ does not vanish for large V; a finite field E is necessary to maintain an infinitesimal current. In finite range models, the barriers Δ_i depends strongly on the local configuration of the CDW, their distribution may be more strongly bounded at large Δ than the distribution of local pinning energies V.

VIII. COMPARISONS WITH OTHER GLASSES

In what follows we compare the behavior of CDW's to other glassy materials and we point out some experimental situations where one can benefit from previous work done on glassy systems.

1. The existence of a large number of metastable states is actually a common feature of disordered materials. This leads in general to anomalous behavior of relaxation at long times whatever the nature of the equilibrium state(s) at zero temperature (ordered, disordered). Different random field systems can be distinguished according to the precise value of the lower critical dimension: $d_\ell = 4$ for the CDW,[5,13] $d_\ell = 2$ for Ising models. Universality of the dynamics is not expected to hold in the case of CDW's, however generic features are still present under typical experimental conditions. In this respect, relaxation experiments are very useful in order to extract both the probability distribution of barriers and the relaxation mechanisms.

   The occurrence of infinite barriers may arise in mean field theories of CDW[9] as well as for spin glass systems[14]. This is the reason why short-range interaction models are believed to behave differently, at least from the dynamic point of view. The importance of infinite range coupling becomes evident only at extremely long times and this is very difficult to reach both in real and numerical experiments. The available time range turns out to be controlled by the distribution of barriers and this is probably the origin of the nonuniversal behavior.

2. As we have shown previously, different regimes appear during the time evolution of the considered system. The initial one is governed by a linear response theory for the transient ("viscous") relaxation towards the first metastable state. In general it is very important to distinguish this regime from thermally-activated processes which dominate at long times. Such a separation between different regimes can be facilitated by the choice of a system with a short characteristic time for the linear response (such as metallic systems, e.g. $NbSe_3$). In CDW's barriers can become very large. In order to allow experimental measurements on laboratory time scales in the hopping regime, it may be appropriate to study systems exhibiting strong pinning where the barriers are smaller.

3. From previous and current studies of spin glasses, it is now clear that history is a very important factor for the time behavior[2] during relaxation. In particular the precise definition of the time origin (ageing process, waiting time) can alter in a sensitive way the fit of data to any prescribed form. Such dependence is also expected for CDW's, which can provide a guide for the study of spin glasses. Simple power law decays can actually be confused with a stretched exponential decay, which can appear as a better fit if the ageing processes are ignored[2]. The main object of the relaxation studies may be the extraction of the barrier distribution and better understanding of dynamics.

REFERENCES

1. L. C. E. Struik, in "Physical Aging in Amorphous Polymers and Other Materials", (Elsevier, New York) (1978); K. L. Ngai, A. K. Rajagopal and C. Y. Huang, J. Appl. Phys. **55**: 1716 (1984).

2. See M. Ocio, M. Alba and J. Hamman, J. Physique Lett. (Paris), **46**: L 1101 (1985) and references therein.

3. A. R. King, J. A. Mydosh and V. Jaccarino, Phys. Rev. **55**: 2525 (1986); R. J. Birgeneau, R. A. Cowley, G. Shirane and H. Yoshizawa, J. Stat. Phys. **34**: 817 (1984).

4. See "Charge Density Waves in Solids", Lecture Notes in Physics, Vol. 217 ed. G. Hutiray and J. Solyom (Springer, Berlin, 1985).

5. H. Fukuyama and P. A. Lee, Phys. Rev. **B17**: 535 (1977), P. A. Lee and T. M. Rice Phys. Rev. **B19**: 3970 (1979).

6. G. Kriza and G. Mihaly, Phys. Rev. Lett. **56**: 2529 (1986).

7. R. M. Fleming, R. J. Cava, L. F. Schneemeyer, E. A. Rietman and R. G. Dunn, Phys. Rev. **B33**: 5450 (1986).

8. P. B. Littlewood and R. Rammal, Phys. Rev. Lett. **58**: 524 (1987).

9. L. Sneddon, Phys. Rev. **B30**: 2974 (1984); D. S. Fisher, Phys. Rev. **B31**: 1396 (1985).

10. P. B. Littlewood, Phys. Rev. **B33**: 6694 (1986).

11. P. B. Littlewood, and R. Rammal, Phys. Rev. B xxx (1987).

12. R. G. Palmer, D. L. Stein, E. Abrahams, and P. W. Anderson, Phys. Rev. Lett. **53**: 958 (1984); J. R. McDonald, Phys. **28**: 485 (1962).

13. L. J. Sham and B. R. Patton, Phys. Rev. **B13**: 2151 (1976).

14. G. Parisi, Phys. Rev. Lett. **50**: 1946 (1983).

NEURAL NETWORKS AND STATISTICAL MECHANICS

Gérard Toulouse

Laboratoire de Physique de l'Ecole Normale
Supérieure 24 rue Lhomond, 75231 Paris, France

The topic of these lectures being outside the mainstream
of this session, the notes are reduced to a sketch. If they
help the reader to gain some perspective on this field and
perhaps entice her/him to join in its exploration, they will
have achieved the desired end.

1. COMPLEX SYSTEMS SHOWING SOME FORM OF INTELLIGENCE

   a) Brains, nervous systems.
       The actors are neurons, interactions via synapses.
       Human brain: ~ 1.300 g, ~ 10^{11} neurons.
       Theory: neural nets.

   b) Immune systems.
       The actors are lymphocytes, interactions via
       antibodies.
       Human immune system: ~ brain in mass, ~ 10^{12}
       lymphocytes.
       Theory: Jerne network.

   c) Evolution, differentiation.
       The actors are genes, interactions via proteins.
       Human genome: ~ 2.10^5 genes.
       Theories: Eigen model, Kauffman model.

2. MOTIVATIONS FOR THE STUDY OF NEURAL NETWORKS

   a) Continuation of statistical physics;
       Study of collective properties of assemblies of
       elements;
       Use of the canonical formalism of statistical
       mechanics (or beyond, with asymmetrical interactions);
       Notions of phase transitions, energy landscapes,
       attractors.

   b) Understand how brains function;
       Biology from neuroanatomy (bottom) to cognitive
       psychology (up);

The brain is an existence proof for many problem-solving abilities.

   c) Design of new devices;
Computers 'that work like the brain', perform pattern-recognition tasks rather than arithmetics, learn by example instead of following explicit rules.

3. THE STANDARD MODEL

It is based on the following ingredients:

- the neuron as a two-state element,

- the network state at time t, defined by the configuration of activity of all component neurons,

- the firing rule, input-output relation for each neuron, which behaves as a linear threshold automaton,

- the dynamics, specified by the firing rule plus an updating procedure; differences in updating lead to different dynamics.

Interpretation: the attractors of the dynamics are meaningful. In particular, persistent firing patterns correspond to 'percepts' or 'concepts'. The emphasis is on memory nets, that function as content-addressable, associative memories.

Neural network theory as inverse problem of statistical mechanics. Two-state elements S_i, interacting via a dynamic rule which involves the intercouplings J_{ij}, are the basic ingredients in both domains. In statistical mechanics, the J_{ij} are given and one looks for the persistent states S_i^μ (plus resistance to thermal noise, phase transitions, etc). In neural network theory, especially for memory nets, one is interested in the inverse problem. Given a set of states S_i^μ, i.e., a set of patterns to be stored in memory, find the couplings J_{ij} which will stabilize them best.

4. STANDARD HOPFIELD MODEL

Hopfield[2] introduced the hypothesis of symmetric interactions ($J_{ij}=J_{ji}$), in order to define an energy function, i.e., the cornerstone of the canonical formalism of statistical mechanics. He added the hypothesis of a fully, self-coupled net (each neuron interacts with every other neuron), that brings with it the simplifications of mean field theory. Given an initial configuration of activity (induced by external stimulation), the network evolves according to its internal dynamics and reaches a decision (settles into an attractor) very rapidly . It thus functions as a memory exhibiting properties of associativity and restoration (tolerance to incomplete data), and robustness (gracious degradation). This is an instance of parallel distributed processing[3]. Although each neuron is a slow element, a massively interconnected architecture gives to the human brain a processing power, around 10^{15} instructions per second, as compared to 10^8 for a Cray-1. Extrapolation of past computer

performances into the future gives a curve that crosses the human brain level about thirty years from now! The last ingredient for the standard Hopfield model is the Hebb learning rule, that has the following properties: one-shot (as opposed to slow), local (as opposed to non-local), symmetric in i,j (as opposed to asymmetric).
Question is: What is the memory capacity as a function of the number of neurons N, and for uncorrelated patterns (to start with)?

If one imposes a constraint of strict stability for pattern storage, the capacity is limited to[4]

$$p ~ N / lnN \tag{4}$$

whereas if one accepts some error tolerance in the positioning of the valleys, the capacity is[5]

$$p ~ 0.14 N \tag{5}$$

The surprises found in the analysis[1] are the existence and role of spurious states, the richness of the phase diagram, the existence of a first order transition from retrieval phase to spin glass phase (confusion catastrophe), the existence of a mixed phase with an ultrametric structure of valleys, and the dependence og updating dynamics on the size and shape of the basins of attraction. This last observation suggests that the notion of energy landscape, in a high dimension space, has to be taken with some care, because many points are saddle points .

In summary, the merits of the Hopfield[2] approach are:

a) feedback in the architecture (not just feedforward, as in perceptrons),

b) asynchronous firing dynamics,

c) solvability of the memory model, via statistical mechanics,

d) programming recipe for more general-purpose computation nets, via design of an *ad hoc* energy function[6].

5. CRITIQUES AGAINST THE STANDARD HOPFIELD MODEL

They have been directed toward the following aspects

a) Symmetric synapses,

b) Tabula rasa,

c) Possibility of sign changes of the couplings. Unbounded efficacies,

d) Forgetting catastrophe,

e) Learning rule (on account of optimality or biological plausibility),

f) Uncorrelated patterns,

g) Complete connectivity,

h) Restriction to two-neuron interactions,

i) No sequences, melodies,

j) Only one class of neurons. Same neuron is both
    excitatory and inhibitory.

Let it just be said that the above criticisms have now
been met and circumvented in some way, though much remains to
be done.

k) Only 'on-off' signals,

l) Lack of architecture, i.e., organization of circuits
    into systems.

6. MEMORY WITHOUT ERRORS

With a non-local learning rule[7], it is possible to obtain
a larger capacity

$$p = N/2. \tag{7}$$

A variant leads to an even larger capacity[8]

$$p = N. \tag{8}$$

Note that overloading in this case leads to a degeneracy
catastrophe (all patterns stable, vanishing interactions)
instead of a fall into confusion.

7. RECAPITULATION ON ULTIMATE CAPACITIES

Four scenarios can be distinguished:

a) plasticity freeze: learning stops, under command from
    a supervisor, or spontaneously[9],

b) confusion catastrophe: loss of retrieval quality,

c) degeneracy catastrophe: loss of attractivity,
    associativity,

d) palimpsests[10]: models that keep learning and
    forgetting.

8. TWO IMPORTANT QUESTIONS

a) What is the optimal capacity that can be hoped for?

b) What learning rules will lead to it?

Recent findings[11] suggest that the optimal capacity (for
random uncorrelated patterns, etc.) is:

$$p = 2N \tag{11}$$

with strict stability of the patterns, using a slow (error-correcting) algorithm. In the case of the Hebb rule , error tolerance allowed one to push the capacity from N/lnN up to 0.14N . How much above 2N is it possible to go , allowing for error tolerance? The virtues of the recent developments are that they do away with the restrictions of symmetric interactions and of complete connectivity , opening new vistas to statistical mechanics . And they make contact with powerful past results, such as the perceptron convergence theorem[12].

9. IN SUMMARY

a) What has statistical mechanics brought to neural network theory?
* Thermodynamic limit (N $\rightarrow\infty$)
* Asynchronous dynamics
* Noise-resistance of pattern stability (T $\neq$ 0)
* Existence of spurious states
* Phase transitions
* Careful study of storage capacities and comparison of learning rules

b) What has neural network theory brought to statistical mechanics?
* Interest for non-hamiltonian systems (asymmetric interactions)[13]
* New class of soluble models (dilution, ...)[14]
* Systematic parameter-space exploration[10]
* Sensitive dependence of attractor basins on dynamics[1]
* Progress in multi-valley analysis of configuration spaces.

Acknowledgements for stimulation are due to H.Gutfreund and R.Pynn.

References

1. D. J. Amit, in _Glassy Dynamics and Optimisation_, ed. I. Morgenstern and L. Van Hemmen (Springer, 1987), and references therein.
2. J. J. Hopfield, Proc. Nat. Acad. Sci. U.S.A. _79_, 2554 (1982).
3. _Parallel Distributed Processing_, Vols. 1 and 2, ed. J. L. MacClelland and D. E. Rumelhart (MIT Press, 1986).
4. G. Weisbuch and F. Fogelman, J. Physique _46_, L623 (1985).
5. A. Crisanti, D. J. Amit, and H. Gutfreund, Europhys. Lett. _2_, 337 (1986).
6. J. J. Hopfield and D. W. Tank, Science _233_, 625 (1986).
7. L. Personnaz, I. Guyon, and G. Dreyfus, Phys. Rev. A _34_, 4217 (1986).
8. I. Kanter and H. Sompolinsky, Phys. Rev. _35_, 380 (1987)
9. P. Peretto, in _Computer Simulation in Brain Science_, ed. R. M. J. Cotterill (Cambridge U.P. 1987).
10. M. Mézard, J. P. Nadal, and G. Toulouse, J. Physique _47_, 1457 (1986).
11. E. Gardner, B. Derrida, private communications.

12. S. Diederich and M. Opper, Phys. Rev. Lett. <u>58</u>, 949 (1987).
13. G. Parisi, J. Phys. A <u>19</u>, L675 (1986).
14. B. Derrida, E. Gardner, and A. Zippelius, Europhys. Lett. (1987).

COMPARISON OF DOMAIN GROWTH KINETICS IN TWO AND THREE DIMENSIONS

Gary S. Grest[*], Michael P. Anderson[*] and David J. Srolovitz[#]

[*]Corporate Research Science Laboratory
Exxon Research and Engineering Company
Annandale, New Jersey 08801 USA

[#]Los Alamos National Laboratory
Los Alamos, New Mexico 87545 USA

I. INTRODUCTION

During the past five years, we[1-4] have been investigating the kinetics of systems with a high ground state degeneracy which have been quenched from a high temperature disordered state, $T \gg T_c$, to a final temperature below the transition temperature T_c. After the quench the system spontaneously develops local domains which are highly ordered. The average domain size then grows in order to reduce the excess free energy associated with the domain walls. For the non-conserved Ising model[5], which has a ground state degeneracy of two, it has been known since the work of Lifshitz[6] and Allen and Cahn[7] that the correlation length R grows algebraically as $t^{1/2}$ for all dimensions above one. This result has been well documented by both analytical[5-11] and computer[9,12-14] studies as well as experimental studies on ordered alloys[7] (e.g., FeAl and CuAu). However five years ago little was known about the growth kinetics for more complex models which have a higher ground state degeneracy. Lifshitz[5] was the first to predict slow kinetics in systems with several degenerate equilibrium states. Safran[15] extended these arguments to show that domains may become pinned if the number of degenerate ground states $Q \geq d+1$, where d is the dimension of space. They suggested that the system could become trapped in locally stable metastable states which would then greatly slow down the kinetics. When Safran, our colleague at Exxon, showed us his results, we decided that computer simulations would be a very good method to test his suggestions. In collaboration with P. Sahni, we began at that time a study of the kinetics for the Potts model, for a wide range of Q's from the Q=2 Ising model to Q=64. While we[1,2] found that the growth remained algebraic, independent of the value of Q, we also found that the higher ground state degeneracy had a large effect on the growth kinetics, the

Examples of systems with highly degenerate ground states (Q>2) include the superlattice structures observed for adsorbed atoms on surfaces[16], such as O/W(110), O/Ni(111) and H/W(001) which are chemisorbed and physisorbed systems such as rare-gas atoms adsorbed on graphite. In addition to surface studies, similar structures have also been observed in intercalation compounds such as $SbCl_5$ intercalated into graphite[17], for which Q=7. In addition, we[2-4] observed that large Q was an excellent description grain growth in poly-crystalline metals (e.g., pure Fe, Al and MgO), if it is assumed that all grain boundaries are energetically equivalent. Since there is an abundance of data on the growth kinetics, grain size distribution and topological distributions of grain edges from the metallurgical literature, it possible to make comparisons between our model simulations and experimential grain growth data. It turns out that the Potts model captures all the essential features of grain growth in real materials. We[3] also found that many of properties of the domain growth kinetics, in particular the microstructures and topological distributions, were essentially the same, independent of dimensionality. When we compared the distribution of grain areas from our two dimensional simulations with the cross sections of our three dimensional systems, we found very good agreement. While the shape of the distribution function changes with Q, it is relatively insensitive to d.

One of the essential features of domain growth involves the competition between individual ordered regions or domains, of different orientations Q which share a common boundary. Many analytical theories[18-20] implicitly assume that domains can be described as spherical, and that growth occurs in an average environment. It is easy to understand why they are relatively poor in describing more than the simple qualitative aspects of experimentally observed kinetics and topology, particulary for large Q. In general, most analytic theories have been concerned with the time dependence of the average grain radius

$$\bar{R} \simeq Bt^n \qquad (1)$$

where B is a temperature dependent constant. Many analytic descriptions give n=1/2, as observed experimentally for the Q=2 Ising model[7]. This value for n is in agreement with two experimential studies using low-energy electron diffraction for O on W(110)[21] (Q=4) and O on W(112)[22] (Q=2) which found n≃0.5. However, experimental data[23] for grain growth in polycrystalline materials consistently give n<1/2, although the data have substantial scatter. For the Ising model, analytic theories[5,10,11] have been very important in clarifying the concept of dynamic scaling, in which the structure function $S(q,t)$ can be written as $\bar{R}^d(t)F(q\bar{R}(t))$. Calculations for $F(q\bar{R}(t))$ are in good agreement with experiment. However for Q>2, much less is known. Some theories have tried to predict the domain size distribution function for polycrystalline (Q large) but have largely been unsuccessful. There are no predictions at all for many topological quantities, like the distribution of the number of grain edges. These failures and inabilities may be traced to their treating the growth in terms of a grain in a mean environment. This type of mean field theory averages out much of the important, topological effects associated with the connectedness of the actual microstructure. For this reason, they obtain results which best describe the Ising model (Q=2), for which Q<d+1 and the topology of the network is irrelvant. Generalizations of this type of theory[24-25] to include many interacting spherical grains are also insufficient as they still do not include the simple space-filling requirements which result in a topologically connected structure.

Because the outlook for finding a solution of an analytic model for domain growth, which describes not only the average growth of the grain radius $\bar{R}$ but also the domain size and topological distributions, does not seem hopeful, we have concentrated on obtaining this information from large scale computer simulations. In our early studies[1-4], we observed that the exponent n for growth in two dimensional (2-d) Potts model was dependent on Q. As Q increased from 2, n decreased slowly from 0.5 to approximately 0.41±0.02 for Q greater than 30. As Q increased, the domain size distribution function narrowed as the domains become more compact. For Q larger than approximately 30, we found both n and the domain size distribution function become independent of Q. This result for large Q was confirmed independently by Wejchert et al.[26] and Mouritsen[27]. However, as pointed out by a number of authors including Kaski et al[28], Dasgupta and Pandit[29], Mazenko, Valls and Zhang [30] and Huse [31], it is difficult to determine whether the exponents one obtains from fitting the simulation results are really asymptotic or not. This is true even when the available data give an excellent fit to Eq. (1). It is always difficult to rule out slow transients. Our original simulations[1-4] were carried out on lattices of size 200x200, already much larger than used to study equilibrium critical phenomena. Such large lattices were necessary since we were interested in following the kinetics for long times. However, even for this size system, we had to stop the simulation when the average grain size was on the order of 200 sites, in order ensure that there were enough domains in the system to give representative results. Because the average area grows as $\bar{R}^2 \simeq t^{2n}$, very large systems are needed to extend the time regime significantly. Here we will present results, previously unpublished for Q=48 on a 1000x1000 lattice that show in fact when the simulations are extended to much larger systems and longer times the value of n for large Q appears to increase to n=0.49±.02, consistent with that for the Ising model. However, all of our other results for the domain size and topological distributions remain unchanged.

We[32] have recently extended our simulations to three dimensions (3-d), where we have studied systems as large as 100^3. Here we observe values of n which are significantly less than 1/2 for large Q. We find that for Q larger than approximately 12, n no longer depends on Q and n=0.38±0.02. While we have seen no difference between the value of n for systems of size 60^3 and 100^3, we cannot say for certain that this value of n is asymptotic. What we have been able to show is that the grain size and topological distribution functions for the high-Q Potts model agree very well with grain growth in real polycrystalline materials.

While the growth kinetics are an interesting problem to

study theortically, to make progress experimentially one often has to be concerned with the presence of quenched impurities. These impurities can occur in many different forms; for example, second-phase particles in grain growth in polycrystalline materials[33] and/or surface defects in the ordering of atoms physisorbed on solid surfaces or intercalated into graphite. As a routine matter, second-phase particles are added to metals and ceramics to control grain size. In these cases the final grain size is affected by both the particle concentration and size. We[34] found that in 2-d the domain growth becomes pinned for quenches to T=0 and that the average domain area A_f varies inversely with the concentration of impurities. For quenches to T>0, the asymptotic growth for long times is expected to be slower than a power law. Huse and Henley[35] predicted that in 2-d the long-time growth would be $(lnt)^4$. This is consistent with simulations by Grest and Srolovitz[36] for the Ising model who were unable to distinguish a lnt from a power of lnt, but did find that the growth was not a power law. More recent work by Chowdhury et al.[37] showed the cross-over from the early time power law regime to the late stage logarthmic regime. While simulations for quenches to T>0 have not been carried out for the Potts model, we expect similiar slow growth for long times.

In this paper, we will review our simulation results for domain growth, with particular emphasis on comparing growth in two and three dimensions. In Sec. II, we will describe the model and Monte Carlo procedure. We will then review our findings for the growth kinetics and describe the changes which occur in the microstructure as Q increases in Sec. III. We will also present results for the domain size distribution and compare our large Q results with experimental data for polycrystalline materials. Finally, in Sec. IV, we will give some conclusions and discuss future directions of this work. In this paper, we will consider only models with a ground state degeneracy Q greater than two. The reader is referred to excellent reviews for the domain growth kinetics for both the conserved and non-conserved Ising model by Gunton et al.[5] and Binder[5]. See also the review by Gunton in this volume. In addition, we will only consider the Potts model and not lattice gas models for adatoms ordering on surfaces, for example O/W(110), which have Q>2 but which cannot be described by the Potts model. The reader is referred to the papers by Sadiq and Binder[38] and Viñals and Gunton[39] for Monte Carlo simulations of this interesting problem.

II. MODEL AND METHOD

We studied the Q-component ferromagnetic Potts model with a nonconserved order parameter,

$$H = -J \sum \delta_{S_i S_j} \tag{2}$$

where S_i is the Q state of the spin on site i ($1 \leq S_i \leq Q$) and δ_{ab} is the Kronecker δ function. The sum is over all spin pairs with a specified distance and J>0. For all of the simulations, we started the system in a random state and rapidly quenched to T≃0. The kinetics of the boundary motion were simulated via a Monte Carlo method in which a site is selected at random and re-orientated to a randomly chosen orientation. If the change in energy associated with the re-orientation, ΔE, is less than equal to zero the re-orientation is accepted. However, if $\Delta E > 0$, the re-orientation is accepted with probability $\exp(-\Delta E/k_B T)$. A generalization[1] of Bortz et al.'s "n-fold" or continuous time technique[40] to the Potts model was used to make the simulations more efficient. Without this method, the number of runs and size of the systems which we would have been able to study would have been greatly reduced. This is particularly true for large Q and late times, when most of the spins are not on the boundary. N re-orientations attempts are referred to as 1 Monte Carlo step (MCS), where N is the number of lattice sites. In all of our simulations we used periodic boundary conditions.

In 2-d, we have carried out simulations on both the triangle and square lattice of size 200^2, 380^2 and 1000^2. For the triangle lattice, we include only nearest neighbors (nn) in Eq. (2). We[1] found that on the square lattice if only nn are included, the domains become pinned for quenches to $T \approx 0$ and the growth exponent n is zero for $Q \geq 3$, in accord with the predictions of Lifshitz[6] and Safran[15]. When the range of interaction is extended beyond nn to include next nearest neighbors (nnn) with coupling J equal for nn and nnn pairs, the growth is not pinned and $\bar{R}$ increases algebraically with the same value of n as on the triangle lattice. Viñals and Gunton[41] found that if J is not exactly equal for the nn and nnn interactions, the domains become pinned for quenches to $T \approx 0$. When the final quench temperature $T > 0$, the growth on the square lattice with nn interactions becomes unpinned and we observed an effective growth exponent n which was T dependent rising toward the values observed on the triangle lattice as T approaches T_c. In 3-d, we observed similar effects for the simple cubic (sc) lattice for $Q > 3$. When the sum is only taken over nn in Eq. (2), growth becomes pinned as there is no local driving force on the vertices where four domains meet. However, the introduction of further neighbors in the sum, Eq. (2), unpins the system and growth occurs even for $T \approx 0$. We[32] have studied systems of size 60^3 and 100^3 on the sc lattice, including 6, 18, 26 and 124 neighbors in the sum in Eq. (2). This corresponds to the inclusion of nn sites, nn plus nnn sites, all sites within a cube out to (111) and all sites within a cube out to (222), respectively. We refer to these as cases k=1, 2, 3 and 4. As the number of neighbors increases, the interaction becomes more spherical, eliminating effects due to the underlying lattice. We have only investigated the case in which the coupling J is equal for all pairs within the range of interaction. When the final quench temperature $T > 0$, the growth on the sc lattice with only nn interactions does becomes unpinned, but the effective exponent n increases only slightly before T_c is reached. Since our main aim has been to search for universal features which describe the growth kinetics, we have concentrated on systems which do not become pinned due to local lattice effects when quenched to $T \approx 0$.

In this paper, we will mainly present new simulation results for the square lattice with nn and nnn interaction. In 2-d, all of the results presented are averaged over 5 runs on a 380^2 lattice. We will also present results for the Q=48 model on a 1000^2 triangle lattice averaged over 2 runs. In 3-d, we will present results on a 100^3 sc lattice for k=2 and 3. For low Q, our results are averaged over 2-3 runs, while for high-Q, we typically made only one run. This is for two reasons. The first is computer time. One run for a Q=48 100^3 simulation takes approximately 12-15 hours of CPU time on an IBM 3090/150 computer. Since the continuous time method we use in the simulations is scalar in its design, the amount of CPU time decreases only slightly on a Cray supercomputer. The second is that we see little fluctuation in our results from sample to sample. For high-Q, there is a sufficient number of grains to obtain good statistics for both the grain size and topological distributions from one run. Two runs for the case k=3, Q=48 on the 100^3 lattice gave nearly identical results for all properties studied. However, for low-Q the small linear dimension of our 3-d simulations does begin to play a role, limiting the time we can follow the growth and increasing the size of the fluctuations. For the Potts model, unlike the Ising model, the growth kinetics can easily be monitored by measuring the average grain area or volume directly by using a cluster-enumeration routine. Both of these quantities are strongly self-averaging unlike[42] $\bar{L}(t)=L^d<\phi^2>_t/<\phi^2>_T$, which is not self-averaging, where ϕ is the order parameter and $<>_t$ and $<>_T$ denote the time dependent and equilibrium averages, respectively. Measuring the average area or volume is computationly efficient and gives an accurate measure of the growth without having to average over many samples as is usually necessary for the Ising model[42]. What would be useful is to be able to use larger lattices in 3-d to follow the growth for very long times, but this is not possible at the present time.

III. RESULTS

In Fig. 1, we display the microstructures for Q=3, 6, 12, and 48 simulations on the square lattice with nn and nnn interactions. The boundary is drawn as the perpendicular bisector between misoriented sites. One can clearly see the gross morphogical changes that occur as Q increases from small Q near the Ising value (Q=2) toward the limit of high

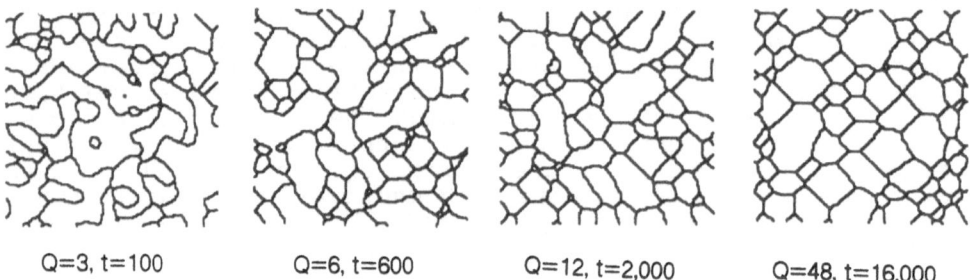

Q=3, t=100   Q=6, t=600   Q=12, t=2,000   Q=48, t=16,000

Figure 1 Domain-boundary configurations for Q=3, 6, 12 and 48 Potts model on a square lattice that were quenched from $T>>T_c$ to T=0. Solid curves represent the boundaries between the regions of different orientations. The times were chosen to yield comparable domain sizes. A 200^2 section of the 380^2 sample is shown.

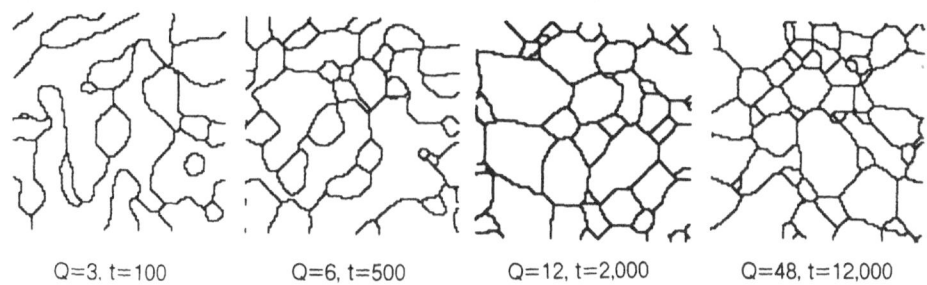

| Q=3. t=100 | Q=6, t=500 | Q=12, t=2,000 | Q=48, t=12,000 |

Figure 2 Planar (100) cross-section of the 3-d microstructure for a 100^3 sc lattice with Q=3, 6, 12 and 48 for case k=3. The times were chosen to yield comparable domain sizes.

Q. The low Q configurations consist of very irregular and asymmetric domains. This irregularity decreases as Q increases. For high Q, the domains are significantly more compact and equiaxed. This dependence of the domain boundaries on Q is very similar in 3-d, as shown in Fig. 2 where we show a (100) planar section on the sc lattice for k=3 (26 neighbors) for the same values of Q. In Fig. 3 we present results for growth on the square lattice with nn and nnn interactions for Q=6 and 30. In Figs. 4 and 5 results for the same values of Q for a (100) planar section on the sc lattice with interaction range k=3 and 2, respectively. Steps are clearly visible in the micrographs in Figs. 2, 4 and 5 due to the small linear dimension of the model (100 lattice sites)$_2$. This is less visible in Figs. 1 and 3, where we show a 200^2 section of a 380^2 simulation. Comparison of these figures indicates that in the low-Q limit, large discontinuous changes in the area of individual grains can occur when one domain meets and coalesces with another domain with the same orientation. The probability of such chance meetings decreases as Q increases, though one can see occurrences of this in the micrographs for Q=30. Since such coalescence events are strictly forbidden in the limit Q→∞, the rarity of such events for Q≳36, indicates that the infinitely degenerate system can be modeled with a large finite Q. An example of a highly degenerate system is the grain structure in polycrystalline materials, where the Q orientations can be associated with the local orientations of the crystalline lattice. It.is interesting to note that the grain morphologies for the two 3-d model are not identical. The boundaries for the k=2 case are flatter; reminiscent of the 2-d square lattice with nn interaction[1], which becomes pinned by "T" shaped vertices where three domains meet. The longer range interaction in the k=3 case which has a more isotropic grain boundary energy seems to eliminate these, producing grain boundaries which on a coarser scale meet at 120^0 even though the sc lattice grain boundaries which meet at triple lines are constrained to have included angles of either 90^0 or 180^0, due to the symmetry of the lattice.

From Figs. 3-5, it appears that the domain size distribution is nearly identical in 2 and 3 dimensions. This correspondence is made more quantitative in Fig. 6, where the grain area distribution function for the 2-d square lattice with nn and nnn interactions is compared with that obtained

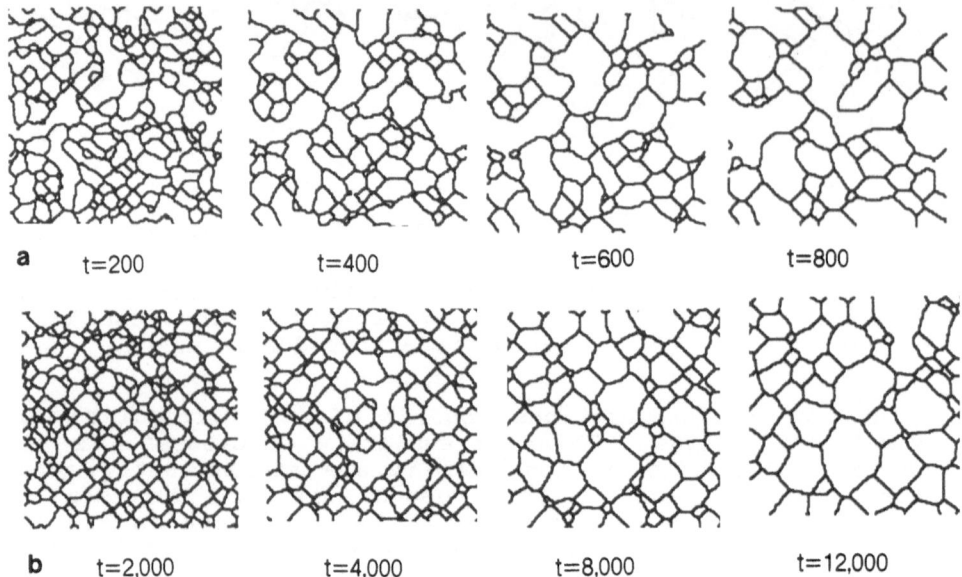

a   t=200   t=400   t=600   t=800

b   t=2,000   t=4,000   t=8,000   t=12,000

Figure 3 Temporal evolution of the domain-boundary for (a) $Q=6$ and (b) $Q=30$ Potts model quenched from $T \gg T_g$ to $T=0$ on the square lattice. A 200^2 section of the 380^2 sample is shown.

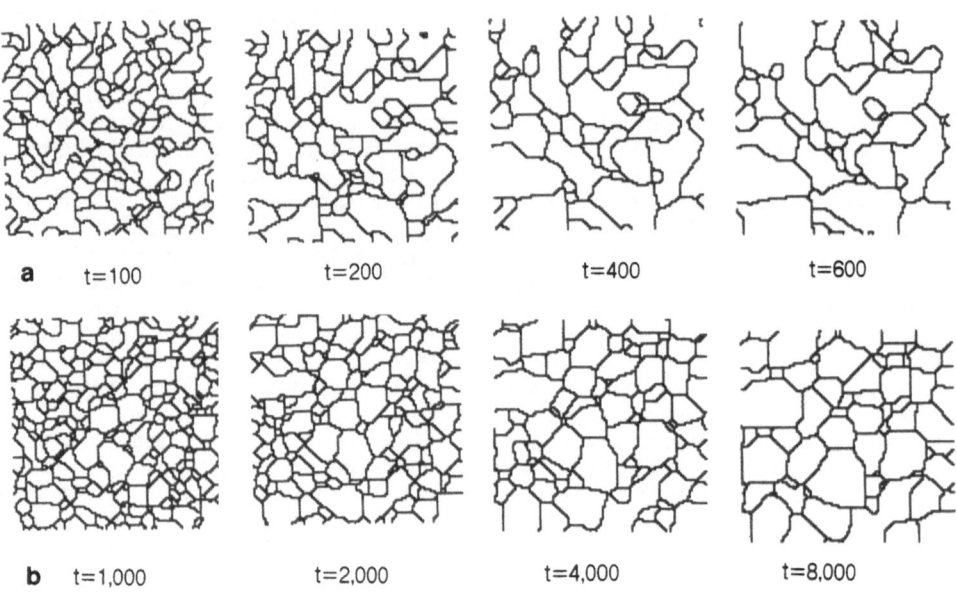

a   t=100   t=200   t=400   t=600

b   t=1,000   t=2,000   t=4,000   t=8,000

Figure 4 Temporal evolution of a (100) planar cross-section from the 3-d microstructure for (a) $Q=6$ and (b) $Q=30$ on a 100^3 sc lattice for case $k=2$.

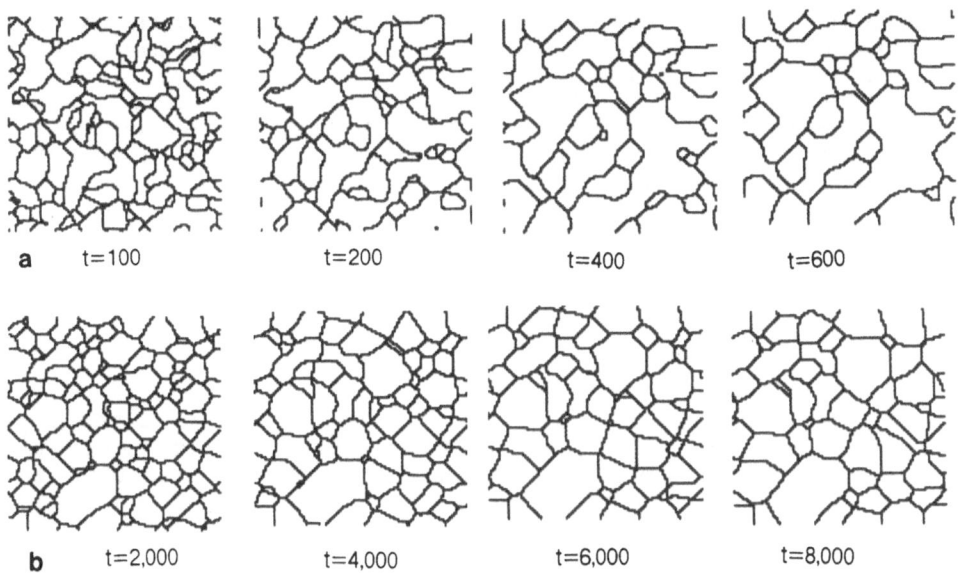

a   t=100   t=200   t=400   t=600

b   t=2,000   t=4,000   t=6,000   t=8,000

Figure 5 Temporal evolution of a (100) planar cross-section from the 3-d microstructure for (a) Q=6 and (b) Q=30 on a 100^3 sc lattice for case k=3.

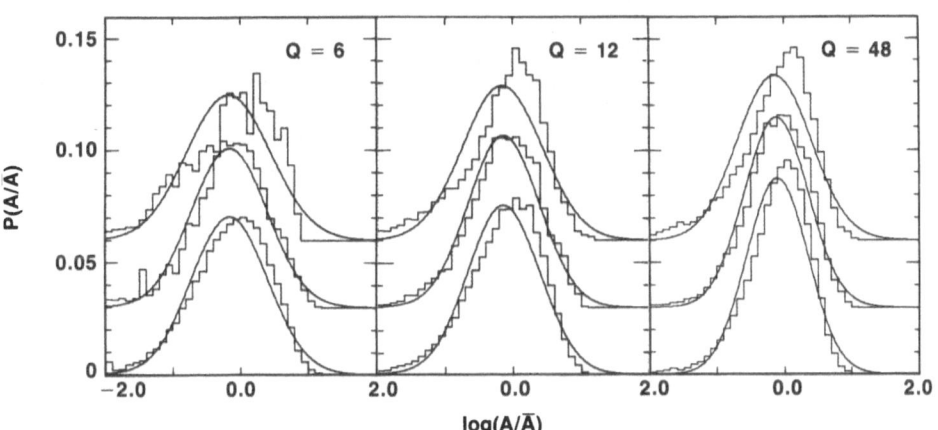

Figure 6 Time averaged domain size distribution function $f(\log_{10}x)$, where $x=A/\bar{A}$ for Q=6, 12 and 48 for the square lattice (bottom curve) and sc lattice with k=2 (middle curve) and k=3 (top curve). The upper two curves have been displaced vertically by 0.03 and 0.06, respectively, for clarity. The solid line corresponds to the log-normal function with appropriate values for the mean and standard deviation as determined from the histograms.

from the (100) planar crosssections of both the k=2 and 3 models on the sc lattice for Q=6, 12 and 48. Results for the triangular lattice with nn interactions are identical to those for the square lattice with nn and nnn interactions. For Q≤4, the 100^3 lattices are too small to obtain good statistics for the volume or cross-sectional area distributions but data for the linear intercept can be obtained. The 2-d results are nearly identical with the k=2 sc results and both differ only slightly from the k=3 sc case. In cross-section there appear to be more small grains for the k=3 sc case than for the other two cases. These distributions were found to be time invariant when normalized by their respective means. This property, which is referred to as statistical self-similarity, was always observed. The only data available to test how well our model compares with experimental data is for the polycrystalline materials. In Fig. 7, we present the cross-sectional area grain size distribution from our simulations for the k=3 model with Q=48 and data for pure Fe. Note that the grain size distribution function for the simulations and experiment agree remarkably well.

Also shown in Fig. 6 as the continuous curve is the log-normal function $f(\log_{10}x)$, where $x=A/\bar{A}$, plotted using the values for the mean μ and standard deviation σ measured for each distribution. Note that the log-normal function is symmetric on the log scale employed and has tails extending to $\pm\infty$ while the data is actually skewed. The log-normal form is a better representation of our data for small Q. The development of the log-normal distribution in domain growth has been rationalized in terms of a probabilistic mechanism in which individual domains are assumed to change area or volume in a random and uncorrelated manner. This argument is not appropriate for high-Q, where the density of vertices is high. The presence of vertices couples each domain to its neighbors, so that changes in area or volume are correlated. In this limit, we find a better fit to the data is Louat's function: $f(x)=\exp(-x)$, where $x=A/\bar{A}$. However, the volume distribution $f(\log_{10}(V/\bar{V}))$ is better fit by a log-normal function. In the low-Q limit, the density of vertices decreases and the correlation between domains becomes smaller so that the log-normal function is expected to give a better fit.

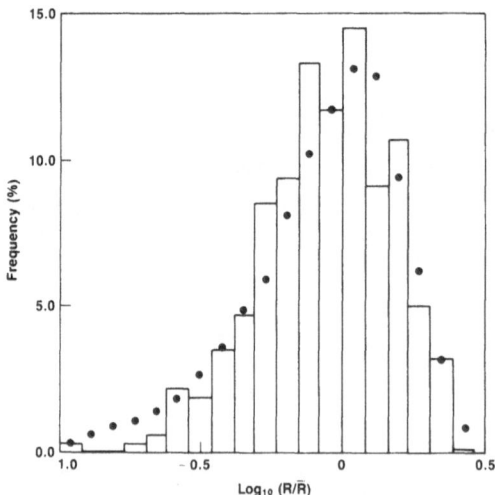

Figure 7 Grain radius distribution as determined from a cross-sectional area analysis of pure Fe (histogram) and from cross-sections of the three-dimensional k=3 lattice model (filled circles).

The kinetics of growth were evaluated by monitoring the time evolution of the mean chord length $\bar{L}$, mean cross-sectional area $\bar{A}$ and mean domain volume $\bar{V}$ (in 3-d). Data for the time dependence of the area $\bar{A}$ are shown in Fig. 8 for four values of Q on the 2-d square lattice with nn and nnn interactions. The lattice is size 380^2 and the data is presented as both a log-log and linear-linear plot. Data for the k=2 and k=3 model in 3-d on a 100^3 lattice are shown in Fig. 9. The data can be fit to the kinetic equation:

$$X(t)^m - X(0)^m = Bt \qquad\qquad (3)$$

where X is equal to either $\bar{L}$, $\bar{A}$, or $\bar{V}$. For X(t)>>X(0), Eq. (3) reduces to $\bar{R}=Bt^n$, Eq. (1), where $\bar{R}=\bar{X}^{1/d}$ and $n=1/(md)$, where d=1 for $\bar{L}$, 2 for $\bar{A}$ and 3 for $\bar{V}$. The average growth exponent, n, determined from fitting $\bar{L}$ or $\bar{A}$ in 2-d or $\bar{L}$, $\bar{A}$, and $\bar{V}$ give the same result within statistical error. Results for growth in 2-d indicate that n=0.49±0.02 for all Q. As can be seen from Fig.8b, the results seem to be quite linear on the plot of $\bar{A}$ versus t. Our new results for Q=48 on a 1000^2 triangle lattice out to 60,000 MCS, give n in agreement with this result for the square lattice. Both of these results give a value of n which is greater than our previous estimates of .41±0.02 for Q≳30. As seen from Fig. 8a, there is significant curvature in the early time regime for high-Q and this is apparently the reason we obtained a lower estimate of n for high-Q from fitting Eq.(3) to our data. If we fit our new data for t≲6000 MCS, we obtain estimates for n which are consistent with our previous results, indicating the rather slow crossover to the asymptotic time regime and the need to do long simulations on large lattices.

In 3-d, we find n=0.38±0.02 for Q≳12 for the k=3 case and n=0.30±0.02 for k=2. It is not possible to determine if this small difference between the k=2 and 3 cases is significant or if the exponent n is time independent. However, a plot of $\bar{A}$ versus t for Q=48, Fig. 10, shows significantly more curvature in 3-d than in 2-d, suggesting that either the crossover to larger values of n occurs for

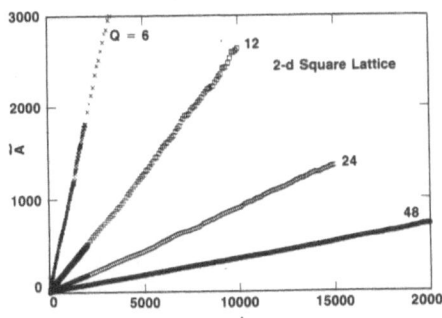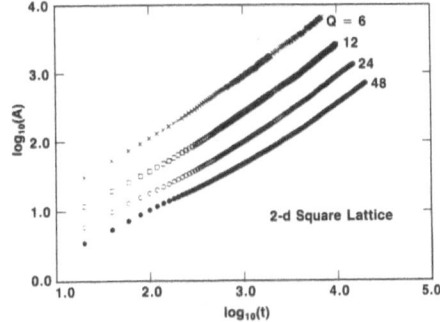

Figure 8 Plot of average area $\bar{A}$ versus t(MCS) for the Q=6, 12, 24 and 48 Potts model following a quench from T>>T$_c$ to T=0. Data is averaged over 5 runs on a 380^2 square lattice with nn and nnn interactions of equal strength. In (a) we show the data on a log-log and (b) on a linear-linear plot. The exponent n is obtained by least square fitting this data to Eq. (3).

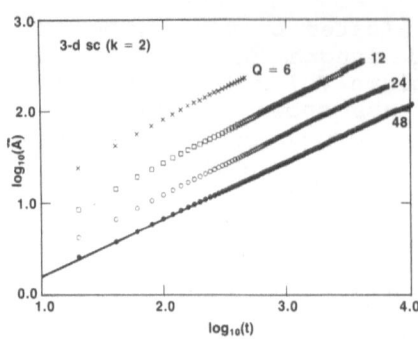 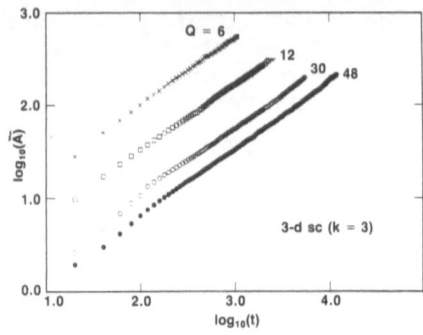

Figure 9 Plot of $\log_{10}(\bar{A})$ versus $\log_{10}(t)$ for the Potts model on a 100^3 sc lattice for four values of Q for (a) k=2 and (b) k=3.

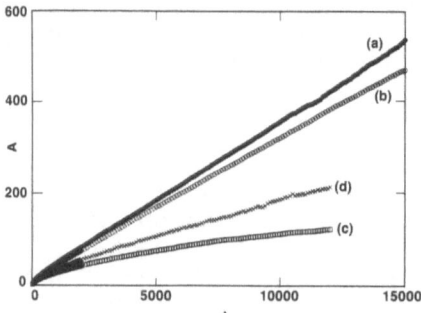

Figure 10 Average area $\bar{A}$ versus t on a linear scale for Q=48 Potts model on (a) 380^2 square lattice with nn and nnn interactions, (b) 1000^2 triangle lattice with nn interactions, (c) 100^3 sc lattice with k=2 and (d) 100^3 sc lattice with k=3.

times much larger than we have been able to study or that n is actually less than 0.5. The lower value of n for k=2 with only 18 neighbors within the range of interaction may be a result of a weak pinning at the vertices which is not strong enough to pin the growth completely as for k=1, but is strong enough to affect the kinetics. For $Q \lesssim 12$, we again find results intermediate between these values and that for the Ising model, n=0.5. From our data, it appears that the effective kinetic exponent n saturates for $Q \gtrsim 12$, while the grain size distributions do not become independent of Q until $Q \gtrsim 30$.

We[34] have studied the effect of quenched random impurities on the growth kinetics for a range of concentration from 0.5% to 20%. The model was initialized by placing Nc impurities with $S_i=0$ on the lattice, where c is the impurity concentration. The remaining sites were assigned a random value from 1 to Q. The impurties were immovable and were noninteracting. These simulations were carried out on the triangle lattice of size 200^2 for Q>4 and 400^2 for Q=3 and 4. Simulations for high-Q in 3-d are presently in progress. In all cases studied, the domain growth becomes pinned after an initial power law regime. This crossover from

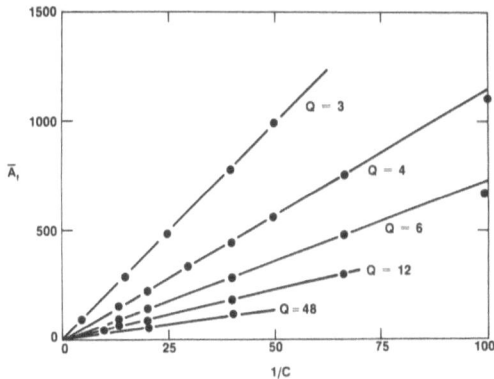

Figure 11 Averaged pinned area A_f versus inverse concentration c^{-1} for five value of Q on the triangle lattice.

a power law regime to pinning occurs at later times with increasing Q and/or decreasing impurity concentration. The averaged pinned domain area A_f varies inversely with the concentration of impurities as shown in Fig. 10. Note that the product cA_f clearly depends on the degeneracy of the model, Q. For high-Q, the domain growth stops when the domain size is comparable with the inter-impurity spacing. Since domain growth occurs when small domains are annihilated and domain annihilation occurs by the meeting of three domain vertices, domain growth stops when domain vertices can no longer meet. Motion of vertices is prevented when the three domain boundaries meeting at the vertex are pinned. Therefore, in 2-d, where the mean number of edges (or vertices) per domain is six (Euler relation), the average number of impurities needed per domain is three: six edges per domain divided by two to account for the fact that each edge is shared by two domains. This is in agreement with the value $cA_f=3$ obtained from Fig. 11 for Q=48. As Q decreases, the domains become increasingly less compact and the value of cA_f increases as the number of impurities to pin the growth increases.

IV. DISCUSSION

One interesting problem which we would like to understand in more detail is why the growth exponent n appeared to be less than 1/2 for high-Q in our earlier 2-d simulations and why it still appears to be less than 1/2 for both the simulations and experiment in 3-d. In 2-d, Mullins[43] has shown that only by assuming statistical self-similarity (which our simulations demonstrate) and local equilibrium that n must equal 1/2. He uses the result, first proven by von Neumann[44] and Mullins[45], that for both bubble growth and idealized grain growth in 2-d, the rate of change of the area A_i of an individual grain (or bubble) depends only on the number of sides,

$$A \equiv \frac{dA_i}{dt} = \frac{\pi k}{3}(N_e-6) \qquad (4)$$

where k is a constant. This equation should be valid for an arbitrarily shaped 2-d grain of N_e-sides under the assumption that u=k/R and the angle between the intersecting grain boundary is 120^0. Here u is the local velocity and R is the signed local radius of curvature lying in the plane, counted positive when it lies along the normal. Thus $\dot{A}_i > 0$ for $N_e > 6$ and is <0 for $N_e < 6$. Writing the statistical self-similarity hypothesis in the form

$$f_{N_e} (A,t) = \phi(A/\bar{A})/\bar{A} \tag{5}$$

where $f_N (A,t)$ is the probability that a grain has N_e edges and area A between A and A+dA. Mullins[43] then proves that if local equilibrium is satisified (i.e. Eq. (4) is valid for all grains at all times) then n=1/2.

Since we have already shown that Eq. (5) is valid, it is of interest to examine Eq. (4) more closely in order to understand the apparent slow cross-over to asymptotic behavior. This is particularly interesting in light of the simulations by Wejchert et al.[26] for 2-d soap bubbles, who using a procedure similiar to ours, introduced the additional constraint that Eq. (4) be satisfied at all times. They[26] found that the growth kinetics followed Eq. (1) with n=1/2 even for very early times. Without this constraint they obtained our previous result, n=0.41. To test the local equilibrium property of the growth, we[46] carried out a simulation for the Q=48 Potts model in which we followed each grain, monitoring the number of edges as a function of time. We first performed a normal domain growth simulation in 2-d for Q=48 on a 200^2 triangle lattice for 1,000 MCS until there were 960 grains remaining. In Fig. 12, we present results for the temporal evolution of the area of several individual grains with the same number of edges. Those grains monitored in Fig. 12 were highly unusual in that their number of edges did not change during the majority of the 10,000 MCS run. For convenience we have reset the clock to 0 after we started monitoring the number of edges and we have plotted a symbol only when the number of edges equal the number N_e specified in the figure. From Fig. 12, we see that the slope of $\dot{A}_i$ is approximately zero for $N_e = 6$ and that for $N_e = 4$ it is approximately of equal magnitude and opposite sign from that for $N_e = 8$. An average slope for $N_e = 5$ and 7 is harder to determine. We must point out that most of the grains change N_e too often to be plotted in this way. Shown in Fig. 13 is the temporal evolution of the area A_i for five more typical grains observed during the 10,000 MCS run. A different symbol is used to label the number of edges at a given time. Note how often N_e changes. The data in Fig. 13 are the norm, while those in Fig. 12, are the exceptions.

Thus it appears that for those grains which keep the same number of sides for a long time, Eq. (4) is satisfied. However, there are large fluctuations around the mean slope and for most grains N_e changes too often for Eq. (4) to be applicable. From these results, we may be able to understand why the the asymptotic growth regime was hard to obtain in our earlier simulations. Since local equilibrium was never established, the apparent grain growth was slower than expected and n appeared to be less than 1/2. Even though the grains were several hundred sites in size, they were continuously changing their number of edges N_e too rapidly for the system to reach local equilibrium. Every time N_e changes, the microstructure rearranges to accommodate the new growth rate, as per Eq. (4). This accommodation process is not instantaneous but requires a finite amount of time τ. If this time had scaled with the grain size, we would have expected that n would remain less than 1/2. Apparently, τ does not scale with the grain size for long times, leading to

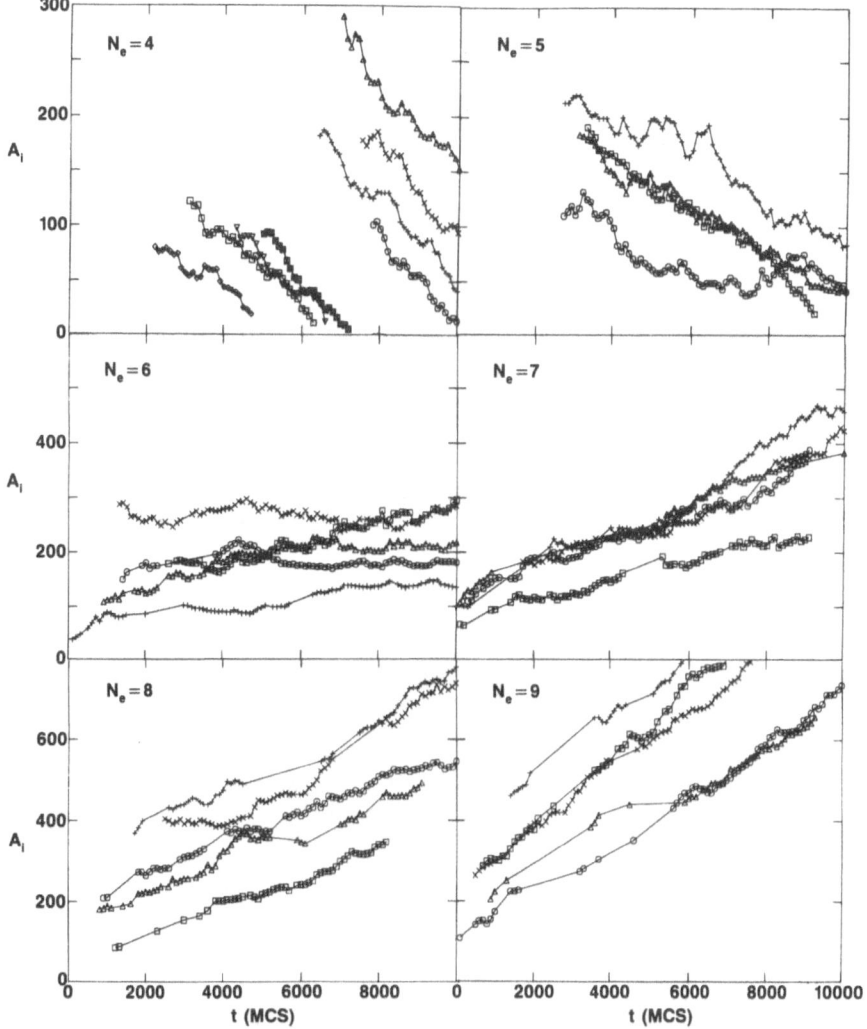

Figure 12 Time dependence of the area, A_i, of individual grains from the 2-d simulation with a specific number of edges, N_e, on the triangle lattice. Symbols are shown every 20 MCS when the grain has the number of edges specified. The simulations were started from a random starting state and run for 1,000 MCS, after which the clock was reset to 0. The data plotted are for grains which had the specified number of edges over unusually large portions of the next 10,000 MCS.

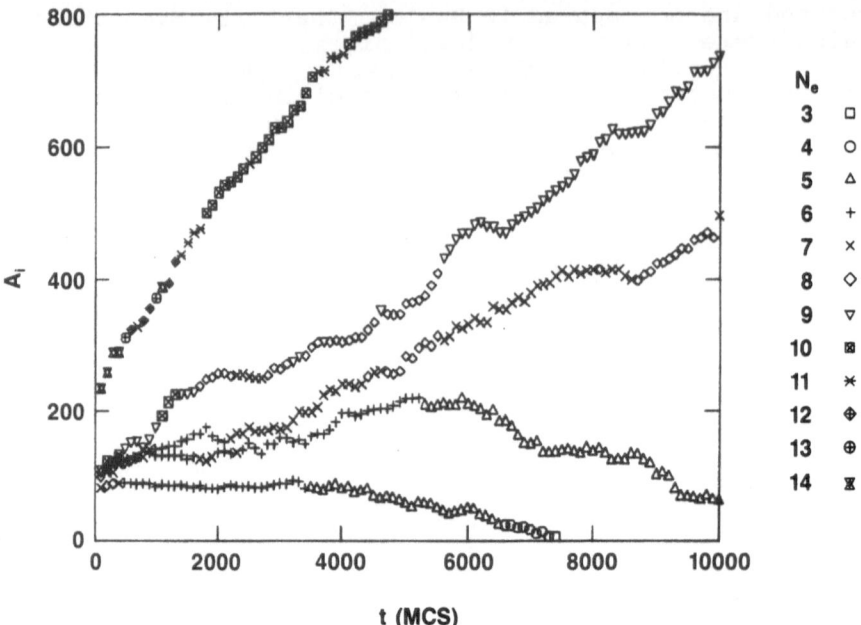

N_e	
3	□
4	○
5	△
6	+
7	×
8	◇
9	▽
10	▨
11	✳
12	⊕
13	⊕
14	⊠

t (MCS)

Figure 13 Time dependence of the area A_i of five randomly chosen grains from the 2-d simulations. The symbols represent the number of edges, N_e. As in Fig. 12, the clock was reset to 0 after an initial run of 1,000 MCS.

the observed cross-over to n≈0.50 kinetics. In the future, we hope to analyze how local equilibrium is attained for longer times in our 1000^2 simulations.

In addition to our studies of domain growth for the Potts model with and without impurities, we[47] have also studied growth when the boundary energy is anisotropic. In 2-d for Q=48, we studied a generalized Hamiltonian, Eq. (2), in which the interaction energy between different orientations depended on their relative mismatch. For quenches to T≈0 for weak anisotropy, n was observed to remain at n=0.41±0.02 and to decrease quickly to n=0.25±0.02 when the anisotropy became large, and to remain there, suggesting[47] two universality classes for the domain growth exponent n. We also observed that the domain size distribution broadened as the anisotropy increased. While the exponent n=0.41 most likely should be interpreted as n=0.5 (in light of the results discussed above) the value n=0.25 we observed for large anisotropy has been found by Mouritsen[48] for a number of other models, including one with Q=2. These models are characterized by their ability to support soft domain walls, in which there exist a spatially smooth gradient of the order parameter across the domain boundary. This is in contrast to systems with hard walls, in which the domain boundary is sharply confined in space. Hard walls result from using discrete Potts-type or lattice gas variables as in Eq. (2). Recent work by Mouritsen and Paestgaard[49] suggest that at least for one Q=2 model with soft walls that gave n≈0.25 for quenches to T=0, n increases as the final quench temperature increases. Thus the fixed point which gives n≈0.25 is not stable for this model[42,50]. At present our anisotropic model has only been studied for quenches to T=0, so we cannot say if there is a similar crossover in n as T increases. In 3-d, the only studies of which we are aware are on discrete models which have hard walls, so that the interesting question of classifying the growth kinetics into universality classes has not been addressed.

References

1. P. S. Sahni, D. J. Srolovitz, G. S. Grest, M. P. Anderson and S. A. Safran, Phys. Rev. B28, 2705 (1983).
2. M. P. Anderson, D. J. Srolovitz, G. S. Grest and P. S. Sahni, Acta Metall. 32, 783 (1984).
3. D. J. Srolovitz, M. P. Anderson, P. S. Sahni and G. S. Grest, Acta Metall. 32, 793 (1984).
4. M. P. Anderson, G. S. Grest and D. J. Srolovitz, Scripta Metall. 19, 225 (1985).
5. For a review for the Ising model, see J. D. Gunton, M. San Miguel, and P. S. Sahni, in Phase Transitions and Critical Phenomena, edited by C. Domb and J. L. Lebowitz (Academic, New York, 1983), Vol. 8, p. 267.; K. Binder, Condensed Matter Research Using Neutrons, edited by S. W. Lovesey and R. Scherm (Plenum, New York, 1985), p. 1.
6. I. M. Lifschitz, Zh. Eksp. Teor. Fiz. 42, 1354 (1962) [Sov. Phys.-JETP 15, 939 (1962)}.
7. S. M. Allen and J. W. Cahn, Acta Metall. 27, 1085 (1979).
8. K. Kawasaki, M. C. Yalabik and J. D. Gunton, Phys. Rev. A 17, 455 (1978); T. Ohta, D. Jasnow and K. Kawaski, Phys. Rev. Lett. 49, 1223 (1982).
9. S. A. Safran, P. S. Sahni and G. S. Grest, Phys. Rev. B 28, 2693 (1983).
10. M. Grant and J. D. Gunton, Phys. Rev. B 28, 5496 (1983).
11. G. F. Mazenko and O. T. Valls, Phys. Rev. B 27, 6811 (1983); G. F. Mazenko, O. T. Valls and F. C. Zhang, Phys. Rev. B 31, 1579 (1985).
12. M. K. Phani, J. L. Lebowitz, M. H. Kalos, and O. Penrose, Phys. Rev. Lett. 45, 366 (1980); P. S. Sahni et al, Phys. Rev. B 24, 410 (1981).
13. E. T. Gawlinski, M. Grant, J. D. Gunton and K. Kaski, Phys. Rev. B 31, 281 (1985).
14. J. Viñals, M. Grant, M. San Miguel, J. D. Gunton and E. T. Gawlinski, Phys. Rev. Lett. 54, 1264 (1985).
15. S. A. Safran, Phys. Rev. Lett. 46, 1581 (1981).
16. M. G. Lagally, G. C. Wang and T. M. Lu, CRC Crit. Rev. Solid State Mater Sci. 7, 233 (1978).
17. H. Homma and R. Clarke, Phys. Rev. Lett. 52, 629 (1984).
18. P. Feltham, Acta Metall. 5, 97 (1957).
19. M. Hillert, Acta Metall. 13, 227 (1965).
20. N. P. Louat, Acta Metall. 22, 721 (1974).
21. P. K. Wu, J. H. Perepezko, J. T. McKinney and M. G. Lagally, Phys. Rev. Lett. 51, 1577 (1983).
22. G.-C. Wang and T.-M. Lu, Phys. Rev. Lett. 50, 2014 (1983).
23. F. Haessner, Recrystallization of Metallic Materials, edited by F. Haessner (Riederer-Verlay, Stuttgart, 1978) p.63.
24. O. Hunderi, N. Ryum and H. Westengen, Acta Metall. 27, 161 (1979); 29, 1737 (1981).
25. V. Yu Novikov, Acta Metall. 26, 1739 (1978); 27, 1461 (1979).

26. J. Wejchert, D. Weaire and J. P. Kermode, Phil Mag. B53, 15 (1986).

27. O. G. Mouritsen, in Annealing Processes - Recovery, Recrystallization and Grain Growth, Seventh Riso Internation Symposium on Metallurgy and Materials Science, 1986 (to be published).

28. K. Kaski, S. Kumar, J. D. Gunton and P. A. Rikvold, Phys. Rev. B29, 4420 (1984).

29. C. Dasgupta and R. Pandit, Phys. Rev. B33, 4752 (1986).

30. G. Mazenko, O. Valls and F. C. Zhang, Phys. Rev. B31, 4453 (1985); B32, 5807 (1985); O. Valls and G. Mazenko (to be published).

31. D. Huse, Phys. Rev. B34, 7845 (1986).

32. M. P. Anderson, G. S. Grest and D. J. Srolovitz, Scripta Metall. 19, 225 (1985).

33. P. A. Beck, M. L. Holzwarth and P. Sperry, Trans. AIME 180, 163 (1949).

34. D. J. Srolovitz and G. S. Grest, Phys. Rev. B32, 3021 (1985).

35. D. Huse and C. L. Henley, Phys. Rev. Lett. 54, 2708 (1985).

36. G. S. Grest and D. J. Srolovitz, Phys. Rev. B32, 3014 (1985).

37. D. Chowdhury, M. Grant and J. D. Gunton, Phys. Rev. B (1987)

38. A. Sadiq and K. Binder, J. Stat. Phys. 35, 617 (1984).

39. J. Viñals and J. D. Gunton, Surf. Sci. 157, 473 (1985).

40. A. B. Bortz, M. H. Kalos and J. L. Lebowitz, J. Comput. Phys. 17, 10 (1975).

41. J. Viñals and J. D. Gunton, Phys. Rev. B33, 7795 (1986).

42. A. Milchev, K. Binder and D. W. Heermann, Z. Phys. B63, 521 (1986).

43. W. W. Mullins, J. Appl. Phys. 59, 1341 (1986).

44. J. von Neumann, in Metal Interfaces (ASM, Metals Park, Ohio, 1952), p. 108.

45. W. W. Mullins, J. Appl. Phys. 27, 900 (1956).

46. G. S. Grest, M. P. Anderson and D. J. Srolovitz, in Computer Simulation of Microstructural Evolution, edited by D. J. Srolovitz (Metallurgical Society, Warrendale, Pa, 1986), p. 21.

47. G. S. Grest, D. J. Srolovitz and M. P. Anderson, Phys. Rev. Lett. 52, 1321 (1984); Acta Metall. 33, 509 (1985).

48. O. G. Mourtisen, Phys. Rev. Lett. 56, 850 (1986); Phys. Rev. 31, 2613 (1985).

49. O. G. Mourtisen and E. Praestgaard (to be published).

50. W. van Saarloos and M. Grant (to be published).

METASTABILITY AND INSTABILITY

IN AN ORIENTATIONAL GLASS

M. Descamps and C. Caucheteux

Laboratoire de Dynamique des Cristaux Moléculaires (U.A. 801)
Université de Lille I
59655 Villeneuve d'Ascq Cedex - France

Molecular crystals offer a large variety of disordered systems. Specially interesting are those in which an average translational order is preserved allowing to
focus on the dynamic orientational disorder. This paper will concentrate on non
equilibrium properties of such a system after a quench. The situation is thus relevant to non-diffusive first-order transformation kinetics and to orientational glassy
behaviour if the quench is deep enough. A temperature time X-Ray analysis[1] will be
reviewed which allows to characterise the kinetics and molecular mechanisms involved both in the aging and the glass like transition of cyanoadamantane (CN-ADM).
Very recent results will be presented in brief at the end of the paper. These concern a new system of a mixed crystal of the CN-ADM and chloroadamantane (Cl-
ADM). Under appropriate time temperature treatment the latter shows apparent
escape and come back to metastability.

The crystal of CN-ADM is in the state of equilibrium disordered phase I above
280K. On its site the molecular dipole randomly tumbles among six orientations
located along the <100> directions of the f.c.c. lattice. By quenching the crystal
quite rapidly to a temperature lower than 180K, a supercooled disordered phase
is obtained[2] which does not transform immediately to the ordered monoclinic
phase II which is the stable phase below 280K. When reheated, a peak in the heat
capacity curve, similar to that obtained when heating glass, is observed at a Tg
in the range 170 - 185 K. The dependence on the thermal history of the sample is
reflected in the modifications of the characteristics of this peak. Dielectric
relaxation[3] shows a freezing of the molecular dipoles in the Tg zone and a Vogel
Fülcher behaviour.

KINETICS OF ORDERING IN CN-ADM

Immediately after the quench, no essential modification of the underlying cubic
order is observed by a Bragg peak inspection. However antiparallel intermolecular correlations slowly set up in the system during aging with characteristic times
of a few dozen hours. They are revealed by the increase of scattered intensity
of superstructure spots at the X-boundary points of the Brillouin zone[4]. The symmetry breaking thus corresponds to a local tetragonal order which develops along
the <100> cubic axis and is not linked in any way to the stable monoclinic order.
This ordering takes place coherently with the cubic matrix. It requires thermally
activated reorientations of the dipoles on such a large time scale that they could
not be probed otherwise. The corresponding order parameter is thus not conserved.
Two distinct regimes for the kinetics are observed both above and below 172K :

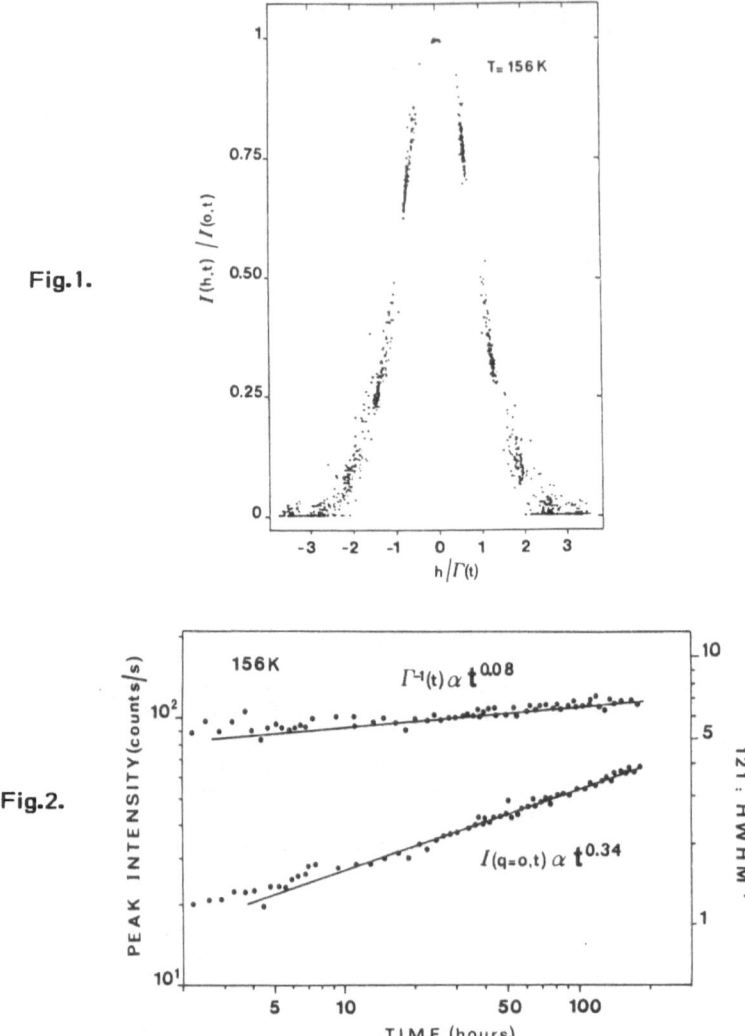

Fig.1.

Fig.2.

* T < 172K. The growth is immediate and gradually slows down. The rate of the kinetics decreases as the temperature decreases which is a probable indication of the primary influence of the molecular mobility in this temperature range. It has been found that the scattering structure function reveals a scaling property within the measured time range : $S(q,t) = L(t)^x S(q.L(t))$. We tried to scale with $L = 2\Gamma^{-1}$, ($\Gamma(t)$: HWHM of the profile) which is a measure of the size of the ordered regions which develop in the system. A normalised scattering function is plotted against $q.\Gamma^{-1}$ in **Fig.1** for 120 scans in the course of a 7 day aging period at 156 K. The result of this scaling analysis is that the universal profile is very close to a Gaussian. This suggests that we are facing clusters, rather well ordered internally, which develop in a disordered matrix. Yet they are of small size ($L \cong 40$ Å at the end of the experiments). The scaling exponent $x \cong 4.4$ ranges between what is expected in 3-dim for free growing [6] and pure coalescence [3] of independant clusters. The integrated intensity shows an increase of the transformed fraction which was found to be about 10% after the 7 days. One of the salient features of the kinetics is its weak time dependence **(Fig.2)**. The peak parameters follow a t^m expression with $m \cong 0.34$ for the peak intensity and $m' \cong 0.08$ for Γ^{-1}. This leads to believe that there are strong constraints which act very early against the expansion. Their origin can be the degeneracy[5] (P = 6) and domain wall softness[6] but we can presume that steric and elastic interactions between clusters are important effects.

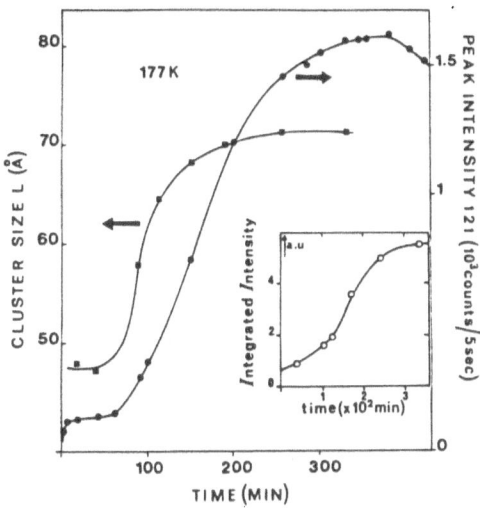

Fig.3.

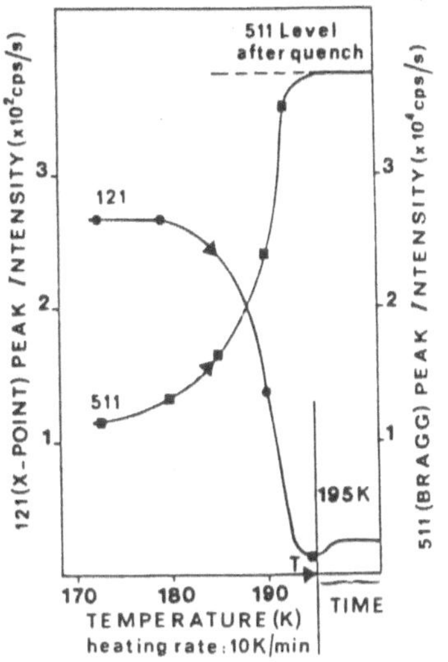

Fig.4.

* **172 K < T < 180 K.** An incubation stage (≅ 1 hour at 177 K) which increases with temperature is noticed. The whole evolution cannot be described by a universal law when scaling the structure function with Γ^{-1}. The evolution seems to stop (after 6 hours at 177 K) to give a metastable array of frozen clusters which are about 70 Å in effective diameter and at most 20% in proportion **(Fig.3).**

The results support the idea of a nucleation and growth process of a metastable phase III. The simplest explanation of the change of regime is that of a transition from possible metastability at high temperature to a regime of instability associated with the drop of the nucleation barrier[7].

TIME DEPENDENT BEHAVIOUR OF THE UNDERLYING LATTICE

The growth of the superlattice reflections is correlated with a decrease of the intensity of the f.c.c. Bragg peaks. The peaks progressively shift towards higher angles which expresses a volume relaxation of the sample. We have also noticed a Q-dependent broadening of the reflections showing up fluctuations of the intermolecular spaces which increase with distance. This reveals coherency strains which probably result from a translational coupling with the orientational ordering inside clusters.

REVERSION OF CLUSTER - GLASS TRANSITION

If, after aging, the specimen is reheated **(Fig.4)** a rapid disappearance of the clusters is observed at about 180K. This comes with a spectacular sharpening of the Bragg peaks. As a result the primary disordered state I is recovered with a high degree of homogeneity. This reversion is clearly the mechanism which accompanies the glass like heat absorption. In several experiments it was noted that the temperature of reversion is a slightly increasing function of the initial size of the clusters. This leads to assume a virtual equilibrium transition I-III at a markedly higher temperature and an interpretation can be based on the classical nucleation theory : the reversion would result from a competition between the rate of reheating and the kinetic ability of clusters to reach the critical size r* at a new temperature. Yet a discussion of cluster stability is rendered uncertain by the unavoidable monoclinic recrystallisation. This point has been recently elucidated by the study of mixed crystals $(CN-ADM)_{1-x}(Cl-ADM)_x$ which show extraordinary Time-Temperature behaviour.

385

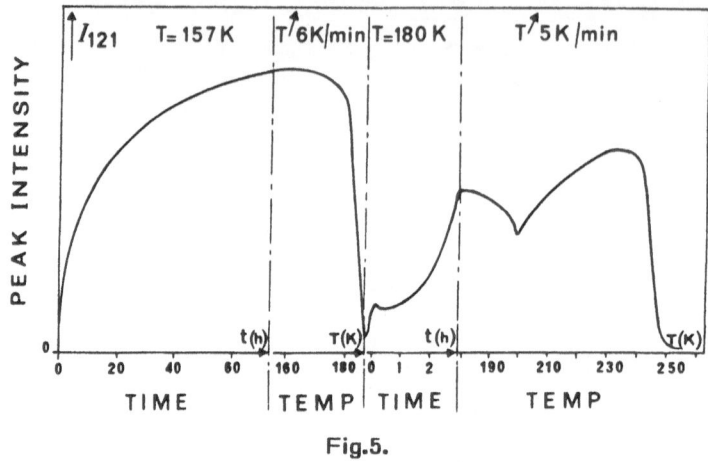

Fig.5.

LEAVING AND RETURNING TO METASTABILITY IN [CN-ADM]$_{1-x}$[Cl-ADM]$_x$

At high temperature the disordered structure of the mixed crystal is isomorphous with that of CN-ADM[8]. This phase can be quenched easily but no monoclinic recrystallisation is detected. As a result states related to the X-point superstructure can be followed in a wider range of time and temperature. An experiment in which the quenched sample is heated from time to time after aging stages at several temperatures gives illuminating information. **Fig.5** sketches the corresponding evolutions of the superstructure peak which reflect clustering. The first two steps – kinetics and reversion – are reminiscent of the pure compound behaviour. Re-aging the sample after reversion, the scattering increases once again after an incubation period and thus clusters of the same structure begin to reform. This stage was not visible with CN-ADM. A second reversion could be observed sometimes at a higher temperature [$\cong$ 196 K] and finally the I-III equilibrium transition could be observed at about 233K. We have thus drawn some kind of staircase in the T-T-T- diagram which shows a repeated leaving of and returning to the metastable state I.

This new system is extremely promising for a careful study of cluster formation and kinetics associated with a non conserved order parameter. Its structure follows such a simple model of disorder that it should deserve theoretical attention.

References

1. M. DESCAMPS, C. CAUCHETEUX [1987] : Dynamics of Molecular Crystals [J. Lascombe Ed] Elsevier. Sci. Pub. Amsterdam, 333 / J. Phys. C. Solid State [1987] in Press
2. M. FOULON, J. P. AMOUREUX, J. L. SAUVAJOL, J. LEFEBVRE, M. DESCAMPS, [1983] J. Phys. C. Solid State 16L, 285
3. J.P. AMOUREUX, G. NOYEL, M. FOULON, M. BEE, L. JORAT [1984] Molecular Phys. 52, n°1, 161.
4. M. DESCAMPS, C. CAUCHETEUX, G. ODOU, J. L. SAUVAJOL, [1984], J. Phys. Lett. 45 L 719
5. S. A. SAFRAN [1981], Phys. Rev. Lett. 46, n°24, 1581
6. O. G. MOURITSEN, [1986], Phys. Rev. Lett. 56, n°8, 850.
7. K. BINDER [1980], Systems far from equilibrium [L. Garido Ed] Springer-Verlag.
8. M. BEE, M. FOULON, J. P. AMOUREUX [1987], Dynamics of Molecular Crystals [J. Lascombe Ed] Elsevier Sci.Pub. Amsterdam 553-558.

RECENT THEORETICAL DEVELOPMENTS IN THE KINETICS OF FIRST ORDER PHASE

TRANSITIONS

J. D. Gunton

Physics Department and Center for Advanced Computational
Science
Temple University
Philadelphia, Pa. 19122

INTRODUCTION

The study of pattern formation in nature is one of the most fascinating fields in modern physics. The subject includes a large number of
disparate phenomena, such as the growth of snowflakes, Rayleigh-Benard
thermoconvections, phase separation in polymer blends and the early stages
of the inflationary universe. Although such problems have important differences, they share several common features, one being the formation of
ordered structures from initially disordered states. As well, the dynamical description of such phenomena typically involves nonlinear equations
of motion.

In these lectures I will describe some of the recent developments in
one area of pattern formation, namely the kinetics of first order phase
transitions. This involves such dynamical processes as nucleation and
growth as well as spinodal decomposition and coarsening. Since a relatively comprehensive review of this field was published[1] in 1983, and several
other reviews have been published recently[2-6] I will focus primarily on
current theoretical and experimental progress. As well, I will discuss
certain directions for investigation which appear to be particularly promising at the moment, although advances in some areas of numerical study may
well require the development of considerably more powerful supercomputers
than currently available. The topics I will summarize include real space
renormalization group methods, interface models, extensions of the Lifshitz-Slyozov theory for droplet growth and numerical solutions of certain
Langevin equations. Where possible I will compare theory and experiment.
Although I will not discuss the extensive literature involving computer

simulation of a variety of microscopic models related to pattern formation (see the article by Grest in this volume for a discussion of this topic), I will briefly describe the pressing need for better statistics in these and other numerical studies.

It will be useful for subsequent discussion to have an example of the type of morphology involved in the kinetics of phase separating systems. In Fig. 1 below, the droplets of a B-rich minority phase are shown growing in an A-rich background of a binary alloy consisting of A and B atoms respectively. Such behavior is observed when a system is quenched from an initially disordered, high temperature equilibrium state to a metastable state below its phase transition temperature. If the initial concentration is close to one of the equilibrium values characteristic of the two phase coexistence at this lower quench temperature, then the state evolves to equilibrium by the birth and subsequent growth of droplets, as shown in Fig. 1. The later stages of this growth are thought to be described by an extended Lifshitz-Slyozov[7] theory which we will describe later. On the other hand, if the initial quench concentration is close to the critical point concentration (such as 50% of each atomic species in an idealized Ising model of a binary system), then the system initially evolves by long wavelength fluctuations in the concentration field, leading to an interconnected structure such as shown in Fig. 2. The early stage of this process is termed spinodal decomposition, with the later stages shown in Fig. 2 called coarsening. Similar phenomena occur in many other systems including binary fluids, quasi-binary glasses and polymer blends. The growth mechanism (and domain growth laws) for equilibration may be system dependent. Thus systems belong to different dynamical universality classes, somewhat analogous to critical phenomena, although a complete theoretical understanding of such classes does not yet exist.

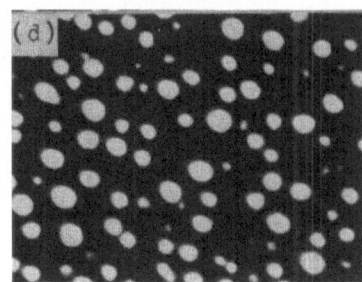

Fig. 1   Domain structure of Fe-Al alloy imaged with B_2 super-lattice reflection in 23.0 atomic % Al alloy. This particular picture shows the droplet morphology 1,000 minutes after the initial quench.

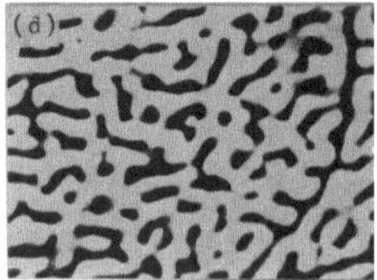

Fig. 2   Same as Fig. 1, except the concentration of Al is 24.7 atomic %.

2. NONEXISTENCE OF SELF-AVERAGING IN KINETICS OF DOMAIN GROWTH

Useful tools for studying microscopic models of ordering include Monte Carlo and molecular dynamics techniques, with the former so far playing a dominant role in the kinetics of domain growth. The simplest models of ordering in binary alloys are the so-called kinetic Ising models, whose Hamiltonian in spin representation is of the form

$$H = - \sum_{i,j} J(\vec{r}_{ij}) \, S_i S_j \, , \, S_i = \pm 1 \tag{2.1}$$

where $J(\vec{r}_{ij})$ denotes the interaction between two spins at sites i and j, separated by the distance $\vec{r}_{ij}$. In many cases the interaction is restricted to nearest-neighbors only. Kinetics is defined for this model by means of a heat bath, with a master equation which involves a transition rate w, which is chosen to satisfy detailed balance. Glauber dynamics involves a transition rate in which a given spin changes from +1 (-1) to -1 (+1) and is used to describe domain growth in systems in which the order parameter is not conserved. Binary alloys which undergo order-disorder transitions have been shown to satisfy the same domain growth law ($\bar{R}(t) \sim t^{1/2}$, where $\bar{R}$ is the average domain size and t is the time following the quench to an unstable state) as the kinetic Ising model with Glauber dynamics. (We often say both systems belong to the same dynamic universality class if their domain growth law is the same.) Another kinetic model which is used to simulate diffusion of atoms in phase separating binary alloys (such as shown in Figures 1 and 2) is known as Kawasaki spin-exchange dynamics, in which the transition rate w now describes a process in which nearest neighbor spins (atoms) exchange states (e.g. + - goes to - + or AB→BA.) In this case the order parameter for a ferromagnetic Ising model is conserved.

Monte Carlo simulations of domain growth for the Glauber and Kawasaki models for both two and three dimensional lattices were very useful in providing significant qualitative insight concerning the kinetics of first order phase transitions during the past decade or so. However, limitations of computing resources and inadequate understanding of the need for very good statistics in such studies resulted in the publication of results for many models in which quantitatively reliable data were probably absent. As this field gradually developed it became clear that it was necessary to make the number n of statistically independent observations of quantities such as the structure factor (scattering intensity) and domain size, $\bar{R}$, as large as possible, and in particular much larger than in previous studies. As well, it became clear that subtle finite size effects could produce incorrect information concerning the later stages of domain growth.

This is not a well understood subject as yet, but some progress has been made recently by Milchev et al.[8] They call a quantity A strongly self-averaging if its error

$$\Delta(n,L) \equiv \sqrt{(<A^2>_L - <A>^2_L)/n} \; , \; n >> 1 \qquad (2.2)$$

satisfies the scaling relation

$$\Delta(n,L) = \Delta(b^{-d} n, bL), \; b > 1 \qquad (2.3)$$

Here $<...>_L$ denotes an average over an appropriate probability distribution function and L is the linear dimension of the system. Eq. (2.3) (with $n = b^d$ where d is the dimensionality) implies that

$$<A^2>_L - <A>^2_L \propto L^{-d} \; . \qquad (2.4)$$

If instead of (2.4) one has

$$<A^2>_L - <A>^2_L \propto L^{-P} \; , \; 0 < P < d \qquad (2.5)$$

then one says that A is self-averaging (but not strongly self-averaging). Lack of self-averaging corresponds to the case in which P = 0. This case is unusual, since then the error Δ of a quantity A does not tend to zero as $L \to \infty$. Extensive quantities such as the magnetization per spin or energy per spin satisfy (2.3) and (2.4) if one is considering equilibrium states not too close to a critical point.

The important observation for us is that Milchev et al.[8] argue that the average domain size $\bar{R}$ (t) for the Glauber model, say, defined as

$$(\bar{R}(t))^d = L^d < \psi^2 >_t / <\psi^2>_T \qquad (2.6)$$

is _not_ self-averaging. (In (2.6) ψ is the order parameter and $< >_t$ and $< >_T$ denote the time dependent and equilibrium averages, respectively.) That is, the error in $\bar{R}$ does _not_ decrease when L increases, but only decreases if one increases the size n of the statistical sample. They further estimate that sample sizes n of the order of 10^4 are necessary if one wants to achieve a relative accuracy of less than one percent. Since most Monte Carlo studies have not usually involved such large values of n, Milchev et al. conclude (in my opinion, correctly) that some of the published estimates of the exponent x describing domain growth ($\bar{R}$ (t)~ t^x) for different models should be considered as only preliminary estimates. Although further theoretical analysis of the issues involving the L and n dependence of the error $\Delta(n,L)$ in nonequilibrium averages is necessary, it seems clear that in general one needs to have large n and to take care to avoid subtle finite size effects before Monte Carlo and molecular dynamics studies of domain growth can yield quantitative results. It should be noted that if L is too small, one can not run for sufficiently long times, since if $R(t) \sim L$, one almost always encounters finite size effects. This means one cannot directly study the asymptotic growth law in such cases.

3. DYNAMICAL SCALING

One quantity which the experimentalists study is the nonequilibrium scattering intensity, which in the absence of multiple scattering is proportional to the so-called structure function $S(\vec{q},t)$. This structure function is the Fourier transform of the nonequilibrium average of the order parameter - order parameter correlation function. A typical behavior of the scattering intensity as a function of the scattering wavenumber $q = |\vec{q}|$ and time t (measured with respect to the initial quench) is shown in Fig. 3 below for the phase separating binary alloy Al Zn.[9] As time increases, a peak develops in the scattering intensity which increases with increasing time. Correspondingly the peak position, $q_m(t)$, decreases with increasing time. One of the theoretical challenges is to calculate $S(\vec{q},t)$.

This function has been found to satisfy dynamical scaling, i.e.

$$S(\vec{q},t) = L(t)^d F(qL(t)), \quad t > t_0 \tag{3.1}$$

where $L(t)$ is a characteristic length scale of the system, such as the average domain size, $\bar{R}(t)$, or $q_m^{-1}(t)$. This scaling has been observed in a variety of systems[1], in both $d = 2$ and $d = 3$ dimensions, for times greater than a transient time, t_0, which is system dependent. The scaling function $F(x)$ for a given system can depend on the quench temperature and concentration, as well as the direction of $\vec{q}$.

Since this scaling is reminiscent of critical phenomena, it has motivated theorists to seek a renormalization group theory of domain growth. Here the problem is more difficult than in critical phenomena, since for one reason the scaling length $L(t)$ is time dependent, e.g., if one assumes a power law behavior,

$$L(t) \simeq At^x. \tag{3.2}$$

One of the goals of the theory is to determine this time dependence, which in some cases could be logarithmic. A more general theoretical description would be to determine the domain growth universality classes, which depend on such factors as (1) whether the order parameter is conserved or nonconserved, (2) the number of components of the order parameter and (3) the number of phases which can coexist at low temperatures. Progress in this direction has been relatively slow, although it is clear that one can find a wealth of different dynamical phenomena involving topological de-

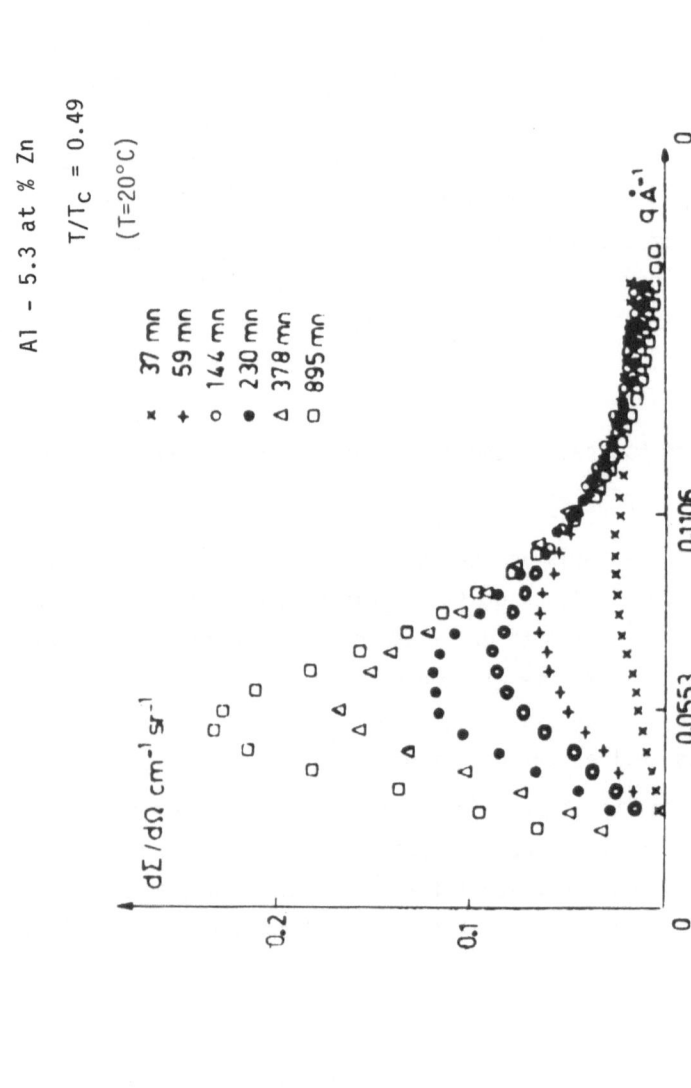

Fig. 3 Scattering intensity as a function of scattering wave number, q, for different times (in minutes), for Al Zn.

fects such as interfaces, vertices and vortices by exploring different physical systems and models.

4. NONLINEAR LANGEVIN EQUATIONS

One starting point for a semi-macroscopic, phenomenological continuum description of the kinetics of ordering is provided by nonlinear Langevin equations. These have been used with great success in the study of critical dynamics[10,11] where one can use an ε-expansion (e.g. ε = 4-d for Ising-like systems, where d is the dimensionality). The simplest examples of such equations are known as models A and B, with

$$\frac{\partial\psi(\vec{r},t)}{\partial t} = -\Gamma(\nabla^a) \frac{\delta H}{\delta\psi} + \zeta \tag{4.1}$$

where $\psi(r,t)$ is the local order parameter (the local concentration field) at spatial position $\vec{r}$ and time t, Γ is a kinetic coefficient and H is the Ginzburg-Landau Hamiltonian

$$H = \int d\vec{r} \{|\nabla\psi|^2 - \tau\psi^2 + g\,\psi^4 + ..\} \tag{4.2}$$

The noise term ζ is assumed to be given by a Gaussian distribution function[13]. These are the continuum analogues of the kinetic Ising model with Glauber or Kawasaki dynamics discussed in section 2, for the cases a = 0 (Model A) or a = 2 (Model B) respectively. The parameter $\tau \propto T_c - T > 0$ for quenches below T_c, so that one is dealing with the dynamics of a double well potential. One can derive the Fokker-Planck equations analogous to these Langevin equations from the master equations for the kinetic Ising models by standard techniques, given certain approximations[12,13].

However, unlike critical dynamics, one has not found a smallness parameter analogous to ε to use to solve the nonlinear Langevin equations in the kinetics of first order phase transitions. Their greatest success to date has been in the formulation of nucleation theories[2,14] and in discussing the early stages of spinodal decomposition[15-18]. The most promising method to solve the nonlinear equations given by (3.1) and (3.2) would seem to be numerical integration.

The first attempt to do so was made by Petschek and Metiu[19] for a two dimensional version of Model B. Their numerical integration of the Langevin equation yielded a qualitatively correct description of the phase separation process, but computer limitations prevented them from developing a quantitative solution. More recently, Model A has been investigated by two different groups. Valls and Mazenko[20] performed a direct numerical integration and verified that the domain growth law was given by $\bar{R}(t) \sim t^{1/2}$ in agreement with earlier theories[21,22]. In addition, they calculated the structure function and obtained a scaling function which is in reasonable agreement with analytical theories[21,23]. Milchev et al.[8] have performed a Monte Carlo simulation of a square lattice version of Model A and have verified the $t^{1/2}$ growth law. They also noted that since this model has "soft walls", the claim by Mouritsen[24] that $\bar{R}(t) \sim t^{1/4}$ for models with soft walls is likely to be incorrect. This conclusion by Milchev et al. that the "softness" of walls is an irrelevant parameter with regard to the asymptotic domain growth law is in agreement with earlier results by Kaski et al.[25] for the herringbone model.

It should be noted, however, that although the numerical results for Model A are encouraging, Model B presents a much greater theoretical challenge. Thus it is natural to extend the earlier study[19] of Model B. This has been done to some extent in a recent Monte Carlo simulation of a lattice version of the nonlinear Langevin equation[26]. Their results for a two dimensional system at critical concentration were consistent with a $t^{1/3}$ growth law. As well, they found that the structure function exhibited an anisotropy consistent with the square lattice geometry.

Another study which is underway is an extension of the numerical integration work of Petschek and Metiu [27]. The preliminary results of the solution of the Langevin equation for Model B are consistent with a $t^{1/3}$ growth law, although better statistics and longer times are needed before a more definitive statement can be made. A typical pattern found from this numerical solution is shown in Fig. 4 below, for a low temperature quench at a critical concentration. As can be seen, the morphology is similar to that of Fig. 2. It seems clear from such numerical studies that as more powerful computers become available, such studies should provide us with a quantitative understanding of the kinetics of such systems. This will hopefully provide a guide for further theoretical and experimental activity.

Finally we should note that there is a real space renormalization group study, discussed later, which predicts that the asymptotic growth law for Model B is ln t, rather than the $t^{1/3}$ mentioned above. Thus the question of the asymptotic behavior for this model is still not settled. However, extensions of the Lifshitz-Slyozov theory (which leads to a $t^{1/3}$ asymptotic growth law in the droplet regime) predict in detail the shape of the droplet distribution function f(R,t). As we discuss in section 5, these theoretical predictions are in reasonable agreement with the experimental data and do not support the ln t behavior.

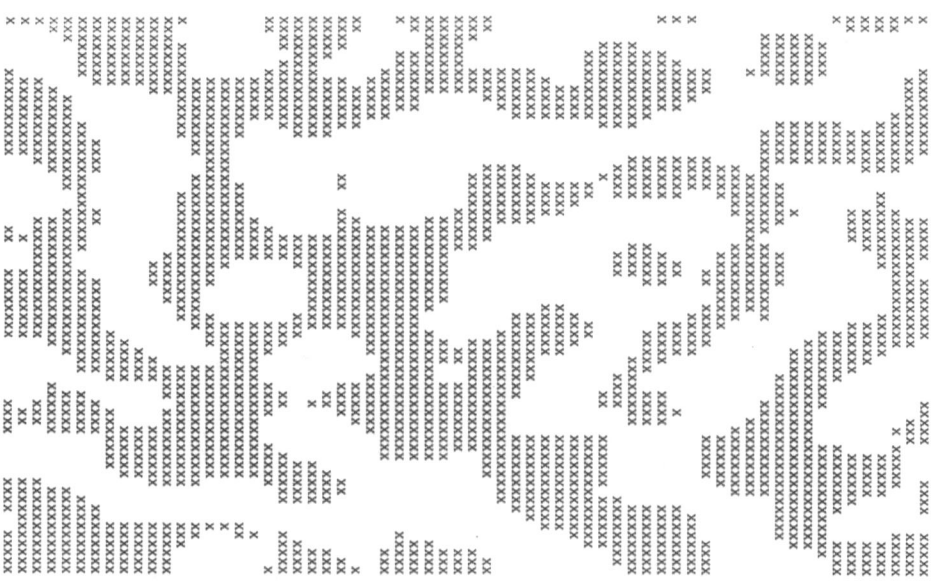

Fig. 4. A typical configuration obtained from a numerical solution of Model B in two dimensions. The regions denoted by x and empty sites correspond to domains of the two possible ordered phases. Periodic boundary conditions are used.

5. EXTENSIONS OF LIFSHITZ-SLYOZOV THEORY

One of the most interesting areas of current theoretical and experimental research involves the late stages of coarsening (Ostwald ripening) of the minority droplet species (Fig. 1). The classic Lifshitz-Slyozov (LS) theory[7] is a mean field kinetic theory for the late stages of this phase separation, in the limit of vanishingly small droplet volume fraction Q. The theory predicted a scaling form for the droplet size distribution function and that the average droplet radius $\bar{R}(t)$ grows in time like $t^{1/3}$, while the droplet number density n decays like t^{-1}. The growth mechanism involves evaporation/condensation, with larger droplets growing at the expense of smaller droplets by diffusion.

In recent years many attempts have been made to go beyond the mean field theory to describe the experimentally relevant case of finite volume fraction.[28-40] This requires a treatment of the effects of interdroplet diffusional interactions. These recent theories start with a monopole approximation to the multidroplet diffusion problem[28], in which the diffusion field in the background matrix is modelled by a collection of point sources or sinks of solute, as discussed below. The droplets are assumed to be immobile and spherical, with the point sources (sinks) located at the centers of the droplets. A recent review by Voorhees[6] covers most of these theoretical developments. We will thus focus on the most recent work by Kawasaki[34-38] and collaborators, which represents the most extensive and, we believe, systematic investigation of this problem. As well, they have recently compared their theoretical predictions with experimental studies of three alloys: a) Ni - 6.5 wt % Si, studied by Rastogi and Ardell[41]; b) Ni - 6.05 wt % Al, studied by Hirata and Kirkwood[42]; c) Al - 1.87% Li, studied by Eguchi et al.[43]. We will discuss this comparison below.

Before discussing their work, we discuss two points. The first has to do with the current state of the extensions of the LS theory. The theories differ in detail, but agree that the interdroplet diffusional interactions do not change the time dependent power laws for $\bar{R}(t)$ and the number density of droplets predicted by LS. The theories also agree that these interactions change the _amplitudes_ of these power laws and alter the _shape_ of the droplet size distribution function. The disagreement between these different theories concerns the detailed nature of these changes. As noted above, Voorhees[6] has given a relatively complete discussion of these differences. The second point is that, as mentioned in the preceding section, there is a real space renormalization group study of the two dimensional kinetic Ising model with Kawasaki dynamics, which is a simple model of the phase separating binary alloy, which claims that the asymptotic growth law is not $\bar{R}(t) \sim t^{1/3}$, but rather $\bar{R}(t) \sim \ln t$.[44] The authors argue that the basic assumptions of the Lifshitz-Slyozov theory are incorrect for binary alloys, such as described by the kinetic Ising model. We will return to this issue in a later section.

We now turn to a summary of the recent work of Kawasaki and collaborators. The starting point of all the theories, including the Tokuyama-Kawasaki (TK) theory[34,35], is that there are $N(t)$ fixed, spherical droplets, with radii $R_i(t)$, $i = 1, 2, \ldots N$, whose centers of mass are located at $\vec{X}_1 , \ldots , \vec{X}_N$ respectively. The basic growth law is described by the multidroplet diffusion equations

$$\frac{d}{dt} \frac{4\pi}{3} R_i^3(t) = 4\pi D B_i(t) \qquad (5.1)$$

$$B_i(t) = [R_i(t)/R_c(t) - 1] - R_i(t) \sum_{\substack{j \neq i}}^{N(t)} B_j(t)/ |\vec{X}_i - \vec{X}_j| \qquad (5.2)$$

In these equations D is the diffusion coefficient of the solute and the critical droplet size $R_c(t) = \alpha/\Delta(t)$. The supersaturation of the solution is $\Delta(t)$ and α is the capillarity length. In addition to these equations there is a conservation law for the solute atoms

$$Q = \sum_{i=1}^{N(t)} \frac{4\pi}{3} R_i^3 (t) + \Delta(t) , \qquad (5.3)$$

where Q is the total initial supersaturation and the first term on the right hand side is the volume fraction of droplets, $q(t)$. In the very late stage scaling regime we can neglect $\Delta(t)$ and (5.3) becomes a conservation law for the volume fraction of the droplets. This can be written as

$$\sum_{i=1}^{N(t)} B_i = 0. \qquad (5.4)$$

Finally, there is a kinetic equation for the droplet size distribution function, $f(R,t)$ which in the TK theory is given by (to order $\sqrt{q(t)}$)

$$\frac{\partial f(R,t)}{\partial t} = \frac{\partial}{\partial R} \frac{\alpha D}{R^2} [\lambda(\rho) + \sqrt{3q(t)} \{V(\rho) + C(R,t)\}]f(R,t) \qquad (5.5)$$

Here $\rho = R/\bar{R}(t)$, and $\lambda(\rho)$ and $V(\rho)$ are drift terms. The term $C(R,t)$ describes the effect of so-called "soft" or distant "collisions" among the droplets, whose origin is the spatial correlations between the droplets generated by the long-range, Coulomb-like interactions in (5.2). Before discussing the numerical solutions of the TK equations, we make a few general remarks. First, the Lifshitz-Slyozov theory is obtained in the zero volume fraction limit ($q(t) \rightarrow 0, Q \rightarrow 0$) of equation (5.1), (5.2), (5.4) and (5.5). That the LS theory is a mean field theory corresponds to the fact that in this limit the second term on the right-hand side of (5.2) disappears. Thus in this limit the growth of a given particle as described by (5.1) depends only on the background supersaturation. However, for the case of interest, $q(t) \neq 0$, the growth of a given droplet is coupled to the other droplets through a long-range interaction, as described by (5.1) and (5.2). Since it is intuitively obvious that the growth of a droplet depends on its local environment of droplets, these interactions play an important role. The second remark concerns the differences between the TK theory and other theories[6]. One important point is the TK theory predicts that there are two distinct coarsening regimes. The first is an intermediate regime in which the droplet volume fraction (and supersaturation) is still changing. In this case the droplet distribution function f(R,t) does not satisfy a scaling law. The second is the late stage scaling regime in which f(R,t) satisfies a scaling form (analogous to the LS scaling). The power law behaviors are the same in both regimes, i.e., $\bar{R}(t) \sim t^{1/3}$ and $n(t) \sim t^{-1}$. In addition, their theory (as well as other theories) involves a second characteristic length, the screening length,

$$1(t) = \frac{1}{\sqrt{4\pi n(t)\ \bar{R}(t)}} \tag{5.6}$$

which behaves like $1(t) \sim t^{1/3}$ in both regimes. A second major difference between the TK theory and other theories is that in addition to the drift terms in (5.5), they include the soft-collision process (described by c(R,t)). This process has not been included in other theories (except in a recent paper by Marder[40]) but has been shown by Kawasaki and collaborators to play an important role in the droplet growth.

Since the technical details of the theories are too complicated to present here, we simply note that the theories differ in their approximations for the statistically averaged source and sink strengths, B_i (for a specified Q) defined in (5.2). We give below three of the approximations in the literature, to give some sense of the differences.

a) Lifshitz-Slyozov :

$$B(\rho) = \rho - 1 \tag{5.7}$$

b) Marqusee-Ross :

$$B(\rho) = [\xi(Q)\rho - 1][1 + \sqrt{3Q/m_3(Q)}\ \rho] \tag{5.8}$$

c) Tokuyama-Kawasaki:

$$B(\rho) = \rho - 1 - \sqrt{3Q/m_3(Q)}\ \rho[m_2(Q) - 1] \tag{5.9}$$

In the above $\xi \equiv \bar{R}/R_c$ and m_2 and m_3 are the second and third moments of the scaled distribution function. Other forms for B(ρ) are given in the review by Voorhees. It should be noted that all of these theories reduce to the LS form as $Q \to 0$.

In Figures 5 and 6 we present some predictions for the late stage scaling behavior of the distribution function f(R,t), for Q = 0.1 and Q = 0.35 respectively. The dots show the result of a computer simulation of equations (5.1), (5.2) and (5.4) by Voorhees and Glicksman[32]. The curves marked by D and D + S are the predictions of the TK theory for the drift terms and the drift plus soft collision terms respectively. The curve marked LSW is the Lifshitz-Slyozov prediction (sometimes called the Lifshitz-Slyozov-Wagner theory).

398

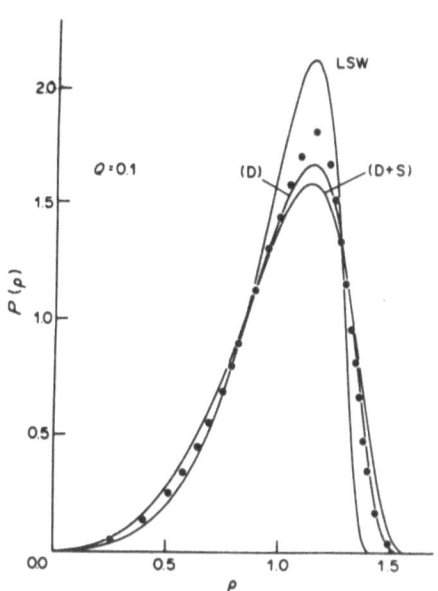

Fig. 5   Plots of the distribution
functions p(ρ) at Q = 0.1.

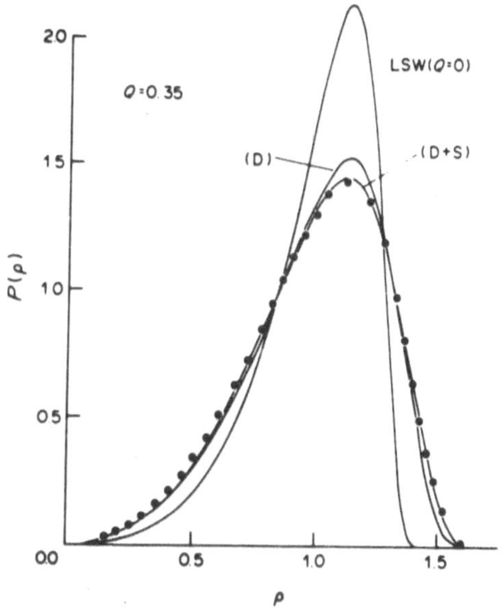

Fig. 6   Plots of the distribution
functions p(ρ) at Q = 0.35.

The scaled distribution function is

$$p(\rho) = \frac{\bar{R}(t)}{n(t)} \ f(R,t) \tag{5.10}$$

It is clear from both figures that the effect of the drift and soft collision processes is to both flatten and broaden the distribution function
as compared with the LS prediction. This is the same qualitative tendency
observed in the experiments. It should also be noted that the TK theory
is in agreement with the computer simulation study of Voorhees and Glicksman at Q = 0.35, but differs from their results at Q = 0.1. A possible
source of this discrepancy is given by Enomoto et al. [36] It should also
be noted the TK theory is in quite good agreement with the results of a
recent computer simulation study of the coarsening equations at Q = 0.1,
as well as with a new simulation of the equations by Enomoto et al. [36].
We will discuss in the next paragraph the comparison between theory and
experiment. There are in fact many experimental studies of f(R,t) and
$\bar{R}(t)$, but one needs a systematic study of these quantities as a function
of volume fraction to really test existing theories. In addition experimental studies are complicated by effects of lattice anisotropy, internal
strain, etc. Before turning to comparison with experiment, however, we
show for completeness in Fig. 7 the behavior of the averaged source and
sink strengths for the results shown in Fig. 5 at Q = 0.1.

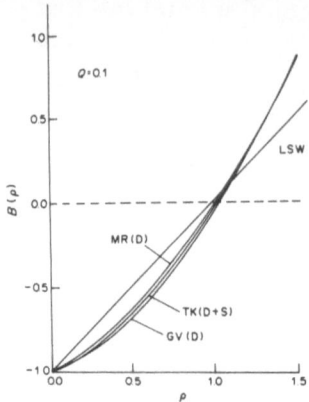

Fig. 7   The averaged source/sink strengths $B(\rho)$ at $Q = 0.1$ given by
Marqusee-Ross (MR), Glicksman-Voorhees (GV) and TK. The
broken line indicates $B(\rho) = 0$.

Finally, we turn to a comparison of the TK theory with experiment.
As noted earlier, they have not only obtained the very late stage scaling
behavior, but as well the "transient" behavior from the intermediate regime into the scaling regime. In Fig. 8 we show the time evolution of the
normalized distribution function for the Al - Li alloy[43], at $Q = 0.1$ and a
quench temperature T = 448 K.   The experimental results are indicated by
histograms. For times greater than 720 ks, both the experimental and TK
results are in the scaling regime. For earlier times the time dependence
of the normalized droplet distribution function is apparent. Similar results
were found for a Ni - Al alloy [42] and a Ni - Si alloy.[41] The agreement in
general seems quite reasonable.

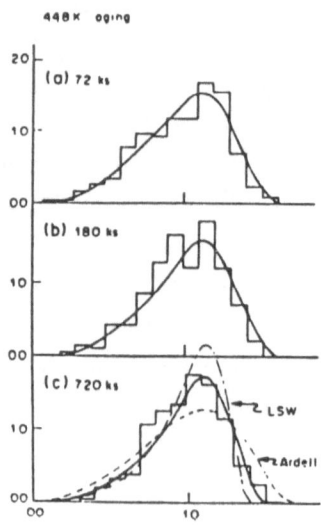

Fig. 8   The time evolution of the normalized droplet size distribution
function for an Al - Li alloy with $Q = 0.1$ and $T_a = 448$ K at
various ageing times: (a) 72 ks. (b) 180 ks and (c) 720 ks.   The
curve labelled Ardell is from the theory of reference 29.

6. DYNAMICS OF INTERFACES FOR MODEL A

As mentioned earlier, focusing on the dynamics of interface motion (such as discussed in the previous section), is a promising direction of theoretical study. An example of such interface dynamics which is much simpler to study theoretically than phase separating binary alloys or binary fluids is the continuum analogue of the kinetic Ising model with Glauber dynamics, i.e., Model A (or the time dependent Ginzburg-Landau model). It is a nonlinear Langevin equation for a nonconserved scalar order parameter. One can derive from the Langevin equation an equation of motion for the interfaces in this case which is just

$$v(a) = L K(a) + \zeta \tag{6.1}$$

where $K(a)$ is the mean curvature of the interface at a point a on the interface. The velocity of the domain wall at this point in the direction normal to the wall is $v(a)$. The noise term ζ will be neglected in our discussion. To begin with, Eq. (6.1) expresses the fact that domain growth for Model A is curvature driven. Furthermore, simple dimensional analysis (taking into account the dimensions of the kinetic coefficient L) leads to the conclusion that the average domain size $\bar{R}(t)$ follows a $t^{1/2}$ growth law. This has been confirmed in a variety of theoretical and Monte Carlo studies[1], as mentioned earlier.

As well, a theory for the nonequilibrium structure function $S(q,t)$ has been developed[23] starting from (6.1). This approximate theory is in reasonable agreement with Monte Carlo studies of the two dimensional Glauber model. In particular, it predicts a scaling form (3.1) which has been seen in studies of this model, as well as in many other physical systems such as binary alloys and binary fluids.[1] The actual form for the function $F(x)$ coincides with that predicted by an earlier theory for Model A which involved a singular perturbation theory.[21] A renormalization group theory for the two dimensional Glauber model has been developed which also predicts the scaling behavior.[45-47]

It is worth noting that the theory developed by Ohta et al.[23] for this interface model has been extended by Grant and Gunton[48] to predict the early time behavior of $S(q,t)$ and $\bar{R}(t)$ for the Ising model in a random magnetic field. This model is interesting for several reasons, one being that the random field competes with the curvature term in (6.1). Also, the random field destroys the scaling form (3.1). This theory is in qualitative agreement with a Monte Carlo study of the early time behavior of this model.[49]

7. INTERFACE EQUATIONS FOR BINARY FLUIDS AND BINARY ALLOYS

A much more difficult class of problems involving interface motion includes binary alloys and binary fluids. (We discussed this problem for the case of late stage droplet growth in section 5). In both cases the order parameter is conserved, which makes the phase separation process much richer than the model discussed in the preceding section. As for Model A, one can derive equations of motion for the interface for binary alloys and binary fluids from their corresponding nonlinear Langevin equations. One obtains the following equation[50-52] for the binary alloy (neglecting the noise term for simplicity):

$$\frac{(\Delta\psi)^2}{\sigma} \int G(\vec{r}(a)-\vec{r}(a'))v(a')da' = L[K(a) - \frac{(\Delta\psi)}{\sigma}h] \qquad (7.1)$$

In this expression $\Delta\psi$ is twice the equilibrium value of the order parameter, σ is the surface tension and $h(t)$ is a measure of the supersaturation (chosen to guarantee the conservation law for the order parameter). The kernel $G(\vec{r})$ is the Green's function of the diffusion equation which satisfies

$$\nabla^2 G(\vec{r}) = - \delta(\vec{r}) \qquad (7.2)$$

where $\vec{r}(a)$ denotes an interface configuration. One can immediately see upon comparing Eq. (6.1) with (7.1) and (7.2) how much more complicated the phase separation of a binary alloy is. The conservation law reflects itself in the fact that what happens at one point on an interface in Fig. 2, say, affects what happens at another point on the interface. In mathematical terms, a kernel $G(\vec{r}(a) - \vec{r}(a'))$ appears in the equation of motion for the velocity $v(a)$. As a consequence of this complication, little progress has been made so far in solving this interface equation for the binary alloy.

An equation analogous to (7.1) can be derived for the binary fluid which we do not write down here. It involves an additional term in the form of the Oseen tensor which reflects the additional effects due to fluid flow not present in binary alloys. The interface equations for the binary alloy and binary fluid have been analyzed to some extent by Kawasaki and Ohta.[52] They were able to show that these equations can reproduce all the late stage processes of spinodal decomposition which has previously been proposed by a variety of authors. For example, after well-defined interfaces are created in the binary fluid the first stages of coarsening occurs by the coalescence of droplets, with a growth law approximately given by $t^{1/3}$. After this initial coalescence regime, there is a break-up of the interconnected structure for concentrated mixtures due to surface-tension driven flow, leading to a linear growth law. (This latter does not seem to occur, however, in two dimensional fluids[53]). There is then a regime in which gravity dominates, as first proposed by Siggia.[54] All of these processes have been seen experimentally in light scattering experiments which indirectly yield the growth laws for binary fluids.[1,55,56]

It would thus seem that the interface equations provide a useful starting point for discussing the later stages of domain growth in binary alloys. These equations are in a sense simpler than the original nonlinear Langevin equations from which they are derived, in that the number of degrees of freedom are considerably reduced if one only focuses on the dynamics of these topological defects. Nevertheless, they are still nonlinear equations and therefore pose a formidable challenge to theorists. It is quite possible that numerical analysis (using the increasingly powerful supercomputers being developed) will be necessary to solve these equations and to obtain predictions for such quantities as the scattering intensity, which are measured in experiments, as we discussed for the Langevin equations in section 4.

8. THEORETICAL CALCULATIONS OF THE SCATTERING INTENSITY FOR BINARY FLUIDS AND BINARY ALLOYS

We now turn to the current status of theoretical calculations of the scattering intensity $S(q,t)$ for binary fluids and binary alloys. There have been a variety of phenomenological calculations[1], which we do not discuss here. Nor do we discuss recent renormalization group calculations for two dimensional kinetic Ising models of binary alloys. Instead we summarize the results of two theories for $S(q,t)$ which are related to the dynamical theories of interface motion discussed here. Both of these calculations are due to Ohta.[57,58] The goal of such calculations is to explain the experimental scattering intensity results for binary alloys, binary fluids and other systems, and in particular to predict the scaling function $F(x)$ defined in Eq. (3.1). A typical measurement of $S(q,t)$ for the binary alloy Al - Zn is shown in Fig. 3.

One of Ohta's theories was based on the work of Tokuyama and Kawasaki[34,35] discussed in section 5. Ohta considered only the lowest nontrivial term in this theory, which as noted earlier involves an expansion in the square root of the volume fraction. His results are thus valid only for small volume fraction of the minority phase. However, he obtained an explicit form for the scaling function in (3.1) which was in reasonable agreement with the Al - Zn results for a concentration of 6.8 atomic % Zn at a quench temperature of T = 293K. However, more precise experimental studies of $S(q,t)$ for small values of the volume fraction are necessary to really test Ohta's predictions. As well, further theoretical work is necessary to take into account higher order terms in the volume fraction expansion for the pair correlation function (neglected by Ohta) to obtain a better theory for S(q,t) in the Lifshitz-Slyozov evaporation-condensation regime.

The second theory by Ohta [58] involved an approximate calculation of the scattering intensity produced by the coalescence of droplets. This coalescence is due to the Brownian motion of the droplets produced by thermal noise. (This effect is negligible in the Lifshitz-Slyozov regime.) Ohta obtained an expression for the scaling function for binary fluids for small volume fractions of the minority phase. His results were in fair agreement with the experimental results of Knobler and Wong. [59] However, it would seem that more detailed experiments are necessary to test the validity of this theory.

9. REAL SPACE RENORMALIZATION GROUP APPROACHES
    A. Callen-Symanzik Type Equations

In a series of papers Mazenko, Valls and collaborators have developed a renormalization group approach to the study of the two dimensional kinetic Ising model with either Glauber or Kawasaki dynamics. In their recent work [61,62] this has been formulated as a differential renormalization group equation of a Callen-Symanzik form. In their approach, as well as a Monte Carlo renormalization group approach discussed below, the fixed points of the renormalization group equation describing the kinetics of domain growth are low temperature fixed points. As both approaches have been successfully applied to the Glauber Ising model, we primarily focus on the more controversial results of Mazenko et al. [61,62] for the Kawasaki model. The basic quantity which they consider is

$$R_M (t, \xi) \equiv M^{-2d} \int_0^M d\vec{r}_j \int_0^M d\vec{r}_i \ \{G(\vec{r},t) - G(\vec{r},0)\} \qquad (9.1)$$

where ξ is the equilibrium correlation length, $\vec{r} = \vec{r}_j - \vec{r}_i$ and $G(\vec{r},t) = \langle S_i S_j \rangle_t$ is the average of the spin correlation function for the Ising model (2.1), when the average is taken with respect to the time dependent probability function given by the master equation. (The typical behavior [60] of $G(\vec{r},t)$ for a critical quench concentration is shown in Fig. 9).

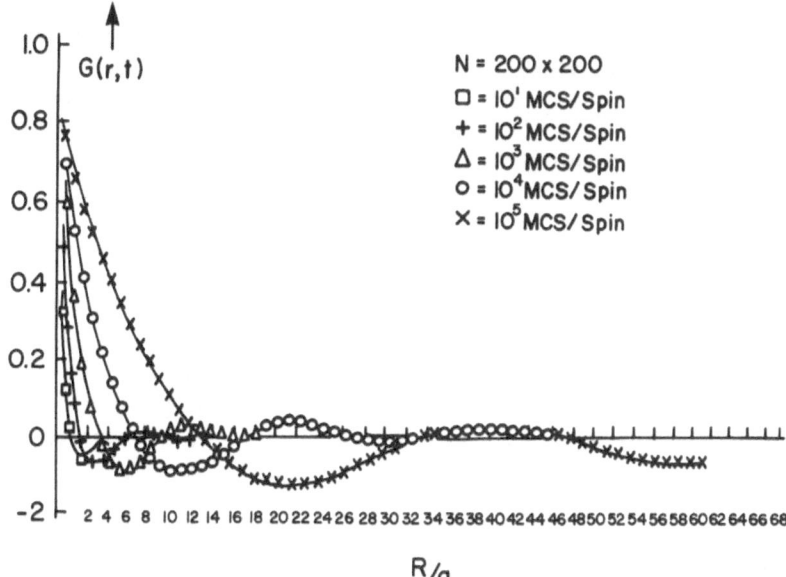

Fig. 9.   The behavior of $G(\vec{r},t)$ as obtained in a Monte Carlo study of the
d = 2 Kawasaki model.

M is the linear size of a cell of M x M spins embedded in a much larger
system of N x N spins, say with N → ∞. Next, introduce a length rescaling factor b > 1, such that M' = M/b, ξ' = ξ/b and <u>define</u> a rescaled time

$$t' = t' (M, b, t, \xi) = \Delta(M, b\ t, \xi)\ t \qquad (9.2)$$

where Δ is a (highly nontrivial!) time rescaling parameter, such that

$$R_M(t, \xi) = R_{M'}(t',\xi'). \qquad (9.3)$$

(In place of ξ one could use the temperature T, but ξ seems to be a more
natural parametrization.)   One can derive a Callen-Symanzik equation from
(9.3) by taking derivatives of the two sides of (2.8) with respect to the
four parameters M, ξ, t and b.   (Note that obviously $\partial R_M/\partial b$ = 0.)   By
eliminating three of the partial derivatives of $R_{M'}$ from these four equations, one then obtains a standard Callen-Symanzik like equation for
$\Delta(M, b, \xi, t)$, say.   This is simply an identity, which we won't write down
here.   A version of it is given in reference 61.   (The above arguments are
following the work reported in reference 62.)

The important issue is how to solve the Callen-Symanzik equation for
the time rescaling factor,Δ , say, which can easily be shown to give the

growth law for the characteristic length scale L(t). Currently this is done by very accurate Monte Carlo simulations of R_M in (9.1), and then by taking appropriate derivatives of this quantity with respect to M, ξ and t. This is done, however, only for very small block sizes (ranging from 6 to 14, say) and very small times (for the Kawasaki model), of the order of a few hundred Monte Carlo steps per spin. The idea is to predict the late time domain growth by studying $R_M(t)$ for early times, but over a range of spatial scales (in practice, a factor of two or so). I consider this idea clever, but at the moment am unconvinced that the Monte Carlo implementation is sufficient to achieve their stated purpose. It would seem most desirable to extend their work to larger values of M and considerably longer times, as well as to carry out a systematic finite size scaling analysis, to check the validity of their growth law predictions.

With this caveat, we present the results of their Monte Carlo simulation for Δ. They have found that their date for a quench at critical concentration can be fit by a parametrization[61,62]

$$\Delta(b, \xi) = \Delta_0(b) \, e^{-\alpha(b)/\xi} \tag{9.4}$$

where Δ_0 and α have been determined as explicit functions of b. Note that in the region studied, there is neither an M or t dependence in (9.4). There is, however, a very strong b and temperature dependence (through ξ). Their interpretation of these results is that for a quench temperature below T_c there is a regime of power-law behavior (with L(t) approximately given by $t^{1/3}$), followed by a crossover to a ln t behavior. Their claim that the asymptotic growth law is ln t, rather than a Lifshitz-Slyozov-like $t^{1/3}$ behavior is, however, controversial. Huse[63], for example, has presented a heuristic argument that the asymptotic growth law at critical concentrations should be $t^{1/3}$ and has carried out a Monte Carlo study that shows a power law behavior for L(t), although much better statistics will be necessary for the late time behavior than currently available. Also, as noted in section 4, the Monte Carlo study of a lattice version of Model B by Milchev et al.[26] is consistent with a $t^{1/3}$ behavior and shows no evidence of crossing over to a ln t behavior. It is clear, however, that neither existing theory nor numerical studies are accurate enough at the moment to resolve this controversy. It should be noted, on the other hand, that the RG theories have not provided predictions for quantities such as f(R,t). At the moment the extended Lifshitz-Slyozov theory fits the experimental data rather well, as discussed in section 5. This seems to indicate that at least for these three dimensional alloys such theories are reasonably accurate.

B. Monte Carlo RG Studies

An alternative approach to that outlined above is to apply the Monte Carlo renormalization group method invented by Ma[64] and developed by Swendsen[65] to the kinetics of domain growth. This has been done successfully for the kinetic Ising model with Glauber dynamics[66,67], in which they found that by a "matching condition" the time rescaling factor was given by

$$\Delta = b^{-2} \tag{9.5}$$

This implies that the growth law is $L(t) = At^{1/2}$, in agreement with the previous theories discussed in section 6. The result (9.5) was also found by Mazenko, Valls and collaborators.[45-47] It should be noted that in these studies of the Glauber model, there is just one zero temperature fixed point. The temperature is an irrelevant variable with respect to domain growth, as is seen from (9.5), (in contrast with the claim of Mazenko and Valls for the Kawasaki model, as given by eq. (9.4)). If one quenches the Glauber model to zero temperature, the system eventually equilibrates. This is in contrast to the behavior of the Kawasaki model. In this case the system begins to order at $T = 0$, but then gets trapped in a metastable state and the system "freezes" (i.e., it does not equilibrate). In the Kawasaki model there is thus a "freezing" fixed point associated with this behavior at $T = 0$, which led Mazenko and Valls to originally predict the ln t growth law.[60] It is possible, however, that the form (9.4) suggests the existence of a second fixed point for the Kawasaki model, although this has not yet been established.

In this regard, it is interesting to note that there is another model which has recently been found to have (at least) two zero temperature fixed points. Namely, a Monte Carlo renormalization group study[68] of the 8-state Potts model[69,70] with both nearest and next-nearest interactions and Glauber dynamics revealed one fixed point associated with freezing behavior and a second associated with equilibration. Viñals and Gunton interpreted their results as implying that the freezing fixed point is only attractive for quenches to zero temperature, while the equilibration fixed point is attractive for quenches to finite temperatures, $T \neq 0$. If true, the asymptotic domain growth law at $T \neq 0$ would be $L(t) \sim t^x$ with $x \simeq \frac{1}{2}$.

10. CONCLUSION

We conclude by listing a few recent studies which are interesting research areas not discussed in this paper. The first consists of a recent experimental study of the kinetics of the liquid-gas phase transition in Langmuir monolayers of stearic acid on water.[71] In the two phase region of this monolayer, a structure resembling a two-dimensional soap foam was observed. Preliminary results were reported for the average linear cell dimension and cell-side distribution as a function of time. Although the authors see a soap foam-like structure, the coalescence growth mechanism observed is not found in soap films. Their work as well as recent studies of adatom ordering on surfaces[72,73] provide interesting new examples of the kinetics of phase transitions in two-dimensional systems.

Another interesting study of two dimensional systems is a numerical simulation of the morphological development occurring during the Ostwald ripening process.[74] The most interesting qualitative result of this study is that the strong interdroplet diffusional interactions which occur at small droplet separations (discussed in section 5) produce motions of the centers of mass of the droplets. This is in disagreement with one of the assumptions of the Lifshitz-Slyozov theory. Nevertheless, the authors find a $t^{1/3}$ growth law. They conclude that such droplet motion might be a general feature of the coarsening process at high volume fractions of the condensing phase.

Finally, we note that recently new (computationally efficient) models of the ordering of quenched phases have been proposed by Oono and Puri.[75] These discrete space-time (coupled map) lattice models involve different dynamical equations than those discussed in this article; they are computationally efficient in terms of obtaining their numerical solutions. The authors claim that their models (the analogues of Models A and B) capture the essential physics of the ordering process, with a considerable reduction in computer time for simulation studies. While it remains to be seen if such models are in the same dynamical universality classes as the authors seek to describe, their preliminary results certainly suggest that their approach merits serious consideration.

ACKNOWLEDGEMENT

This work was supported by NSF Grant No. DMR-8312958.

REFERENCES

1. J. D. Gunton, M. San Miguel and P. S. Sahni, in Phase Transitions and Critical Phenomena, Vol. 8, p. 267 (1983), edited by C. Domb and J. L. Lebowitz (New York, Academic Press).

2. J. S. Langer, in Systems Far From Equilibrium, Lecture Notes on Physics, No. 132, (1980), ed. L. Garrido (Springer Verlag, Heidelberg).

3. K. Binder, Condensed Matter Research Using Neutrons, p. 1 (1985), edited by S. W. Lovesy and R. Scherm (New York, Plenum).

408

4. J. D. Gunton, J. Stat. Phys. $\underline{34}$, 1019 (1984).

5. K. Binder, in Statistical Physics, p. 35 (1986) edited by H. Eugene Stanley (North Holland, Amsterdam).

6. P. W. Voorhees, J. Stat. Phys. $\underline{38}$, 231 (1985).

7. I. M. Lifshitz and V. V. Slyozov, J. Phys. Chem. Solids $\underline{19}$, (1961).

8. A. Milchev, K. Binder and D. W. Heermann, Z. Phys. B$\underline{63}$, 521 (1986).

9. M. Hennion, D. Ronzaud and P. Guyot, Acta. Metl. $\underline{30}$, 599 (1982).

10. P. C. Hohenberg and B. I. Halperin, Rev. Mod. Phys. $\underline{49}$, 435 (1977).

11. J. D. Gunton, in Lecture Notes in Physics, No. 104, Springer-Verlag, Heidelberg (1979).

12. J. S. Langer, Ann. Phys. (N. Y.) $\underline{54}$, 258 (1969).

13. J. D. Gunton and M. Droz, Introduction to the Dynamics of Metastable and Unstable States, Lecture Notes in Physics, Vol. $\underline{183}$ (1983), ed. J. Zittartz (Springer-Verlag, Berlin).

14. M. Grant and J. D. Gunton, Phys. Rev. B$\underline{32}$, 7299 (1985).

15. J. Cahn, Acta Met. $\underline{9}$, 795 (1961); ibid $\underline{10}$, 179 (1962).

16. J. S. Langer, M. Bar-on and H. D. Miller, Phys. Rev. A$\underline{11}$, 1417 (1975).

17. M. Grant, M. San Miguel, J. Vinals and J. D. Gunton, Phys. Rev. B$\underline{31}$, 3027 (1985).

18. C. Billotet and K. Binder, Z. Phys. B$\underline{32}$, 195 (1979).

19. R. Petschek and H. Metiu, J. Chem. Phys. $\underline{79}$, 3343 (1983).

20. O. Valls and G. F. Mazenko, to be published, Phys. Rev. B.

21. K. Kawasaki, M. C. Yalabik and J. D. Gunton, Phys. Rev. A$\underline{17}$, 455 (1978).

22. S. M. Allen and J. W. Cahn, Acta Metl. $\underline{27}$, 1085 (1979).

23. T. Ohta, D. Jasnow and K. Kawasaki, Phys. Rev. Lett. $\underline{49}$, 1223 (1982).

24. O. G. Mouritsen, Phys. Rev. Lett. $\underline{56}$, 850 (1986).

25. K. Kaski, S. Kumar, J. D. Gunton and P. A. Rikvold, Phys. Rev. B$\underline{29}$, 4420 (1984).

26. A. Milchev, D. W. Heermann and K. Binder, unpublished.

27. E. T. Gawlinski, J. D. Gunton and C. Miller, unpublished.

28. J. J. Weins and J. W. Cahn, in Sintering and Related Phenomena, p. 151 Plenum, New York (1973).

29. A. J. Ardell, Acta. Met. $\underline{20}$, 61 (1972).

30. A. D. Brailsford and P. Wynblatt, Acta. Met. $\underline{27}$, 489 (1979).

31. P. W. Voorhees and M. E. Glicksman, Acta. Met. $\underline{32}$, 2001 (1984).

32. P. W. Voorhees and M. E. Glicksman, Acta Met. $\underline{32}$, 2013 (1984).

33. J. A. Marqusee and J. Ross, J. Chem. Phys. $\underline{80}$, 536 (1984).

34. M. Tokuyama and K. Kawasaki, Physica $\underline{123}$A, 386 (1984).

35. M. Tokuyama, K. Kawasaki and Y. Enomoto, Physica $\underline{134}$A, 323 (1986).

36. Y. Enomoto, M. Tokuyama and K. Kawasaki, Acta Met. $\underline{34}$, 2119 (1986).

37. Y. Enomoto, K. Kawasaki and M. Tokuyama, to be published, Acta Met. (1987).

38. Y. Enomoto, K. Kawasaki and M. Tokuyama, "Computer Modelling of Ostwald Ripening", Kyushu University preprint (1986).

39. C. W. J. Beenakker, Phys. Rev. $\underline{33}$A, 4482 (1986).

40. M. Marder, Phys. Rev. Lett. $\underline{55}$, 2953 (1985).

41. P. K. Rastogi and A. J. Ardell, Acta Metl. $\underline{19}$, 321 (1971).

42. T. Hirata and D. H. Kirkwood, Acta Met. $\underline{25}$, 1425 (1977).

43. T. Eguchi, Y. Tomokiyo and S. Matsumura, in Phase Transitions (1987), to be published.

44. G. F. Mazenko and O. T. Valls, Phys. Rev. $\underline{33}$, 1823 (1986).

45. G. F. Mazenko and O. T. Valls, Phys. Rev. B$\underline{27}$, 6811 (1983).

46. G. F. Mazen o and O. T. Valls, Phys. Rev. B$\underline{30}$, 6732 (1984).

47. S. R. Anderson, G. F. Mazenko and O. T. Valls, J. Stat. Phys. $\underline{41}$, 17 (1985).

48. M. Grant and J. D. Gunton, Phys. Rev. B$\underline{29}$, 6266 (1984).

49. E. T. Gawlinski, S. Kumar, M. Grant, J. D. Gunton and K. Kaski, Phys. Rev. B$\underline{32}$, 1575 (1985).

50. K. Kawasaki and T. Ohta, Progr. Theoret. Phys. $\underline{67}$, 147 (1982).

51. K. Kawasaki and T. Ohta, Progr. Theoret. Phys. $\underline{68}$, 129 (1982).

52. K. Kawasaki and T.Ohta, Physica A$\underline{118}$, 175 (1983).

53. M. San Miguel, M. Grant and J. D. Gunton, Phys. Rev. B$\underline{31}$, 1001 (1985).

54. E. D. Siggia, Phys. Rev. A$\underline{20}$, 595 (1979).

55. Y. C. Chou and W. I. Goldburg, Phys. Rev. A$\underline{23}$, 858 (1981).

56. N. C. Wong and C. M. Knobler, Phys. Rev. A$\underline{24}$, 3205 (1981).

57. T. Ohta, Prog. Theor. Phys. Supplement No. $\underline{79}$, 141 (1984).

58. T. Ohta, Ann. of Phys. (N. Y.) $\underline{158}$, 31 (1984)

59. C. M. Knobler and N. C. Wong, J. Phys. Chem. **85**, 1972 (1981).

60. D. Chowdhury, K. Kaski and J. D. Gunton (unpublished).

61. G. F. Mazenko, O. T. Valls and F. C. Zhang, Phys. REv. B**32**, 5807 (1985).

62. G. F. Mazenko, invited talk, presented at the March meeting of the APS, Las Vegas, Nevada (1986).

63. D. Huse, Phys. Rev. B**34**, 7845 (1986).

64. S. K. Ma, Phys. Rev. Lett. **37**, 461 (1976).

65. R. H. Swendson, in Real Space Renormalization, edited by T. W. Burkhardt and J.M.J. van Leeuwen (Springer-Verlag, New York), p. 57 (1982).

66. J. Vinals, M. Grant, M. San Miguel and J. D. Gunton and E. T. Gawlinski, Phys. Rev. Lett. **54**, 1264 (1985).

67. S. Kumar, J. Vinals and J. D. Gunton, Phys. Rev. B**34**, 1908 (1986).

68. J. Vinals and J. D. Gunton, Phys. Rev. B**33**, 7795 (1986).

69. P. Sahni, D. J. Srolovitz, G. S. Grest, M. P. Anderson and S. A. Safran, Phys. Rev. B**28**, 2705 (1983).

70. K. Kaski, J. Nieminen and J. D. Gunton, Phys. Rev. B**31**, 2998 (1985).

71. B. Moore, C. M. Knobler, D. Broseta and F. Rondelez, J. Chem. Soc., Faraday Trans. 2, **82**, 1753 (1986).

72. K. Heinz, to be published in Proceedings of Campobello Meeting on Surface Science, Springer-Verlag (1987), editors M. Grunze and J. Kreuzer.

73. J. D. Gunton, "Kinetics of Adatom Ordering on Surfaces", ibid ref. 72.

74. P. W. Voorhees, G. B. McFadden, R. F. Boisvert and D. I. Meiron, submitted to Acta Met. (1986).

75. Y. Oono and S. Puri, Phys. Rev. Lett. **58**, 836 (1987).

A NEUTRON SCATTERING STUDY OF THE KINETICS OF PHASE SEPARATION IN $Mn_{1-x}Cu_x$

B. D. Gaulin,* S. Spooner* and Y. Morii[+]

*Solid State Division
Oak Ridge National Laboratory
Oak Ridge, Tennessee 37831, U.S.A.

[+]Physics Division
Japan Atomic Energy Research Institute
Tokai, Ibaraki, 391-11
Japan

Although the kinetics of first order phase transitions[1] have been of keen interest to the theoretical and numerical simulation physicist for at least ten years, the experimental status of this field is not mature. The most informative experiments must measure the appropriate time dependence of spatial correlations in a system which has been rapidly quenched from a temperature well above its ordering temperature to a temperature at which its equilibrium state displays order. Such time resolved studies are inherently difficult.

We have undertaken a time resolved small-angle neutron scattering study of $Mn_{1-x}Cu_x$ for x = .4, .33, and .2 to examine the kinetics of phase separation. Both early and late time behavior were investigated for x = 0.33 so as to establish a comprehensive kinetic description of the process. A preliminary report of some of this work has been submitted for publication elsewhere.[2]

The Cu-Mn alloy system has a miscibility gap in Mn rich alloys.[3] The variation of solid solution lattice parameters[4] in the range of phase separation is small which leads to minimal strain energy effects. The concentrations x = 0.33 and 0.2 both lie near the middle of the miscibility gap while x = 0.4 is closer to the side. With a decomposition observation temperature of 450 C, none of the

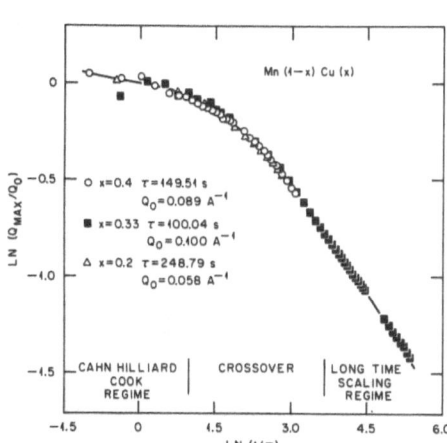

Fig. 1. The $LN(Q_{max}/Q_0)$ is plotted against the $LN(t/\tau)$ for the $Mn_{1-x}Cu_x$ system. The solid line is the theoretical curve based on Huse's arguments as described in the text.

three alloys' position on the phase diagram would be close to the phase coexistance line, although the x = 0.2 and 0.33 alloys are more likely to be within the classical spinodal instability regime.

The experiments were carried out on the 30 meter small-angle scattering instrument of the National Center for Small-Angle Scattering Research at the Oak Ridge National Laboratory. The experimental details will be described elsewhere.[2,5] All of the materials were polycrystaline and the phase separation boundaries for $Mn_{1-x}Cu_x$ are ~ 560 C for x = 0.33, ~ 570 C for x = .2, and ~ 535 C for x = .4. The samples were mounted in a specially designed rapid quench furnace and solution treated at 800 C for at least 45 minutes prior to quenching. The quench to 450 C required ~ 30 seconds from initiation until control was established at 450 C and the kinetics of the phase separation was followed in situ.

The measured structure function, $S(Q)$, directly determines the equal time concentration correlation function. It displays a peaked form, where the wave vector at which $S(Q)$ is a maximum, Q_{max}, is proportional to the average linear domain size, $R(t)$. Each data set was fit to a general form for the purpose of determining $Q_{max}(t)$. The results of the fits can be extracted from Fig. 1.

The linear theory of phase separation due to Cahn and Hilliard[6] was extended by Cook[7] and makes a sharp distinction between states which evolve via spinodal decomposition and those which evolve via nucleation and growth. It only describes spinodal decomposition. The Cahn-Hilliard-Cook (CHC) structure function is

$$S(Q,t) = \bar{S}(Q)\exp[RQ^2(Q_c^2 - Q^2)t]$$

$$- \frac{\alpha T}{(Q_c^2 - Q^2)} \left\{ 1 - \exp[RQ^2(Q_c^2 - Q^2)t] \right\}$$

where T is the final state temperature.

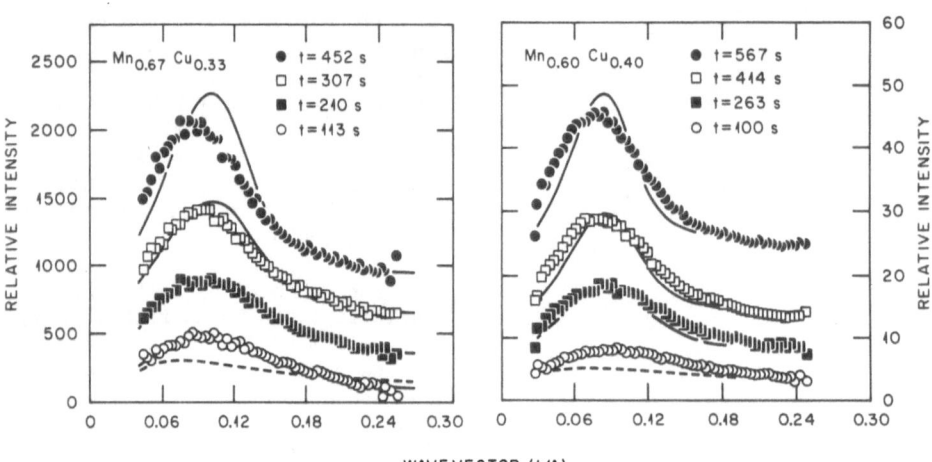

Fig. 2. Relatively early time data sets of S(Q) for $Mn_{.67}Cu_{.33}$ and $Mn_{.60}Cu_{.40}$ are shown along with solid lines representing their time evolution predicted by CHC theory, as discussed in the text. For clarity, the S(Q) zero level is offset for each data set.

Figure 2 shows our early time data for $Mn_{.67}Cu_{.33}$ and $Mn_{.60}Cu_{.40}$. Using an assumed form of $\tilde{S}(Q)$, our early time data could be well represented by the CHC structure factor. The form of $\tilde{S}(Q)$ used to propagate the structure functions is shown at the bottom of each panel as the dashed line (the dashed line in the $Mn_{.60}Cu_{.40}$ panel actually displays $.88\ \tilde{S}(Q)$). Our best description of these data used $Q_c = .139\ A^{-1}$, $R = 32\ (A^4/sec)$ and $\alpha T = 2.8$ in units of $\tilde{S}(Q)A^{-2}$ for $Mn_{.67}Cu_{.33}$; and $Q_c = .104\ A^{-1}$, $R = 69\ (A^4/sec)$ and $\alpha T = .037$ in units of $\tilde{S}(Q)A^{-2}$ for $Mn_{.60}Cu_{.40}$. The solid lines through the data show the excellent description of the data by the theory for times less than an appropriate t_{max} value. For $Mn_{.67}Cu_{.33}$, this t_{max} value is between 250 and 350 seconds, while for $Mn_{.60}Cu_{.40}$, it is between 350 and 450 seconds. In both alloys for times larger than t_{max}, the peak in $S(Q)$ moves down to smaller wave vectors while the peak intensity lags behind its predicted value. This failure of the CHC theory at later times is due to its linear nature.

At late times a dynamic scaling of the time dependent structure is predicted. The expected form of the structure function is

$$S(Q,t) = \left(Q_{max}(t)\right)^{-d} F\left(Q/Q_{max}(t)\right)$$

where d is the dimensionality of the system, while $F\left(Q/Q_{max}(t)\right)$ is the universal scaling function.

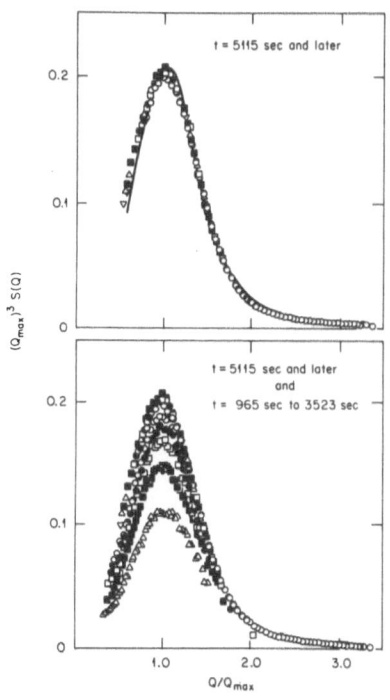

Fig. 3. The expected form of the scaling function, $F(Q/Q_{max})$, is shown for $Mn_{.67}Cu_{.33}$ at times exceeding 5000 seconds in the top panel. The breakdown of the scaling relation at earlier times can be followed in the bottom panel. The solid line in the top panel is the best fit of this late time data to theory as described in the text.

Figure 3 shows our results for $(Q_{max})^3 S(Q)$, the expected form of $F(Q/Q_{max}(t))$, at later times for $Mn_{.67}Cu_{.33}$. Data sets of $F(Q/Q_{max}(t))$ for times from 5115 through 8429 seconds are shown in the top panel of Fig. 3. These data sets fall almost precisely on top of one another as do still later data sets from 12582 seconds through 21109 seconds. This indicates that at these late times, the dynamic scaling relation is followed exactly. At earlier times, the data sets follow the scaling form to a systematically lesser and lesser extent, as can be seen in the bottom panel of Fig. 3. Hence we can explicitly follow the breakdown of the late time dynamic scaling in $Mn_{.67}Cu_{.33}$ for times less than 5000 seconds. The data sets for $Mn_{.60}Cu_{.40}$ and $Mn_{.80}Cu_{.20}$ do not exhibit dynamic scaling over the time range of observation.

The form of $F(Q/Q_{max}(t))$ has been of interest to several theoretical groups. We have fit Furukawa's[8] form of $F(Q/Q_{max}(t))$ to our late time data and the best fit is shown as the solid line in the top panel of Fig. 3. As previously reported,[2] the fitted value of Furukawa's γ value is 4.8.

The time dependence of the average linear domain size, R(t), is currently a controversial theoretical point. Lifshitz and Slyozov[9] originally argued $R(t) \sim t^{1/3}$ at late times. Very recently Huse[10] has argued for this same result; however, he has also produced earlier time corrections to this behavior. The competing argument[11] predicts $R(t) \sim \ln(t)$ and is valid only at late times. A logarithmic time dependence is not observed over any convincing time range for any of the alloys. In addition for the alloy where the behavior was followed over the widest kinetic range, $Mn_{.67}Cu_{.33}$, the power law we obtain in fitting the late time data is slightly below $- 1/3$; it is $-.37 \pm .03$.

We have modified Huse's earlier time arguments in such a way as to make a comparison with our data. The relation

$$\left(\frac{Q_{max}}{Q_0}\right) = \left(\frac{t}{\tau}\right)^{-n}$$

with

$$n = \frac{1}{3} - \frac{A}{(t/\tau)^{1/3} - A\ln(t/\tau)}$$

was produced.[2] We have fit this form of $Q_{max}(t)$ to data sets of all three alloys with A, Q_0, and τ as adjustable parameters. The fits of each data set independently produced values of A between A = .285 and .293 for the three alloys. Consequently, the sets were refit with the common value A = .289 and the results are shown in Fig. 1. All three data sets fall on the same curve and the fit to the theoretical curve (the solid line in Fig. 1) is excellent for all times observed. The fitted τ and Q_0 values are also given in Fig. 1 for each alloy.

The behavior of all three alloys is then consistent with dividing the kinetics of phase separation in this system into three kinetic regimes: an early time regime where CHC theory holds, a late time regime where dynamic scaling applies, and a crossover regime linking the early and late time regimes. The values of the maximum time, t_{max}, over which CHC theory applies fall in the ratio $\sim 1.3 \pm .3 : 1$ for $Mn_{.60}Cu_{.40}$ as compared with $Mn_{.67}Cu_{.33}$, whereas the fitted τ values for the two concentrations are in the ratio 1.49 : 1. Also $Mn_{.80}Cu_{.20}$ and $Mn_{.60}Cu_{.40}$ do not exhibit dynamic scaling over the kinetic range of the experiment, which is consistent with the assigned boundary separating the late time scaling regime from the crossover regime.

This research was carried out under the U.S.-Japan Cooperative Program on Neutron Scattering, and supported in part by the Division of Materials Sciences, U.S. Department of Energy under contract DE-AC05-840R21400 with Martin Marietta Energy Systems, Inc.

REFERENCES

1. J. D. Gunton, M. San Miguel, and P. S. Sahni in "Phase Transition and Critical Phenomena," C. Domb and J. L. Lebowitz (editors), Vol. 8, Academic Press, London (1983).
2. B. D. Gaulin, S. Spooner, and Y. Morii (submitted for publication).
3. J. M. Vitek and H. Warlimont, Metal Science, 7, January (1976).
4. N. Cowlam, Metal Science, 483, October (1978).
5. B. D. Gaulin, S. Spooner, and Y. Morii (to be published).
6. J. W. Cahn and J. E. Hilliard, J. Chem. Phys. 28 , 258 (1958).
7. H. E. Cook, Acta. Metal. 18, 297 (1970).
8. H. Furukawa, Physica A 123, 497 (1984).
9. I. M. Lifshitz and V. V. Slyozov, J. Phys. Chem. Solids 19, 35 (1961).
10. D. A. Huse, Phys. Rev. B 74, 7845 (1986).
11. G. F. Mazenko, O.T. Valls, and F.C. Zhang, Phys. Rev. B 32, 5807 (1985).

ON DENSE BRANCHING PHASE SEPARATION

R. Hilfer[*], S. Alexander[*†] and R. Bruinsma[*]

[*]Department of Physics
University of California, Los Angeles
Los Angeles, CA 90024, USA

[†]Racah Institute of Physics
Hebrew University
Jerusalem, Israel

Recently Deutscher and Lareah[1] discovered a new mode of phase separation in thin films of Al/Ge alloys. They observe the growth of circular "colonies" whose densely packed appearance has been called "dense branching morphology"[2]. The colonies consist of a highly branched starlike "island" of polycrystalline Ge inside a "lake" of monocrystalline Al which is only slightly larger than the Ge island. Thus the Al forms a thin but essentially uninterrupted rim around the Ge peninsulas. The whole colony is embedded in the amorphous phase having an overall composition of 40% Al and 60% Ge. As these colonies grow into the metastable amorphous surrounding they preserve their more or less circular shape.

This immediately raises the question why on the one hand the Al/Ge-interface shows an instability, while on the other the Al/amorphous boundary does not. We investigate this question first. We then present the theoretical description of the new growth morphology. We outline the solution of our equations and indicate how a unique growth velocity is selected. We finally compare our results with experiment.

We first turn to the stability problem for the Al/amorphous interface: During ordinary solidification from a melt the excess foreign atoms have to be cleared away from the solid/liquid interface by chemical diffusion. If this diffusion process has to occur into the metastable phase and over large distances it will be more effective if the interface area is increased. This leads to the Mullins-Sekerka instability and a dendritic morphology for the growing crystal[3].

The MS-instability is more formally derived from the diffusion equation together with two boundary conditions for the moving interface. The first boundary condition demands energy conservation across the boundary while the second condition is a statement of thermodynamic equilibrium including the Gibbs-Thomson correction due to surface tension. A steady state solution is given by a flat interface moving with velocity $\vec{v}$ into the metastable phase in front of which the diffusion field decays exponentially. One then adds a small perturbation ζ to the moving front

$$\zeta(\vec{r}_\perp, t) = \zeta_k \exp (i\vec{k} \cdot \vec{r}_\perp + \omega t)$$

where $\vec{r}_\perp$ denotes the position in the plane perpendicular to $\vec{v}$. The problem becomes to determine the amplification rate ω for a perturbation with wavevector $\vec{k}$. The sign of ω will decide whether the perturbation is enhanced or suppressed. Carrying out the linear stability analysis along the lines of Langer[3] one obtains

$$\omega = -\frac{D}{2\xi^2} + D \left(\frac{1}{\xi} - 2d_0 k^2 \right) \left(\frac{\omega}{D} + k^2 + \frac{1}{4\xi^2} \right)^{1/2}$$

as the relation between ω and k. In the above relation $\xi = D/v$ is the diffusion length and d_0 is the so called capillary length which is determined by the surface tension. The relation is valid for the symmetric model, i.e. when the diffusion constant D in the solid equals that in the melt. The usual argument[3] proceeds with two assumptions: 1. $k\xi \gg 1$ assumes that the diffusion length is very large (usually macroscopic). 2. $\omega \ll Dk^2$ assumes that the time scale for diffusing a distance on the order of the wavelength of the perturbation is much shorter than the time scale on which growth occurs (quasistationarity). With these two assumptions one finds the MS-instability for wavelengths above $\lambda = (2d_0\xi)^{1/2}$.

What distinguishes our case from solidification in a melt is the fact that the diffusion constant in the amorphous phase is much smaller than in a melt. Thus ξ is no longer macroscopic and we have to consider the case $k\xi \ll 1$. At the same time the quasistationarity assumption breaks down because the term ω/D under the root can no longer be neglected. One finds that $\omega=0$ at $k=0$ while $\omega = -D/2\xi^2 = -v^2/2D$ at the wavelength governing the MS-instability. Thus in this case perturbations are damped out. For very long wavelengths the damping becomes weak. This effect can also be observed in the experiment.[4]

We now proceed to develop a set of equations describing the growth. Compelled by the experimental evidence we arrived at the following central features of our description: 1. The dominant diffusion is that of atomic Ge backward from the Al/amorphous interface into the crystalline Al. 2. Atomic diffusion in the amorphous phase is very slow compared with the crystalline phase. 3. Nucleation and growth of Ge crystallites occurs only at the interface between Al and Ge. 4. Nucleation of Al crystals in the amorphous phase is much more frequent than that of Ge but still rare; it controls the initiation of new colonies. Of these assumptions the first one is at the heart of our understanding. It stems from the observation that the Al/Ge interface is separated from the amorphous phase by an Al rim and can only grow if Ge atoms diffuse across this rim. We argue that the interplay between the Al/amorphous boundary and the Al/Ge boundary which act respectively as source and sink for Ge atoms sets the length scale of the problem. Note that the length scale is strongly temperature dependent.[1]

To approach the problem mathematically we concentrate on the smooth Al/amorphous boundary and describe the highly irregular Al/Ge interface in an averaged fashion. We replace the local concentration of atomic Ge by its radial average $c(r)$ and that of crystalline Ge by its radial average $\rho(r)$. Then we have

$$\frac{\partial c}{\partial t} = D_L \nabla^2 c - \frac{\partial \rho}{\partial t} \tag{1a}$$

on the "left" inside the Al and

$$\frac{\partial c}{\partial t} = D_R \nabla^2 c \tag{1b}$$

in the amorphous phase on the "right" of the interface. D_L resp. D_R is the diffusion constant on the left resp. right. At the interface on has $c = c_L$

on the Al side ($r=R_-$) and $c = c_R$ on the amorphous side ($r=R_+$). Taking excluded volume effects into consideration in these equations somewhat complicates the analysis but does not change its essential features. We describe the growth process in its simplest form as

$$\frac{\partial \rho}{\partial t} = Bc\rho \tag{2}$$

The phenomenological rate constant B describes the growth of the branched Ge structure and thus incorporates nucleation and growth of Ge crystallites. At the Al/amorphous interface, $r = R(t)$, the diffusion field must obey mass conservation:

$$\frac{d}{dt} R(t) \left(\Delta c + \rho(R) \right) = D_L \left. \frac{\partial c}{\partial r} \right|_{r=R_-} + D_R \left. \frac{\partial c}{\partial r} \right|_{r=R_+} \tag{3}$$

Here $\rho(R)$ is a small seed concentration of crystalline Ge at the boundary which is seen in the experiment. $\Delta c = c_R - c_L$ denotes the discontinuity in the concentration across the interface (miscibility gap).

The ramified Al/Ge boundary close to the Al/amorphous boundary acts as a sink for the diffusing Ge and from Eq. (3) this implies a finite concentration gradient and thus a finite velocity for the moving front. Searching for constant velocity profiles we write

$$c(r,t) = c_0 f(z) \qquad \text{resp.} \qquad \rho(r,t) = c_0 g(z) \tag{4}$$

where $R = vt$, $z = (r-R)/\xi$, c_0 is the concentration of the in the amorphous phase and $\xi = D_L/v$ is the basic length scale in the problem. For sufficiently long times ($v^2 t/D_L \gg 1$) the curvature of the interface can be neglected and one obtains a closed nonlinear equation for f [5]

$$f' + f'' = \beta f (1 - f - f') \tag{5}$$

with the boundary conditions

$$f(-\infty) = f'(-\infty) = 0 \tag{6a}$$

$$f(0) = c_L/c_0 \tag{6b}$$

$$f'(0) = 1 - c_L/c_0 - \varepsilon \tag{6c}$$

where $\beta = c_0 B D_L/v^2$ is a dimensionless control parameter and $\varepsilon = g(0)$ is the small seed concentration at the interface introduced in Eq. (3). We display the solutions to Eq. (5) in f-f'-space in Fig. 1 . Trajectories fulfilling the boundary conditions at $z=-\infty$ emerge from the origin with a slope $f'/f = 1/\zeta = [-1 + (1+4\beta)^{1/2}]/2$. The straight line $1-f-f'=0$ is a separatrix. The boundary condition Eq. (6c) determines a straight line parallel to the separatrix. If we choose a value for β we must follow the associated flow line starting from $(0,0)$ until it intercepts this straight line and read off the corresponding value $f(0) = c_L/c_0$. This determines β and thus v as a function of c_L (see inset of Fig. 1).

After demonstrating how the boundary conditions effect a velocity selection we compare the results with experiment.[4] The growth velocity is found to be constant as predicted. For slow velocities, i.e. $\beta \gg 1$, one derives the relation $v \approx (c_L/c_0) D_L/\xi\zeta$. It has been checked experimentally by comparing the temperature dependence of $D/\xi\zeta$ with that of v and seems to be in quantitative agreement. If the temperature dependence of v were given by that of D_L one would expect a temperature independent length scale. Instead we have $v \propto (BD_L)^{1/2}$ which shows that both B and D_L contribute. In a first check

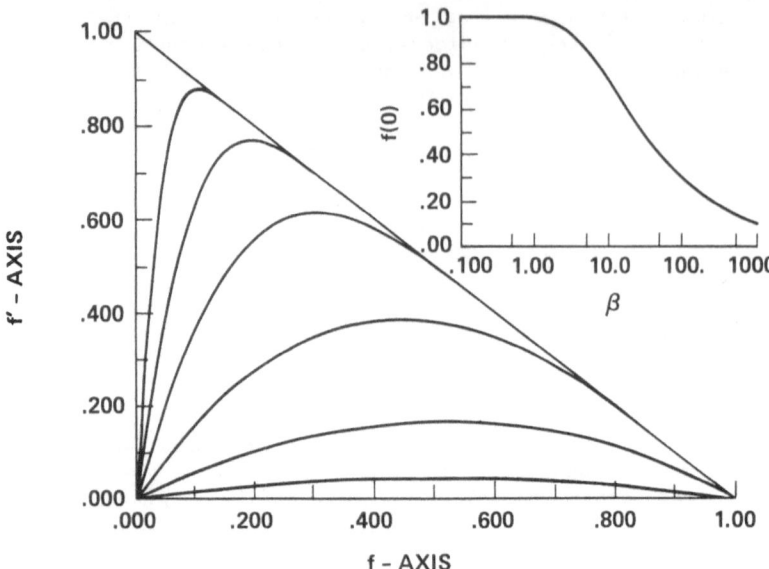

Fig. 1 : Trajectories fulfilling the boundary conditions at $z=-\infty$ for selected values of β (β = .2,1,5,25,100,500).
Inset: Dependence of $f(0)=c_L/c_0$ on β for trajectories fulfilling all boundary conditions with ε = .001

this relation has been found consistent with the values of the activation energies of the quantities involved.[4] In addition the values of c_L and c_R have been checked against the Al/Ge phase diagram.[5] In conclusion we have shown that our simple analysis agrees qualitatively with all known experimental facts. Further theoretical work has to concentrate on a microscopic model for the nucleation and growth process. More detailed experimental data are needed to explore the limits of quantitative agreement.

ACKNOWLEDGMENTS

We wish to thank G. Deutscher for extremely stimulating discussions. We gratefully acknowledge financial support from the National Science Foundation under Grant No. 84-12898 and from the Deutsche Forschungsgemeinschaft.

REFERENCES

1. G. Deutscher and Y. Lereah, Physica 140A (1986), 191
2. E. Ben-Jacob, G. Deutscher, P. Garik, N. Goldenfeld and Y. Lereah, Phys. Rev. Lett. 57 (1986), 1903
3. J.S. Langer, Rev. Mod. Phys. 52 (1980), 1
4. G. Deutscher and Y. Lereah, preprint
5. S.Alexander, R.Bruinsma, R.Hilfer, G.Deutscher and Y.Lereah, preprint

CRYSTALLIZATION KINETICS OF METALLIC GLASSES $Co_{78-x}Fe_xSi_9B_{13}$ PROBED BY ELECTRICAL RESISTIVITY AND MAGNETIZATION

K. Pękała, M. Pękała*, P. Jaśkiewicz, and R. Trykozko

Institute of Physics, Warsaw Technical University
*Department of Chemistry, University of Warsaw

INTRODUCTION

Metallic glasses are metallic alloys distinguished by lack of a long range order of atoms. Their unique physical properties are caused by both the topological and chemical disorder in amorphous structure [1]. Namely, the interplay of exchange interactions and casted in anisotropy along the ribbon length involves the asperomagnetic structure [2]. On the other hand, the disorder modified structure factor changes electron scattering leading to increase in electrical resistivity [3].

Metallic glasses are cast in metastable state and are subject to structural relaxation which in turn involves evolution of their magnetic and electron transport properties. Generally, rate of this evolution is faster at higher temperature which initiates more thermally activated processes of atomic rearrangements. Therefore, the actual state and physical properties of metallic glass sample are determined by its "curriculum vitae", i.e. the summarized exposure for available relaxation processes occuring during sample storage at temperatures T_i for times t_i [4].

The present experimental study is aimed to elucidate the time variation of magnetization and electrical resistivity of metastable alloys of series $Co_{78-x}Fe_xSi_9B_{13}$ (x = 0 to 66) when they are annealed at temperature just below the crystallization threshold temperature. Knowledge of the stability against crystallization is of crucial importance to establish the working range and to adjust a proper heat treatment in order to achieve controlled modification of physical parameters.

EXPERIMENTAL

The amorphous structure of alloys obtained by melt spinning method with rapid quenching rate of 10^6 K/s as well as products of crystallization were checked by X-ray diffraction and differential scanning calorimetry. Variations of electrical resistivity and magnetization during either the isochronal anneal with rate of 5 K/min or the isothermal anneal at temperatures below the crystallization threshold temperature were recorded by means of the dc four probe and Faraday balance methods, respectively.

RESULTS AND DISCUSSION

Time variation of magnetization and electrical resistivity may be recognized as the indirect probes of crystallization kinetics in metallic glasses since both quantities are sensitive to atomic rearrangements [4]. The principal effect of crystallization processes is the gradual restoring of translational symmetry within volume roughly increasing with time. This process influences electron transport phenomena in a simple way causing for example an abrupt drop in electrical resistivity occuring in a narrow time or temperature intervals [3,4]. For magnetization the same effect is just as abrupt, however, the sign of change depends on the ratio of magnetization in amorphous and crystalline phases which not necessarilly have to be smaller than one.

Analysis of crystallization occuring by nucleation and growth processes becomes burdensome since the polymorphous crystallization, i.e. a direct transition from amorphous to crystalline phase of the same composition, is very seldom. More often crystallization of a new phase is followed by crystallization of the remaining amorphous matrix.

During the first runs of isochronal anneal the number of crystallization stages and crystallization threshold temperatures were determined from the temperature dependence of magnetization and electrical resistivity. All the alloys studied start to crystallize at temperature rising almost monotonically with iron content from 700 to 780 K (Fig.1). The second stage appears between 795 and 815 K. Alloys containing 58% or more iron exhibit also the third stage starting between 825 and 860 K.

Magnetization and electrical resistivity recorded during isothermal anneals lasting up to 1000 min show that during the first period, typically 1 to 100 min, the slow structural relaxation of amorphous phase occurs [5]. Then in a next period the abrupt variation of magnetization σ and electrical resistivity ρ could be fitted to expressions $\ln(\sigma/\sigma_0) = t/\tau$ and $\ln(\rho/\rho_0) = t/\tau$, respectively, resulting in values of time constants τ. Since crystallization rate is controlled mainly by diffusion it may be described as thermally activated process:

$$\tau = \tau_0 \exp(Q/kT) \qquad (1)$$

where Q and k are an activation energy and a Boltzmann constant, respectively.

Activation energies Q_1 for the first crystallization stage spread between 0.6 and 3.7 eV (Fig. 1). Values for the second stage Q_2 lie remarkably higher ranging between 2.4 and 4.6 eV whereas values of Q_3 span from 3.3 to 3.5 eV. The mentioned values of Q_i coincide with activation energies reported by Luborsky [6] for series of transition metal-metalloid glasses.

Concluding one may state that crystallization of metallic glasses studied here is not of polymorphous type. It should be rather described as a sequence of precipitation and transformation processes. Comparing the present results with structural studies of Załuska and Matyja [7] one may ascribe Q_1 energy to precipitation of the bcc phase based on Co-Si or

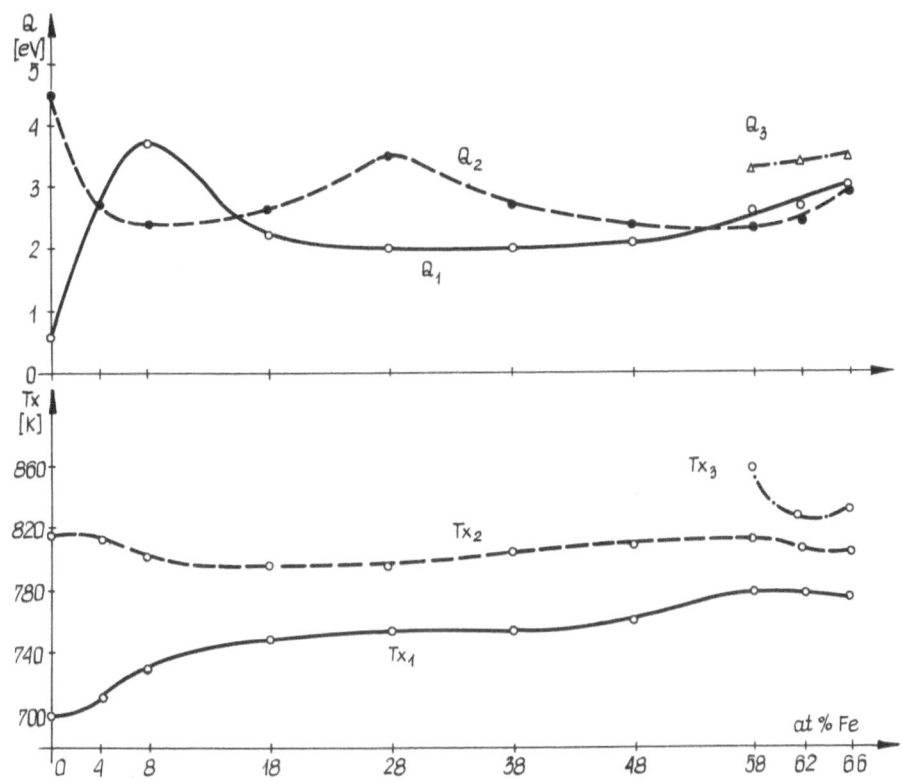

Fig.1 Crystallization threshold temperatures T_{xi} and activation energies Q_i for various crystallization stages of metallic glasses $Co_{78-x}Fe_xSi_9B_{13}$ versus iron content x. Lines are guides for eyes only.

Fe-Si respective to the alloy composition. At higher Q_2 energies borides, mainly Co_2B and Fe_3B, are formed while the latter one transforms to Fe_2B during the third stage with Q_3.

REFERENCES

1. F.E. Luborsky in Amorphous Metallic Alloys, ed. F.E. Luborsky, Butterworth, London, 1983
2. K. Moorjani, J.M.D. Coey, Magnetic Glasses, Elsevier, Amsterdam-New York-Tokyo, 1984
3. K. Pękała, M. Pękała, J. Latuszkiewicz, J.J. Bara, B.F. Bogacz, P. Jaśkiewicz, R. Trykozko, IEEE Trans. Magn. MAG 20, 1338 (1984)
4. U. Köster, U. Herold, in Glassy Metals I, ed. H.-J. Güntherodt, H. Beck, Topics in Applied Physics, vol.46, Springer Verlag, Berlin-Heidelberg, 1981
5. M.R.J. Gibbs, J.E. Evetts, J.A. Leake, J. Mater. Sci. 16, 278 (1983)
6. F.E. Luborsky, Materials Sci. Eng. 28, 139 (1977)
7. A. Załuska, H. Matyja, preprint

POLYMER DYNAMICS AND GELATION

James E. Martin[*]

Sandia National Laboratories
Albuquerque, New Mexico 87185

INTRODUCTION

In this talk we will review some of the recent advances in the study of the sol-gel transition, and in the dynamics of the sol clusters. The primary emphasis will be on the experimental literature, but theory will be discussed where it is needed to interpret the data. The first part of this lecture will focus on the kinetics of growth of the sol beneath the gel point, and on a few purely static issues, such as the fractal structure of the sol and the observation of a spatial correlation length in undiluted gels.

The second part of this lecture will discuss the relaxation of concentration fluctuations from the diluted and undiluted sol. Much of this research on the dynamics of gels is new, so the theoretical ideas are either qualitative or, in some cases, essentially nonexistent. I hope that presenting some of the uninterpreted data will spark theoretical interest into the dynamics of sol clusters.

Although gelation has been an active area of research for many years, there are a number of fundamental questions that have not yet been satisfactorily answered. An issue of particularly high current interest is, "What is the importance of kinetic effects on the critical behavior observed near the gel point?" At this time the answer to this question is unknown, since many of the theories that have been developed to describe the gelation process- e.g. the Flory-Stockmayer theory, bond percolation, correlated percolation- describe equilibrium systems. However, since much of the experimental data is taken on irreversible chemical gels, where the bond energy is very high, a kinetic description of the gelation process may be necessary. In this talk we will try to emphasize this aspect of the gelation problem.

[*]This work performed at Sandia National Laboratories, Albuquerque, NM and supported by the U.S. Department of Energy under Contract No. DE-AC-04-76DP00789.

THE KINETIC RATE EQUATION

There are several theoretical approaches to gelation, but perhaps the simplest is the Smoluchowski equation[1], often referred to as the kinetic rate equation. The kinetic rate equation gives a mean field description of the gelation process, and so is only qualitatively correct, but it provides a useful framework in which to classify a variety of growth processes. This classification helps to clarify the difference between gelation and the various types of aggregation.

The Smoluchowski equation expresses the time evolution of the number of m-mers, $N(m)$, in terms of a reaction kernel K_{ij}, which gives the probability of an i-mer reacting with a j-mer.

$$\frac{dN(m)}{dt} = \frac{1}{2} \sum_{i+j=m} N(i)K_{ij}N(j) - N(m) \sum_{j} K_{mj}N(j) \quad (1)$$

The first term accounts for the creation of m-mers through binary collisions of i-mers and (m-i)-mers; the second term represents the 'annihilation' of m-mers due to binary collisions with other clusters. Although the structure of this equation appears simple enough, it is difficult to determine the form of the reaction kernel for a given physical system.

Van Dongen and Ernst[2] recently introduced a classification scheme for *homogeneous* kernels, based on the relative probabilities of large clusters sticking to large clusters, and small clusters sticking to large clusters. Depending on which of these processes dominates, one might expect quite different growth kinetics and size distributions. For example, if small-large interactions dominate, then large variations in cluster mass are discouraged, and the size distributions will tend to be tightly bunched, like a bell-shaped curve. On the other hand, if large-large interactions dominate, then the small clusters tend to get left behind in the scramble and a monotonically decreasing size distribution (with an algebraic decay) is obtained. The discussion which follows is a brief summary of the pertinent results of van Dongen and Ernst[2].

To define the growth classes we need to introduce two exponents. Let the probability that a j-mer reacts with a j-mer (large with large) be $K_{jj} \sim j^{\lambda}$, and let the probability that a j-mer reacts with a monomer (large with small) be $K_{1j} \sim j^{\nu}$. Based on these exponents, a few general comments can be made about homogeneous kernels. First, kernels with either $\lambda > 2$ or $\nu > 1$ are unphysical, since the reactivity cannot increase more rapidly than the cluster size. Second, if λ is greater than 1 the Smoluchowski equation predicts the formation of an infinite cluster in finite time- a gel point. Finally, kernels with $\lambda \leq 1$ give nongelling behavior, that is, an infinite cluster is formed at infinite time. Clearly, much of the qualitative behavior of the Smoluchowski equation is controlled by λ.

Class I Growth

In *class I* growth large-large interactions dominate, i.e. $\lambda > \nu$. In the nongelling regime, where $\lambda < 1$, the number distribution decays algebraically like $N(m) \sim 1/m^{\tau}$, with $\tau = 1 + \lambda$, and the weight-average mass, M_w, grows as a power of time, $M_w \sim t^z$, with $z = 1/(1-\lambda)$. We note that since $\lambda < 1$, the *polydispersity* exponent τ is less than 2, in contrast to systems which exhibit a gel point, where τ is greater than 2. Also, as λ approaches 1 the growth exponent z approaches ∞ - in this limit the growth becomes exponential in time.

In the gelling regime of class I growth, where $\nu < 1 < \lambda \leq 2$, the number distribution again decays algebraically, $N(m) \sim m^{-\tau}$, with $\tau = (\lambda+3)/2 > 2$. Below the gel point this algebraic decay is valid only for clusters smaller than the typical cluster size (the z-average mass, M_z). However, above t_{gel} this algebraic decay extends to arbitrarily large clusters, and the prefactor of the decay decreases monotonically with time, as clusters bond to the infinite cluster. Near the gel point M_w diverges like $|t_{gel}-t|^{-\gamma}$, where $\gamma = 1/(\lambda-1)$, and the mean cluster size, M_z, diverges like $|t_{gel}-t|^{-1/\sigma}$, where $\sigma = (\lambda-1)/2$. Thus for homogeneous gelling kernels the Smoluchowski equation quite generally predicts $M_w \sim M_z^{1/2}$.

In the Flory-Stockmayer[3,4] approach to gelation it is assumed that all sites on a cluster are equally reactive. This gives the prototypical class I kernel $K_{ij} = ij$. The homogeneity exponents for this kernel are $\lambda = 2$ and $\nu = 1$, which gives $N(m) = m^{-5/2}$, $M_w \sim |t_{gel}-t|^{-1}$, and $M_z \sim |t_{gel}-t|^{-2}$. These are the well-known predictions of the classical theory of gelation.

Class III Growth

Class III growth is defined by the domination of small-large interactions. Since ν must always be less than 1 and by definition λ must be less than ν, a gel point is not possible in this class of growth. Instead, the size distributions tend to be bell-shaped curves. The original Smoluchowski kernel[1], $K_{ij} = (R_i + R_j) \times (D_i + D_j) \propto (i^{1/3} + j^{1/3}) \times (i^{-1/3} + j^{-1/3})$, developed to describe *diffusion-limited* aggregation, is in this class.

Class II Growth

Class II growth is a complex intermediate ground between class I and class III growth. In class II growth $\lambda = \nu \leq 1$, so neither small-large nor large-large interactions are dominant. Again, no gel point is possible, since $\lambda \leq 1$. Sum kernels of the form $(i^a + j^a)^b$ are examples of Class II kernels. Although little is known about the size distributions for class II kernels, for $\lambda = 1$ the growth of the average mass is known to be exponential in time, which suggests that class II kernels are relevant to aggregation under *reaction-limited* conditions.

In summary, the kinetic rate equation predicts three qualitatively distinct behaviors of growth:

power-law growth:	$M_w \sim t^z$	$z=1/(1-\lambda)$
exponential growth:	$M_w \sim e^{ct}$	
gel growth:	$M_w \sim |t_{gel}-t|^{-\gamma}$	$\gamma=1/(\lambda-1)$

In silica systems experimental evidence exists for all three types of growth. For example, when an orthosilicate is reacted under alkaline conditions, small colloidal particles are formed. The colloidal suspension can then be destabilized by adding salt and adjusting the pH, so that the colloidal particles aggregate. If the aggregation process proceeds rapidly, power-law growth is observed[5], as shown in Fig. 1. On the other hand, if the aggregation process proceeds slowly, exponential growth of the average cluster mass is observed[6,7], as in Fig. 2. By changing the initial reaction conditions of the orthosilicate, covalent gels are formed[8] which exhibit the power-law divergence shown in Fig. 3. In order to understand these last data, we will now discuss various equilibrium models of gelation.

EQUILIBRIUM GELATION

Any discussion of equilibrium gelation runs the risk of being *very* long. Rather than attempt to give a broad review of the current theoretical state of affairs, we will restrict ourselves to those issues which are germane to the experimental work to be described. (Those wishing a more expansive review of gelation might consult the excellent article by Stauffer, Coniglio, and Adam[9]; a very readable introduction to percolation theory has been recently published by Stauffer.[10]) These experimental issues include;

1. The divergence of the average cluster mass, M_w, and average radius, R_z.

2. The fractal dimension, D, of a 'typical' sol cluster, and the effective fractal dimension of the sol ensemble.

3. The asymptotic form of the scattering function, I(q), of the diluted sol.

4. The algebraic decay of the number distribution of the sol, N(m).

Since most of the experimental data to be described were taken in the pregel stage, we will largely neglect post-gelation issues, such as the gel fraction and the growth of the modulus.

Flory-Stockmayer Theory

The earliest theory to attempt to describe the divergences observed in gelation is due to Flory[3] and Stockmayer[4], who formulated a mean field theory for the polycondensation of bi- and poly-functional monomers. In this model bonds are randomly formed between adjacent nodes on an infinite Cayley tree, or Bethe lattice. The Flory-Stockmayer theory was

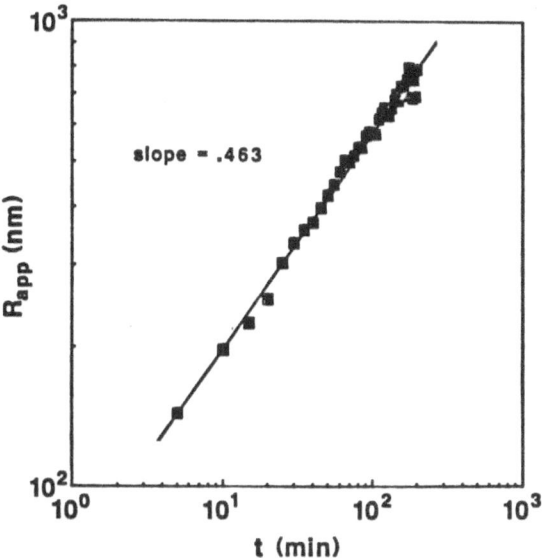

Fig. 1: Colloidal silica reacted under rapid, diffusion-limited conditions[5] exhibits power-law growth, $M_w \sim t^z$, where the exponent z is experimentally determined to be about .8.

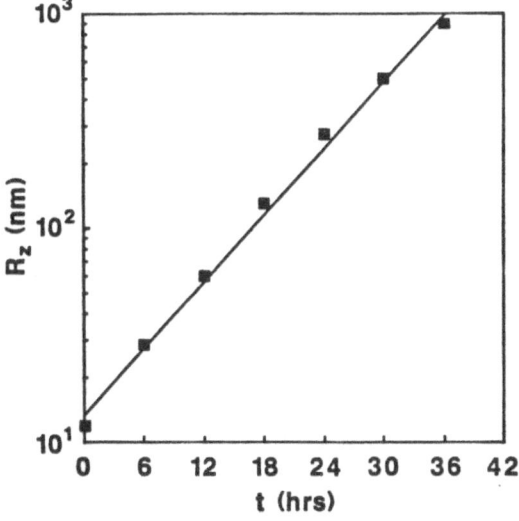

Fig. 2: Colloidal silica reacted under slow, reaction-limited conditions[6] exhibits exponential growth, $M_w \sim e^{ct}$.

a huge success, since it properly described the salient features of gelation, e.g. the emergence of an infinite cluster at some critical extent of reaction p_c, and gave quite reasonable predictions for the gel point. Additionally, the FS theory correctly predicts power-law singularities in the mean cluster size and z-average radius as $\varepsilon=|p_c-p|$ vanishes from above or below. These results are

$$M_w \sim \varepsilon^{-\gamma}$$

(2)

$$R_z \sim \varepsilon^{-\nu}$$

with $\gamma=1$ and $\nu=1/2$.

Despite these qualitative successes, the FS model has been criticized because of the unphysical nature of the Cayley tree. On a Cayley tree the notion of spatial dimension is utterly lost, and essentially all of the vertices are monofunctional, surface nodes. That is, in

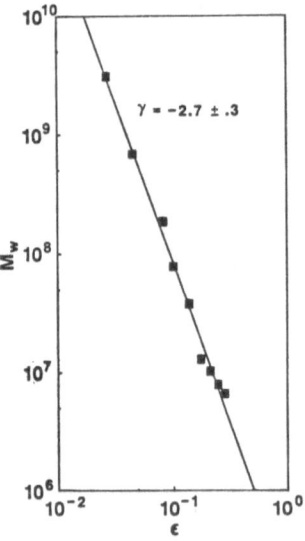

Fig. 3: Gels made from tetramethoxysilicon and water in a 1:4 molar ratio with a two-step, acid-base hydrolysis exhibit critical gel growth[8], $M_w \sim |t_{gel}-t|^{-2.7\pm0.3}$, near the sol-gel transition.

an f-functional Cayley tree the number of nodes in the m-th cycle (counting the root as zero) is $\approx (f-1)^m$. For a lattice with L cycles, the fraction of surface sites is thus $(f-1)^L/\Sigma^L(f-1)^m$, which is a constant that depends *only* on the functionality, *not* on the lattice size. Thus, the fraction of surface sites does not vanish as the lattice size goes to infinity, in contrast to Euclidean lattices. This leads to the principle short-coming of the FS approach- the purely branched clusters formed on the Cayley tree are predicted to have a fractal dimension of 4 ($M \sim R^4$). This result is overtly unphysical in three dimensions[9], since the mean cluster density, $\rho \equiv M/R^3 \approx R$, diverges with the radius. More subtle arguments[11] show that the FS approach is internally inconsistent in less than six dimensions.

Bond Percolation

In light of these difficulties, Stauffer[12] and de Gennes[11] promulgated the use of bond percolation as a description of polycondensation. In the percolation model, bonds are placed at random between adjacent nodes of a regular or random d-dimensional lattice. Clusters containing circuits are thus admitted, excluded volume effects are directly accounted for, and the spatial dimension becomes an explicit part of the problem. As a result, the exponents that describe the divergence of the weight-average mass and z-average radius become dimension dependent for d<6, which is the upper critical dimension for percolation, above which the mean field theory is exact.

Many of the results of percolation theory can be summarized by the similarity form of the size distribution introduced by Stauffer[13] and Essam[14];

$$N(m)=m^{-\tau} f(m/M_z). \qquad (3)$$

Here M_z is the 'typical', or z-average, cluster mass and f(x) is a cut-off function which is a constant for x<<1, but decays rapidly for x>>1 [for the purpose of calculating exponents[9] it will suffice to set f(x)=0 for x>1.] A second exponent σ can be defined by $M_z \sim \varepsilon^{-1/\sigma}$, and the moments of this distribution can then be computed in terms of τ and σ. For example, the weight-average mass is $M_w \equiv \Sigma m^2 N(m)/\Sigma m N(m) \sim \varepsilon^{-(3-\tau)/\sigma}$; comparing this to eq. 2 gives[9] $\gamma=(3-\tau)/\sigma$.

To compute the z-average radius an additional aspect of the sol ensemble is important[9]: *sol clusters whose radius is smaller than the correlation length R_z (m<M_z) have the same fractal dimension D as the infinite cluster at the gel point.* The z-average radius is then $R_z^2 \equiv \Sigma R^2 m^2 N(m)/\Sigma m^2 N(m)$, where $R^D \sim M$. Thus $R_z^D \sim M_z \sim \varepsilon^{-1/\sigma}$, and comparing this to $R_z \sim \varepsilon^{-\nu}$ gives[9] $\nu=1/\sigma D$. Combining our results gives $R_z^{D(3-\tau)} \sim M_w$. In effect, D(3-τ) is a fractal dimension for the polydisperse sol ensemble.

At this point it is appropriate to discuss the scattered intensity measurements, $I(q)$, from which much of the experimental data is derived. The variable q, which has the dimensions of inverse length, is loosely referred to as the 'momentum transfer'. In terms of the scattering angle q and the wavelength in the scattering medium λ, $q=4\pi\sin(q/2)/\lambda$. However, before we discuss the form of the scattering function, a basic point needs to be clarified to avoid confusion amoung those accustomed to thinking about thermal phase transtions. The *random* percolation model is a divergence in *connectivity*, not in density fluctuations. Therefore, the density-density correlation function is a simple constant, regardless of the extent of reaction. In particular, nothing interesting happens to the scattering function as the system crosses the gel point[9] (more on this later). However, if one freezes the percolating network at some extent of reaction and dilutes the clusters, the connected bits will stay connected, and this connectivity will give rise to spatial correlations. Thus, scattering from the *diluted* system allows an investigation of the connectivity.

In general, the intensity scattered from a *monodisperse* solution of clusters of mass M can be written $I(q)/c \sim MF(qR)$, where c is the mass concentration. On long length scales, where $qR \ll 1$, the well-known Guinier expansion of the scattering function is $I/c \sim M(1-q^2R^2/3+...)$, so the mass and radius are determined by the intercept and slope of I/c versus q^2. However, on short length scales, where $qR \gg 1$, we expect the scattered intensity to be independent of the cluster mass since the scattering from chunks of the cluster of size $1/q$ add incoherently[15]. Using $R^D \sim M$ we deduce that the function $F(qR)$ must be a power law; $I/c \sim M(qM^{1/D})^x \sim M^0$. This is solved by $x=-D$, which gives the standard relation for the scattering from a monodisperse ensemble of fractals, $I/c \sim q^{-D}$ for $qR \gg 1$.

This form of the scattering function is not useful, however, unless the sol is fractionated before measurements are made. To compute the scattering function for the unfractionated sol, we must average the scattering over the number distribution function $N(m)$. This is readily accomplished by noting that the polydispersity-averaged scattering function can be written in the form $I(q)/c \sim M_w F(qR_z)$. Again, the Guinier expansion is $I/c \sim M_w(1-q^2R_z^2+...)$, so the *weight-average* mass and *z-average* radius are obtained from the intercept and slope of I/c versus q^2. On length scales smaller than the correlation length ($qR_z \gg 1$) the intensity will become independent of the correlation length, and therefore of M_w. Using $R_z^{D(3-\tau)} \sim M_w$ then gives[15,16] $I/c \sim M_w(qM_w^{1/D(3-\tau)})^x \sim M_w^0$, so $x=-D(3-\tau)$ and $I/c \sim q^{-D(3-\tau)}$ for $qR_z \gg 1$. Note that for a system with a sol-gel transition ($\tau > 2$) the scattering exponent is $D(3-\tau)$, whereas for a system which does not exhibit a sol-gel transition ($R_z^D \sim M_w$) the scattering exponent is D.

In summary, our results for quantities that can be measured in a scattering experiment are:

$$\text{divergences:} \qquad R_z \sim \varepsilon^{-1/\sigma D} \qquad\qquad M_w \sim \varepsilon^{-(3-\tau)/\sigma}$$

$$\text{structure:} \qquad M_w \sim R_z^{D(3-\tau)} \qquad\qquad I/c \sim q^{-D(3-\tau)}$$

These expressions are valid for both the percolation model and the FS theory: in order to make comparison to experiments, we need only the numerical values of these exponents. For the percolation model the exponent τ can be computed from the *hyperscaling* relation[9] $\tau = 1 + d/D$ (this relation does not apply to the dimension-independent FS theory). For three-dimensional percolation $D \cong 2.5$, so $\tau \cong 2.2$. The exponent σ is known to be ca. 0.45. Thus the *strict* percolation predictions are $R_z \sim \varepsilon^{-0.88}$, $M_w \sim \varepsilon^{-1.76}$, $R_z^{2.0} \sim M_w$, and $I/c \sim q^{-2.0}$. In contrast, the Flory-Stockmayer theory gives $R_z \sim \varepsilon^{-1/2}$, $M_w \sim \varepsilon^{-1}$, $R_z^2 \sim M_w$, and $I/c \sim q^{-2}$. It would seem that an intermediate scattering experiment would be unable to distinguish between these theories, since in each case the exponent $D(3-\tau) \cong 2$. However, it is not clear that these are the exponents which would actually be measured in an experiment on the diluted sol, since upon dilution it is thought that the sol clusters might swell.

Swelling of the Diluted Sol

Using mean field, Flory-type arguments, Isaacson and Lubensky[17], and de Gennes[18], have computed the fractal dimension of both diluted and undiluted sol clusters. In the following, we summarize these results. The Flory formula for the free energy is

$$\frac{F}{k_B T} = \frac{R^2}{R_0^2} + \frac{v}{N_w} \rho^2 V \qquad (4)$$

The first term, R^2/R_0^2, is a purely elastic term that depends on the radius of the cluster in its unperturbed state, where two-body interactions vanish. The second term counts screened, repulsive, two-body interactions in terms of the mean monomer density of a typical cluster, $\rho = N/V$, where the cluster volume is taken to be R^d. The energy of an interaction is v/N_w, where v is the 'bare' two-body interaction parameter and N_w is the weight-average molecular weight of the surrounding medium. In the undiluted sol, N_w will be proportional to some power of the typical cluster size, i.e. $N_w \sim N^b$; in a sol diluted by a simple solvent, however, $N_w \sim$ const. and $b=0$. This screening of two-body interactions is not intuitively obvious, but it is well-founded theoretically.[19]

The equilibrium radius is determined by minimizing the free energy given by eq. 4. However, this equation is very general- it can be applied to linear polymers in melts, good solvents etc.- so we must somehow specify some *additional* parameter of our gelling system in order to apply this equation. The additional parameter we supply is the fractal dimension

433

of a single cluster at (or above) the critical dimension, D_c. We can then use eq. 4 to compute the upper critical dimension, d_c, for our system and from d_c we can find the fractal dimension of our clusters as a function of the spatial dimension.

At the upper critical dimension two-body interactions effectively vanish, so the equilibrium radius is equal to the unperturbed radius, $R=R_0$, where $R_0^{D_c}$~N. Using this in eq. 4 it can be shown that $d_c=D_c(2-b)$. Minimizing the free energy for dimensions less than the upper critical dimension then gives the dependence of the fractal dimension on d.

$$D = \frac{d_c}{(2-b)} \frac{(d+2)}{(d_c+2)} \qquad (5)$$

To apply this to percolation clusters ('branched' polymers) we note that at the critical dimension the FS theory is exact, i.e. $D_c=D_{FS}=4$. In the case of *undiluted* percolation clusters we use the mean field theory result, N_w~$1/\varepsilon$~$N^{1/2}$, to obtain b=1/2. This gives an upper critical dimension $d_c=4(2-1/2)=6$. For *dilute* clusters N_w is a constant (b=0), and $d_c=4(2-0)=8$.

The fractal dimension of undiluted percolation clusters, $D=(d+2)/2$ $(d \leq d_c)$, is then obtained from eq. 5 with b=1/2, $d_c=6$. In 3 dimensions and higher this expression is in very good agreement with simulations (D=2.5 for d=3), but in 2 dimensions this Flory-type formula predicts nonfractal clusters, in contrast to the accepted value D=1.89. We can combine this expression with the hyperscaling formula $\tau=1+d/D$ to obtain $\tau=(3d+2)/(d+2)$, from which the FS result $\tau=2.5$ is obtained for $d=d_c=6$.

Finally, the fractal dimension of clusters diluted by a *good* solvent, $D=2(d+2)/5$, may be obtained from eq. 5 by setting b=0 and using $d_c=8$. In 3 dimensions this gives D=2, in marked contrast to the undiluted, unswollen value of 2.5. Using D=2, $\tau=2.2$, and $\sigma=0.45$ then gives the following percolation (FS in parenthesis) predictions for the divergences of size parameters[20], and the intermediate scattering[16] from the swollen, diluted sol:

divergences:	R_z ~ $\varepsilon^{-1.11(1/2)}$	M_w ~ $\varepsilon^{-1.76(1)}$
structure:	M_w ~ $R_z^{1.6(2)}$	I/c ~ $q^{-1.6(2)}$

We will now compare these predictions to experimental measurements on various chemical gels.

SCATTERING MEASUREMENTS

Intermediate Scattering

Scattering measurements have been reported for the diluted sol of the tetramethoxysilicon/H_2O/methanol system by Martin and Keefer.[21] In this study the reaction conditions were varied by changing the initial TMOS concentration, and by changing the method of catalyzing both the hydrolysis and condensation to form Si-O-Si bonds. Typical results for the scattered intensity are shown in Fig. 4. For large values of ε the correlation length is small and $qR_z < 1$ over the experimental q regime. In this limit the scattering function is described by $I/c \sim M_w(1-q^2R_z^2+...)$. However, close to the gel point the correlation length becomes much larger than the largest experimental length scale and pure power-law scattering is observed[8], with $I/c \sim q^{-1.67 \pm 0.06}$. This exponent is in good agreement with the percolation prediction of 1.60, but does not support the mean field prediction of $I/c \sim q^{-2}$.

These light scattering measurements are also in good agreement with the intermediate neutron scattering measurements of Bouchaud, Delsanti, Adam, Daoud, and Durand,[22] on polyurethanes. In the polyurethane study, scattering measurements were made on the diluted sol both before and after fractionation. The unfractionated sol gave a scattering exponent of $D(3-\tau)=1.59 \pm 0.05$, whereas the fractionated sol gave $D=1.98 \pm 0.03$. From this it can be concluded that $\tau = 2.2 \pm 0.04$, in agreement with percolation.

Radius versus Mass

Measuring the divergence of the average mass and radius is difficult, due to uncertainties in determining the gel point. However, if R_z is plotted against M_w, the exponent $D(3-\tau)$ can be determined without any knowledge of the gel point. This experiment has been reported by a number of research groups.

From light scattering measurements in the 'Guinier' regime ($qR_z < 1$) Schmidt and Burchardt[23] were able to measure R_z and M_w as functions of ε for styrene anionically copolymerized with divinylbenzene. They found $M_w \sim R_z^{2.0}$, in very good agreement with the FS prediction $D(3-\tau)=2$ (or the *unswollen* percolation model), but in poor agreement with the swollen percolation prediction of 1.6. Schosseler and Leibler[24] reported low-angle light scattering measurements of R_z and M_w for the γ-radiation induced 'vulcanization' of polystyrene. From these data they found $D(3-\tau)=1.72$, again in good agreement with the swollen percolation model. Martin, Wilcoxon and Adolf[8] used low-angle light scattering on the acid-catalyzed TMOS/H_2O/methanol system, Fig. 5, and found $D(3-\tau)= 1.79 \pm 0.1$, in experimental agreement with the intermediate scattering exponent of 1.67 ± 0.06 for the same system.

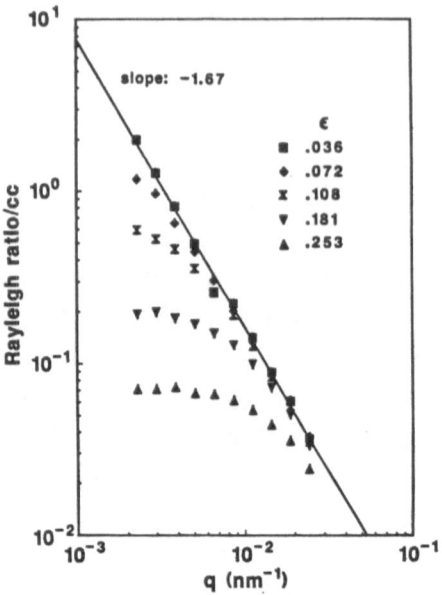

Fig. 4: The light scattering intensity of sols diluted at various extents of reaction[8,21] is shown to approach a power law near the gel point. The scattering exponent, 1.67±0.06, is in good agreement with the swollen percolation prediction of 1.60.

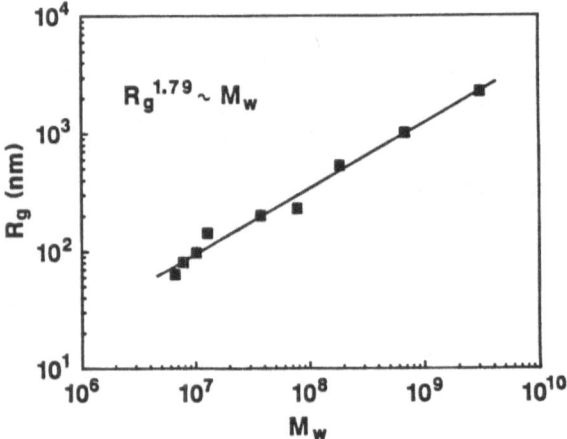

Fig. 5: Light scattering measurements of the z-average radius, R_z, and weight-average mass, M_w, of a dilute silica sol, are plotted to verify the relation $R_z^{D(3-\tau)} \sim M_w$. The exponent $D(3-\tau)$ is found to be 1.79±0.1, in agreement with the light scattering data for the same system, shown in Fig.4.

436

Divergence of the Radius and Mass

Thus far, the experimental data have supported the percolation model, however, we have not yet discussed measurements of the divergences of M_w and R_z. Relatively little data is in the literature, since the errors in these measurements can be substantial. However, Martin, Wilcoxon and Adolf[8] have have recently reported low-angle static light scattering measurements of M_w and R_z, and *dynamic* light scattering measurements of the z-average hydrodynamic radius, R_h. All of these measurements were made on the TMOS/H_2O/methanol system, using either acid or base catalysis. The largest source of error in the measured exponents is due to uncertainties in the gel time. All of the quoted errors reflect *this* uncertainty, and not the relatively small error obtained from the least squares fit to the data. We have already mentioned the divergence of M_w, Fig. 3, which gave $M_w \sim \varepsilon^{-2.7\pm0.3}$. This exponent is in disagreement with either the percolation prediction of $\gamma=1.76$, or the mean field prediction of $\gamma=1$, and there does not seem to be any obvious experimental source of this disagreement. Likewise, the z-average radius of gyration, Fig. 6, was found to diverge as $\varepsilon^{-1.52\pm0.2}$, again in disagreement with either the *swollen* percolation prediction of $\nu=1.11$, or the mean field prediction of $\nu=1/2$. The ratio of these exponents gives $D(3-\tau)=1.76$, which of course agrees with Fig.5.

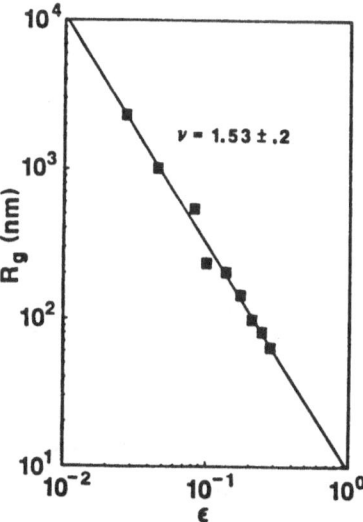

Fig. 6: A critical plot of the z-average radius of the TMOS/H_2O/methanol system, obtained from static light scattering, gives $\nu=1.52\pm0.2$, in disagreement with the percolation prediction of 1.11 or the FS prediction of 1/2.

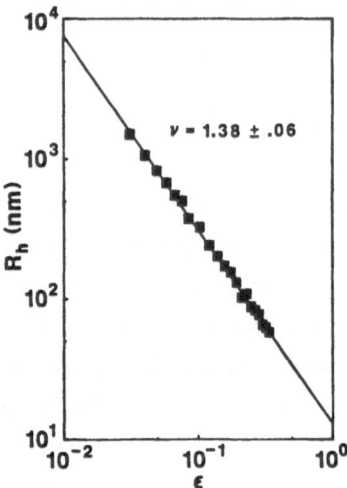

Fig. 7: The z-average hydrodynamic radius of diluted TMOS sols is seen to give a critical exponent[8] of 1.38±0.06, in agreement with the divergence shown for the static radius (Fig. 6), but in disagreement with the percolation prediction of 1.11, or the FS prediction of 1/2.

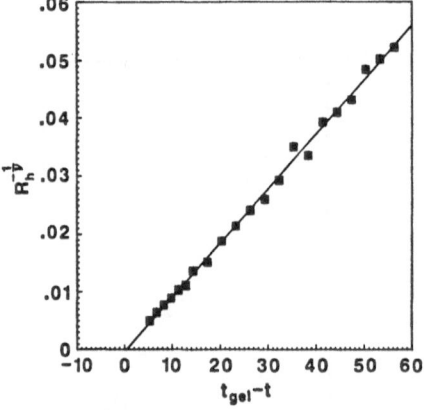

Fig. 8: The data in Fig. 7 are linearized[8] using the experimental value of v, so that a linear extrapolation of the data is possible. The extrapolated gel point is found to be within .34 minutes of the estimated gel time, which is well within the estimated uncertainty of .75 minutes for t_{gel}.

Data for the hydrodynamic radius[8], Figs. 7&8, are in accord with the static measurements. The details of the dynamic light scattering measurements and their interpretation will be deferred until later, suffice it to say that it is expected that $R_h \propto R_z$. Experimental measurements bear this out: the hydrodynamic radius diverges as $R_h \sim \epsilon^{-1.38 \pm 0.06}$, in good agreement with the value $\nu = 1.52 \pm 0.2$ obtained from the static measurements. As a matter of interest, the hydrodynamic radius is linearized and plotted against t_{gel}-t in Fig. 8. The point of this exercise is to demonstrate that a linear extrapolation of these data comes within .34 minutes of the estimated gel time. This is within the estimated experimental uncertainty of .75 minutes for t_{gel}, demonstrating that the estimated error of ± 0.06 in our exponent is very reasonable.

Size Distribution

We have already mentioned the neutron scattering measurements of Bouchaud et al.[22] on fractionated and unfractionated sols. These measurements, we recall, gave a polydispersity exponent of $\tau = 2.2 \pm 0.04$, in excellent agreement with the percolation model ($\tau = 2.2$). Taking a direct approach, Leibler and Schosseler[25] used size exclusion chromatography to fractionate γ-irradiated polystyrene sols. They were able to demonstrate the scaling form of the number distribution, eq. 3, and found $\tau = 2.3 \pm 0.1$. Again, this exponent is in reasonable agreeement with percolation, but is somewhat smaller than the FS value $\tau = 5/2$.

SOLVENT EFFECTS

Correlated Percolation

Thus far the effect of solvent has been completely neglected, in order to clarify the discussion. However, in the TMOS/H_2O/methanol system the concentration of monomer can be quite low (less than 0.3 %), so the effect of the solvent cannot be ignored. In fact, in a completely random system at these concentrations, the monomers themselves will not *site* percolate. Therefore, in the absence of monomer-monomer attractions (e.g. an *athermal* solvent), there will not be a connected path of infinite extent of neighboring silica monomers, and the system cannot *bond* percolate without the introduction of spatial correlations in the final gel.[11] Thus, on general grounds we are led to conclude that in the presence of a solvent there are two correlation lengths in the final gel: a connectivity correlation length R_z, and a spatial correlation length ξ. Since bond percolation is a transition in connectivity alone, it cannot account for the presence of a correlation length (larger than a monomer) in an undiluted gel; percolation must be modified to describe low-density equilibrium gels.

One approach to this problem has been to correlate the occupied sites[9,11,26] by introducing an exchange energy J. In the simplest form of this model, adjacent sites are assumed to be bonded. The qualitative behavior of this model is as follows; (1) At high 'temperatures' $(J \ll k_B T)$ sites are randomly distributed, so the number of bonds is $O(Np^2)$, where N is the size of the lattice and p is the occupation probability. (2) As the temperature is progressively reduced, spatial correlations will increase, until at some point an infinite cluster of connected sites emerges. At this temperature there is a sol-gel transtition with percolation exponents, but the spatial correlation length will still be finite, and the pair correlations will be described by the Ising model. (3) Finally, as the temperature is further reduced the system will reach the phase boundary. Note that if the site concentration is too low, the system can phase separate without passing through a sol-gel transition.

If the sol-gel transition occurs reasonably close to the phase boundary it should be possible to observe Ising exponents in the gel, e.g. $I \sim q^{-2}$. Experimentally, one can move the sol-gel transition close to the phase boundary by reducing the initial concentration, however, there is no real guaranty that this will be sufficiently close to the critical concentration to observe critical exponents.

Kinetic Description

To describe dilute irreversible gelation we may adopt a kinetic description of the gelation process. At early times, we imagine a thermodynamically stable dilute solution of monomers which diffuse, collide, and just generally have a good time. This friendly state of affairs is abruptly terminated by the addition of a catalyst, whereupon diffusing monomers, of radius b, collide and react to form clusters. Depending on the activation energy of the forward reaction, this clustering process will be either reaction or diffusion limited. If the bond formation energies are high, the reaction is essentially irreversible (no fragmentation) and the randomness will be *quenched* on the experimental time scale. If we assume that in this early stage of the reaction fractal clusters of dimension D are formed, then the volume fraction of these clusters, $\phi = (R/b)^{d-D}\phi_0$, increases from the initial monomer volume fraction ϕ_0 as the average radius increases. Eventually the volume fraction approaches unity, at a critical overlap radius $\xi = b/\phi^{1/(d-D)}$, and fractal *spatial* correlations no longer develop. Thus, a scattering experiment on an undiluted gel should find a correlation length proportional to $\phi^{-1/(d-D)}$, and should demonstrate intermediate scattering of the form $I \sim q^{-D}$.

In the second stage of this kinetic growth process the *connectivity* correlation length increases as the clusters react to form very large *overlapping* clusters. Eventually, there is a sol-gel transition. A number of questions arise; (1) What is the fractal dimension of the overlapping clusters, made of 'monomers' of radius ξ? (2) If the system gels, what is the critical behavior near the gel point? A kinetic model of this type has been proposed by Kolb and Herrmann,[27] who extended two-dimensional, diffusion-limited cluster-cluster

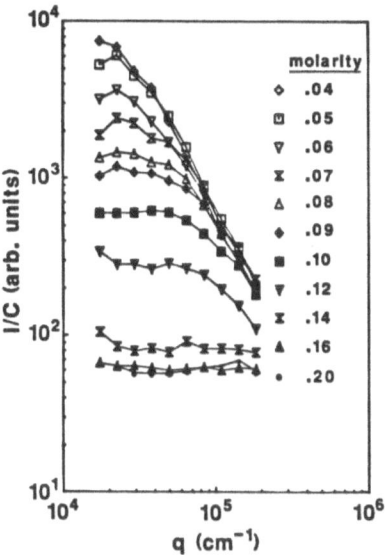

Fig. 9: Light scattering data taken from gels made from very dilute solutions of TMOS in methanol demonstrate the growth in the spatial correlation length as the molar concentration of TMOS is diminished. At the lowest concentrations, the correlation length approaches 1 micron.

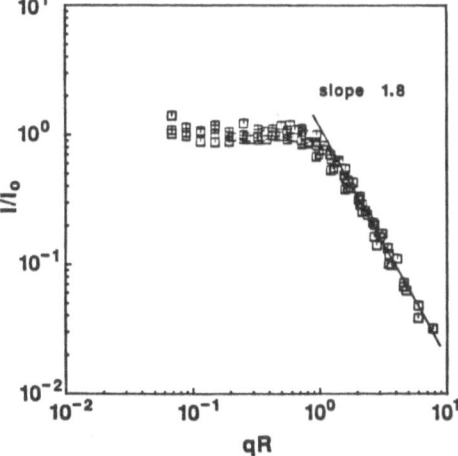

Fig. 10: The light scattering data in Fig. 9 are plotted on master axes, to find the dependence of the correlation length on the initial monomer concentration. These data are collapsed by a fractal dimension of 2.3, although the measured scattering exponent is 1.8.

aggregation simulations to high initial particle densities. In the second stage of growth, they found a fractal dimension of 1.75±0.07, which is much larger than the low-density value for diffusion limited aggregation, D=1.42±0.05, and somewhat smaller than the percolation value of 1.89. More important, for the physically reasonable case where the cluster mobility depends on the inverse radius they did not observe a gel at finite time. It is noteworthy that reaction-limited aggregation, with $\tau=2$, is much closer to gelation. Therefore it seems possible that reaction-limited aggregation might give a sol-gel transition at finite time, and might give a physically reasonable description of irreversible gelation.

Measurement of Correlations

In order to investigate the effect of solvent, very dilute TMOS gels were made, in the range of .03 to .2 molar TMOS. The static light scattering data from the undiluted gels are shown in Fig. 9. At the highest concentrations, the scattered intensity is independent of q over the experimental range 25 nm $< q^{-1} <$ 400 nm. At low TMOS concentrations, however, the spatial correlations start to enter the experimental length scale window, and roughly power-law scattering becomes evident. A data collapse, Fig. 10, can be made by plotting on the axes $I/(\phi\xi^D)$ versus $q\xi$, with $\xi=\phi_0^{-1/(d-D)}$. The collapse is achieved with a fractal dimension of 2.3, but the apparent slope of the scattering curve is closer to 1.8, so these data do not support the very simple kinetic description of growth given above. Neither do they agree with the correlated percolation prediction, $I\sim q^2$, although this may be due to the fact that we are too far from the critical point.

DYNAMIC ASPECTS

We have seen that the interpretation of the static properties of sol-gels, obtained from elastic scattering data, is reasonably straightforward, and that the observed *scattering* exponents are reasonably well accounted for by the swollen percolation cluster model. We will now address the more complex issue of using quasielastic light scattering to understand the dynamics of these clusters. These *dynamic* scattering data are sensitive to less universal aspects of the sol- e.g. the flexibility of the clusters- so the interpretation of these data is somewhat more involved.

Before entering a more detailed discussion of the quasielastic scattering experiment, it is instructive to give a rough description of what is measured. A static light scattering experiment probes density correlations on a length scale q^{-1}; a dynamic light scattering experiment probes the relaxation times of processes which relax on some length scale q^{-1}. For example, suppose one is scattering from a particle undergoing anomalous diffusion in a fractal space, i.e. $R^{d_w}\sim t$, where R is the r.m.s excursion of the particle after some time t. The observed relaxation time will just be the time it takes the particle to move a distance q^{-1},

$\tau_o \sim q^{-d}w$. Thus in this simple case the exponent on q gives the dimensionality of the walk (e.g. in the Brownian case the relaxation *rate* is $D_t q^2$).

In a quasielastic light scattering experiment the autocorrelation function of the scattered field, $S(q,t) \propto <E_s(q,t)^* E_s(q,t)>$, is determined. All of the dynamical processes of the clusters- translational and rotational diffusion, and internal modes- contribute to some degree to this correlation function. The difficulty lies in determining the dominant relaxation processes that give rise to the highly nonexponential correlation functions observed for the sol. In the regime $qR_z \ll 1$ this is simple enough; here translational diffusion dominates and the *dynamic structure factor* for a cluster of mass m is[16] $S_m(q,t) = S_m(q) \exp(-q^2 D_t t)$, where D_t is the translational diffusion coefficient, $S_m(q)$ is the static structure factor [$S_m(0)=1$], and $S_m(0,0)=1$. Under the assumption of strong hydrodynamic interactions the diffusion coefficient is ζ/R_h, where $\zeta = k_B T/6\pi\eta_o$, η_o is the solvent viscosity, and R_h is the hydrodynamic radius. However, as $qR \rightarrow 1$ this simple diffusion analysis breaks down since internal modes can contribute to the relaxation. In fact, for large qR internal modes may completely dominate the observed relaxation.

A full description of the internal modes of the sol clusters would be very complex. However, even though detailed models are not yet available, there are a few general comments one can make concerning the dynamics of these clusters, since many of the interesting features of the observed relaxation are a simple result of polydispersity. Further, we will show that the effects of rotations and flexibility can be taken into account for systems which exhibit a sol-gel transition. In the following, we first discuss the dynamics expected for a sol comprised of rigid, nearly spherically symmetric clusters. We then consider the case of flexible clusters.

Rigid Clusters

Although many physical systems exhibit a spectrum of relaxation times, very often correlation functions of these systems can be expressed as a function of a nondimensioned time, t/τ_o, where τ_o is an average relaxation time for the system. For the dynamic structure factor, it is often useful to ask how this time scale depends on the product of momentum transfer times a length. The most surprising feature of correlation functions of rigid power-law polydisperse systems is the presence of *three* different time scales[6], each of which depends on a different power of the momentum transfer when $qR \gg 1$. Because of these multiple time scales it is not possible to write the dynamic structure factor as a function of a single nondimensioned time, t/τ_o, as it is for flexible systems, such as linear polymers. In other words, for $qR_z \gg 1$ the *shape* of the correlation function will depend on the momentum transfer. (In the following, subscripted τ's will always refer to relaxation times, whereas an unscripted τ will continue to denote the polydispersity exponent.)

It is convenient to refer to the three time scales as short, τ_s, intermediate, τ_i, and long, τ_l. The short and intermediate relaxation times can be obtained from the dynamic structure in a model-independent way. The short time is just the inverse of the Rayleigh linewidth, Γ, which is obtained as the first cumulant of the dynamic structure factor,

$$\tau_s^{-1} = \Gamma = \frac{d \ln S(q,t)}{dt}\bigg|_{t=0} \qquad (6)$$

This linewidth, which is just the *harmonic* average of the spectrum of relaxation times, $\langle 1/\tau_k \rangle$, describes the exponential time decay, $S(q,t) = S(q)e^{-t/\tau_s}$, of the correlation function on times short compared to τ_s.

The intermediate time scale[27] is the integral of the correlation function,

$$\tau_i = \int_0^\infty \frac{S(q,t)}{S(q,0)} \, dt \qquad (7)$$

This time scale is the *arithmetic* average of the relaxation times, $\langle \tau_k \rangle$, so τ_i is always larger than τ_s. Finally, the longest time scale[6] is $\tau_l = 1/q^2 D_z$ where D_z is the z-average diffusion coefficient, ζ/R_z^{d-2}. This time dominates the long-time behavior of the correlation function, where we may write $S(q,t \gg \tau_l) = f(t/\tau_l)$.

For *rigid*, power-law polydisperse scatterers a rich behavior is found for the short and intermediate time scales. Specifically, these time scales are found to depend on fractional powers of the momentum transfer. This behavior is obtained by applying eqs. 6&7 to the experimentally observed structure factor,

$$S(q,t) = \frac{\displaystyle\int_1^\infty m^2 S_m(q,t) N(m) \, dm}{\displaystyle\int_1^\infty m^2 N(m) \, dm} \qquad (8)$$

In this way Martin and Leyvraz[28] find the relations $\tau_s = \tau_l f(qR_z)$ and $\tau_i = \tau_l h(qR_z)$, where the functions $f(x)$ and $h(x)$ are simple constants for $x \ll 1$ but are the power laws $f(x) \sim x^{2-\alpha}$ and $h(x) \sim x^{2-\beta}$ for $x \gg 1$. The exponents α and β depend on both the fractal dimension of the scatterers and the polydispersity exponent, Fig. 11. In the $qR_z \gg 1$ regime $\tau_s \sim R_z^{d-\alpha}/\zeta q^\alpha$ and $\tau_i = R^{d-\beta}/\zeta q^\beta$, where

$$\alpha = \begin{cases} 2 & \tau < 2-(d-2)/D \\ d-D(2-\tau) & 2-(d-2)/D < \tau < 2 \\ d & \tau > 2 \end{cases} \qquad \beta = \begin{cases} 2 & \tau < 2 \\ 2+D(\tau-2) & 2 < \tau < 2+(d-2)/D \\ d & \tau > 2+(d-2)/d \end{cases} \qquad (9)$$

If these results are applied to *rigid* percolation clusters one obtains $\tau_s \sim 1/\zeta q^d$ and $\tau_i \sim R_z^{d-D+\eta}/\zeta q^{D-\eta}$. Since the exponent η is very nearly zero for percolation in 3 dimensions, the q-dependence of τ_i is essentially determined by the fractal dimension of a single cluster.

In the colloidal silica system an anomalous fractional scaling of the linewidth, with $\Gamma \sim q^{2.7}$, has been experimentally observed[29,6] under slow aggregation conditions. From the above analysis, this indicates a polydispersity exponent of ~1.9 for this system. In contrast, under rapid aggregation conditions[5] $\Gamma \sim q^2$, indicating that the polydispersity exponent must be less than ~1.5. These estimates for the polydispersity of rapidly and slowly aggregating systems are in agreement with simulations.

In short, we have seen that the behavior of the short and intermediate relaxation times is dominated by the usual assortment of exponents, and is independent of the forms of the various cut-off functions mentioned in passing. In contrast, the long-time tail of the correlation function tends to be dominated by the form of the cut-off function for the size distribution. On reflection, this is not too surprising; at long times only the very slowest processes contribute to the correlation function. These slow processes are due to the diffusion of the 'exponentially rare' clusters, whose mass is larger than the typical mass. For example, if the cut-off function $f(m/M_z)$ for the size distribution is of the form $f(x) \sim \exp(-x^b)$, then at very long times the correlation function will be a stretched exponential,[6]

$$S(q,t) \approx A(t/\tau_i)\, e^{-(ct/\tau_i)^\delta} \tag{10}$$

with $\delta = bD/(bD+1)$, and $A(x) = x^{\delta[(2-\tau)/b - 1/2]}$. Van Dongen and Ernst[30] have recently

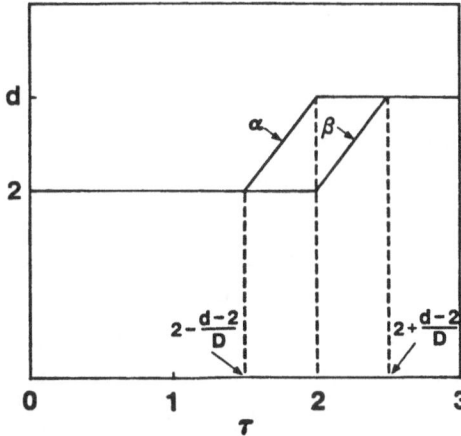

Fig. 11: Exponents for the short and intermediate time scales of the dynamic structure factor for rigid scatterers ($qR_z \gg 1$) are plotted against the polydispersity exponent. At $\tau=2$ a maximum separation of time scales occurs.

shown that for the Smoluchowski equation the cut-off function is universally an exponential, thus $\delta = D/(D+1)$ for rigid systems described by the kinetic rate equation. This stretched exponential behavior has been observed for the colloidal silica system.[6] For the percolation model, the cut-off function appears to be a Gaussian, so $\delta = .8$.

Flexible Clusters and Rotations

For systems that exhibit a sol-gel transition, a few general comments can be made concerning the effects of flexibility and rotational diffusion. If we assume the rotational and internal modes are only weakly coupled to translational diffusion, then the dynamic structure factor of a cluster of mass m may be written in the form[16,28]

$$S_m(q,t) = e^{-q^2 D_t t} H(qR; \Omega t) \qquad (11)$$

where Ω is a characteristic frequency for the scatterer. For rigid rotators this characteristic frequency is the rotational diffusion coefficient $\theta \propto D_t/R^2$. It is somewhat surprising that for completely flexible systems- where the relaxation time of a mode of length $1/q$ is independent of the radius of the scatterer- the characteristic frequency ($\sim$ the longest relaxation time) has the same form, $\Omega \propto D_t/R^2$. Linear polymers (free- or non-draining, theta or good solvent) are prototypical flexible systems if the persistence length is small compared to the radius. Of course, a flexible system will have many internal modes, but the relaxation rates of these modes may be written as a product of the characteristic frequency times some function, often a power, of the mode index, and since this index is summed over in calculating $S_m(q,t)$, eq.11 is still valid. Thus we conclude that in the case of strong hydrodynamic interactions, where the diffusion coefficient is $D_t \sim \zeta R^{2-d}$, the characteristic frequency is $\Omega \sim \zeta R^{-d}$.

This $\Omega \sim D_t/R^2$ scaling of the characteristic frequency has a very simple physical interpretation; it is the dynamical analog of dilation symmetry. Dilation symmetry may be defined as the inability to determine a length scale from the static structure. The analogous dynamical symmetry is the inability to determine a length scale from the dynamical structure- the particle trajectories. For example, suppose one is given a detailed history- in the form of a movie- of the dynamics of a sol ensemble. Can one determine the length scale? In the absence of any absolute time or length scale, the best we can do is analyze the trajectories in a relative manner. For example, we define our fundamental clock as the time it takes a cluster with a radius equal to the frame size (our *only* standard of length) to rotate through unit angle. From the rotational diffusion equation we know this time is θ^{-1}. We then ask how far this cluster will diffuse in this time. Using $D_t \sim R^2$ gives $R^2 \sim D_t/\theta \sim L^2$. Thus in the time it takes a cluster to move through a unit angle it diffuses a distance proportional to its own length. In other words, the complete cluster trajectories, including all normal modes, are self similar; the trajectory of a small cluster, of radius R, is statistically identical to the trajectory of a cluster of size R'.

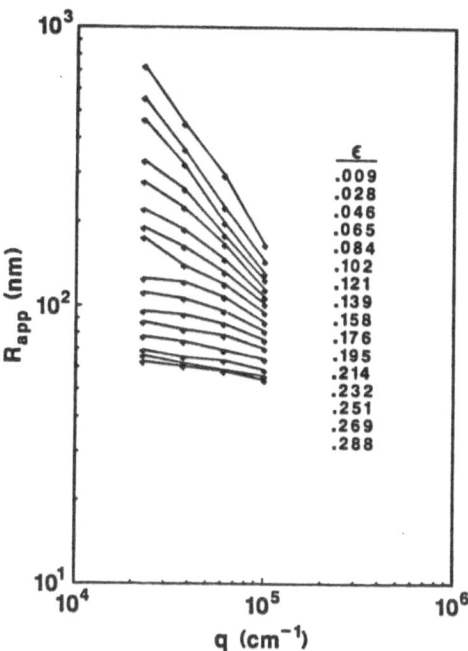

Fig. 12: The apparent hydrodynamic radius is shown as a function of the momentum transfer for dilute silica sol clusters at various times before the gel point.[8] The highest curve is the closest to the gel point. A gradual crossover can be seen from $R_{app} \sim R_z$, far from t_{gel}, to $R_{app} \sim 1/q$, close to t_{gel}.

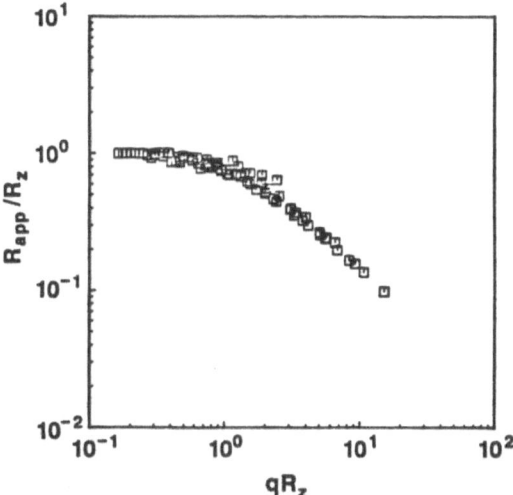

Fig. 13: The dynamics data in Fig.12 are collapsed[8] by plotting on the nondimensioned axes R_{app}/R_z versus qR_z. This data very clearly illustrates the crossover from $R_{app} \sim R_z$ to $R_{app} \sim 1/q$ as qR_z becomes $\gg 1$.

For a dilute sol of flexible clusters, the momentum transfer dependence of the mean relaxation rate $\Gamma=<1/\tau_k>$ can be obtained by a simple argument. From eqs. 6&11 the linewidth can be shown to be of the form $\Gamma=\Omega h(qR)$. Now suppose we examine the system at two different length scales, $1/q$ and $1/q^*$ (imagine these as 'windows' of size $1/q$), both chosen to satisfy $qR_z>>1$. The static structure of the system will be invariant because the individual clusters are fractal, and the size distribution is a power law (i.e. the relative population of m-mers and n-mers depends only on n/m). In effect, clusters of radius R have simply been replaced by statistically identical clusters of radius $R^*=Rq/q^*$. However, although the static properties are scale invariant, the dynamic properties are not, since smaller clusters give rise to shorter relaxation times. At q^*, the linewidth will be $\Gamma(q^*)=\Omega^*h(q^*R^*)$, and using our expression for the characteristic frequency, $\Omega\sim\zeta R^{-d}$, gives $\Gamma(q^*)=(q^*/q)^d\Gamma(q)$. Thus when we have a change of length scale, the observed relaxations dynamic processes simply change in proportion to the volume, i.e. $\Gamma\sim\zeta q^d$!

Alternatively, one may proceed by directly averaging $S_m(q,t)$ over the size distribution. In this way it can be shown that if there is a sol-gel transition (i.e. $\tau>2$) then the Rayleigh linewidth is given by[16] $\Gamma\sim\zeta q^d$ for $qR_z>>1$. Furthermore, for a flexible system it can be demonstrated that regardless of the polydispersity there is a single time scale for the correlation function if $qR_z>>1$, so the correlation function at fixed q can be written as a function of a single nondimensioned time Γt, where $\Gamma\sim\zeta q^d$. Thus the primary difference between a flexible and rigid polydisperse sol is in the momentum transfer dependence of the arithmetic-average relaxation time, τ_i. For flexible clusters $\tau_i\sim q^{-d}$, whereas for rigid systems $\tau_i\sim q^{-D}$.

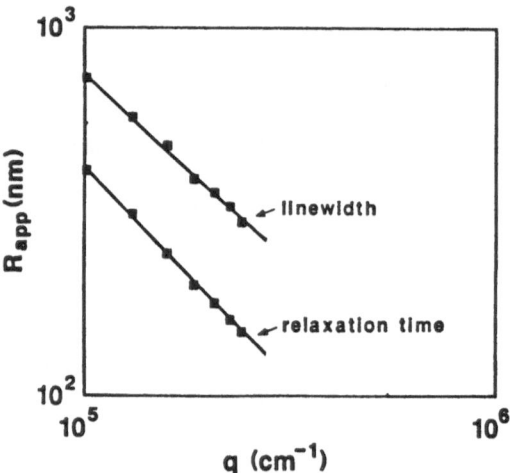

Fig. 14: Dynamics data[21] for the harmonic (top line) and arithmetic average relaxation time of dilute silica clusters show that for large q the apparent hydrodynamic radius scales as 1/q. This implies that the sol clusters are flexible.

Dynamic light scattering measurements were taken on the diluted sol of the TMOS/H_2O/methanol system.[8,21] The experimental data for the harmonic average relaxation time, τ_s, are shown in Fig.12 for various values of $|t_{gel}-t|$. In this figure the data are plotted as an *apparent* hydrodynamic radius, $R_{app} \equiv \zeta q^2 \tau_s$. From our discussion on the dynamics we know that R_{app} has the scaling behavior $R_{app}=R_z G(qR_z)$, where $G(x<<1)=1$ and $G(x>>1)=1/x$ (see eq. 9 with $\tau>2$). For example, for $\varepsilon>0.2$ the correlation length is small and $qR_z<<1$; in this regime the apparent radius is completely determined by the correlation length, $R_{app}=R_z$. However, very close to the gel point ($\varepsilon<0.05$), where $qR_z>>1$, R_{app} becomes nearly independent of the correlation length and we approach the limit $R_{app}\sim 1/q$. In Fig.13 these data are collapsed on the nondimensioned axes R_{app}/R_z versus qR_z. The data collapse illustrates the scaling behavior more clearly; the shift factors for the data collapse were actually used to make the critical plot in Fig.7.

Finally, in Fig.14 we show the q dependence of both the harmonic and arithmetic average relaxation times, again in terms of an apparent hydrodynamic radius. It is observed that both of these relaxation times scale like q^{-3}, in support of the interpretation of the sol clusters as flexible.

CONCLUSIONS

Although a variety of aspects of gelation have been discussed, we have not been able to paint a definitive picture of the sol-gel transition. In part, this is due to the paucity of experimental data in this field. But at least a few tentative conclusions can be drawn;

1. The static structure of the fractionated and unfractionated sol seems to be fairly well accounted for by the percolation model if swelling is taken into account.

2. The divergences of the average cluster mass and radius are stronger than predicted by the percolation model, and much stronger than the Flory-Stockmayer predictions.

3. Measurements of the polydispersity exponent are in good agreement with percolation.

4. The static structure observed in the undiluted, low-density silica gels are in reasonable agreement with either an equilibrium model, such as correlated percolation, or a kinetic growth model. Discrepancies exist in either case.

5. The dynamics data indicate that the sol clusters are flexible.

At this point both theoretical and experimental work are needed to understand the sol-gel transition. For example, a kinetic model which describes polycondensation, yet gives a gel point in finite time, might give a useful description of the gel point. And experimental work needs to be extended to a wide variety of systems, to test for universality at the gel point.

OPTICAL INVESTIGATIONS OF AGGREGATION PROCESS IN AQUEOUS NOBLE

METAL COLLOID SYSTEMS

U. Kreibig, M. Quinten, and D. Schönauer

FB 11-Physik, Universitätdes Saarlandes
D-6600 Saarbrücken/Germany

INTRODUCTION
Small,spheroidal inclusions in an otherwise homogeneous matrix yield simple model substances for investigations of inhomogeneous/disordered materials.
Two-component-materials consisting of metallic particles in dielectric embedding media are of special interest, both, for basic and for technical
purposes (e.g. Cermets).
Even in such apparently simple systems, physical properties vary over extreme ranges (e.g. the electric conductivity may change for 10 decades),
depending on the amount of particles, the particle sizes and the particle
topology.
So, investigation of disordered materials mainly means to investigate influences of the sample topology. We present experimental and theoretical
results of the effect of variations of topology on the optical extinction
of such small particle systems (1).

SAMPLE PREPARATION
We prepared our samples starting from diluted aqueous Zsigmondy-type noble
metal colloids (2). The particles are nearly spherical, of narrow size distributions, and typical mean diameters range from 3 to 100 nm.
The topography of these primary samples is the diluted statistically disordered geometrical arrangement of the particles.
The sample topography can extremely be modified if the particles are enab-
led to form
- coagulation aggregates with electrically separated, close packed particles (3), and
- coalescence aggregates with grain boundaries formed between neighboring
  particles in an aggregate (4).
The formation of coagulation aggregates can be released in the primary
samples by various ways. All are due to complex concurrences of attractive
v.d.Waals forces, repulsive Debye-Hückel forces and diffusion processes.
The main advantage of these samples, compared, e.g., to thin island films,
is that during formation of the aggregates, the individual particles remain unchanged and so, the amount of aggregation can be varied over wide
ranges as the only parameter of the system (3).
The state of coagulation aggregation may be stable if
- particles, with bound overlayers of atoms/ions/molecules at the interfaces, touch each other, or if
- particle pairs develop a Hamaker-Derjaguin-Verwey-Overbeek-like secondary potential minimum.

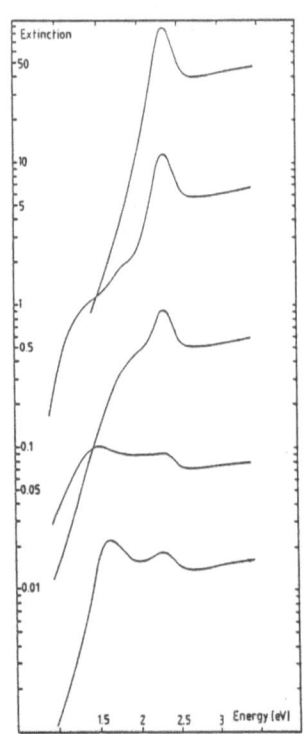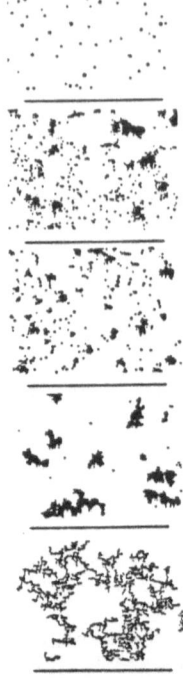

Fig.1 Extinction spectra
and TEM micrographs
of Au-particle samp-
les with varying
amount of aggregation

(2R = 38 and 56 nm)

In the primary systems, repulsion is high enough to keep the particle distances in the mean markedly higher. If, however by addition of ions to
the suspension, the Debye-Hückel-parameter is caused to increase, then particles can approach each other sufficiently during their thermal random
walk, and eventually particle aggregates are formed with shapes varying
from pseudo-fractal (i.e. fractal in a limited size region) chainlike
structures to massive lumps. The time scale of this aggregation process
varies from seconds to weeks, depending on
- the amount of added counter-ions, and
- the presence of stabilizing agents, as, e.g., gelatin.
Adding, after some time, stabilizing agents to sufficient amounts, the coagulation process can completely be stopped. If, even, the water content is
subsequently extracted and, thus, the packing density of the aggregates
as a whole is markedly increased in the remaining solid gelatin, the size
and structure of the aggregates remain roughly unchanged, the only effect
being the ordering in a more or less two-dimensional array. So, it is
possible to obtain information from TEM characterization of solidified
thin layers. Fig.1 shows such samples.

SAMPLE CHARACTERIZATION
Treating, both, ion-concentration and time of aggregation as independent
parameters we obtained series of samples with different aggregate structures and aggregate size distributions. These samples are better defined than
samples obtained by centrifugation of strongly aggregated systems. Presently, it is important to have small aggregates since only these can be
treated theoretically. The aggregates grow by addition of single particles and by merging of two or more aggregates. So, the probability P_N of
aggregates containing a given number of N particles at a time t is determined by
- the creation processes, taking place from aggregates of smaller N and
- the loss processes due to the production of larger aggregates from
  N-particle aggregates.
We found this probability P_N to follow a power law $P_N = P_o \cdot N^{-1/A}$ (1) where

452

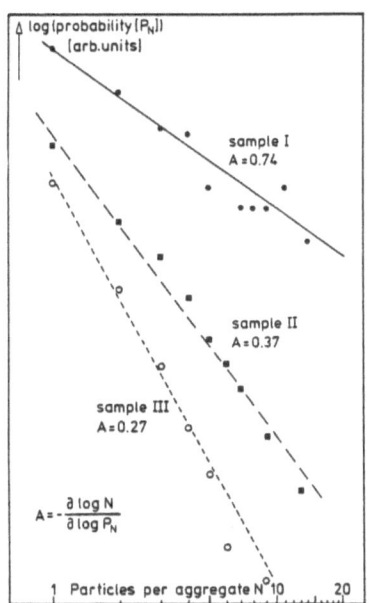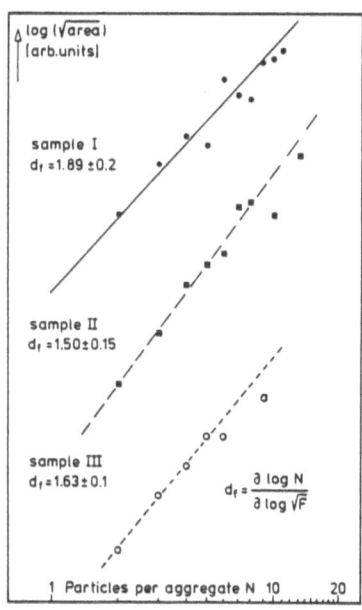

Fig.2 Left: Probability P_N of aggregates consisting of N particles.
Right: Compactness of the aggregates: mean linear size versus N.
(Two-dimensional samples of Au particles with 2R = 1o nm)

we call A the "aggregation number". A proves, however, not to be an univer-
sal exponent but is sensitive to details of the aggregation process.
Fig. 2a shows results for samples with small aggregates.
The packing density or compactness of the aggregates can be varied also.
This is shown in Fig. 2b, where a mean linear dimension L of the aggregates
(here: (area on the TEM-micrograph)$^{1/2}$) is plotted versus particle number
N. Again a power law is obtained: $L = L_1 N^{-1/d_f}$ (1). In the limit of extended aggregates, the exponent is the Hausdorff dimension or (in fractal
systems) the fractal dimension. Extended numerical evaluations of 15 samples with 4 ooo aggregates and 3o ooo particles yielded values of d_f between 1.4 and 2.o for two-dimensional aggregates. d_f proves to depend on
N for equal aggregate structure (e.g. for linear chains) if N is small.
Therefore, an aggregate dimension was recently developed which is independent of N (5).

OPTICAL PROPERTIES
(A) Coagulation Aggregates
The optical extinction (i.e. absorption plus scattering losses) of such
samples is determined, both, by
- properties of the individual particles and
- collective effects due to interaction among the particles densely packed
  in the aggregates.
As clearly shown by Fig.1, where only the topography of otherwise unchanged particles is varied, the appropriate extinction spectra are very sensitive to these collective effects.
The individual properties are properly described by the phenomenological
theory of Mie and Debye (6) and only minor corrections due to ABC and finite size effects of the dielectric material function have been added
later. The electromagnetic excitation is expanded into "electric" and "magnetic" multipole contributions, the former being due to spherical plasmon
polariton modes, the latter being due to eddy currents.
If aggregation occurs, the collective effects in our systems are beyond
effective medium ansätzen, which only hold for special sample topologies
(pure statistical and cubic arrangements) and were developed only for
quasi-static conditions.

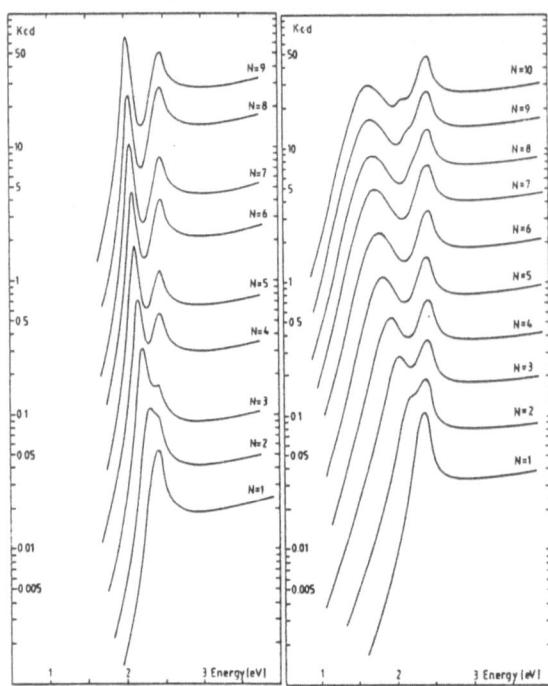

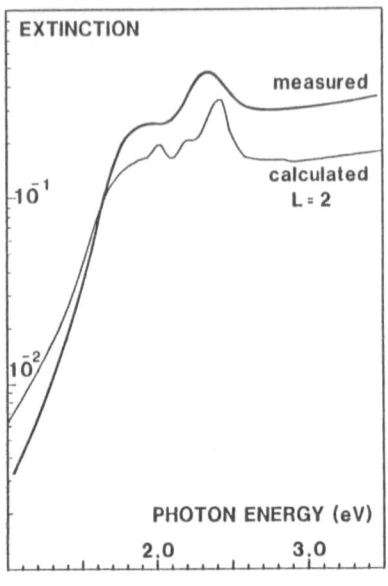

Fig.3 Calculated extinction spectra of linear aggregates of N Au-particles. Left: 2R=1o nm; Right: 2R=56 nm. Particles nearly touching. Embedding matrix: gelatin. Orientation averaged.

Fig.4 Quantitative comparison of the measured and computed extinction spectrum of a sample with small Au-particle aggregates (2R= 38 nm). Computed spectrum shifted down by factor o.5. Computations include retarded dipolar and quadrupolar excitations.

To describe samples containing coagulation aggregates properly, a more detailed description is demanded, which has recently been developed (7, 1). This theory extends Mie's theory and is based upon the assumption of a direct coupling between all particles of an aggregate via the near-field multipole scattering fields which change the effective field at a given particle. This effect has been numerically computed including retardation for the combination of dipolar and quadrupolar excitations and coupling. Linear aggregates could be treated up to N = 3o and nonlinear aggregates up to N = 1o (8). Some selected extinction spectra, obtained after aggregate orientation averaging, are shown in Fig. 3. It is obvious that the narrow single-peaked dipolar plasmon polariton band of the well isolated single particles splits by the aggregation into mainly two peaks, one of them shifting to low frequencies when N and/or 2R is increased.

These features are qualitatively also observed in the measured spectra of Fig.1. Recently, Quinten et al. (8) performed the first quantitative comparison between the measured and the calculated extinction spectrum of an Au-particle sample containing small, well separated particle aggregates. We present these two spectra in Fig 4 and refer to (8) for further details.

(B) Coalescence Aggregates

The numerical computations prove that the splitting of all N-particle aggregates is largest for the linear chain and reaches an asymptotic value with increasing N (Fig.5). Hence, coagulation aggregates cannot have extinction peaks beyond a certain low frequency limit ω_g.

Fig.6 shows a quasifractal Au-particle sample, measured extinction spectra of which are given in Fig.7.

Curve (a) gives the plasmon-polariton band of the primary sample before aggregation.

Curves (b) and (c) give the spectra after coagulation has taken place and

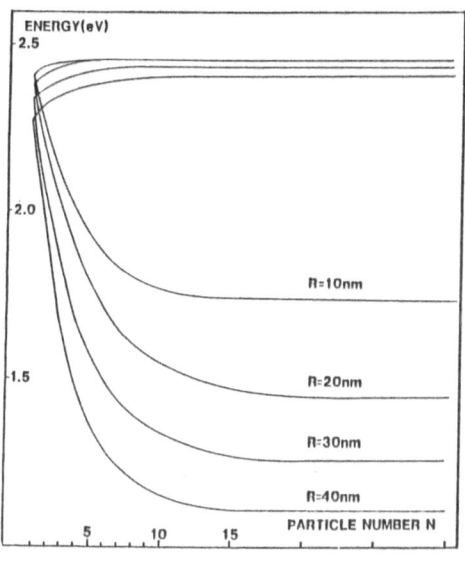

Fig.5 Computed extinction peak positions of linear Au-particle chains with varying number of particles N. R: particle radius. Calculations include retarded dipolar and quadrupolar excitations and coupling.

has been stopped by stabilization. (b) was obtained from the aqueous colloidal system with 3-dimensional aggregates and (c) from the according 2-dimensional aggregates produced by extracting the water.
The limiting frequencies ω_g are almost reached by the samples, since the aggregates mainly consist of chainlike structures.
In addition to the rapid coagulation we observed a slow change of the spectra within several weeks. As a result, the low frequency band was reduced, and far beyond ω_g additional extinction occured (curves (d) and (e)) which cannot be attributed to coagulation aggregates.

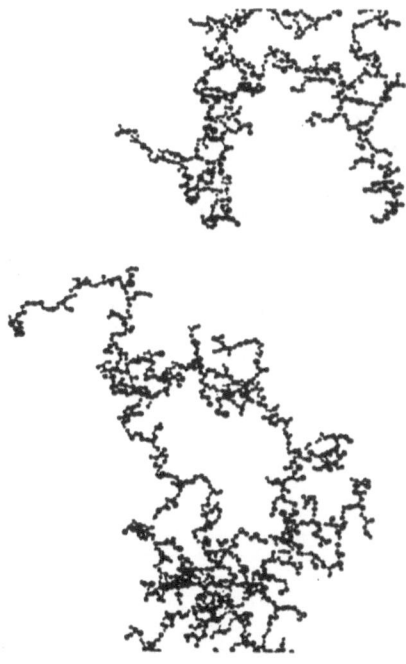

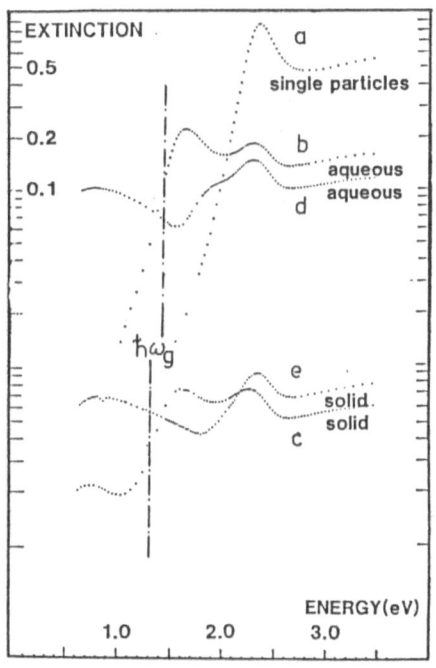

Fig.6 Au-particle sample with quasi-fractal aggregates.
2R = 38 nm.

Fig.7 Measured extinction spectra of the sample of Fig.6
a): before aggregation. b),d): 3-dim. aggregates. c),e):2-dim. aggregates.

455

Electron micrographs proved that the slow process was coalescence, i.e. the formation of grain boundaries between neighboring particles in the otherwise unchanged aggregates. This effect has been described by Thölen (9). The particles transmute into electrically connected larger metallic entieties, which are pre-states of percolation paths in the sample. From the resulting new extinction maximum, a mean length of such paths of about 5 particles can roughly be estimated, in correspondence to the electron micrographic informations. Thus, the optical extinction allows to observe the formation of percolation, not on the length scale of the whole sample, as as does the static conductivity but on a smaller scale of the order of the wavelength of light (4).

Analogously, it should be possible to observe the formation of nanocrystalline material when starting with more compact coagulation aggregates.

As a resumé, we have shown that optical spectral analysis of aggregates in small particle samples is possible. On the other hand, as a consequence, properties of the individual particles can be extracted from optical spectra only if no aggregates are present in the samples. Otherwise they may be more or less veiled by collective interaction effects.

REFERENCES

(1) U. Kreibig, M. Quinten, D. Schoenauer : physica scripta T13 (1986),84

(2) R. Zsigmondy : "Das kolloide Gold" Leipzig, Deuticke 1925

(3) U. Kreibig, A. Althoff, M. Pressmann : Surface Sci. 106 (1981), 308
    M. Quinten, U. Kreibig: Surface Sci. 172 (1986), 557
    D. Schoenauer, U. Kreibig : Surface Sci.156 (1985), 100

(4) M. Quinten, U. Kreibig : Proceedings of the DPG-Frühjahrstagung 1987,
    DS-13.2

(5) D. Schoenauer, U. Kreibig : Proceedings of the DPG-Frühjahrstagung
    1987, DS-13.1

(6) e.g. in: M. Born "Optik" Berlin, Springer

(7) J. Gerardy, M. Ausloos : Phys.Rev. B22 (1980), 4950; B25 (1982), 4204;
    B27 (1983), 6446.

(8) M. Quinten, U. Kreibig : Proceedings of the Symposium on "Optical
    Particle Sizing: Theory and Practice"
    Rouen, France 1987

(9) A. Thölen in "Contribution of Cluster Physics to Material Science
    and Technology" ed. J.Davenas,P.Rabette; Nijhoff,
    Dordrecht 1986, p.601

PATTERN FORMATION IN LIPID MEMBRANES

Aspects of fractal solidification, grain-boundary kinetics, interfacial melting, and wetting

O.G. Mouritsen

Department of Structural Properties of Materials
Technical University of Denmark, Building 307
DK-2800 Lyngby, Denmark

H.C. Fogedby and E. Schwartz Sørensen

Institute of Physics, University of Aarhus
DK-8000 Aarhus, Denmark

M.J. Zuckermann

Department of Physics, McGill University
Montreal, Quebec, Canada H3A 2T8

I. INTRODUCTION

This paper draws attention to a class of pseudo-two-dimensional materials which are exceptional candidates for observing a variety of time-dependent effects, including fractal growth during crystallization, grain-boundary formation and dynamics, interfacial melting, and wetting. These materials are planar macromolecular aggregates formed by amphiphiles in contact with water.[1] Appropriate amphiphiles could be simple detergents and soaps or the more complicated lipids which we shall concentrate on in this paper.

A typical class of lipids is the phospholipids which are made up of a polar moiety, the head, and two flexible hydrocarbon chains, the tail. It is the amphiphilic nature of such molecules which is responsible for their spontaneous self-organization into planar aggregates, so-called lipid membranes, in exposure to water.[2] When spread on an air/water interface, lipid molecules form condensed two-dimensional phases in the presence of a lateral pressure (Π), Fig. 1a. Similarly, when mixed with water, lipid molecules may form a bimolecular sheet, the lipid bilayer, Fig. 1b, which constitutes the background matrix in every biological cell membrane. Lipid membranes have a very rich phase structure in terms of lateral pressure and temperature. This is because there are several mechanisms available for the ordering of lipid molecules, e.g. (i) the conformational ordering of the hydrocarbon chains via rotational isomerism and van der Waals interactions between the chains, and (ii) two-dimensional crystallization involving translational degrees of freedom. The spectacular interfacial properties of lipid membranes to be discussed in Secs. III and IV are caused by a subtle interplay between these two sets of internal and crystalline degrees of freedom.

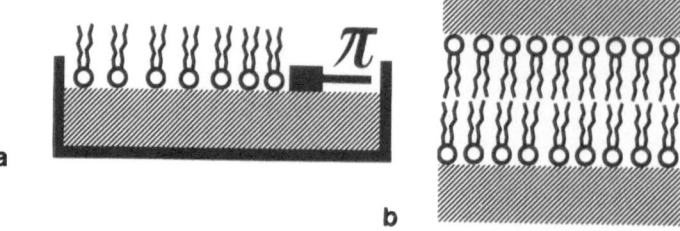

Fig. 1. Schematic cross-section of pseudo-two-dimensional lipid
membranes. (a) Lipid monolayer on an air/water interface.
(b) Lipid bilayer in bulk water. The lipid molecules are
indicated by a polar head and two flexible hydrocarbon chains.

This paper gives a short overview of results which have been obtained
recently from theoretical model studies of pattern formation in lipid membranes using mainly computer simulation, with an emphasis on the general
aspects of the phenomena. Brief reference is given to the current experimental situation.

II. FRACTAL GROWTH IN IMPURITY-CONTROLLED SOLIDIFICATION

A model has recently been proposed[3] to describe solidification processes
in liquid systems where the growth-limiting factor is diffusion of a second
component, such as an impurity or a solvent which is miscible in the liquid
phase only. This model is believed to describe the pattern formation in
certain rapidly compressed lipid monolayers[4] containing a fluorescent dye
impurity which is only miscible in the fluid lipid phase.

The model is formulated as a microscopic aggregation model on a lattice
with particles. The growth process in the model is assumed to be governed
by two mechanisms: (i) irreversible solidification of fluid particles at the
solid-liquid interface, and (ii) chemical diffusion of impurities (or solvent
molecules) in the liquid phase only. It is assumed that the heat of solidification is negligible and that there is no tension associated with the interface.
Hence, the growth model is characterized by two intrinsic time scales, that
of liquid-solid conversion (τ_S) and that of impurity diffusion (τ_D). The two
basic model parameters are therefore the time scale ratio τ_S/τ_D and the impurity concentration C. The growth model may be considered as a generalized Eden model[5] in which the growth is limited by the local impurity concentration near the solidification front. It is the impurities and their diffusional characteristics which cause the interfacial stability of the present
growth model and are responsible for the morphology of the growing solid.
The model differs from previous theoretical approaches to aggregation processes in two principle ways: (i) it accounts microscopically for a second
component, and (ii) the aggregating particles are all present simultaneously,
in contrast to models of diffusion-limited aggregation (DLA).[6]

Figure 2 shows a solid aggregate consisting of about 35000 particles
grown according to the model described above. The analysis in Fig. 3
demonstrates that this aggregate is fractal and scale invariant characterized
by a Hausdorff dimension of $D \simeq 1.8$, where D is defined[7] by $N(R) \sim R^D$.
$N(R)$ is the number of particles in a region of linear extension R. Aggregates grown for other values of τ_S/τ_D and C are also of fractal nature with
the following non-universal trends: For a fixed value of τ_S/τ_D, D decreases
as C is increased. For a fixed value of C, D has a minimum at a value of
τ_S/τ_D of the order of unity. For slower diffusion, D increases towards 2
and the aggregates have uniform density but are very tenuous. For faster

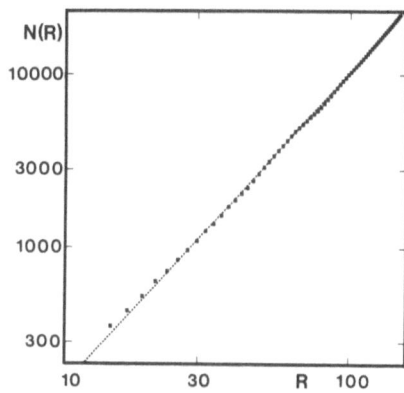

Fig. 2. Fractal solid aggregate consisting of about 35000 particles grown at τ_S/τ_D = 1 and C = 0.7. The solid bar indicates 200 lattice units.

Fig. 3. Double logarithmic plot of the particle content, N(R), vs linear extension, R, of the aggregate in Fig. 2. Low-R and high-R cut-offs have been introduced. The fractal dimension is $D \simeq 1.8$.

diffusion, D also increases towards 2 and the aggregates become compact. The minimum of D gets more shallow and tends towards the DLA-value, $D \simeq 1.7$, for large C. There is an obvious physical explanation of this observation: For large C, fluid-fluid particle correlations can be neglected and the individual aggregating fluid particle can be considered as isolated. This is precisely the DLA-limit.

The actual fractal-forming mechanism of the present model of solidification is revealed from analysis of the spatial distribution of impurities as a function of time. This distribution shows that the active growth zone has elevated levels of impurities which are driven as a halo by the solidification front. This halo is the fractal former of the solidification process and it thus plays a role analogous to that of the well-known screening effect in DLA.[6] Ref. 3 discusses these results in relation to other model studies of fractal and dendritic growth.

Fractal solid domains have been detected experimentally by fluorescence microscopy[4] on lipid monolayers of dimyristoylphosphatidyl ethanolamine in the presence of a dye impurity. The fractal dimension is around $D \simeq 1.5$. Furthermore, the experimental study also found increased levels of dye between the arms of the fractal aggregate and in the active growth zone, suggesting the same fractal-forming mechanism to be operative as in the model discussed here.

III. GRAIN-BOUNDARY KINETICS AND INTERFACIAL MELTING

Lipid monolayers and bilayers undergo a first-order phase transition, the so-called gel-to-fluid transition, which takes the membrane from a low-temperature state of high chain-conformational order to a high-temperature state of low chain-conformational order. This transition is very entropic, $\Delta S \sim 15 k_B$ /molecule, and it is accompanied by a dramatic expansion of the membrane area. The transition is a solid-liquid transition, though it should be stressed that the membrane is a highly anisotropic system and the transition resembles more that of smectic melting except that it is driven by intramolecular excitations of a polymeric character.[8] It is conventionally assumed that at this transition, condensation takes place in terms of translational and conformational degrees of freedom simultaneously. Recently, however, synchrotron x-ray studies[9] of *in situ* dimyristoyl phosphatidic acid monolayers have shown that the chain-ordering transition and the crystallization of the monolayer need not take place at the same lateral pressure.

We shall here outline the results of a model study geared towards an understanding of coupled transitions in lipid membranes.[10-12] We discuss the case of a bilayer, but the general idea and results carry over to the case of monolayers as well. The study is based on a microscopic triangular lattice gas Potts model with a variable vacancy concentration. This model is a hybrid of Pink's multi-state model,[13] which accounts for the interaction between the internal degrees of freedom, and the high-q-state Potts model which in an approximate way accounts for crystallization in terms of positional variables.[14] In terms of the characteristic energy constants of the two parts of the Hamiltonian, J_0 and J_p respectively, the phase diagram of the model can be described as follows: For large J_p/J_0, a single phase line separates a crystalline solid from a fluid phase. For small J_p/J_0, an intermediate phase intervenes which is a kind of disordered solid phase. Thus, for small J_p/J_0 the two phase transitions decouple in accordance with the synchrotron x-ray experiments.[9]

Time-dependent effects associated with the ordering processes in this model[10-12] have been studied by Monte Carlo computer simulation.[15] The time is measured in units of Monte Carlo steps per site (MCS/S). Figures 4 and 5 give snapshots of non-equilibrium microconfigurations characteristic of a thermal cycle through the phase diagram. The equilibration time at each temperature is 1000 MCS/S.

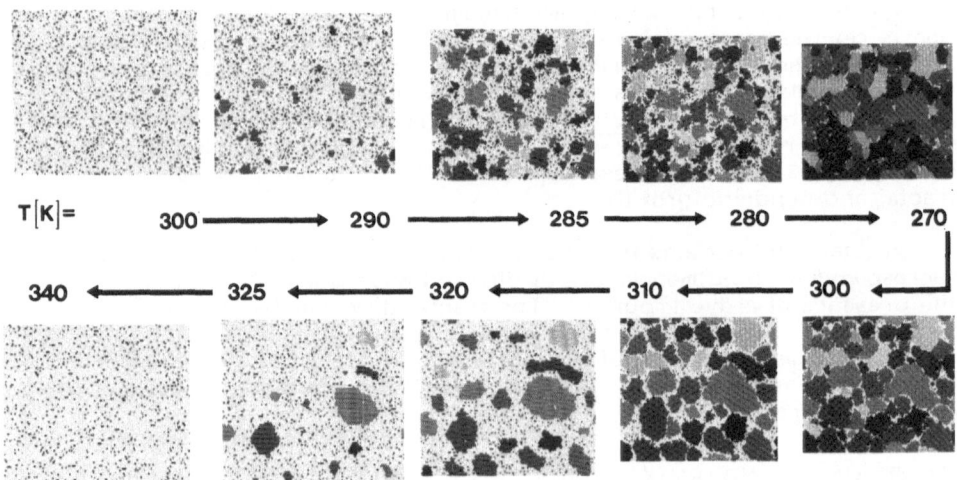

Fig. 4. Snapshots of microconfigurations illustrating the non-equilibrium processes of grain-boundary formation and subsequent interfacial melting in a lipid bilayer model with coupled transitions. The system contains 10000 lipid chains. White areas indicate fluid domains and symbols indicate gel lipid molecules, with each symbol labelling a crystal-orientational state (Potts state). The lattice parameter has been scaled to display the thermally induced area contraction and expansion during the thermal cycle.

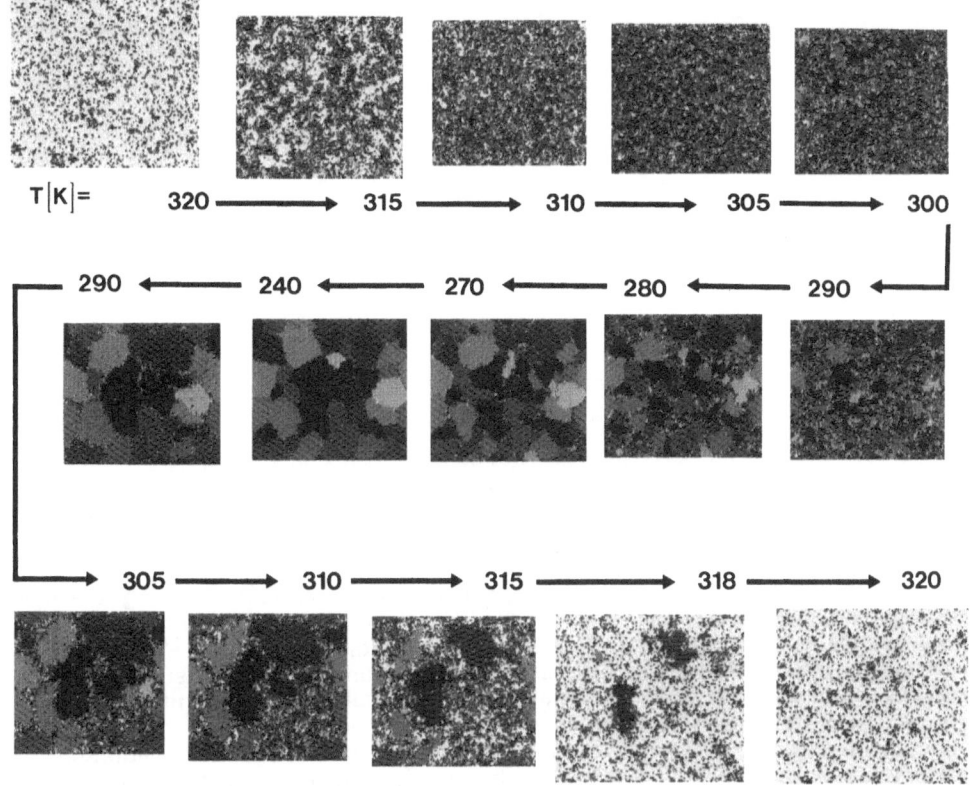

Fig. 5. Same as Fig. 4 but for uncoupled transitions.

In the case of coupled transitions, Fig. 4 shows the formation of crystalline domains floating in a fluid sea as the temperature is lowered. At low temperatures, the solid is composed of a polycrystalline array which anneals very slowly in time (actually according to a power law $R(t) \sim t^{0.41}$).[16,14] When the temperature is increased again, this polycrystalline array melts at the domain boundaries at a time scale which is much faster than that associated with grain growth. This is the phenomenon of non-equilibrium interfacial melting. In the case of uncoupled transitions, Fig. 5, an intermediate glassy solid is formed, and as the temperature is lowered further, the poly-crystal appears. When this polycrystal is heated up through the intermediate phase, a sort of interfacial 'glassification' takes place where the crystal-lites gradually disorder at the domain boundaries.

A quantitative description of the interfacial melting process may be obtained by calculating the interfacial energy and interfacial density,[12] as shown in Fig. 6, as functions of the distance, t, from the non-equilibrium melting point. The interfacial density, $\eta(t)$, measures the melted fraction relative to the solid fraction and thus diverges at $t = 0$. This is substantiated by the data in Fig. 6 which are accurately described by $\eta(t) \sim t^{-\gamma}$, with $\gamma \simeq 2$.

Lipid membranes are exceptional candidates for pronounced interfacial melting behaviour since the flexibility of the hydrocarbon chains facilitates the formation of fluid interfaces between crystalline grains of solid lipid domains.[17] Recent real-time synchrotron low- and wide-angle x-ray studies of lipid bilayer suspensions show a kinetic behaviour of the transition which is consistent with interfacial melting behaviour.[18]

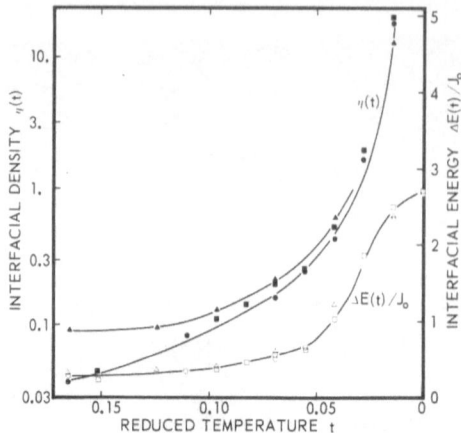

Fig. 6. Interfacial density and interfacial energy as functions of
reduced temperature. Results are given for different system
sizes, $N = 30^2$ (triangles), 50^2 (squares), and 100^2 (circles).
The equilibration time is 2000 MCS/S.

Although difficult to observe experimentally, interfacial melting is a process which in general should take place in a large number of materials. It
has been observed indirectly in Bi[19] and in certain bicrystals.[20] Furthermore, neutron scattering of the melting transition of incommensurate K and
Rb layers in quasi-two-dimensional layers intercalated in graphite show that
the melting starts at the discommensuration domain boundaries of the modulated two-dimensional lattice.[21] Finally, a number of computer simulation
studies have been reported of grain-boundary melting in systems where the
interface is subject to a constrained thermal equilibrium[22] (see also Sec. IV).

IV. WETTING

The previous section described the non-equilibrium melting process in a
statistical ensemble of crystalline grains. The pronounced interfacial melting
process there was a consequence of a competition between grain growth and
melting which, on the chosen time scales, favoured the melting process at
the tense grain boundaries. Here we consider a slightly different situation
where an interface mediates an equilibrium melting process which takes place
between two grains of fixed orientations (Potts states). The model is the
same as that for lipid membranes discussed in Sec. III.

Figure 7 displays snapshots of microconfigurations typical of equilibrium

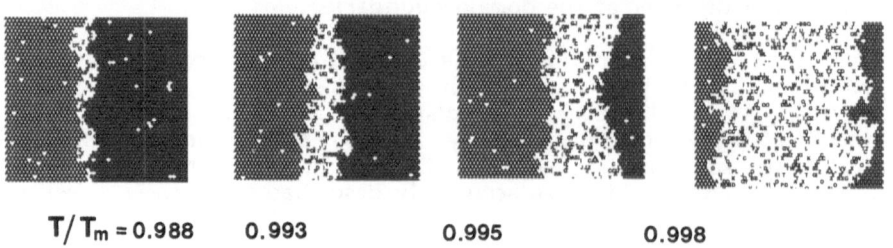

$T/T_m = 0.988$ 0.993   0.995   0.998

Fig. 7. Wetting phenomena at a domain boundary. A fluid region is
formed between two differently oriented crystalline domains.
The system contains 50^2 particles on a triangular lattice.

for different temperatures below the uniform bulk first-order melting point,
T_m. These configurations are derived from Monte Carlo simulations on a cell

with fixed rows of particles in the horizontal direction and periodic boundaries in the vertical direction. As T_m is approached from below, a fluid region is seen to wet the grain boundary. The original sharp first-order transition now appears to be continuous, i.e. the symmetry-breaking weak field induced by the fixed rows of particles smear the transition.

A preliminary analysis shows that the width, W, of the fluid interface diverges at T_m according to a power law $W \sim |T-T_m|^{-w}$ with $w \approx 0.90 \pm 0.15$. The statistics needed to sustain an accurate analysis of the interface as a function of temperature and system size is enormous, at least $5 \cdot 10^5$ MCS/S are required for system sizes up to about 50 x 50.

These findings are in accordance with similar studies of interfacial adsorption by Selke[23] carried out on various multi-state models, including the standard q-state Potts model. In particular, Selke found $w \approx 1.1 \pm 0.1$ for multi-state models with bulk first-order transitions. All these model studies are special cases of general wetting phenomena[24] where a third phase of macroscopic thickness intervenes between two other phases of different symmetry. This intervening phase shows interesting critical wetting behaviour close to or possible at the bulk transition temperature.

V. CONCLUSIONS

This brief overview of pattern formation in lipid membranes has demonstrated that lipid membranes exhibit a variety of fascinating time-dependent effects. In many cases these effects are quite general in nature but may be specially pronounced in lipid systems. Lipid membranes thus offer themselves as useful experimental systems to study interfacial phenomena in an advantageous setting.

From the point of view of biology, interfacial regions in lipid bilayers and biological membranes may be extremely important for supporting a variety of biological and physiological functions.[25] In fact, it is likely that the laterally differentiated regions of a biological membrane with soft interfacial regions play a role for regulation of e.g. protein and enzymatic activity, growth, transmembrane transport, and nervous activity.

Acknowledgements : This work was supported by the Danish Natural Science Research Council under grants J.nr. 5.21.99.72 and J.nr. 11-5593, NSERC of Canada, and Le FCAC du Quebec.

REFERENCES

1. For a recent review, see O.G. Mouritsen in "Physics and the Living Matter" (M. Droz et al., eds.) Springer-Verlag, Heidelberg, 1987.
2. J.N. Israelachvili, S. Marčelja, and R.G. Horn, Q. Rev. Biophys. 13:121 (1981).
3. H.C. Fogedby, E. Schwartz Sørensen, and O.G. Mouritsen (submitted to Phys. Rev. Lett. 1987).
4. A. Miller, W. Knoll, and H. Möhwald, Phys. Rev. Lett. 56:2633 (1986).
5. M. Eden, "Proc. Fourth Berkeley Symp. Math. Stat. and Prob." (J. Neuman, ed.) Vol. 4, p. 233 (1961).
6. T.A. Witten and L.M. Sander, Phys. Rev. B 27:5686 (1983).
7. B.B. Mandelbrot, "The Fractal Geometry of Nature" (Freeman, San Francisco, 1982).
8. S. Doniach, in "Ordering in Two Dimensions" (S.K. Sinha, ed.) Elsevier North Holland, Amsterdam, p. 67 (1980).
9. K. Kjær, J. Als-Nielsen, C.A. Helm, L.A. Laxhuber, and H. Möhwald (Phys. Rev. Lett. 1987).
10. O.G. Mouritsen and M.J. Zuckermann, Chem. Phys. Lett. (1987).

11. M.J. Zuckermann and O.G. Mouritsen, Eur. Biophys. J. (1987).
12. O.G. Mouritsen and M.J. Zuckermann, Phys. Rev. Lett. 58:389 (1987).
13. D.A. Pink, T.J. Green, and D. Chapman, Biochemistry 19:345 (1980).
14. P.S. Sahni, G.S. Grest, M.P. Anderson, and D.J. Srolovitz, Phys. Rev. Lett. 50:263 (1983).
15. O.G. Mouritsen, "Computer Studies of Phase Transitions and Critical Phenomena" (Springer Verlag, Heidelberg, 1984).
16. O.G. Mouritsen, in "Annealing Processes - Recovery, Recrystallization and Grain Growth" (N. Hansen et al., eds.) Proceedings of the 7th Risø International Symposium on Metallurgy and Materials Science, p. 457.
17. O.G. Mouritsen, Biochim. Biophys. Acta. 731:217 (1983); O.G. Mouritsen and M.J. Zuckermann, Eur. Biophys. J. 12:75 (1985).
18. M. Caffrey, Biochemistry 24:4826 (1985).
19. M.E. Glicksman and C.L. Vold, Surf. Sci. 31:50 (1972).
20. P.S. Ho, T. Kwok, T. Nguyen, C. Nitta, and S. Yip, Scr. Metall. 19:993 (1985); P. Deymier and G. Kalonji, Scr. Metall. 20:13 (1986); K.E. Sickafus and S.L. Sass, Acta Metall. (1987).
21. H. Zabel, S.E. Hardcastle, D.A. Neumann, M. Suzuki, and A. Magerl, Phys. Rev. Lett. 57:2041 (1986).
22. J.Q. Broughton and G.H. Gilmer, Phys. Rev. Lett. 56:2692 (1986); T. Nguyen, P.S. Ho, T. Kwok, C. Nitta, and S. Yip, Phys. Rev. Lett. 57:1919 (1986).
23. W. Selke, Surf. Sci. 144:176 (1984); W. Selke, in "Static Critical Phenomena in Inhomogeneous Systems" (Springer Verlag, Heidelberg, 1984) p. 191.
24. M.E. Fisher, Faraday Symp. Chem. Soc. 20 (1986).
25. E. Sackmann, in "Biological Membranes" Vol. 5 (Academic Press, London, 1984) p. 105.

DIRECT SPECTROSCOPY OF MICROEMULSION DROPLET FLUCTUATIONS

Dieter Richter*, Bela Farago*, and John S. Huang+

* Institut Laue-Langevin, Grenoble, France
+ Exxon Research and Engineering Corp., Annandale, N.J., USA

Microemulsions are homogeneous mixtures of oil, water and surfactants[1]. They contain large internal interfacial areas that separate the hydrocarbon domains from the aqueous regions. This interfacial area is very large and e.g. amounts to about 100 m^2/cm^3 for a microemulsion containing 10% surfactant. Thus, the properties and structure of microemulsions will crucially dependent on the nature of these surfactant layers. Two alternative ideas have been brought forward : (i) it was argued that the surface tension must be extremely low such that the droplet formation is not inhibited by its large contribution to the free energy[2,3]; (ii) alternatively one may think, that droplet formation results mainly from the natural bending tendency of the surfactant layer[4]. Then, the observed microemulsion structures are due to a minimalization of the bending elastic energy. The only way to distinguish between these two basic mechanisms for microemulsion formation is the study of the dynamics of the surfactant layer.

In the case of the 3-component microemulsion containing the surfactant AOT (sodium di-2-ethylhexyl sulfusuccinate), water and decane, the existence of a single phase droplet structure is well established[5,6]. Furthermore, it is known that in this system, the mean radius of the water droplets depends linearly on the surfactant to water ratio[6].

The aim of the dynamic experiment is to measure the thermal shape fluctuations of the single droplet. As indicated by the relatively small effective polydispersity the amplitudes of these fluctuations from the spherical form are expected to be small. Therefore, it is important to reduce the scattering volume to the actually fluctuating part. This can be done by contrast matching the neutron scattering length density of the internal droplet phase to that of the external continuum. Under these conditions the scattering originates solely from the fluctuating shell of surfactant layer. From the dispersion relation of the relaxation frequency, we can ascertain whether surface tension or elastic bending energy drives the dynamics.

The time and momentum dependent scattering intensity $I(Q,t)$ from a fluctuating droplet of radius r_0 can be written as :

$$I(Q,t) = e^{-D_t Q^2 t} V_s^2 (\Delta\rho)^2 \left\{ f_o(Qr_o) + \frac{1}{4\pi} \sum_{l=2}^{\infty} (2l+1) f_l(Qr_o) \right.$$

$$\left. \frac{\langle a_l(0) a_l(t) \rangle}{r_o^2} \right\}$$

(1)

Eq. (1) can be derived by expanding the shape fluctuation into spherical harmonics. Thereby f_o is the static form factor of the shell :

$$f_o = \left(\frac{\sin Qr_o}{Qr_o}\right)^2 \text{; the inelastic form factors of the shell are}$$

$$f_l = [(1 + 2)j_l(Qr_o) - Qr_o\, j_{l+1}(Qr_o)]^2, \; V_s \text{ is the scattering volume,}$$

$\Delta\rho$ the scattering length contrast and j_l are spherical Bessel function of order l. For the shell it is assumed that $Q\Delta \ll 1$ (Δ is the surface layer thickness). The time dependent amplitudes of the fluctuations $\langle a_l(o) a_l(t)\rangle \sim e^{-t/\tau_l}$ are expected to be overdamped. Finally $D_t = kT/6\pi\eta r_o$ is the translational diffusion coefficient, where η is the viscosity of the solvent.

The form factor of the lowest order fluctuation mode ($l = 2$) at small (Qr_o) is effectively given by j_2^2. Its first maximum occurs in the vicinity of the first zero of the form factor of the shell j_o^2, while in the case of the sphere the elastic form factor $(3\, j_1/Qr_o)^2$ still overhelmingly dominates the spectrum in this Qr_o range. Therefore, we expect a much better visibility of the lowest order fluctuation for the shell labelling.

The dynamic experiment was carried out on the neutron spin echo spectrometer (IN11) at Institut Laue-Langevin in Grenoble, France. The energy change of the scattered neutrons is measured directly through the Larmor precessions of the neutron spin in an external guide field. The intermediate scattering function $S(Q,t) = I(Q,t)/I(Q,o)$ is directly given by the final polarization of the scattered neutrons[7].

Five shell like microemulsion samples were used for this study. The water-to-AOT molar ratios ranged from 8 to 40 giving mean droplet radii (center to the outer edge of the surfactant tails) from 70 to 24 Å[8]. The dispersed volume was 5% and constant. The continuous phase was a per-deuterated decane which matches the neutron scattering length density of the internal D_2O phase almost exactly such that the surfactant coated droplet appears to be only an empty shell to the scattering beam. Fig. 1 presents a set of relaxation curves determined on a droplet with a radius of 49 Å. The solid line represents a fit with a single exponential decay yielding a relaxation rate $\Gamma(Q)$. Fig. 2 displays the effective diffusion coefficient Γ/Q^2 as a function of Q for 4 different droplet sizes. For each sample a pronounced peak is observed the position of which occurs at $Qr_o = 3.2 \pm 0.1$ for all samples measured. For small Q Γ/Q^2 approaches the value of the center of mass diffusion estimated for each of the droplet sizes. Both the low-Q limit and the occurrence of a peak in our spectra are qualitatively understood on the basis of the scattering function of Eq. (1) which holds independently of any specific model of the dynamics. From the size dependence of the fluctuation rates we can distinguish between modes driven by elastic bending forces and those dominated by surface tension. From dimensional analysis we immediately realise that for the case of splay elasticity the relaxation rate $1/\tau$ should scale with $K_c/\eta r_o^3$, where K_c is the curvature elastic modulus, while for a dominating surface tension σ $1/\tau \sim \sigma/\eta r_o$ should hold. Since the line broadening due to translational diffusion at the peak position ($Qr_0 \simeq 3.1$) DQ^2_m scales as r_0^{-3}, we obtain for the ratio of peak height and translational diffusion coefficient $S = \Gamma/DQ^2_m \sim$ const (r_0) for elastic forces and $\sim r_0^2$ for surface tension. According to table I the measurements performed on samples with radii from 70 to 30 Å give S = const consistent with a dominating splay elasticity.

TABLE 1 : Water to surfactant ratio, mean radius r_0 observed peak position and relative peak height S for the studied microemulsions

$\dfrac{[D_2O]}{[AOT]}$	Radius r_0(Å)	$Q_m\, r_0$	$S = \Gamma/DQ^2_m$
40.8	70	3.15	2.4
32.6	59.2	3.2	2.3
24.5	48.7	3.30	2.3
16.3	38.3	3.16	2.0
8.2	27.5	3.16	2.2

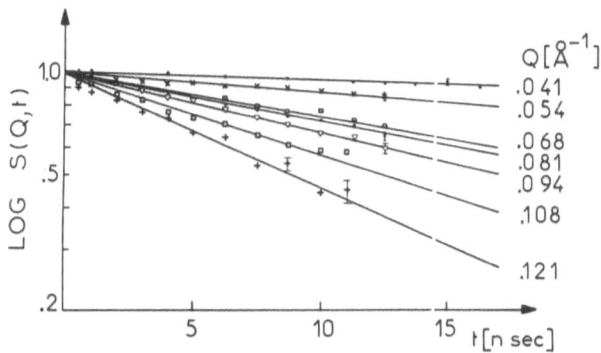

Figure 1 : Measured spin echo relaxation spectra for different Q values using the r_0 = 49 Å sample. The solid lines are fits to an exponential decay.

To calculate the peak height beyond simple scaling arguments, we extend and modify the work of Schneider et al[9] in order to include non-zero spontaneous curvature. According to the equilibrium prediction of Safran[10] the mean square fluctuations give rise to a mean excess surface area (the excess area is the difference between the droplet area and that of a sphere with equivalent volume). In order to calculate the relaxation frequencies from the expression for the total bending energy this excess area has to be taken into account as an additional constraint. For the relaxation frequency of the lth mode we get[10] :

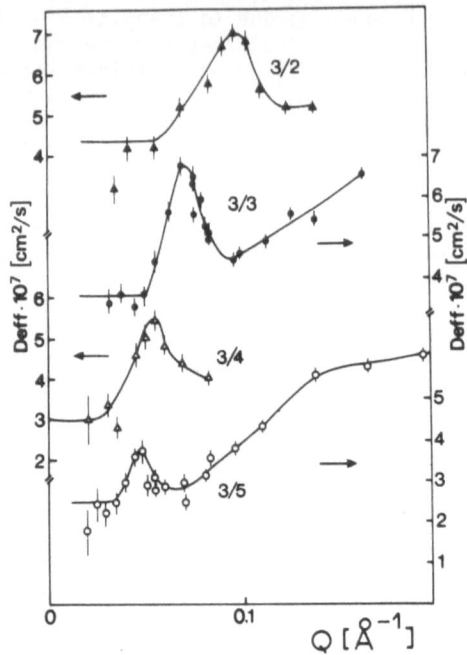

Fig. 2. The effective diffusion coefficient Γ/Q^2 is plotted vs Q for
microemulsions of different droplet size. 3/2 : r_0 = 38 Å;
3/3 : r_0 = 49 Å; 3/4 : r_0 = 59 Å; 3/5 : r_0 = 70 Å.

$$\omega_l = \frac{1}{\tau_l} = \frac{K_c}{\eta r_0^3} \frac{l(l+1)-6+4r_0/r_s}{Z(l)} \qquad (2)$$

and the amplitude of the l-th mode is

$$a_l^2 = \frac{kT}{K_c} \frac{1}{(l+2)(l-1)[l(l+1)-6+4r_0/r_s]} \qquad (3)$$

where $Z(l) = (2l+1)(2l^2 + 2l-1)/[l(l+1)(l+2)(l-1)]$, r_s is the radius of
spontaneous curvature and kT is the thermal energy. At finite r_s, a
small temperature-dependent entropy of mixing term can be neglected in
the above expressions.

Since the experimental spectra show a smooth decay even in the peak
region, the droplet relaxation times have to be of the same order as the
relaxation times due to translational diffusion. Therefore, to a good
approximation, the experimentally measured decay times are identical to
the initial slope of $S(Q,t)$:

$$\Gamma(Q) = - \frac{\delta \ln S(Q,t)}{\delta t}\bigg|_{t=0} = DQ^2 + \frac{\sum_{l>1} \omega_l f_l(Qr_0)(2l+1)\langle|a_l^2|\rangle}{4\pi j_0^2(Qr_0) + \sum_{l>1}(2l+1)f_l(Qr_0)\langle|a_l^2|\rangle} \qquad (4)$$

In the peak region ($Qr_0 \simeq 3.1$) polydispersity crucially influences the
value of $\Gamma(Q)$. For a further evaluation we consider it self-consistently
by introducing Safrans result[10] for the mean l = 0 fluctuation of the
ensemble.

$$\langle \delta R^2 \rangle / R^2 = kT / [8K_c (3 - 2 \; r_0/r_s)] \qquad (5)$$

Replacing $j_0^2 (Qr_0 = \pi)$ by $\langle \delta R^2 \rangle / R^2$ and summing up Eq. (4) we obtain a simple semi-quantitative result for $\Gamma(Qr_0 = \pi)$

$$\Gamma(\pi) = D(\frac{\pi}{r_0})^2 + \frac{0.3K_c}{\eta r_0^3} \; / \; \left[\frac{\pi/2}{3/2 - r_0/r_s} + \frac{0.15}{r_0/r_s} \right] \qquad (6)$$

For r/r_s intermediate between 0 and 3/2 (which are the limits of stability of droplet microemulsions), the r-dependence of the denominator is weak, giving $\Gamma \sim r_0^{-3}$.

Using Eq. 6 we can estimate the value for the bending elastic constant K_c. For $r_0/r_s = 0.5$

$$\frac{K_c}{k_T} = \frac{\pi}{0.9} \left[\frac{\Gamma}{DQ_m^2} - 1 \right]. \qquad (7)$$

For $\Gamma/DQ^2 = 1.5$ and $Qr_0 = \pi$, we have $K_c = 5$ kT. With Eqs. 2, 3 and 5 we finally calculate the relaxation time for the $l = 2$ mode $\tau_2 = 7$ns ($r_0 = 50$ Å); its amplitude $\langle a_2^2 \rangle^{1/2} \simeq 16\%$ and the $l = 0$ dispersion $[\delta R^2/R^2]^{1/2} \simeq 11\%$. The resulting total dispersion of 20% agrees well with the SANS result of 25%[5].

In summary, we have successfully used quasielastic neutron scattering to probe the dynamics of the shape fluctuation of the microemulsion droplets. It is found that these thermal fluctuations can be described by the overdamped bending elastic modes of the surfactant layer while the surface tension seems to play very minor roles, if at all. The bending elastic constant for this layer is estimated to be about 5 kT.

We acknowledge theoretical advice by S. T. Milner.

REFERENCES

1. For a general survey, see for instance Surfactants in Solutions edited by K. Mittal and B. Lindman (Plenum, New York, 1984)
2. J.H. Schulman, W. Stoeckenius and L. Prince, J. Phys. Chem. 63, 1677 (1959)
3. P.G. DeGennes, C. Paupin, J. Phys. Chem. 86, 2294 (1982)
4. S.A. Safran and L.A. Turkevich, Phys. Rev. Lett., 50, 1930 (1982)
5. J.S. Huang and M. Kotlarchyk, Phys. Rev. Lett., 57, 2587 (1986)
6. M. Kotlarchyk, S.H. Chen, J.S. Huang and M.W. Kim, Phys. Rev. A29, 2054 (1984)
7. F. Mezei, Neutron Spin Echo, Lecture Notes in Physics, Vol. 128, Springer Verlag (1979)
8. See also J.S. Huang, S.T. Milner, B. Farago, D. Richter, submitted for publication
9. M.B. Schneider, J.T. Jenkins and W.W. Webb, J. Phys. Paris, 45, 1457 (1984)
10. S.A. Safran, J. Chem. Phys. 78, 2073 (1983).

RENORMALIZATION GROUP RESULTS FOR RANDOM RESISTOR NETWORKS

A. B. Harris[+]

School of Physics and Astronomy
Raymond and Beverly Sackler Faculty of Exact Sciences
Tel Aviv University, Tel Aviv 69978 ISRAEL

The object of this review is to provide a starting point for those unfamiliar with the techniques underlying renormalization group (RG) (or other) approaches to dynamic properties of random networks. Since it is not practical to review here the RG technique itself, the present discussion is confined to a qualitative description of the formulation required for the RG. In this way it is hoped that we may separate the difficulties in approaching this subject into those distinct from other problems (discussed here) and those common to any RG treament, for which the reader is referrred to standard references.[1] The outline of this paper is as follows.

A. Very Brief Review of Percolation
B. Relations between Models of Resistance, Diffusion, and Spin Waves
C. Two-point Resistance, $[R(x,x')]_{av} \sim r^{\zeta/\nu}$ (where $r=|x-x'|$) Related to Conductivity, $\Sigma(p) \sim \Delta p^t$, where $\Delta p=|p-p_c|$. Scaling Form for $[R(x,x')]_{av}$.
D. Hamiltonian and Correlation Function for $[R(x,x')]_{av}$ - Mean-Field Theory
E. Resistance Noise (See Rammal's Lecture). Fluctuations due to Noise versus Those due to Geometry. Distribution of Bond Currents. Multifractality.
F. Non-Linear Resistors, $V \sim I^\alpha$. Identification of Special Values of α. The Case $\alpha \sim 1$.
G. Resistance of Walks, Granular Superconductors, Continuum Percolation.
H. Open questions - "Superconducting" exponent s, Elastic Networks and Self-Avoiding Walks on Percolating Clusters.

A.   BRIEF REVIEW OF PERCOLATION THEORY[2,3]

We consider bond percolation in which each bond is randomly occupied with probability p and vacant with probability 1-p. Let $\nu(x,x')$ be unity if sites x and x' are in the same cluster of occupied bonds and zero otherwise and [ ]$_{av}$ denote an average over random configurations. Then $[\nu(0,r)]_{av}$ decays to its r=∞ value over a length scale ξ, with $\xi \sim \Delta p^{-\nu}$, as shown in Fig. 1a. Here p_c is the critical percolation concentration such that for p>p_c an infinite cluster exists and $[\nu(0,\infty)]_{av}=P_\infty(p)^2$, where $P_\infty(p)$ is the probability that a given site belong to the "infinite cluster," and $P_\infty(p) \sim \Delta p^\beta$ for p→p_c^+ as shown in Fig. 1b. We may also consider the probability P(iϵn) that a given site i belong to a cluster of n sites. As discussed by Stauffer[2]

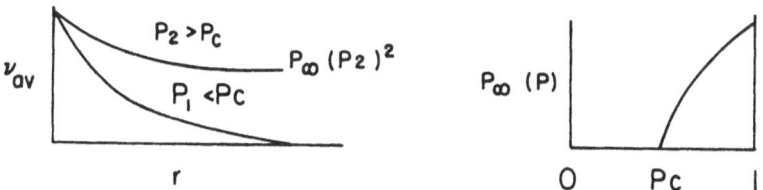

Fig. 1. Left: (a) $[\nu(0,r)]_{av} \equiv \nu_{av}$ vs r for two values of p. Right: (b) $P_\infty(p)$ vs p.

$$P(i\epsilon n) \sim n^{1-\tau} \, f(n \, \Delta p^{\beta+\gamma}) \, , \tag{1a}$$

where γ is the exponent for the percolation susceptibility. All the percolation exponents can be expressed in terms of any two independent ones. We may write (1a) as

$$P(i\epsilon n) \sim n^{1-\tau} \, f(n/\xi^{(\beta+\gamma)/\nu}) \, , \tag{1b}$$

which indicates that the typical size, n, of a cluster depends on ξ as

$$n \sim \xi^{(\beta+\gamma)/\nu} \equiv \xi^{D_f} \, , \tag{2}$$

where $D_f = (\beta+\gamma)/\nu$ is the fractal dimension of the cluster over length scales $r < \xi$. Many people have been seduced by the simplicity of the fractal picture[3] to talk in terms of D_f and ν. I usually talk in terms of the older exponents β, γ, ν, etc. These two exponent languages are completely equivalent, at least as far as dominant behaviors are concerned: $\beta = (d-D_f)\nu$ and $\gamma = (2D_f - d)\nu$, for instance.

The fact that $P(i\epsilon n)$ depends on only a single scaling variable or is "unifractal" leads to "constant gap" scaling for the moments of the cluster size, e. g.

$$\int n \, P(i\epsilon n) \, dn \sim \Delta p^{-\gamma}; \qquad \int n^2 \, P(i\epsilon n) \, dn \sim \Delta p^{-\gamma - D_f \nu} \tag{3a,b}$$

$$\int n^3 \, P(i\epsilon n) \, dn \sim \Delta p^{-\gamma - 2D_f \nu} \, . \tag{3c}$$

II. MODELS OF DIFFUSION, RESISTANCE, AND SPIN WAVES

Following deGennes[4] we consider models of the "ant in the labyrinth." At each time step the "blind ant" blindly selects a direction in which to hop. If the bond in that direction is present, the ant hops, otherwise it waits for the next time step. In Fig. 2a one sees that hopping is symmetric in this model. Both the probability to hop from site 3 to site 2 and that to hop from 2 to 3 are equal to $1/z$, where z is the coordination number of the underlined lattice. The corresponding master equation for the probability, $P_i(t)$, that the ant be at site i at time t is

$$dP_i/dt = \Sigma_{\delta, \, occ} \, (P_{i+\delta} - P_i) \, , \tag{4}$$

where the sum is over occupied bonds intersecting site i. In contrast, the "myopic ant" can see which neighboring bonds are occupied and selects randomly from these a bond over which to hop. Thus the myopic ant always hops at each time step and is described by the equation

$$dP_i/dt = \Sigma_{\delta, occ} \, [P_{i+\delta}/z_{i+\delta} - P_i/z_i] \, , \tag{5a}$$

where z_i is the coordination number of site i in the diluted lattice. In Fig. 2a one sees that the myopic ant hops with probability 1 from site 2 to site 3 and with probability 1/3 from 3 to 2. This process is thus not symmetric, but if we set $P_i/z_i = u_i$, we can write

$$z_i \, du_i/dt = \Sigma_{\delta, occ} \, (u_{i+\delta} - u_i) \, . \tag{5b}$$

Kirchhoff's equations for a network in which a) each occupied bond has unit resistance, and b) each site is connected by a capacitor C to ground are[5]

Fig. 2. Two small clusters, Left: (a), Right: (b).

$$C \, dV_i/dt = \Sigma_{\delta,occ} \, (V_{i+\delta} - V_i) + I_{ext}(i) \, , \tag{6}$$

where $I_{ext}(i)$ is the external current put into the system (if $I_{ext}(i)>0$) at site i.

Spin waves[6] in a Heisenberg spin system are described by the Hamiltonian $H=-(J/2) \, \Sigma_{\delta,occ} \, \vec{S}_i \, \vec{S}_{i+\delta}$ from which the equations of motion are

$$i \, dS_j^+/dt = \Sigma_{\delta,occ} \, J \, S_j^+ S_{j+\delta}^z - S_{j+\delta}^+ \, S_j^z \, , \tag{7a}$$

where $S_j^+ = S_j^x + iS_j^y$ (Here $i=e^{i\pi/2}$). At low temperature this equation may be linearized by setting $S_j^z = S_j$, where S_j is the magnitude of the spin at site j. Setting $u_j=S_j^+/S_j$ we have

$$iS_i \, du_i/dt = \Sigma_{\delta, \, occ} \, J \, S_i \, S_{i+\delta} \, (u_i - u_{i+\delta}) \, . \tag{7b}$$

These models are clearly quite similar and we make the following statements:
a)   The blind ant and the resistor network are equivalent: $P_i(t)$ with the initial condition that the ant is known to start from site k is equivalent to the voltage $V_i(t)$ of a network with $I_{ext}(i)=0$ for all i if the initial voltage is zero at all sites except k where it is unity.
b)   By comparing (5b) and (7b) we see that the myopic ant and spin waves are equivalent.
c)   It is widely accepted that the blind and myopic ants have the same asymptotic long time or long distance behavior. This equivalence is established analytically by Harris et al.[7]
d)   Accepting c) and comparing (5b) and (6) we can state that "weak randomness" in the capacitances $C_i=C+\delta C_i$, with $|\delta C_i| < C$, does not change the asymptotic properties of the network.
e)   For $q\xi \langle\langle 1$ spin waves obey the dispersion relation $\omega=D(p)q^2$ and for $r\rangle\rangle\xi$ diffusion obeys $r^2 \sim D(p)t$. In the fractal regime ($q\xi>1$ or $r<\xi$) these relations break down and one has anomalous diffusion[8] $r^{2+\theta} \sim t$ and fracton excitations[9] $\omega \sim q^{d_s}$. Here θ and the spectral dimension d_s are related to the conductivity exponent t defined by $\Sigma \sim \Delta p^t$, where $\Sigma(p)$ is the bulk conductivity:

$$\theta\nu = (t-\beta); \qquad\qquad d_s\nu = (\beta+\gamma)/(t-\beta+2\nu) \, . \tag{8a,b}$$

f)   A relation equivalent to the Einstein relation is[10]

$$D(p)/D(1) = [\Sigma(p)/ \, \Sigma(1) \,] - [(P_\infty(p))/(P_\infty(1) \,) \,] \tag{9}$$

or $D(p) \sim \Sigma(p)/P_\infty(p)$. To see the physical content of this relation consider the simple circuits a and b of Fig. 2. Considering sites 1 and 2 we can say that $\Sigma(a)=\Sigma(b)$. For these "infinite" clusters $P_\infty(a) > P_\infty(b)$, so that (8) predicts that $D(a) < D(b)$. But this is correct, because the diffusing particle spends some time in the dead end which does not affect the conductivity.

C.   TWO-POINT RESISTANCE $[R(x,x')]_{av}$

Consider a section of the diluted lattice shown in Fig. 3a and imagine measuring the resistance between all pairs of points i and f at a fixed separation $\mathbf{r}$, as indicated in Fig. 3b. As can be seen from Fig. 3a the measured resistance will vary for different choices of sites i and f and the resulting probability distribution $P(R,r)$ that the resistance assume the value R gives rise to the histogram[13] represented schematically in Fig. 4a. Note that we allowed a box in this histogram at $R=\infty$. From this histogram one can deduce the probability that two sites at separation $\mathbf{r}$ be in the same cluster: this quantity is just the fraction of sites having finite resistance: $P(R<\infty,r) = [\nu(0,r)]_{av}$. To characterize the resistance between sites we introduce

Fig. 3. Left: (a) A cluster of resistances. Right: (b) Resistance probe between terminals i and f. In (a) we indicate possible choices for locations of the terminals i and f.

473

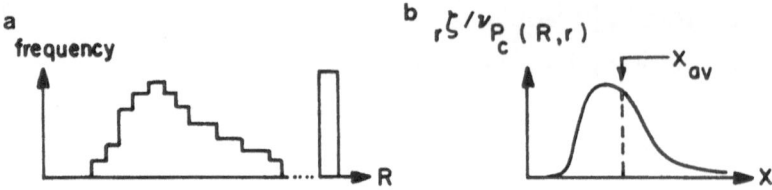

Fig. 4. (a) Resistance histogram. (b) $P_c(R,r)$ for fixed r vs $x \equiv R/r^{\zeta/\nu}$.

the conditional probability $P_c(R,r)$ that the resistance at separtion r assume the value R if the two sites are known to be in the same cluster: $P_c(R,r)=P(R,r)/[\nu(0,r)]_{av}$. It has been shown by several techniques[11-13] that this function depends on a single scaling variable $R/R_{av}(r)$, where $R_{av}(r)$ is the average resistance between two points <u>on the same cluster</u> at a separation r: $R_{av}(r) \sim r^{\zeta/\nu}$, for $r<\xi$. Thus $P_c(R,r)=r^{-\zeta/\nu} f(R/r^{\zeta/\nu})$, as shown in Fig. 4b. This scaling form shows that the resistance distribution is "unifractal" and gives rise to constant gap scaling,[14]

$$\int P_c(R,r) \, R^q \, dR \sim r^{q\zeta/\nu} . \tag{10}$$

Next we consider how to get the conductivity exponent t, assuming we have calculated ζ. For this purpose we use the node-link argument of Skal and Shklovskii[15] and deGennes[16] illustrated in Fig. 5. Here one imagines a d dimensional regular lattice consisting of nodes separated by a distance ξ connected by resitances $R(\xi)\sim\xi^{\zeta/\nu}$. The current density J, i. e. the current per unit area, is I/ξ^{d-1}, and the electric field is $E=dV/dx=V/\xi$. Thus the conductivity is

$$\Sigma = J/E = (I/\xi^{d-1}) \, (\xi/V) = (I/V)\xi^{2-d} . \tag{11}$$

But we identify V/I as $R(\xi) \sim \xi^{\zeta/\nu}$. Now using $\xi\sim\Delta p^{-\nu}$ we get

$$\Sigma(p) \equiv \Delta p^t \sim \Delta p^{(d-2)\nu + \zeta} . \tag{12}$$

This result seems to rely on the validity of the node-links picture. However, it can be derived[14] using scaling and spin-wave hydrodynamics[17]. For $p>p_c$ the zerofrequency spinwave response function G(q) for wavevector q in the hydrodynamic regime ($q\xi<<1$)is $M/(Dq^2)$, where M is the magnetization. We set $M=P_\infty(p)$, since finite clusters do not order, and use scaling to extend this result into the regime where $q\xi$ is no longer small by writing $G(q)=P_\infty(p)D(p)^{-1}q^{-2}f(q\xi)$. This form was derived for $p>p_c$, but it will also hold for $p<p_c$. In that case, however, since we do not have spin wave excitations, $f(q\xi)$ will vary as $(q\xi)^2$ at small $q\xi$. Thus, for $p<p_c$

$$G(q) \sim P_\infty(p) \, D(p)^{-1} \, \xi^2 \sim \Delta p^{2\beta-t-2\nu} . \tag{13}$$

But since the spin-wave problem is essentially the same as the resistor network, G(q) is equal to the spatial Fourier transform of the voltage response function, $\langle V V\rangle_q$, which for $q\to0$ is

$$\langle V V\rangle_{q\to0} = \Sigma_j \, [\nu(i,j) \, R_{ij}]_{av} . \tag{14}$$

In this sum the dominant contribution comes for $r_{ij}\sim\xi$, so that

$$\langle V V\rangle_{q\to0} = R(\xi) \, \Sigma_j \, [\nu(i,j)]_{av} = \Delta p^{-\gamma-\zeta} . \tag{15}$$

Equating the right-hand sides of (13) and (15) and using $2\beta+\gamma=d\nu$, we recover (12): $t=(d-2)\nu+\zeta$.

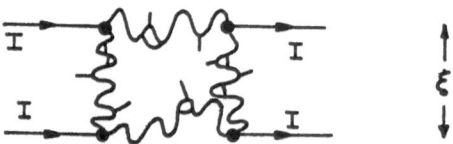

Fig. 5. Node-link picture of resistor network. The resistance between nodes is $R(\xi)$.

D. HAMILTONIAN FORMULATION[5,13] FOR $[R(x,x')]_{av}$

Now we discuss how to calculate ζ. As we will see, the following correlation function $\chi_\lambda(x,x')$ contains the desired information:

$$\chi_\lambda(x,x') \equiv [\exp (- \lambda^2 R(x,x')/2]_{av} , \tag{16}$$

where λ is a parameter at our disposal. For instance, when λ is very small, the quantity inside the square bracket of (16) is zero when the sites are disconnected and the resistance is infinite and is essentially unity when the sites are connected. Thus we see that

$$\chi_{\lambda \to 0}(x,x') = [\nu(x,x')]_{av} , \tag{17}$$

i. e. χ_λ is equal to the percolation connectedness susceptibility when $\lambda \to 0$. For small, but non-zero λ, this correlation function still has no contributions when the sites x and x' are disconnected. When they are connected, the factor inside the square bracket of (16) is approximately $1 - (1/2) \lambda^2 R(x,x')$. Thus we write

$$\chi_\lambda(x,x') = [\nu(x,x')]_{av} - (1/2)\lambda^2[\nu(x,x') R(x,x')]_{av} \tag{18a}$$

$$= [\nu(x,x')]_{av} [1 - (1/2)\lambda^2 R_{av}(x-x')] . \tag{18b}$$

Summing this over x' gives the $q=0$ Fourier component of χ:

$$\chi_\lambda(q=0) = \chi_{perc}(q=0) [1 - (1/2) \lambda^2 R_{av} (\xi)] \tag{19a}$$

which is of the form

$$\chi_\lambda(q=0) = \chi_{perc}(q=0) \left[1 - \frac{1}{2} \frac{\lambda^2}{\Delta p^\zeta} \right] . \tag{19b}$$

The strategy is to calculate χ_λ and compare the result to (19b) in order to determine ζ.

In what sense is the above correlation function similar to those used for the Ising model? The answer to that question lies in the following mathematical relation

$$\exp\left(- \frac{1}{2} \lambda^2 R(x,x') \right) = Z^{-1} \int_{-\infty}^{+\infty} \pi DV \; e^{i\lambda V(x)} \; e^{-i\lambda V(x')} \; e^{-H(V)} , \tag{20}$$

where $Z = \int \pi DV \exp[-H(V)]$, πDV indicates integration over all $V(x)$'s, and

$$H(V) \equiv (1/2) \Sigma_{x,x'} \sigma_{x,x'} [V(x) - V(x')]^2 . \tag{21}$$

Thus we interpret $H(V)$ as the "Hamiltonian" for the network of conductances $\sigma_{x,x'}$ and the right-hand side of (20) as the order parameter-order parameter correlation function where $\exp[i\lambda V(x)]$ is the order parameter at x. I still have not told you how you are to evaluate (20). To do that involves taking account of the randomness via the replica trick.[5,13]

Ignoring such intricacies, we consider the mean-field result

$$\chi_\lambda(q=0) = \left[\Delta p + (1/2)\lambda^2 \right]^{-1} \tag{22a}$$

$$= \Delta p^{-1} \left[1 - \frac{1}{2} \frac{\lambda^2}{\Delta p} ... \right] . \tag{22b}$$

and by comparison with (19b) we see that $\zeta = 1$ in mean field theory. This implies that

$R_{av}(r) \sim r^{\zeta/\nu} \sim r^2$. However, in high spatial dimension d, one can view the percolating cluster connecting two points as a random walk. In that case, the number of steps in the walk, i. e. its resistance, will be proportional to the square of the separation, as we just found. For d>6, this result holds. However, for d<6 the RG gives[13]

$$\zeta = 1 + \epsilon/42 + \dots \quad \epsilon = 6 - d . \tag{23}$$

Apparently ζ increases slowly with decreasing dimension to its value for d=2, ζ=1.30.[18] "Who cares about this?" you may say. Well, the Alexander-Orbach conjecture,[9] d_s=(4/3), implies that

$$\zeta = (\beta+\gamma)/2 = \{[1-(\epsilon/7)] + [1+(\epsilon/7)]\}/2 = 1 , \tag{24}$$

which disagrees with (23). Thus the Alexander-Orbach relation is good but not exact.

From (16) it is clear that χ_λ contains information on averages of moments M_q of resistances. Since these obey constant gap scaling, ratios like $M_3 M_1 / M_2^2$ are universal. The values of such ratios from the RG[13] have been verified quite nicely by series work.[19]

E.   RESISTANCE NOISE[20] AND CURRENT DISTRIBUTION[20,21]

Here the conductance of each bond has a probability distribution

$$P(\sigma_b) = (1-p) \delta(\sigma_b) + p g(\sigma_b) , \tag{25}$$

where $g(\sigma)$ is sharply peaked for $\sigma \sim 1$. There are now two sources of fluctuations in R(x,x'):
a)   Fluctuations in geometry as in the histogram of Fig. 4;
b)   The width of $g(\sigma)$ (i. e. resistance "noise") even when the geometry is fixed.
How should we characterize effect b)? Averages over $g(\sigma)$ at fixed cluster geometry, denoted $\langle \ \rangle_g$, such as

$$\langle R^2(x,x') \rangle_g - \langle R(x,x') \rangle_g^2 , \tag{26a}$$

measure just this effect, and would be zero without resistance "noise." The quantity in (26a) tells us about the width in the resulting distribution of R(x,x'). In general we study the higher cumulants, as for example

$$\langle R^3(x,x') \rangle_g - 3\langle R^2(x,x') \rangle_g \langle R(x,x') \rangle_g + 2\langle R(x,x') \rangle_g^3 \tag{26b}$$

and so forth. A problem now arises: How can we obtain averages over g independently from that over cluster geometry? To answer this, note that the average over cluster geometry (not discussed explicitly here, see Ref. 5) required the introduction of a replica index. Now that we have two such averages, we introduce two replica indices,[22] one associated with each average.

More insight as to the meaning of the above quantities comes from the relation[20,22] between cumulants, such as in (26) and moments of the current distribution. For this purpose let us consider injecting a unit current into the network at site x and removing it at site x'. In this process we will generate an approximately dipolar array of currents in the bonds in a region centered on x and x'. Let $i_b(x,x')$ denote the current in bond b in this situation. For an arbitrary network it can be shown[20] that $dR(x,x')/dr_b = i_b(x,x')^2$. Thus, fluctuations in resistance are closely linked to the bond currents. More precisely to lowest order in the width of the distribution g one has

$$\langle R(x,x')^q \rangle_{g,cum} = \langle (\sigma_b)^q \rangle_{g,cum} \Sigma_{bonds} i_b(x,x')^{2q} , \tag{27}$$

where "cum" indicates a cumulant average [as in (26a) for q=2 and (26b) for q=3]. Thus the cumulants of the resistance are related to the moments of the current distribution, $n(i_b)$, shown schematically in Fig. 6 for fixed values of x and x'. Note that we only get the full current (i_b=1) in the singly-connected or "red" bonds. Thus n(i_b=1) scales as $r^{1/\nu}$, where we have used Coniglio's theorem[23] for the scaling of singly connected bonds. On the other hand, the set of bonds through which current flows comprise the backbone. Consequently $\int n(i_b)di_b$ scales like $r^{D_{BB}}$, where D_{BB} is the fractal dimension of the backbone.[3] Also, the total power for unit input current is the resistance $R(x,x')$, so that $\int n(i_b)i_b^2\, di_b$ scales like the resistance, i. e. as $r^{\zeta/\nu}$. Thus different parts of $n(i_b)$ scale differently.[21] This is the essence of multifractality.[24]

A convenient way to characterize multifractality is to specify how the moments of the current distribution scale with separation $r\equiv x-x'$:

$$\Sigma_b \left[\, i_b(x,x')^{2q}\,\right]_{av} \sim r^{[-\gamma-\psi(q)]/\nu}\ .\tag{28}$$

(These exponents $\psi(q)$ are related[22] to those introduced by Rammal et al.[20]) The RG result[22] is

$$\psi(q) = 1 + \frac{\epsilon}{7(q+1)(2q+1)}\ .\tag{29}$$

Had the current distribution been "unifractal" we would have had constant gap scaling, i. e. $d\psi(q)/dq$ would have been constant. In most applications of multifractality one considers the qth moment of the probability distribution for q over the entire range $-\infty$ to $+\infty$. From (29) one sees that sufficiently negative values of q ($q<q_c$, with $q_c=-1/2$ according to the RG) are not allowed. A rigorous bound for q_c for diluted resistor networks is given by Blumenfeld et al.[25] This pathology for negative q probably does not occur for diffusion limited aggregation.

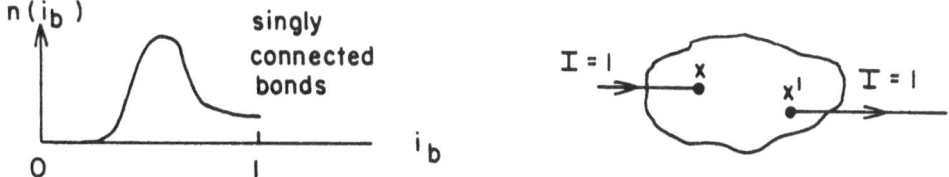

Fig. 6. Distribution of bond currents (Left) for fixed terminals x and x' (Right).

F. NON-LINEAR RESISTORS

One may consider networks of non-linear resistive elements[26] each of which obeys a generalized Ohm's Law with a non-linearity exponent α:

$$\Delta V = r\,\left|I^\alpha\right|\,\text{sign}(I) \qquad \text{or} \qquad I = \sigma\,\left|\Delta V\right|^{1/\alpha}\,\text{sign}(\Delta V)\ .\tag{30}$$

One can define[27] the exponents ζ and $\psi(q)$ for arbitrary value of α. Special values of α are related[28-30] to interesting geometrical quantities. Consider first a parallel circuit in which two resistors r_1 and r_2 are subjected to a voltage drop $\Delta V=1$. The combined current flowing in both resistors is $(1/r_1)^{1/\alpha} + (1/r_2)^{1/\alpha}$ so that the effective resistance for parallel resistors is

$$\left[\,1/R\,\right]^{1/\alpha} = \left[\,1/r_1\,\right]^{1/\alpha} + \left[\,1/r_2\,\right]^{1/\alpha}\ .\tag{31}$$

For resistors in series one can easily verify that the usual rule $R=r_1+r_2$ still holds for non-linear elements. Following Blumenfeld and Aharony[28] we consider the special cases $\alpha=\infty$ and $\alpha=0^+$. For $\alpha=\infty$ we see that the right-hand side of (31) is 2, so that $R\to2^{-\alpha}\to0$. Thus only bonds not in parallel contribute, i. e. the total resistance is due to singly connected bonds. According to Coniglio's theorem[23] we therefore have $R_{av}(r)\sim r^{1/\nu}$, i. e. $\zeta(\infty)=1$. For $\alpha\to0^+$, the right-hand side of (31) is dominated by the smaller of the resistors in parallel, $r_<$, so that $R=r_<$. Thus the current always flows in the path with the least resistance, i. e. the shortest path or the "chemical" distance, R_{min}. Thus $R_{min}(x,x')\sim r^{\zeta(0+)/\nu}$. For $\alpha=0-$ $\zeta(\alpha)$ becomes the exponent associated with the longest self-avoiding walk (on finite clusters) and $\zeta(\alpha=-1)$ is the fractal dimension of the backbone and $\zeta(-\infty)$ gives the entropy of domain walls.[29] In Fig. 7 $\zeta(\alpha)$ is shown for d=2.

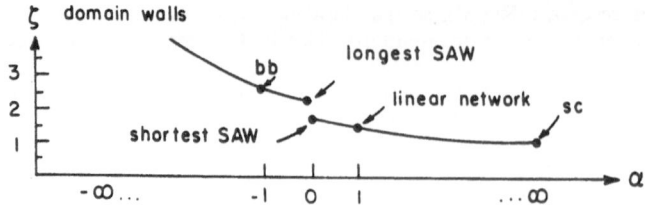

Fig. 7. $\zeta(\alpha)$ vs α for two dimensions.[29] Here "bb" (at $\alpha=1$) denotes backbone and sc (at $\alpha=\infty$) singly connected bonds.

The RG results[27] for ζ and $\psi(q)$ for general α are too cumbersome to give here. Instead a simplified version of a perturbative argument[27] relating these two exponents for general α will be given. It is unusual to be able to derive such a relation using elementary methods, but since this is the case here, it is worth repeating the argument. To simplify the algebra we consider the case of a slightly non-linear network, that is we set $\alpha=1-\delta$ with δ small. The more general case is treated in Ref. 27 and, by a different method, in Ref. 29. For small δ we write

$$I_{ext}(i) = \Sigma_j \; \sigma_{ij} \; \Delta V_{ij} \; \left| \Delta V_{ij} \right|^\delta \sim \Sigma_j \; \sigma_{ij} \; \Delta V_{ij} \; [\; 1 + \delta \; \ln \left| \Delta V_{ij} \right| \;] \; . \tag{32}$$

For simplicity we take the external current to be of unit magnitude, into the system at x and out of the system at x'. We view these equations as those of a linear system with bond conductances

$$\sigma_{ij} = \sigma_0 + \sigma_0 \; \delta \; \ln \left| \Delta V_{ij} \right| = \sigma_0 + \sigma_0 \; \delta \; \ln \; \left| I_{ij} \right| \; , \tag{33}$$

where $\sigma_0 \equiv r_0^{-1} \equiv 1$ is the conductance of a single occupied bond b. If R_0 represents the linear resistance and R_α the non-linear resistance, then correct to first order in δ we have

$$R_\alpha(x,x') = R_0(x,x') + \Sigma_b \; [\partial R_(0(x,x')/\partial\sigma_b] \; \sigma_b \; \delta \; \ln \left| I_b \right| \; . \tag{34}$$

For unit external currents, we have $R_0(x,x')=\Sigma_b \; I_b^2 r_0$ and $\partial R_0(x,x')/\partial\sigma_b = -I_b^2 r_0^2$. Thus (34) is

$$R_\alpha(x,x') = \Sigma_b \; [\; r_0 I_b^2 - \delta \; R_0 I_b^2 \; \ln \left| I_b \right|] \tag{35a}$$

$$= \Sigma_b \; r_0 \; I_b^{2-\delta} \; , \tag{35b}$$

which shows that $\zeta(1-\delta)=\psi(1-\delta/2)$ to first order in δ. Thus for $\alpha=1$ we have

$$2 \; \partial\zeta(\alpha)/\partial\alpha = \partial\psi(q)/\partial q|_{q=1} \; . \tag{36}$$

This relation can be generalized to arbitrary α[27,29] and differs from that of de Arcangelis et al.[30] which holds only for networks with high symmetry.

G.   OTHER TOPICS

We mention briefly three other related problems. Firstly, John and Lubensky[31] have treated a model of granular Josephson-coupled superconductors for which the Hamiltonian is

$$H = - \; \Sigma_{ij} \; J_{ij} \; \cos\left(\phi_i \; - \; \phi_j \; + \; \int_{r_i}^{r_j} \vec{A} \; (r) \; d\vec{l} \right), \tag{37}$$

where J_{ij} is zero if the bond (i,j) is vacant and is unity if it is occupied. This model reduces to an x-y model when the vector potential A(r) is zero, i. e. when the magnetic field is zero. For an x-y model at low temperature, $\cos(\phi_i-\phi_j) \sim (\phi_i-\phi_j)^2$, in which case the Hamiltonian is isomorphic to that for the resistor network given in (21). Now consider the effect of non-zero magnetic field. In the case of dangling ends (not part of loops) one can "gauge away" the vector potential. For loops there will be field induced frustration because a uniform variation of the superconducting phase can not accomodate loops of randomly varying size and connectivity. The reader is referred to Ref. 38 and the references therein for more details.

Fluid flow in random pores[32] resembles current flow in a random resistor network. Due to narrow necks in the pores, the bonds have a distribution of σ's with $g(\sigma)\sim\sigma^{-\alpha}$. This model, introduced by Kogut and Straley,[33] has complicated cross-overs which have been elucidated by the RG work of Lubensky and Tremblay,[34] but the results are too complicated to give here.

Finally, Harris and Christou[35] have considered the end-to-end resistance, R, of several random walk models. The first, shown in Fig. 8a, is one[36] in which each step of a random walk is associated with a conductance σ. If a bonds is traversed m times, its conductance is $m\sigma$. They find $R\sim N^x$, with x=1 for d>4 and x=1-ϵ/4 for d=4-ϵ. For a similar model,[37] shown in Fig. 8b, of "bridged" self-avoiding walks in which sites which are nearest neighbor are connected by bridges of conductance σ, they find R~N, as given previously by Ball and Cates.[38]

From these examples one sees that the techniques used to handle resistance problems have potentially quite wide application. They might be applied to charge density waves, the model for which was compared to the x-y model by Rammal in his lectures. Indeed, any phase fluctuation model on percolation clusters with an energy of the form $|\nabla\phi|^2$ is a candidate for these methods.

Fig. 8. Left: (a) Random walk with end-to-end resistance $13\sigma/5$. Right: (b) self-avoiding walk with bridges (dashed lines) whose end-to-end resistance is $11\sigma/4$.

H.   OPEN QUESTIONS

Brief mention may be made of three open questions. Although much numerical information on them is available, there is as yet no convincing scaling picture or RG treatment to provide a firm basis for placing their critical behavior in a more general context.

The first of these concerns the mixture of normal (σ=1) and superconducting (σ=∞) elements. Here the conductivity diverges as $\Sigma(p)\sim\Delta p^{-s}$ as p→p_c. The relation s=2ν-β, given by Stephen,[5] disagrees violently with the known result s=t$\simeq$1.3 in two dimensions.

The second problem is to give a Stephen-like field theory for the diluted elastic network and to thereby apply the RG to this problem. For central-forces[39] the elastic threshold does not coincide with the percolation threshold, so there are bound to be associated difficulties. An attempt[40] has been made to invoke a different kind of percolation in this case. For general bond bending forces the two thresholds do coincide and one might expect progress in this case. However, the analog of the resistive susceptibility does not seem to have a simple interpretation.

Finally there is the problem of self-avoiding walks (SAW's) on percolation clusters.[41] As we have said, the resistor network is equivalent to the (n=3 component) Heisenberg model. In fact, it is equivalent to the n-component spin model with n$\geq$2. Also, the n=1 model is just the Ising model, which has been treated at the percolation threshold.[42] Since SAW's can to be represented by the n→0 limit of the n-component spin model,[43] one could well ask why this problem has not yet been solved. The explanation lies in the fact that SAW's correspond to a high-temperature expansion of the partition function, whereas for n$\geq$1 the cross-over at the percolation threshold involves the assumption of low temperatures where the spins are nearly parallel.

In conclusion, for slightly more detail, the reader is referred to another review.[44]

I am grateful to Tel Aviv University for its hospitality and acknowledge support from the US-Israel Binational Science Foundation, the Israel Academy of Sciences and Humanities, and the Israel AEC. I thank A. Aharony and T. C. Lubensky for many helpful discussions.

REFERENCES

+Permanent address: Department of Physics, University of Pennsylvania, Philadelphia, PA 19104

1.   C. Domb and M. Green, Phase Transitions and Critical Phenomena, Vol. 6, Academic Press, 1982, New York.
2.   D. Stauffer, Introduction to Percolation Theory, Taylor and Francis, London, 1985).
3.   A. Aharony in Directions in Condensed Matter Physics, ed. G. Grinstein and G. Mazenko (World Scientific, Singapore, 1986), p1.
4.   P. G. de Gennes, La Recherche 7, 919 (1976).
5.   M. J. Stephen, Phys. Rev. B 17, 4444 (1978).
6.   L. R. Walker in Magnetism, ed. G. T. Rado and H. Suhl (Academic Press, New York, 1963) Vol. I, Ch. 8.
7.   A. B. Harris, Y. Meir, and A. Aharony, Diffusion on Percolating Clusters, preprint.
8.   Y. Gefen, A. Aharony, and S. Alexander, Phys. Rev. Lett. 50, 77 (1983).
9.   S. Alexander and R. Orbach, J. de Phys. (Paris) Lett. 43, 625 (1982); R. Rammal and G. Toulouse, J. de Phys. (Paris) Lett. 44, L13 (1983).
10.   S. Kirkpatrick, Phys. Rev. Lett. 27, 1722 (1971); A. B. Harris and S. Kirkpatrick, Phys. Rev. B 16, 542 (1977).
11.   R. B. Stinchcombe and B. P. Watson, J. Phys. C 9, 3221 (1976).
12.   R. Rammal, M. A. Lemieux, and A.-M. S. Tremblay, Phys. Rev. Lett. 54, 1087(C) (1985).
13.   A. B. Harris and T. C. Lubensky, Phys. Rev. B 35, 6964 (1987).
14.   A. B. Harris and R. Fisch, Phys. Rev. Lett. 38, 796 (1977).
15.   A. S. Skal and B. I. Shklovskii, Fiz. Tekh. Poluprovodn. 8, 1582(1974) [Sov. Phys. Semicond. 8, 1029 (1975)].
16.   P. G. de Gennes, J. Phys. (Paris) Lett. 37, L1 (1976).
17.   B. I. Halperin and P. C. Hohenberg, Phys. Rev. 188, 898 (1969).
18.   Y. Meir, R. Blumenfeld, A. Aharony, and A. B. Harris, Phys. Rev. B 34, 3424 (1986).
19.   Y. Meir, A. B. Harris, and A. Aharony, unpublished.
20.   R. Rammal, C. Tannous, and A.-M. S. Tremblay, Phys. Rev. A 31, 2662 (1985).
21.   L. de Arcangelis, S. Redner, and A. Coniglio, Phys. Rev. B 31, 4725 (1985).
22.   Y. Park, A. B. Harris, and T. C. Lubensky, Phys. Rev. B 35, 5048 (1987).
23.   A. Coniglio, Phys. Rev. Lett. 46, 250 (1981).
24.   T. C. Halsey, M. H. Jensen, L. P. Kadanoff, I. Procaccia, and B. I. Shraiman, Phys. Rev. B 33, 1141 (1986).
25.   R. Blumenfeld, Y. Meir, A. Aharony, and A. B. Harris, Phys. Rev. B 35, 3524 (1987).
26.   S. W. Kenkel and J. P. Straley, Phys. Rev. Lett. 49, 767 (1982).
27.   A. B. Harris, Phys. Rev. B 35, 5056 (1987).
28.   R. Blumenfeld and A. Aharony, J. Phys. A 18, L443 (1985).
29.   R. Blumenfeld, Y. Meir, A. B. Harris, and A. Aharony, J. Phys. A 19, L791 (1986).
30.   L. de Arcangelis, S. Redner, and A. Coniglio, J. Phys. A 18, L805 (1985).
31.   S. John and T. C. Lubensky, Phys. Rev. B 34, 4815 (1986).
32.   B. I. Halperin, S. Feng, and P. N. Sen, Phys. Rev. Lett. 54, 2391 (1985).
33.   P. M. Kogut and J. P. Straley, J. Phys. C 12, 2151 (1979).
34.   T. C. Lubensky and A. M.-S. Tremblay, Phys. Rev. 34, 3408 (1986).
35.   A. B. Harris and A. Christou, to be published.
36.   J. R. Banavar, A. B. Harris, and J. Koplik, Phys. Rev. Lett. 51, 1115 (1983).
37.   J. S. Helman, A. Coniglio, and C. Tsallis, Phys. Rev. Lett. 53, 1195 (1984).
38.   R. C. Ball and M. E. Cates, J. Phys. A 17, 2531 (1984).
39.   S. C. Feng and P. N. Sen, Phys. Rev. Lett. 52, 216 (1984).
40.   J. Wang and A. B. Harris, Phys. Rev. Lett. 55, 2459 (1985).
41.   J. W. Lyklema and K. Kremer, Z. Phys. B 55, 41 (1984).
42.   M. J. Stephen and G. S. Grest, Phys. Rev. Lett. 38, 567 (1977).
43.   P. G. de Gennes, Phys. Lett. A 38, 339 (1972).
44.   A. B. Harris, Phil. Mag., to be published.

HAMILTONIAN FORMULATION FOR THE CONFORMATION AND RESISTANCE OF RANDOM WALKS

A. Christou and A. B. Harris [†]

Department of Theoretical Physics, University of Oxford
1 Keble Rd. Oxford OX1 3NP, UK.

Recently there has been great interest in obtaining detailed information concerning the geometrical and dynamical properties of random structures. Here we present a general technique for developing Hamiltonian formulations for such problems and illustrate them for various models of stochastic walks. Having developed a Hamiltonian formulation one can then discuss the desired properties via the usual methods, e.g. mean-field theory, the renormalization group (RG) etc. The steps involved in such a program are shown in Fig. 1. Here we describe in detail the first step, developing a Hamiltonian, and summarize briefly the other steps, since they involve well-known methods.[1]

I. Hamiltonian Formulation — Correlation Functions

Our aim is to develop a Hamiltonian H that is to be expressed in terms of spin variables $\{S\}$ such that $\text{Tr}\{\exp(-H)\}$ is a generating function from which the desired properties can be obtained. The main idea is to define the spin variables via their <u>trace rules</u> so as to achieve this end. As a simple example we reproduce the well-known[2] result that self-avoiding walks (SAWs) are described by the $n \to 0$ limit of the n-component Heisenberg model.

Consider the average

$$\chi(x, x') \equiv \langle S_1(x)S_1(x') \rangle \equiv \frac{\text{Tr}_S\left\{S_1(x)S_1(x')\exp\left(-H(S)\right)\right\}}{Z}, \tag{1}$$

where Z is the "partition function": $Z = \text{Tr}_S \exp\left(-H(S)\right)$ and $H(S)$ is an n-component Hamiltonian

$$H(S) = -\frac{1}{2}\sum_{x,x'}\gamma_{x,x'}\sum_{\alpha=1}^{n}S_\alpha(x)S_\alpha(x'), \tag{2}$$

where $\gamma_{x,x'} = 1$ if x and x' are nearest neighboring sites and is 0 otherwise. We would like $\chi(x, x')$ to be the generating function for SAWs, i.e. that $\lim_{n\to 0}\chi(x, x') = \sum_N c_N^{SAW}(x, x')K^N$ where $c_N^{SAW}(x, x')$ is the number of N-step SAWs starting at x and ending at x'. For this purpose we introduce the following <u>trace rules</u>. Spin operators at different sites are independent of one another. For any site x the trace rules are

$$\text{Tr}_S(1) = 1, \quad \text{Tr}_S(S_\alpha) = 0, \quad \text{Tr}_S(S_\alpha S_\beta) = \delta_{\alpha,\beta}, \quad \text{Tr}_S(S_{\alpha_1}S_{\alpha_2}\cdots S_{\alpha_k}) = 0, \quad k > 2. \tag{3}$$

[†] *Permanent Address: Department of Physics, University of Pennsylvania, Philadelphia, PA 19104 USA.*

Fig. 1. Flow chart for the calculation of scaling exponents in random systems.

For $n \to 0$ any free sum over replica indices vanishes, so that $Z \to 1$. Also, because of the trace rules (3), the numerator of (1) can be interpreted as giving rise to all possible diagrams each of which consist of a direct product of a walk from x to x' with an arbitrary set of closed polygons. However, since each polygon carries a free sum over a replica index, only walks <u>without</u> polygons survive the $n \to 0$ limit. Thus in the $n \to 0$ limit, $\chi(x, x')$ is the desired generating function for SAWs. Although the spin operators defined by (3) are not the usual Heisenberg ones, the resulting Hamiltonian can easily be shown to be in the same universality class as that obtained for Heisenberg spins.

Gaussian random walks (GRWs) can be handled in a similar way. We set

$$H(u) = -\frac{1}{2}K \sum_{x,x'} \gamma_{x,x'} u(x)u(x') \tag{4}$$

and define independent traces for each site x via

$$\mathrm{Tr}_u \{A(x)\} \equiv \left(\frac{1}{2\pi}\right)^{\frac{1}{2}} \int_{-\infty}^{+\infty} du(x) \{A(x)\} \exp\left(-\frac{1}{2}u^2(x)\right). \tag{5}$$

Then, if $\mathbf{I}$ and γ are matrices in the site labels, with $I_{x,y} = \delta_{x,y}$, one has

$$\chi^{GRW}(x_1, x_2) \equiv \langle u(x_1)u(x_2) \rangle = \{\mathbf{I} - K\gamma\}^{-1}_{x_1,x_2}. \tag{6}$$

Expanding the right-hand side of (6) one gets

$$\chi^{GRW}(x, x') = \sum_N c_N^{GRW}(x, x')K^N, \tag{7}$$

where $c_N^{GRW}(x, y)$ is the number of N-step GRWs which start at x and end at y.

So far we have not produced anything new. However, now consider

$$H(u, S) \equiv -\frac{1}{2}\sum_{\alpha=1}^{n}\sum_{x,x'} K\gamma_{x,x'} S_\alpha(x)S_\alpha(x')u_\alpha(x)u_\alpha(x'), \tag{8}$$

where the u_α's obey (5) for each value of α and now the S_α's obey the trace rules

$$\mathrm{Tr}_S(1) = 1, \quad \mathrm{Tr}_S\left(S_{\alpha_1}S_{\alpha_2}\ldots S_{\alpha_{2k+1}}\right) = 0 \tag{9a}$$

$$\mathrm{Tr}_S\left(S_{\alpha_1}S_{\alpha_2}\ldots S_{\alpha_{2k}}\right) = w_k \quad \text{if} \quad \alpha_1 = \alpha_2 \ldots = \alpha_{2k}$$
$$= 0 \quad \text{otherwise}. \tag{9b}$$

The trace rules (9) are designed to assign a fugacity w_k to each site visited k times in the walk, which we call a <u>valence</u> walk. As we discuss in detail elsewhere[3] the Su-Su correlation function is the generating function for such valence walks. Specifically

$$\lim_{n\to 0} \langle u_1(x)S_1(x)u_1(y)S_1(y) \rangle = \sum_N \sum_{\gamma \subset \gamma_N(x,y)} w(\gamma), \tag{10}$$

482

Fig. 2. a) A seven step Gaussian random walk. If each step represents a unit conductance then the end-to-end resistance of this walk is $\frac{5}{2}$. b) A five step "bridged" self-avoiding walk whose associated resistance is $\frac{11}{4}$.

where γ is summed over all N-step GRWs starting at x and ending at y and each such walk carries a fugacity factor $w(\gamma)$ which is a product over all sites of the fugacity factor for each site. The fugacity factor for a given site is w_k if the site has associated $2k$ spin operators as in (9b), i.e. if it is visited k times in the walk γ. To get the correct fugacity weighting it was necessary not only to incorporate spin variables obeying (9) into the Hamiltonian (4), but also to replicate the u variables n times in the limit $n \to 0$. In essence this procedure is necessary because one needs a quenched average over walks.[3]

The same idea of "marrying" walk variables to other variables can be used to obtain resistance properties of walks. To do this we use the relation[4] for the resistance $R(y, z)$ between two nodes y and z in an arbitrary network of conductances $\{\sigma_{x,x'}\}$

$$\lim_{J \to \infty} \lim_{q \to 0} \langle \mathbf{v}(y) \cdot \mathbf{v}(z) \rangle = \nu_{y,z} \left\{ 1 - \frac{R(y, z)}{J} + O(J^{-2}) \right\} \tag{11}$$

where $\nu_{x,x'} = 1$ if sites x and x' are connected by non-zero conductances and is zero otherwise, and $\langle \mathbf{v}(y) \cdot \mathbf{v}(z) \rangle$ is evaluated with respect to the q-state Potts Hamiltonian

$$H \equiv -\frac{1}{2} J(q-1) \sum_{x,x'} \sigma_{x,x'} \mathbf{v}(x) \cdot \mathbf{v}(x') \equiv -\frac{1}{2} \sum_{x,x'} \sigma_{x,x'} H_{x,x'}^{(P)}(\mathbf{v}), \tag{12}$$

where $\mathbf{v}(x)$ is a unit vector which can assume any one of q states such that if $\mathbf{v}(x)$ and $\mathbf{v}(x')$ are in the same state, $\mathbf{v}(x) \cdot \mathbf{v}(x') = 1$ and otherwise $\mathbf{v}(x) \cdot \mathbf{v}(x') = -(q-1)^{-1}$.

We now consider the resistance of random walks for which each step of the walk is associated with a conductance σ_0. In this model[5] a bond which is traversed k times in a walk has an associated conductance $k\sigma_0$, as is illustrated in Fig. 2. To treat this model we must[3] replicate the u variables and include for each bond (x, x') the appropriate Potts interaction, $H_{x,x'}^{(P)}(\mathbf{v})$, replicated in the limit $m \to 0$:

$$H(u, \mathbf{v}) = -\frac{1}{2} K \sum_{x,x'} \gamma_{x,x'} \sum_{\alpha=1}^{n} u_\alpha(x) u_\alpha(x') \exp \left\{ \sum_{\beta=1}^{m} H_{x,x'}^{(P)}(\mathbf{v}_\beta) \right\} \tag{13a}$$

$$= -\frac{1}{2} K \sum_{x,x'} \sum_{\alpha} \gamma_{x,x'} u_\alpha(x) u_\alpha(x') \prod_{\beta=1}^{m} \left\{ 1 + \frac{(q-1)}{1 + (J\sigma_0)^{-1}} \mathbf{v}_\beta(x) \cdot \mathbf{v}_\beta(x') \right\}. \tag{13b}$$

In the limit when first m, second n, third q, and finally $1/J$ go to zero, we have[3]

$$\langle u_1(x) \mathbf{v}_1(x) \cdot \mathbf{v}_1(y) u_1(y) \rangle = \sum_N K^N \sum_{\gamma \subset \gamma_N^{GRW}(x,y)} \left\{ 1 - \frac{1}{J} R(\gamma) \right\}$$

$$= \sum_N (zK)^N \left\{ 1 - R_{av}(N)/J \right\} \tag{14}$$

where z is the coordination number of the lattice.

483

For SAWs we "marry" Potts variables to spins and use the Hamiltonian H^{SAW}:

$$H^{SAW}(S, \mathbf{v}) = -\frac{1}{2}K \sum_{x,x'} \gamma_{x,x'} \sum_{\alpha=1}^{n} S_\alpha(x)S_\alpha(x') \exp\left(-\sum_\beta H_{x,x'}^{(P)}(\mathbf{v}_\beta)\right), \quad (15)$$

where the $S_\alpha(x)$ obey the trace rules in (3) and the Potts interaction is that of (12). This Hamiltonian will give the generating function for the end-to-end resistance of SAWs for which each step has an associated conductance σ_0. This problem is rather trivial in that every SAW of length N steps has a resistance $\frac{N}{\sigma_0}$. More interesting is a model[6] in which each step of the walk has a conductance, which for convenience we take to be $\sigma_0 + \sigma_{br}$, and between sites on the walk which are nearest neighbors on the lattice, but not adjacent sites along the path of the walk, there is a conductance σ_{br}. We call this network a "bridged SAW" or BSAW and its properties are obtained from the Hamiltonian $H^{BSAW}(S, \mathbf{v})$ with

$$H^{BSAW}(S, \mathbf{v}) = H^{SAW}(S, \mathbf{v}) - \frac{1}{2}J \sum_{x,x'} \gamma_{x,x'} \sum_{\beta=1}^{m} T(x)T(x')H_{x,x'}^{(P)}(\mathbf{v}_\beta). \quad (16)$$

As we have seen, $H^{SAW}(S, \mathbf{v})$ will endow each step of the SAW with a conductance σ_0. We would like the second term in (16) to connect nearest neighboring sites (whether or not they are successive steps along the walk) by parallel conductances σ_{br}. This will be the case if the T obey <u>trace rules</u> which make $T(x) = 1$ if accompanied by an $S_\alpha(x)$ and $T(x) = 0$ otherwise. (Without this restriction, the T's would generate a parallel set of conductances between <u>all</u> nearest neighboring sites on the lattice, irrespective of whether sites were on the SAW or not.) This restriction is obtained by setting

$$\mathrm{Tr}\ \{T^p\} = 0$$

$$\mathrm{Tr}\ \{T^p S_{\alpha_1} S_{\alpha_2} \cdots S_{\alpha_k}\} = 1 \ \ \text{if} \ \ \alpha_1 = \alpha_2 = \cdots = \alpha_k, \ \quad \text{for } k \geq 1. \quad (17)$$
$$= 0 \ \ \text{otherwise},$$

where p is an integer. With these trace rules and those previously introduced we have[3]

$$\langle S_1(x)\mathbf{v}_1(x) \cdot \mathbf{v}_1(y)S_1(y)\rangle = \sum_N K^N \sum_{\gamma \subset \gamma_N^{SAW}(x,y)} \{1 - R(\gamma)/J\}, \quad (18)$$

where we take the limit when m, q, n, J^{-1} all go to zero as in (14).

II. Field Theories [9]

The above Hamiltonians can be converted into a field theoretic form by using the Hubbard-Stratanovich transformation involving a complete set[3,4,7] of auxiliary fields conjugate to each combination of local operators in the bilinear Hamiltonians we have constructed. Thus we introduce fields $\Psi_\alpha(x)$ conjugate to $u_\alpha(x)$, $\Psi_{\alpha\beta}^\mu(x)$ conjugate to $u_\alpha(x)v_\beta^\mu(x)$, etc., where v_β^μ is the μth cartesian component of the Potts vector $\mathbf{v}_\beta$ in the βth replica. Thus the general Ψ has a <u>scalar index</u> α and t <u>Potts</u> indices (β_1, μ_1), (β_2, μ_2), ..., (β_t, μ_t). As in the dilute Ising[7] or Potts[4] models, one can classify fields according to the number, t, of Potts indices they involve, with $t = 0, 1, \cdots, m$. Then the field-theoretic free energy density for GRWs is schematically of the form

$$F^{GRW} = \frac{1}{2} \sum_{t=0}^{m} \sum_x \sum_\Psi \Psi^{(t)}(x) \{r_t - \nabla^2\} \Psi^{(t)}(x) + uV_c^{(4)}(\{\Psi\}) + vV_s^{(4)}(\{\Psi\}), \quad (19)$$

where $\Psi^{(t)}$ indicates a Ψ with t Potts indices and $\sum_\Psi$ indicates that all upper and lower indices on the Ψ's are summed over. Also, $V_s^{(4)}$ represents a potential fourth order in the Ψ's, which has "spherical" symmetry and $V_c^{(4)}$ a potential which has cubic symmetry. Unlike the pure Potts model, this field theory has no third order terms because the trace over the u or S variables ensures an even number of operators. The symmetries of the quartic potentials are analogous to those one encounters in treating the Heisenberg model in a cubic crystalline environment,[8] where they reduce to

$$V_s^{(4)} = \left(\sum\nolimits_{\alpha=1}^{n} S_\alpha^2 \right)^2 ; \quad V_c^{(4)} = \sum\nolimits_{\alpha=1}^{n} S_\alpha^4 . \tag{20}$$

Also $r_t = (K - z^{-1} + tJ^{-1})$, i.e. for $J^{-1} = 0$ the critical value of K is $1/z$. The model represented by (19) has non-trivial critical properties for $d < 4$ and so we treat $d = 4 - \epsilon$.

For BSAWs the above procedure yields a free energy density of the form of (19) except that in this case additional, more complicated, fourth order potentials emerge. These new potentials are irrelevant and too complicated to discuss here.[9]

The RG recursion relations in $4 - \epsilon$ dimensions can be obtained in the usual way.[8] The meaning of these recursion relations can be most easily understood from the potential flow in the u-v plane shown in Fig. 3. For the GRW model one starts with, and always stays on the line, $u + v = 0$. Thus the R fixed point describes GRWs. For $(1/J) = 0$, the geometrical properties of GRWs are described by the new R fixed point. These properties differ from those of the G or Gaussian fixed point only insofar as one calculates: a) statistics of interactions between two or more GRWs, or b) the end-to-end resistance of GRWs. The usual Gaussian statistics for a single GRW are reproduced by the fixed point R. Corrections to Gaussian behavior do appear, but because they are proportional to $(u + v)$, they vanish, as they must, for GRWs. These properties of the fixed point R indicate that it represents a satisfactory description of conformation and resistance of GRWs, and gives us confidence in our results. The average end-to-end resistance $< R_N >$ of N-step walks is determined from the t-dependence of r_t in (19). Recursion relations for r_t yield the result for GRWs,

$$< R_N > \sim N^x, \quad N \to \infty. \tag{21}$$

For $d > 4$, $x = 1$ and for $d = 4 - \epsilon < 4$,

$$x = 1 - \epsilon/4 + \epsilon^2/16 \ldots . \tag{22}$$

Note that the line $u + v = 0$ is an unstable ridge line. That is, if the system is started slightly away from this line, it flows elsewhere. Leaving the ridge line corresponds to adding self-avoidance. In other words, even a small amount of self-avoidance, or repulsion, is relevant and takes the walk from the GRW universality class to the SAW universality class, as is well-known. Then the system is described by the stable fixed point H, which corresponds to the Heisenberg $n \to 0$ fixed point, which is known[2] to describe SAWs. We now have to repeat the calculation leading to (21) and (22) for the fixed point H. Since the additional fourth-order potential due to bridges is irrelevant, one sees that SAWs and BSAWs have the same value of x, which clearly must be unity. Indeed, an explicit calculation does give x=1, in agreement with the previous results of Ball and Cates.[10] We can also study the corrections to scaling for BSAWs. These do

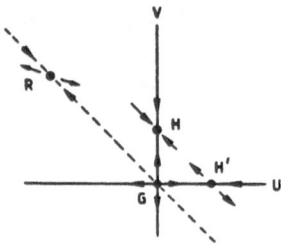

Fig. 3. The flow diagram for the coefficients u and v of the spherical and cubic symmetry potentials respectively. The random walk fixed point (R) is singly unstable and lies on the ridge $u + v = 0$. The Heisenberg fixed point (H), with only v non-zero, is stable and corresponds to the SAW fixed point. The other 'Heisenberg' fixed point (H'), with only u non-zero, is unstable and represents a multicritical point. G is the Gaussian fixed point.

depend on the degree of irrelevance of the above-mentioned new fourth-order potentials attributable to bridges. We find

$$< R_N > \sim N[1 + AN^{-\Delta}] \qquad (23)$$

with $\Delta = \epsilon/4$. Thus BSAWs have the same resistance exponent x as SAWs, but the corrections to scaling may be important enough to make numerical checks of this identity difficult. In fact the numerical situation is not entirely clear.

In summary, we have developed a general approach whereby one can express essentially arbitrary statistical properties of walks in terms of "spin" operators. The advantage of doing this is that such a formulation leads naturally, via the Stratanovich transformation, to a field theory whose analysis may be undertaken. At present we are exploring the application of these techniques to spiral and related walks.

Acknowledgments
AC acknowledges support from the SERC in the form of a Research Studentship. ABH wishes to thank the Department of Theoretical Physics for its hospitality and acknowledges partial support as a Visiting Fellow from the SERC and from the National Science Foundation under grant DMR 82-19216.

References

1. C. Domb and M. S. Green, *Phase Transitions and Critical Phenomena*, Vol 6 (Academic Press, 1982, New York).
2. P. G. de Gennes, Phys. Lett. A **38**, 339 (1972).
3. A. B. Harris and A. Christou, J. Phys. A, to be published.
4. C. Dasgupta, A. B. Harris, and T. C. Lubensky, Phys. Rev. B **17**, 1375 (1978).
5. J. R. Banavar, A. B. Harris, and J. Koplik, Phys. Rev. Lett. **51**, 1115 (1983).
6. J. S. Helman, A. Coniglio, and C. Tsallis, Phys. Rev. Lett. **53**, 1195 (1984).
7. M. J. Stephen and G. S. Grest, Phys. Rev. Lett. **38**, 567 (1977).
8. P. Pfeuty and G. Toulouse, *Introduction to the Renormalization Group and Critical Phenomena*, J. Wiley, New York (1977).
9. A.B. Harris and A. Christou, J.Phys. A , to be published.
10. R. Ball and M. E. Cates, J.Phys. A., **17**, L531 (1984).

RESONANT SCATTERING AND CLASSICAL LOCALIZATION

Carlos A. Condat

Institute for Physical Science and Technology
University of Maryland
College Park, Maryland 20742

INTRODUCTION

After the work of John, Sompolinsky, and Stephen[1], the study of localization in classical systems has received renewed interest. Most calculations have been carried out using a Gaussian noise representation for the disorder. Kirkpatrick[2] was the first to study localization in the presence of finite-size inhomogeneities. A "pseudosphere" approximation was then used to obtain more explicit results for the localization of acoustical waves (AW) by a random array of hard spheres and of electromagnetic waves (EW) by a random array of conducting spheres.[3] Related models were analyzed for the case of AW in a two-component composite[4] and of vector EW.[5] It is our purpose to investigate how the nature of the inhomogeneities affects the localization processes.

THE MODEL AND ITS ANALYSIS

Consider a collection of uncorrelated, identical, randomly distributed spheres (disks if d=2) immersed in a uniform lossless fluid. The volume fraction occupied by the spheres is n* and their radius is a. The velocity potential in the fluid, ϕ, satisfies the wave equation

$$\left(\frac{\partial^2}{\partial t^2} - c^2 \nabla^2\right) \phi(\vec{x},t) = 0 \quad , \tag{1}$$

with $c = (1/\rho\kappa)^{1/2}$, ρ and κ being, respectively, the density and compressibility of the fluid. The boundary conditions depend on the nature of the scatterers; we will consider hard, soft, and permeable (density ρ_s and compressibility κ_s) spheres. In the first two cases we must use, respectively

Neumann $\left(\frac{\partial \phi}{\partial r} = 0\right)$ and Dirichlet ($\phi=0$) boundary conditions at the surface of each sphere. For the permeable scatterers we need continuity conditions involving the velocity potential ϕ_s inside of the spheres:

$$\rho_s \phi_s = \rho \phi \quad \text{and} \quad \frac{\partial \phi_s}{\partial r} = \frac{\partial \phi}{\partial r} \quad \text{(at } r = a) \quad .$$

The study of localization involves the creation of a disturbance at $t = 0$ and $\vec{x} = 0$ and the observation of the evolution of the energy density at long times and large distances. The appropriate magnitude to calculate is the energy-density propagator $\langle P_E(\vec{k},\omega)\rangle$, which is the Fourier-Laplace transform of the squared Green's function. The brackets represent an average over the random distribution of scatterers. While the external frequency ω is related to the overall time evolution (since we are interested in long times we will take $\omega \to 0$), the internal frequency E is the one to be detected in an experiment probing the microscopic excitations.

It can be shown[2] that $\langle P_E \rangle$ has a diffusive hydrodynamic pole,

$$\langle P_E(\vec{k},\omega)\rangle \sim \frac{1}{-i\omega + D(E,\omega)k^2} \quad . \tag{2}$$

The diffusion coefficient $D(E,\omega)$ is calculated using the self-consistent theory of localization.[6] This yields an algebraic equation for $D(E,\omega)$:

$$D(E,\omega) = D_B(E)\left[1 + f_d \left(\frac{c}{E}\right)^d S(E) \int_0^Q \frac{d\vec{q}}{i\omega D^{-1}(E,\omega) - q^2}\right] \quad . \tag{3}$$

Here $D_B(E)$ is the Boltzmann diffusion coefficient, which results from considering only incoherent processes, f_d is a constant that depends on the dimensionality d, and $Q = 2\pi c D_B^{-1}(E)$ is a cut-off. (The importance of the choice of Q has been considered elsewhere.[7]) The total cross section $S(E)$ for the individual scatterers is a monotonously increasing (decreasing) function of the frequency E for hard (soft) spheres. In the most interesting case, $\kappa_s > \kappa$, it can have a peak structure due to partial wave resonances. This resonant behavior can be used to optimize the conditions for the observation of localization.

RESULTS

For three-dimensional systems, Eq. (3) yields two types of solutions

(A) $\quad D(E,\omega\to 0) = -i\omega\xi^2(E) \quad .$

This implies $P_E(\vec{x},t) \sim \exp[-|\vec{x}|/\xi(E)]$: the energy density is localized at long times in a region whose size is given by the localization length $\xi(E)$.

(B) $D(E,\omega \to 0) = D(E)$

This implies that the energy density is diffusing and the excitations of frequency E are extended.

The curve in the (E,n*) plane separating solutions of type A and B is a phase boundary between localized and extended states. A typical phase diagram for permeable scatterers is represented in Fig. 1. The oscillations in the phase boundary are due to the scattering resonances of the individual spheres. For example, as indicated in Fig. 1, a sweep in frequencies at n* = .175 would encounter four mobility edges. Delocalization occurs at low frequencies ($S(E) \sim E^{d+1}$ and the scattering is inefficient) and at high frequencies (geometrical optics limit). A partial smearing out of the oscillations should be expected in an experiment, because some scattering events will occur in the near field from previous scatterers. For this reason the conditions for localization are not necessarily much improved by a large increase in M. Localization effects are much weaker if M < 1 or if $\rho_s \neq \rho$ but $\kappa_s = \kappa$, and localization is not possible for hard scatterers. From Fig. 1 we also see that for soft scatterers the low frequency excitations are localized [$S(E=0) = 4\pi a^2$].

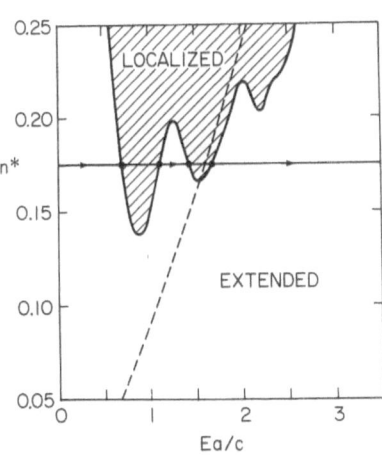

Fig. 1. Phase diagram for $\rho_s = \rho$ and $\kappa_s = (1.75)^2\kappa$. The dashed line is the phase boundary for soft (Dirichlet) scatterers.

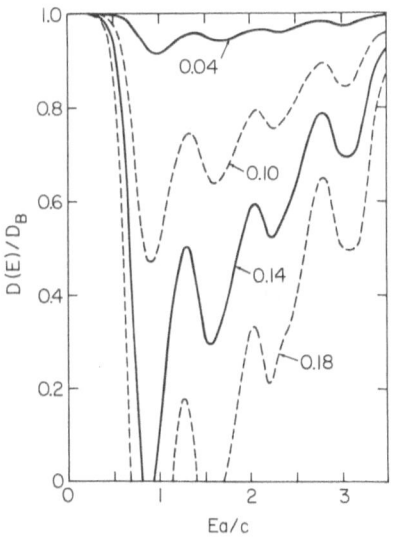

Fig. 2. Diffusion coefficient for $\rho_s = \rho$ and $\kappa_s = (1.75)^2\kappa$. The numbers next to the curves are the volume fractions n*.

489

In Fig. 2 we have plotted $D(E)/D_B(E)$, which is a measure of the strength of the coherent effects, which decrease the diffusion of energy. We note the existence of two (four) mobility edges for $n^* = 0.14$ (0.18). Calling the transport mean free path ℓ_T, it is possible to evaluate at what values of the Ioffe-Regel parameter $E\ell_T/c$ we find the mobility edges.[7] They are usually found in the range $1 \lesssim E\ell_T/c \lesssim 3$.

For two-dimensional systems we find that all excitations are localized, although $\xi(E)$ can be quite large. Except for the soft-sphere case, $\xi(E)$ is shortest when $Ea/c \simeq 1$ (and hence the wavelength is of the order of the scatterer perimeter) and diverges at high and low frequencies.

A one-dimensional localization model, in which third-sound waves are scattered by randomly located identical parallel strips was described in Ref. 8. Resonant transmission by the individual scatterers gives rise to a discrete, but experimentally meaningful, set of extended states.

We have taken all the inhomogeneities to be identical. A different model, describing a related physical situation, is that of fractons, the localized excitations in a self-similar structure.[9]

ACKNOWLEDGMENT

Most of this work was done in collaboration with T. R. Kirkpatrick.

REFERENCES

1. S. John, H. Sompolinsky, and M. Stephen, Phys. Rev. B 27, 5592 (1983).
2. T. R. Kirkpatrick, Phys. Rev. B 31, 5746 (1985).
3. C. A. Condat and T. R. Kirkpatrick, Phys. Rev. Lett. 58, 226 (1987).
4. P. Sheng and Z. Q. Zhang, Phys. Rev. Lett. 57, 1879 (1986).
5. K. Arya, Z. B. Su, and J. L. Birman, Phys. Rev. Lett. 57, 2725 (1985).
6. D. Vollhardt and P. Wölfle, Phys. Rev. B 22, 4666 (1980).
7. C. A. Condat and T. R. Kirkpatrick, unpublished.
8. C. A. Condat and T. R. Kirkpatrick, Phys. Rev. B 33, 3102 (1986).
9. O. Entin-Wohlman, S. Alexander, and R. Orbach, these Proceedings.

QUANTUM PERCOLATION: MEANING AND RESULTS

Vipin Srivastava

School of Physics
University of Hyderabad
Hyderabad - 500 134, India

INTRODUCTION

Simple ideas of percolation have been applied widely in physics problems. We like to draw attention here to the application of percola-tion ideas to understand the difficult problem of electron localization[1] in a disordered array of potentials. This was first attempted by Ziman[2] at a stage when the significance of localization phenomenon was realized but its comprehension was still posing a challenge to the theoretical physicists. As it became clear that localization was a purely quantum mechanical phenomenon, its analogy to percolation ideas became dubious. However, study of percolation under quantum mechanical conditions -- now known as quantum percolation -- became interesting in its own right. The aim here is to explain the meaning of the term 'quantum percolation' and to report some of the interesting results which are generally not expected in a typical localization problem and, therefore, pertain to this specific problem only.

MEANING OF QUANTUM PERCOLATION

Just as in the classical percoltion problem one studies the flow of a fluid through a porous medium, in quantum percolation (q.p.) one investigates diffusion of an electron in a potential field created by atoms thrown randomly on an empty lattice. If the fraction of occupied lattice sites exceeds a critical value, x_c, the percolation threshold, a connected network of sites extends from one end of the system to another and this is understood as the onset of classical conduction. However, in the microscopic quantum mechanical problem, scattering and quantum interference of the electron wave may preclude electron conduction through the percolation channel. Conduction, in quantum mechanical sense, requires the formation of an extended state in the percolation channel. This should presumably occur at a concentration higher than x_c as will become clear in the following.

It is apparent that q.p. is the same as localization problem in a binary system but it should be emphasised that it demands the system to be infinitely disordered, i.e. there should be an infinite potential barrier between two sites occupied by unlike atoms. This is explained in Fig.1 -- If we start with a binary alloy with finite disorder,

491

at certain concentration of minority A atoms extended states can be created in the A-subband. Now imagine a mechanism whereby disorder (i.e. the disparity between the potentials offered by A and B atoms) can be increased continuously. At some stage it will convert the extended states in the A- subband into localized states. However, an increase of the A-concentration can futher produce extended states in the A-subband. Such a competition between concentration and disorder can alternately produce and destroy extended states in A-subband until in the limiting situation of infinite disorder, B atoms are formally removed from the picture and an extended state can spread along the connected network of A-sites. This extended state marks the onset of indestructable conduction along a percolation channel and the concentration of A atoms is defined as the q.p. threshold, x_q. This is summarised in Fig.2 in terms of divergence of average cluster-size and average size of localized states with varying concentration.

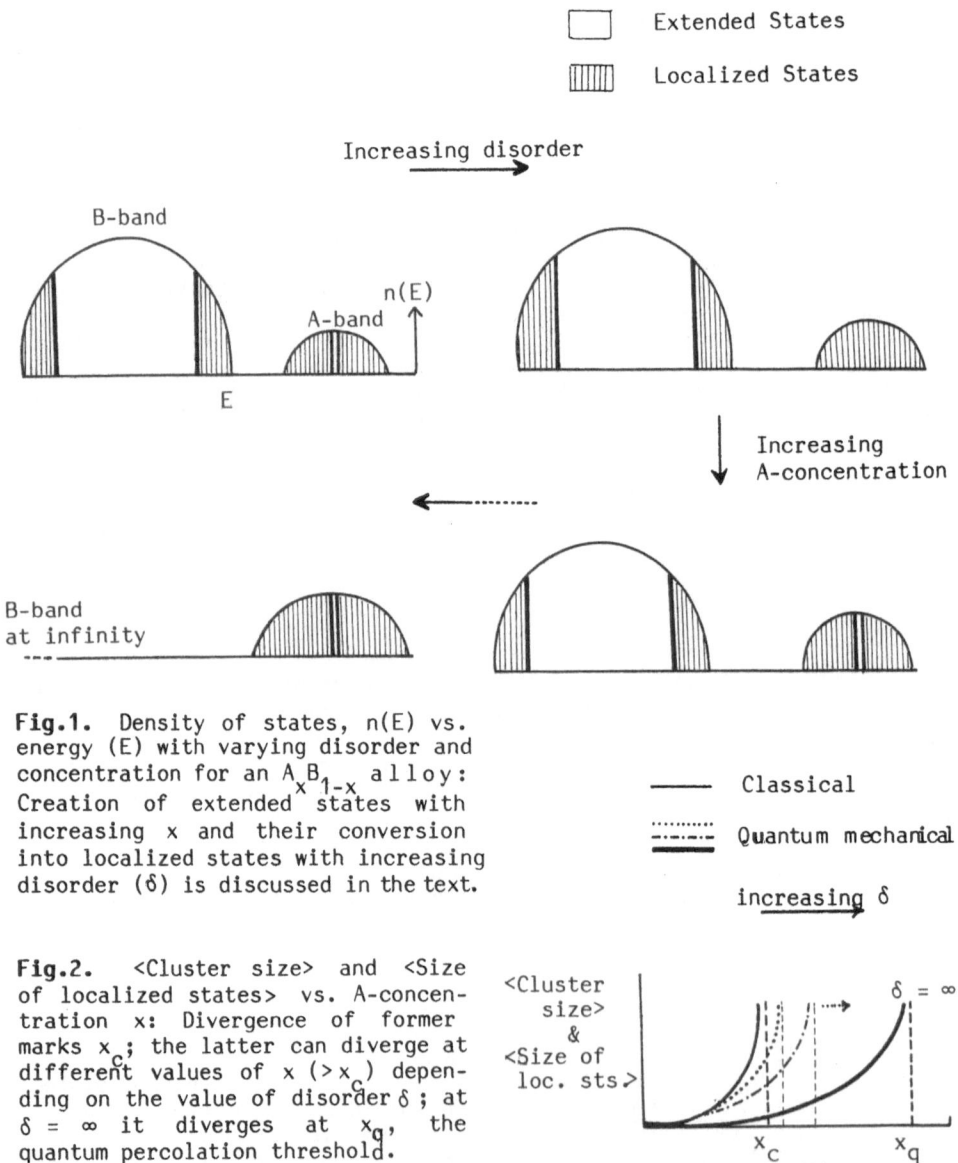

Fig.1. Density of states, n(E) vs. energy (E) with varying disorder and concentration for an $A_x B_{1-x}$ alloy: Creation of extended states with increasing x and their conversion into localized states with increasing disorder (δ) is discussed in the text.

Fig.2. <Cluster size> and <Size of localized states> vs. A-concentration x: Divergence of former marks x_c; the latter can diverge at different values of x (> x_c) depending on the value of disorder δ ; at $\delta = \infty$ it diverges at x_q, the quantum percolation threshold.

RESULTS

Quantum Percolation Threshold

Earlier estimates of x_q made by Economou and Cohen[8] showed $x_q < x_c$ which can not be correct because, as discussed here, q.p. pertains to an infinitely disordered binary system described by Hamiltonian,

$$H_{AA} = V \Sigma |i><j|, \text{ (V=constant for A atoms on nearest} \tag{1}$$
$$\text{neighbour sites; zero, otherwise),}$$

where tunneling through B sites, which may be thought to connect finite clusters below x_c, can evidently not occur for B sites form an impenetrable boundary.

In one dimensional (d=1) systems it is clear that x_q should be one in view of the fact that all states are localized in presence of any disorder.[3] In d=3 x_q has been reported to have values between 0.32 and 0.7[4,5,6,7] depending on whether one studies site problem or bond problem. The d=2 case has proved controversial. This is discussed in some detail below.

A direct measurement of x_q can be made in terms of participation ratio or the fraction of sites over which a wave function spreads. One expects the average participation ratio $< IP >$ (averaged over energy) to exhibit a monotonic behaviour as a function of x, with asymptotic divergence at x_q. It is more practical to calculate $< IP >^{-1}$ which is expected to decrease monotonically with x and show a kink at $x=x_q$ from whereon the contribution to $< IP >^{-1}$ remains small. Contrary to the expected monotonic behaviour novel features were found to appear at $x=x_c$ and $x=x_q$.[9] This is shown in Fig. 3. The dips at x_c and x_q are attributed[5] to the divergence of mean cluster size and size of localized states, respectively. The subsequent peaks are reckoned to be caused by the downward shift of the size-distribution of finite clusters and that of localized states for $x > x_c$ and $x > x_q$.

For d=2, x_q is found to be 0.74 (determined for square lattice).[5,9,10] The real space renormalization group calculation gives the values 0.86 and 0.76 respectively for site and bond problems. It is disturbing that x_q is found to be less than one, in view of the general understanding[11] that in d=2 any amount of disorder will cause complete localization of states. Since quantum percolation pertains to infinite disorder one actually expects 'exponential' localization which should keep $< IP >^{-1}$ very high down to x=1 as shown by dash-dot line in Fig.3. It appears difficult to reconcile the great disparity between the expected and the obtined results.

Fig.3. Schematic plot of $< IP >^{-1}$ versus x drawn with the help of numerical results of refs. 5 and 9. Attention should be paid to the dips at x_c and x_q. The dash-dot line should represent the situation where states in the percolation channel may be always assumed to be <u>exponentially</u> localized at $\delta = \infty$ in d = 2.

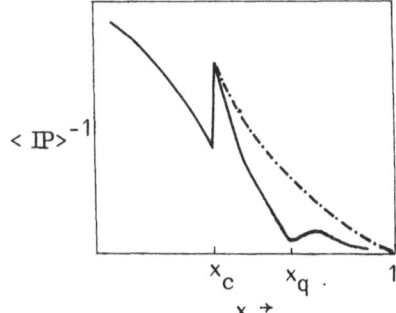

Hop-Scotch States

A new kind of localized states found typically in strongly disordered binary alloys are shown in Fig.4. They arise due to quantum interference of electrons after they are reflected from the B atoms placed in special positions as shown.[9,12] What is interesting is that these very strongly localized states appear at and around the centre of the band. This presents a wild departure from the Mott-CFO model[13] which predicts localization in band tails and does not allow mixing of localized and extended states.

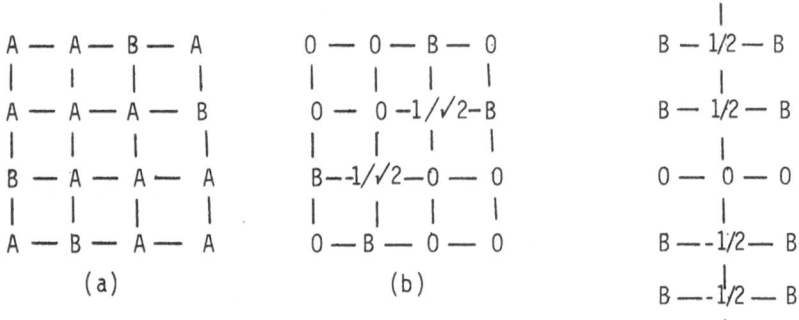

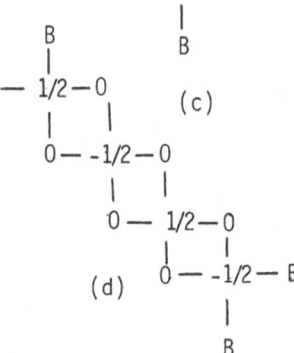

Fig.4. Sketches of Hop-scotch type localized states in strongly diordered A_xB_{1-x} alloy. (a) A local configuration of A-atoms which is not isolated but supports a localized eigenstate (with wave function shown in (b)) of eigenenergy E=0. In (c) and (d) wave functions of larger localized states are shown whose energies are close to E=0.

SUMMARY

In summary we have defined here the quantum mechanical version of the prcolation problem and have shown its conncection with the localization problem. It is found to have some novel features that reveal new physics in the diluted binary systems.

REFERENCES

1. P.W. Anderson, Phys. Rev. 109: 1492 (1958).
2. J.M. Ziman, "Models of Disorder", Cambridge University Press, Cambridge (1979).
3. N.F. Mott and W.D. Twose, Adv. Phys. 10:107 (1961).
4. T. Odagaki and K.C. Chang, Phys. Rev. B30: 1612 (1984).
5. V. Srivastava and M. Chaturvedi, Phys. Rev. B30: 2238 (1984).
6. R. Raghavan and D.C. Mattis, Phys. Rev. B23: 47(1981).
7. Y. Shapir, A. Aharony, and A.B. Harris, Phys. Rev. lett. 49, 486 (1982).
8. E.N. Economou and M.H. Cohen, Phys. Rev. B5: 2931 (1972).
9. V. Srivastava and D. Weaire, Phys. Rev. B18: 6635 (1978).
10. G.M. Scher, J. Non-Cryst. Sol. 59 and 60: 33 (1983).
11. E. Abrahams, P.W. Anderson, D.C. Licciardello and T.V. Ramakrishnan, Phys. Rev. lett. 42: 673 (1979).
12. S. Kirkpatrick and T.P. Eggarter, Phys. Rev. B6: 3598 (1980).
13. N.F. Mott, Adv. Phys. 16: 49 (1967); M.H. Cohen, H. Fritzsche and S.R. Ovshinski, Phys. Rev. Lett. 22: 1065 (1969).

PARTICIPANTS

Marc Aertsens
    Limburgs Universitair Centrum, Universitaire Campus
    B-3610 DIEPENBEEK, Belgium

Amnon Aharony
    School of Physics and Astronomy, Tel Aviv University
    69978 TEL AVIV, Israel

Preben Alstrøm
    NORDITA, Blegdamsvej 17, DK-2100 COPENHAGEN Ø, Denmark

Arne F. Andresen
    Institute for Energy Technology, P.O.B. 40, N-2207 KJELLER, Norway

Javid Ashraff
    Dept. of Theoretical Physics, 1, Keble Road, Oxford OX1 3NP, UK

Trond Aukrust
    Inst. for Theoretical Physics, University of Trondheim-NTH
    N-7034 TRONDHEIM-NTH, Norway

Jörgen Axell
    Dept. of Theoretical Phycis, The Royal Institute of Technology
    S-10044 STOCKHOLM, Sweden

Paul D. Beale
    Dept. of Physics, University of Colorado, BOLDER, CO 80309, USA

Collin L. Broholm
    Risø National Laboratory, DK-4000 Roskilde, Denmark

Patrice Bujard
    CIBA-GEIGY AG, Forschungszentrum KA, CH-1701 FRIBOURG, Switzerland

Silvia Celi
    Dept. of Physics, University of Pavia
    Via Bassi 6, I-27100 PAVIA,Italy

Anna Maria N. Chame
    Centro Brazileiro des Pesquisas Fisicas
    R. Xavier Sigaud 150, RIO DE JANEIRO RJ,Brazil

Henry Chou
    Physics Department, City College of New York, NEW YORK, N.Y. 10031, USA

Alexis Christou
    Dept. of Theoretical Physics, 1, Keble Road, OXFORD OX1 3PN, UK

Carlos A. Condat
    University of Maryland, IPST, COLLEGE PARK, MD 20742, USA

Marc Descamps
    Dept. of Physics, University of Lille, F-59655 VILLENEUVE-D'ASCA, France

Edgardo R. Duering
    School of Physics, Tel Aviv University, 69978 TEL AVIV, Israel

Wolfgang G.M. Dultz
    Fakultät für Physik, Universität Regensburg, 8400 REGENSBURG, West-Germany

Josette Dupuy
    Dept. de Physique, Université Claude Bernard
    69622 LYON VILLEURBANNE, France

Sir Sam Edwards
    Cavendish Laboratory, Madingley Road, CAMBRIDGE CB3 OHE, UK

Hans-Friedrich Eicke
    Inst. for Physical Chemistry, University of Basel
    Klingelbergerstrasse 80, CH-4056 BASEL, Switzerland

Ora Entin-Wohlman
    School of Physics and Astronomy, Tel Aviv University
    69978 TEL AVIV, Israel

Thorbjörn Farestam
    Dept. of Theoretical Physics, Chalmers University of Technology
    S-41296 GOTHENBURG, Sweden

Jens G. Feder
    Dept. of Physics, University of Oslo
    POB 1048 Blindern, N-0316 OSDLO 3, Norway

Vidar Frette
    Dept. of Physics, University of Oslo
    POB 1048 Blindern, 0316 OSLO 3, Norway

Jan Frøyland
    Fysisk Institutt, University of Oslo
    POB 1048, Blindern, 0316 OSLO 3, Norway

Liv Furuberg
    Dept. of Physics, University of Oslo
    POB 1048 Blindern 0316 OSLO 3, Norway

Bruce D. Gaulin
    Solid State Division, Oak Ridge National Laboratory
    OAK RIDGE, TN 37831, USA

Gary S. Grest
    Exxon Research and Engineering Co. Route 22 East,
    ANNANDALE, N.J. 08801 USA

Göran Grimvall
    Dept. of Theoretical Physics, The Royal Institute of Technology
    S-10044 STOCKHOLM, Sweden

Jan Petter Hansen
    Dept. of Physics, University of Aarhus, AARHUS, Denmark

A. Brooks Harris
    University of Philadelphia, pt. Dept. of Physics
    University of Tel Aviv, 69978 TEL AVIV, Israel

Geir Helgesen
    Dept. of Physics, University of Oslo
    POB 1048 Blindern, 0316 OSLO 3, Norway

Emilio L. Hernandez-Garcia
    Dpt. Fisica, Universitat de les Illes Balears
    E-07071 PALMA DE MALORCA, Spain

Marc Hewitt
    Dept. of Physics, University of Edinburgh
    Mayfield Road, EDINBURGH EH9 3JZ, UK

Paul G. Higgs
    TCM, Cavendish Laboratory, CAMBRIDGE CB3 0B3, UK

Rudolf Hilfer
    Dept. of Physics, University of California
    405 Hilgard Ave., LOS ANGELES, CA 90024, USA

Daniel Hong
    Inst. for Theoretical Physics, University of California
    SANTA BARBARA, CA 93106, USA

Jean-Pierre Hulin
    Labo ESPCI, 10, rue Vauquelin, F-75231 PARIS, France

Ragnvald Høier
    Dept. of Physics and Mathematics,
    Technical University of Norway, 7034 TRONDHEIM-NTH, Norway

Mogens Høgh Jensen
    NORDITA, Blegdamsvej 17, DK-2100 COPENHAGEN Ø, Denmark

Torstein Jøssang
    Dept. of Physics, University of Oslo
    POB 1048 Blindern, 0316 OSLO 3, Norway

Jørgen K. Kjems
    Risø National Laboratory, DK-4000 ROSKILDE, Denmark

Ger J.M. Koper
    Lorentz Institut, Niewsteeg 18, 2311 SB LEIDEN, Netherlands

Uwe A. Kreibig
    Fachbereich 11 - Physik, Universität des Saarlandes
    D-6600 SAARBRÜCKEN, West-Germany

Werner Köhler
    EP IV, Universität Bayreuth
    D-8586 BAYREUTH, West-Germany

Laurent Limat
    Lab. HMP-ESPCI, 10, rue Vauquelin, F-75231 PARIS, France

Per-Anker Lindgård
    Risø National Laboratory, DK-4000 ROSKILDE, Denmark

Herbert M. Lindsay
    Exxon Research & Engineering Co., Route 22-E, ANNANDALE, NJ 08801, USA

Kenneth B. Lyons
    RM1a126, AT+T, Bell Laboratories,
    600 Mountain Avenue, MURRAY HILL, N.J. 07974, USA

Paul Manneville
    DPh-G/PSRM, CEN Saclay, F-91191 GIF-SUR-YVETTE, France

James E. Martin
    Division 1152, Sandia National Laboratories, ALBUQUERQUE, N.M. 87185 USA

Vicent J. Martinez-Garcia
    Dept. de Mathematica Aplicada, Universitat de Valencia
    E-46100 BURJASSOT (Valencia), Spain

Joseph L. McCauley
    Dept. of Physics, University of Houston, HOUSTON, Tx. 77004, USA

Paul Meakin
    Bldg. 356, Rm 251, Experimental Station
    E.I. du Pont de Nemours & Co., WILMINGTON, DA 19868, USA

Yigal Meir
    School of Physics, University of Tel Aviv, 69978 TEL AVIV, Israel

Sushil K. Mendiratta
    University of Aveiro, 3800 AVEIRO, Portugal

Ole G. Mouritsen
    Dept. of Structural Properties of Materials, Technical University
    of Denmark, Dldg. 307, DK-2800 LYNGBY, Denmark

Michael Murat
    School of Physics, University of Tel Aviv, 69978 TEL AVIV, Israel

Knut J. Måløy
    Dept. of Physics, University of Oslo
    POB 1048 Blindern, 0316 OSLO 3, Norway

Jan Naudts
    Dept. of Physics, University of Antwerpen
    Universiteitsplein 1, B-2610 ANTWERPEN, Belgia

Ursula J. Nicholls
    Dept. of Physics, University of Edinburgh
    Mayfield Road, EDINBURGH EH9 3JZ, UK

Gunnar A. Niklasson
    Physics Department, Chalmers University of Technology
    S-41296 GOTHENBURG, Sweden

Mark Novotny
    IBM Bergen Scientific Centre
    Allégt. 36, N-5000 BERGEN, Norway

Kaare Otnes
    Institute for Energy Technology, P.O.B. 40, N-2007 KJELLER, Norway

498

Unni Oxaal
    Dept. of Physics, University of Oslo
    POB 1048 Blindern, 0316 OSLO 3, Norway

Marek Pekala
    Dept. of Chemistry, University of Warsaw
    Al. Zwirki i Wigury 101, 02-089 WARSAW, Poland

Iveta Pimentel
    Dept. of theoretical Physics, 1, Keble Road, OXFORD OX1 3NP, UK

Jean-Jacques Préjean
    CNRS-CRTBT, BP 166, F-38042 GRENOBLE-CEDEX, France

Jean Rajchenbach
    Lab. d'Optique Physique, ESPCI
    10, rue Vauquelin, F-75005 PARIS, France

R. Rammal
    CNRS-CRBT, BP 166, F-38042 GRENOBLE-CEDEX, France

Dieter Richter
    Institute Laue-Langevin, 156X, F-38042 GRENOBLE-CEDEX, France

Risto K. Ritala
    Finnish Pulp and Paper Research Institute
    POB 136, SF-00101 HELSINKI, Finland

Jacques Schneck
    CNET, 196 Rue Henri Ravera, F-92220 BAGNEUX, France

Peter Schofield
    Theoretical Physics Division, AERE Harwell, OXFORDSHIRE OX11 ORA, UK

Roger Serneels
    Limburgs Universitair Centrum, Universitaire Campus
    B-3610 DIEPENBEEK, Belgium

France Sevsek
    Institut "Josef Stefan", POB 199, Yu-61001 Ljubljana, Yugoslavia

Stephen M. Shapiro
    Lab. Leon Brillouin, CEN Saclay
    F-91191 GIF-SUR-YVETTE CEDEX, France

David Sherrington
    Dept. of Physics, Imperial College, LONDON SW7 2BZ, UK

Pado Sibani
    Dept. of Physics, University of Odense
    Campusvej 55, DK-5230 ODENSE M, Denmark

José M. Nunes da Silva
    Dept. of Physics, University of Porto 4000 PORTO, Portugal

Vipin Srivastava
    School of Physics, University of Hyderabad
    500134 HYDERABAD, India

H. Eugene Stanley
    Center for Polymer Studies, University of Boston,
    590 Commonwealth Avenue, BOSTON, Mass. 02215, USA

Olav Steinsvoll
    Institute for Energy Technology, POB 40, N.2007 KJELLER, Norway

Robin B. Stinchcombe
    Dept. of Theoretical Physics, 1, Keble Road, OXFORD OX1 3NP, UK

Erik Schwartz Sørensen
    Fysisk Institut, Aarhus Universitet, DK-Aarhus C, Denmark

Joseph M. Thijssen
    University of Nijmegen, Toernooiveld, 6525 ED NIJMEGEN, Netherlands

Harry Thomas
    Dept. of Physics, University of Basel
    Klingelbergerstrasse 82, CH-Basel, Switzerland

Gerard Toulouse
    Lab. de Physique, Ecole Normale Supérieure
    24, Rue Lhommond, F-75005 PARIS Cedex 05, France

Peder A. Tyvand
    Dept. of Physics, Agricultural University of Norway, POB 19
    1432 ÅS-NLH, Norway

Paul J. Upton
    Dept. of Theoretical Physics, 1, Keble Road, OXFORD OX1 3NP, UK

René Vacher
    U.S.T.L., Lab. Science Matériaux Vitreux
    F-34060 MONTPELLIER CEDEX, France

David Weitz
    Exxon Research & Engineering Co., Route 22-E ANNANDALE, N.J. 08801, USA

ORGANIZING COMMITTEE

Roger Pynn, director
    LANSCE MS-H805, Los Alamos National Laboratory
    Los Alamos, N.M. 87545, USA

Tormod Riste, co-director
    Institute for Energy Technology, POB 40, N-2007 KJELLER, Norway

Arne T. Skjeltorp
    Institute for Energy Technology, POB 40, N-2007 KJELLER, Norway

Gerd Jarrett, secretary
    Institute for Energy Technology, POB 40, N-2007 KJELLER, Norway

KTaO$_3$, 297

Langevin equation
 non-linear, 393-395
Laplace equation, 23, 54-55
Levy flight, 37
Lifshitz-Slyozov theory, 387, 395-
 400, 416
Light scattering, 1, 435-439
 dynamic, 437, 449
 inelastic, 297
 quasi-elastic, 97-101
Lipid membrane, 457-463
Localisation, 276
 Anderson, 244
 classical, 487-490
 length, 246
 weak, 235

Map
 circle, 177
 Henon, 196
 Poincare, 197
 sine, 190
 tent, 185
Markov chain, 217
Master equation, 234
Melting, 13-15
 interfacial, 1, 459-462
Memory effects, 293-294
Metastability, 383-386, 388
Microemulsion
 droplet fluctuations in, 465-469
 water-in-oil, 219-223
Microparticles, 1-16, 71
Microstructures, 366, 370, 379
Modes
 local, 348
 vibrational, 235
Molecular crystal, 383
Molecular dynamics, 389
Monte Carlo, 389, 407-408
Morphology, 388
Multifractals, 27-29, 52, 124,
 145-198, 209, 270, 477

Network
 elastic, 259
 non-linear, 272
 neural (see neural network)
 resistor (see resistor network)
Neural networks, 318, 336, 359-363
Neutron scattering, 1, 251, 255,
 413-416, 439, 462
Neutron spin echo, 466-467
Noise
 diffusion, 264-267
 1/f, 263-276, 311
 Nyquist, 264-267
 reduction of, 19, 31-34, 36, 39
Nucleation, 387, 393, 422
N-vector model, 36

Optimization, 331
Ostwald ripening, 395

Pattern
 formation, 36, 93, 387, 457-463
 recognition, 360
Peclet number, 39, 81, 227
Percolation, 1, 21, 27, 103, 220,
 263, 305, 471
 bond, 425, 431
 clusters, 209-212, 311
 continuum, 271
 correlated, 439
 invasion, 132, 141, 164
 multifractality of, 163-171
 quantum, 491-494
 site, 22
 threshold, 200, 219-223, 233,
 259
Period doubling, 173, 176, 182
Phase separation, 413-420
Phase transition
 first order, 387-408
Pinning
 energy, 345, 355
 weak, 345, 376
Polydispersity, 443-445
Polymer
 branched, 434
 cooperative motion in, 289-291
 dynamics of, 425-449
 glass transition in, 279-295
 melt, 10
Polystyrene
 spheres of, 2
Porous media, 46, 111-143, 199
 tracer dispersion in, 225-228
Potts model, 36, 365, 368-374, 460

Quasiperiodicity, 178
Quenched impurities, 368, 376
Quenched variables, 318
Quenching, 1

Radius of gyration, 5, 7, 24, 45,
 52-53, 94, 116
Random walk, 5, 23-24, 28, 45,
 47, 55, 73, 93, 107, 213, 481-
 486
 Gaussian, 482
 self-avoiding, 166, 209-212, 285,
 479, 481
Rate equation, 426-428
Rayleigh-Benard
 flow, 82-86
 problem, 71, 75, 173, 178-180
Reaction kernel, 63, 66
Relaxation
 dynamics of, 311-313
 glassy, 343-357
Renormalisation group, 471-480
Replica
 broken, 318
 method, 319, 334, 476
Resistance fluctuations, 166
Resistor network, 217
 non-linear, 168, 209, 477-478
 random, 155, 471-480
Ridgefield sandstone, 227

Scaling, 8, 21, 60, 398, 449
  crossover, 234
  dynamical (*see* dynamical scaling)
  finite size, 27
Screening, 52
Sedimentation, 71
Self averaging, 370, 389-391
Silica, 3
  aerogel, 239, 255-258
  colloidal, 429
  smoke-particles, 251-255
Smoluchowski equation, 8, 61, 64,
    426
Snowflake, 5, 35-38
Sol, 425-449
Solidification
  dendritic, 19-43
Spectral dimension, 265
Spin dynamics, 305-314
Spin glass, 317-337
  1/f noise in, 263, 272-275
Spinodal decomposition, 387-388, 393,
    402
Spin waves, 235
Stability analysis, 418
Standard model, 360
Sticking probability, 15, 37, 93
Strain
  coherency, 385
Stretched exponential, 356
Structure factor, 414
  for binary fluids, 403-404
Surface tension, 15
Susceptibility
  non-linear, 299

Thermal conductivity, 243-248
Thermodynamic limit, 333
Tip splitting, 33, 76, 89
Transmission electron microscopy,
    452
Turbulence, 154
Two-level system, 243, 299, 301

Ugelstad technique, 2, 75
Ultrametric
  inequality, 326
  structure, 361
Ultrasonic attenuation, 244
Universality, 199, 356
  class, 380, 388, 391

Viscous fingers, 20, 25, 29-31, 71
    in Hele-Shaw cells, 115-116
    on percolation clusters,
      199-207
    in porous media, 111-143
  fractal dimension of, 25, 114-
  117
  structure of, 131-135, 206
Vogel-Fulcher law, 280, 292, 298,
    383

Wetting, 139, 462-463
Winding number, 177

X-ray scattering, 1, 383, 460
XY model, 19, 22